Caves in Context

The Cultural Significance of Caves and Rockshelters
in Europe

Edited by

Knut Andreas Bergsvik and Robin Skeates

Oxbow Books
Oxford and Oakville

Published by
Oxbow Books, Oxford, UK

© Oxbow Books and the authors, 2012

ISBN 978-1-84217-474-6

A CIP record for this book is available from the British Library

This book is available direct from:

Oxbow Books, Oxford, UK
(Phone: 01865-241249; Fax: 01865-794449)

and

The David Brown Book Company
PO Box 511, Oakville, CT 06779, USA
(Phone: 860-945-9329; Fax: 860-945-9468)

or from our website

www.oxbowbooks.com

Library of Congress Cataloging-in-Publication Data

Caves in context : the cultural significance of caves and rockshelters in Europe / edited by Knut Andreas Bergsvik and Robin Skeates.
  p. cm.
Includes bibliographical references.
ISBN 978-1-84217-474-6
1. Caves--Europe--History. I. Bergsvik, Knut Andreas. II. Skeates, Robin.
GB608.42.C34 2012
551.44'7094--dc23
                                2012008862

Front cover: Kirkehelleren at Træna, Norway. Photo:Knut Andreas Bergsvik
Back cover: inside the cave La Garma at Cantabria, Spain. Photo: Pablo Arias

Printed in Great Britain by
Short Run Press
Exeter

Contents

List of contributors.. v
Preface..vii

Chapter 1: Caves in context: an introduction 1
    Knut Andreas Bergsvik and Robin Skeates

The British Isles and Scandinavia

Chapter 2: From Assynt to Oban: some observations on prehistoric cave use in western Scotland 10
    Clive Bonsall, Catriona Pickard and Graham A. Ritchie

Chapter 3: Mesolithic caves and rockshelters in Western Norway 22
    Knut Andreas Bergsvik and Ingebjørg Storvik

Chapter 4: Rockshelters in central Norway: long-term changes in use, social organization
        and production 39
    Anne Haug

Chapter 5: On the outer fringe of the human world: phenomenological perspectives
        on anthropomorphic cave paintings in Norway 48
    Hein Bjartmann Bjerck

Iberia and France

Chapter 6: On the (l)edge: the case of Vale Boi rockshelter (Algarve, Southern Portugal)................................65
    Nuno Bicho, João Cascalheira and João Marreiros

Chapter 7: The use of caves and rockshelters by the last Neandertal and first Modern Human
        societies in Cantabrian Iberia: similarities, differences, and territorial implications 82
    Javier Ordoño

Chapter 8: La Garma (Spain): long-term human activity in a karst system...101
    Pablo Arias and Roberto Ontañón

Chapter 9: Shedding light on dark places: Deposition of the dead in caves and cave-like
        features in Neolithic and Copper Age Iberia 118
    Estella Weiss-Krejci

Chapter 10: The Bronze Age use of caves in France: reinterpreting their functions
        and the spatial logic of their deposits through the *chaîne opératoire* concept 138
    Sébastien Manem

The Central Mediterranean

Chapter 11: Caves in Context: the late medieval Maltese scenario..153
    Keith Buhagiar

Chapter 12: Caves in need of context: prehistoric Sardinia 166
    Robin Skeates

Chapter 13: Discovery and exploratory research of prehistoric sites in caves and rockshelters
in the Barbagia di Seulo, South-Central Sardinia 188
Giusi Gradoli and Terence Meaden

Chapter 14: Notes from the underground: caves and people in the Mesolithic and Neolithic Karst 199
Dimitrij Mlekuž

Central and Eastern Europe

Chapter 15: Cave Burials in Prehistoric Central Europe 212
Jörg Orschiedt

Chapter 16: Late Caucasian Neanderthals of Barakaevskaya cave: chronology,
palaeoecology and palaeoeconomy 225
Galina Levkovskaya, Vasiliy Lyubin and Elena Belyaeva

Chapter 17: Interstratification in layers of unit III at Skalisty rockshelter
and the origin of the Crimean final Palaeolithic 254
Valery A. Manko

Contributors

PABLO ARIAS
Instituto Internacional de Investigaciones Prehistóricas de Cantabria
(Unidad Asociada al CSIC)
Edificio Interfacultativo de la Universidad de Cantabria
Av. de Los Castros s/n
39005 Santander
Spain
pablo.arias@unican.es

ELENA BELYAEVA
Russian Academy of Sciences
Institute of History of Material Culture
Dvortsovaya nab., 18
St.-Petersburg
191186 Russia
biface@mail.ru

KNUT ANDREAS BERGSVIK
University of Bergen
Department of Archaeology, history, Cultural Studies and Religion
P.O. Box 7805
5020 Bergen
Norway
knut.bergsvik@ahkr.uib.no

NUNO BICHO
FCHS – University of Algarve
Campus de Gambelas
8005-139 Faro
Portugal
nbicho@ualg.pt

HEIN BJARTMANN BJERCK
Norwegian University of Science and Technology (NTNU)
Museum of Natural History and Archaeology
NO-7491 Trondheim
Norway
hein.bjerck@vm.ntnu.no

CLIVE BONSALL
University of Edinburgh
School of History, Classics, and Archaeology
Old High School
Infirmary Street
Edinburgh EH1 1LT
United Kingdom
C.Bonsall@ed.ac.uk

KEITH BUHAGIAR
University of Malta,
Department of Classics and Archaeology
Msida MSD 2080
Malta
keithbuhagiar@onvol.net

JOÃO CASCALHEIRA
FCHS – University of Algarve
Campus de Gambelas
8005-139 Faro
Portugal
jmcascalheira@ualg.pt

JAVIER ORDOÑO DAUBAGNA
Department of Geography, Prehistory and Archaeology
University of País Vasco – Euskal Herriko Unibertsitatea
c/o Tomás y Valiente s/n
01006 Vitoria-Gasteiz (Álava)
Spain
javier.ordono@ehu.es

M. GIUSEPPINA GRADIOLI
COMET Valorizzazione Risorse Territoriali and ISSEP Sardinia
Via Pitzolo 20
09126 Cagliari
Sardinia
Italy
issepsardegna@gmail.com

ANNE HAUG
Norwegian University of Science and Technology (NTNU)
Museum of Natural History and Archaeology
NO-7491 Trondheim
Norway
Anne.Haug@vm.ntnu.no

GALINA M. LEVKOVSKAYA
Russian Academy of Sciences
Institute of History of Material Culture
Dvortsovaya nab., 18
St.-Petersburg
191186 Russia
ggstepanova@yandex.ru

VASILIY LYUBIN
Russian Academy of Sciences
Institute of History of Material Culture
Dvortsovaya nab., 18
St.-Petersburg
191186 Russia
biface@mail.ru

SÉBASTIEN MANEM
Institute of Archaeology
University College London
31-34 Gordon Square
London WC1H 0PY
United Kingdom
s.manem@ucl.ac.uk

VALERY MANKO
Institute of Archaeology
National Academy of Sciences of Ukraine
12 Geroiw Stalingradu ul.
Kyiv
Ukraine
valery_manko@yahoo.com

JOÃO MARREIROS
FCHS – University of Algarve
Campus de Gambelas
8005-139 Faro
Portugal
jmmarreiros@ualg.pt

G. TERENCE MEADEN
Oxford University
Archaeology Section
Department of Continuing Education
Rewley House
1 Wellington Square
Oxford, Oxfordshire
OX1 2JA
United Kingdom
terence.meaden@torro.org.uk

DIMITRIJ MLEKUŽ
Ghent University
Faculty of Arts and Philosophy
Department of Archaeology and ancient history of Europe
Sint-Pietersnieuwstraat 35 UFO
9000 Gent
Belgium
dmlekuz@gmail.com

ROBERTO ONTAÑÓN
Instituto Internacional de Investigaciones Prehistóricas de Cantabria
(Unidad Asociada al CSIC)
Edificio Interfacultativo de la Universidad de Cantabria
Av. de Los Castros s/n
39005 Santander
Spain
ontanon_r@gobcantabria.es

JÖRG ORSCHIEDT
University Hamburg
Archaeological Institute
Edmund-Siemers-Allee 1, Fluegel West
20146 Hamburg
orschiedt@uni-hamburg.de

CATRIONA PICKARD
University of Edinburgh
School of History, Classics, and Archaeology
Old High School
Infirmary Street
Edinburgh, EH1 1LT
United Kingdom
Catriona.Pickard@ed.ac.uk

GRAHAM A. RITCHIE
University of Edinburgh
School of History, Classics, and Archaeology
Doorway 4, Teviot Place
Edinburgh, EH8 9AG
United Kingdom

ROBIN SKEATES
Durham University
Dept of Archaeology
South Road
Durham, DH1 3LE
United Kingdom
robin.skeates@durham.ac.uk

INGEBJØRG STORVIK
University of Bergen
Department of Archaeology, history, Cultural Studies, and Religion
P.O. Box 7805
5020 Bergen
Norway
Ingebjorg.storvik@student.uib.no

ESTELLA WEISS-KREJCI
University of Vienna
Department of Social and Cultural Anthropology
Universitätsstraße 7
A–1010 Wien
Austria
estellawk@hotmail.com

Preface

Caves and rockshelters are found all over Europe, and have frequently been occupied by human groups, from prehistory right up to the present day. Some appear to have only traces of short occupations, while others contain deep cultural deposits, indicating longer and multiple occupations. Above all, there is great variability in their human use, both secular and sacred. The aim of this book, then, is to explore the multiple significances of these natural places in a range of chronological, spatial, and cultural contexts across Europe.

The majority of the chapters published here were originally presented in a conference session on 'Caves in Context: The Economical, Social, and Ritual Importance of Caves and Rockshelters', held at the 14th Annual Meeting of the European Association of Archaeologists in Malta in September 2008. This session was organized by Knut Andreas Bergsvik, and the discussant was Robin Skeates.

The session was well attended and there was a great deal of discussion on the subject. Several key questions were raised, and many of them were related to caves themselves. How uniform or diverse are the categories 'cave' and 'rockshelter'? To what extent should we separate 'economic', 'social', and 'ritual' aspects of cave use? How do caves' internal and external environments change over time? What kinds of caves are not selected for human use? In what ways do people transform natural caves? What do people do in caves, what do caves mean to those people, and what do caves do to them? How are practices such as dwelling, production, and ritual performance experienced in caves by different kinds of person through all of their senses? What numbers of people are involved in different cave activities? In what different ways are the light and dark zones of caves used?

Other questions dealt with the relationships between caves and the outside world. What kinds of materials do people bring into caves? How does the architecture of caves used as residential or burial places compare to that of houses or megalithic monuments? Are caves connected to other culturally significant places in the landscape through the movement of people and things as part of rites of passage? When do caves become marginal places? How have caves and cave dwellers been represented through the history of archaeology? Are there recognizable patterns of cave-use that vary according to basic economic and cultural differences? For example, do hunter-gatherers use caves differently from farmers, or do egalitarian societies use them in different ways to hierarchical groups? To what extent are they situated adjacent to natural resources, hunting grounds, pastures or farmland?

Methodological issues were also brought up: Why have archaeological archives relating to cave archaeology been neglected? What is the significance of cave names? Why are caves still relevant today? How does European cave archaeology compare to that practiced in other parts of the world?

Because of the interest and lively discussion generated by the presented papers, we decided to ask the session participants to expand and elaborate their contributions into articles for this volume, in order to benefit a wider audience. The result is this anthology, which will primarily be of interest to archaeologists, and to cave archaeologists in particular. However, it is also of relevance to other scholars working in the related fields of speleology, earth sciences, landscape studies, and anthropology, which together comprise the thriving inter-disciplinary field of cave studies. We would like to acknowledge the generous funding provided by the Department of Archaeology, History, Cultural Studies and Religion at the University of Bergen which facilitated the publication of this volume. We would also like to thank Clare Litt, Julie Gardiner and Sam McLeod of Oxbow Books for their support in seeing this volume through from start to finish.

Knut Andreas Bergsvik and Robin Skeates
Bergen and Durham, July 2011

Chapter 1

Caves in context: an introduction

Knut Andreas Bergsvik and Robin Skeates

Caves in context

The premise for this volume is that the archaeology of caves in Europe needs to be more consciously and comprehensively studied 'in context', for the benefit of both speleology and archaeology. This 'necessity of adopting a contextual approach to the study of the human use of caves' was first emphasized in the 1990s (Tolan-Smith and Bonsall 1997, 218; cf. Skeates 1994), but still needs reiterating today. One problem is that cave studies are so well established as a specialized field of research that it is now possible to investigate caves in relative isolation, including as a sub-discipline of archaeology (e.g. Inskeep 1979; CAPRA 1999–2007; Gunn 2004). Another problem is that cave archaeology is now dominated by scientific data collection and analysis, to the detriment of interpretative approaches to their social and cultural significance. As a consequence, archaeological cave studies can be accused of a loss of meaning and relevance to the social sciences in general and to archaeology in particular, especially in contrast to their dynamic development in the mid-nineteenth century, when they were entangled in some key scholarly debates. In this volume, then, we hope to demonstrate, through a diversity of European archaeological approaches and examples, that cave studies, whist necessarily focussed, can also be of significance to wider, contemporary, archaeological research agendas, particularly when a contextual approach is adopted.

Meanings and the search for them lie at the heart of scholarly uses of the term 'context'. In general, 'context' is used to refer to the ambience, arena, background, circumstances, conditions, environment, framework, habitus, relations, situation, or surroundings that determine or clarify the meaning of a thing. And so, for linguists, 'context' refers to the 'parts [of a text] that precede or follow a passage and fix its meaning' (*OED*); hence when that passage is taken 'out of context' it can be misleading; while for 'contextualist' philosophers, attributions of knowledge are determined by the specific contexts in which they occur. In field archaeology, 'context' refers to any discrete stratigraphic unit identified and recorded during the excavation of an archaeological site, such as a layer or a pit – whose stratigraphic relationships over space and time are of fundamental importance to reconstructing the historical development of that site (e.g. Harris 1979; Drewett 1999, 107). By contrast, in theory-led post-processual (or interpretative) archaeology, this idea of context is extended to refer to the whole web of associations of a particular material thing or practice being studied, with the goal of 'contextual archaeology' being the weaving together of a rich interpretative network of associations and contrasts within which to situate that past thing or practice's particular, 'context-dependent', meanings (e.g. Butzer 1982; Hodder 1986, 118–146; 1987; Barrett 1987; Conkey 1997; Barrowclough and Malone 2007). This takes us back to the Latin root of the term 'contextual', meaning woven together, closely connected, or continuous; but the approach is also informed by Clifford Geertz's (1973) ethnographic method of 'thick description', in which both a human behaviour and its context are explained, so as to make the behaviour meaningful to an outsider.

Not all of the contributors to this volume use the concept of context in a theoretically explicit manner. Nevertheless, together, they do help to define at least six contextual dimensions of relevance to cave archaeology. First, there is the 'architectural' context of cave and rockshelter structures, their natural and cultural formation processes (or 'speleogenesis'), and their typological relation to other architectural forms (such as megalithic tombs), all of which frame and add significance to the various human activities carried out in and around them, which in turn affect the culturally diverse values and names ascribed to caves. Second, the caves themselves may offer exceptionally good contexts in terms of detailed stratigraphic resolution and their sometimes favourable conditions for preservation of organic material. Third, there is the spatial context of caves, both as architectural spaces and as meaningful places in the landscape, connected to (or maginalized from) other landforms, resources, and patterns of human behaviour. Fourth, there is the temporal context of the human use of caves, including the history of their occupation, transformation, and remembrance (or forgetting) both seasonally and over the long-term of centuries and millennia. Fifth, there is the (overlapping) socio-economic context of caves: the meaningful place of caves within wider cosmologies, ritual actions, economic strategies, social practices, power relations, identities, and memories. And, sixth, there is the scholarly context of cave archaeology, in relation to the dynamic history of science and of archaeology.

Approaches to cave archaeology in Europe

The history of cave archaeology in Europe can be traced in some regions back to the first half of the nineteenth century, when a few prominent scientists began to explore caves and their deposits as part of broader geological, palaeontological, and antiquarian research agendas, including the great debate over the antiquity of humankind (e.g. Daniel 1981, 48–55; Grayson 1983; Simek 2004; Trigger 2006, 138–156; Ahronson and Charles-Edwards 2010; Prijatelj 2010). Caves soon came to be regarded as significant archaeological resources, valued in particular for the common stratification of their deposits which facilitated the relative chronological ordering of faunal remains and cultural material, and for their often protected and non-acid sedimentary environments which ensured the relatively good preservation of inorganic and organic materials – values which have endured to this day. Over the years, a wide range of professionals and enthusiasts have made archaeological discoveries in hundreds of caves across Europe. In Sardinia, for example, over 100 cave excavations have been undertaken since 1873, not only by the state-funded staff of archaeological superintendencies and university departments, but also by members of regional speleological societies and by local archaeology enthusiasts (Skeates – this volume). In Norway, priests and other educated men excavated caves and published the results during the latter part of the nineteenth century until the enterprise was taken over by professional archaeologists in 1907 (Bergsvik 2005; Bergsvik and Storvik – this volume). Sometimes, discoveries of caves have been accidental. But, since the mid 1990s, systematic archaeological field surveys – some specifically focussed on caves – have added significantly to our contextual understanding of the place of archaeological caves in present-day and ancient landscapes (e.g. Bicho et al.; Bonsall et al. – this volume; Holderness et al. 2006). Cave studies are, consequently, now seeing a resurgence in various parts of Europe, particularly as part of larger multi-disciplinary studies of natural and cultural landscapes.

Over this long history, a diversity of theoretical and methodological approaches to cave archaeology has developed, in part related to wider traditions of archaeological research associated with different periods and regions of study (cf. Watson 2001; Kornfeld et al. 2007). In this volume, we acknowledge and accept this diversity, whilst also promoting a contextual approach. Indeed, to deny this diversity would be to misrepresent the scholarly context within which cave archaeology is practiced in Europe today.

In terms of theory, the big three paradigms of archaeological thought – commonly labelled as 'culture-historical' (or 'traditional'), 'processual' (or 'cognitive-processual'), and 'postprocessual' or 'interpretive' (e.g. Shanks and Hodder 1995) – all remain very much alive in interpretations of cave archaeology in Europe. For example, Manko's approach to caves and their deposits is fundamentally culture-historical, being characterised by an interest in the 'when' and 'where' of past cultures, based on the identification of distinctive assemblages of material remains and their attribution to individual archaeological cultures. Manko consequently argues – based upon detailed categorizations and comparisons of cave stratigraphies, stone artefact types, and faunal remains – that Skalisty rockshelter in the Crimean mountains was used in the Final Palaeolithic primarily as a long-term base camp by hunters using a forest-based economic strategy and the Shankobien lithic industry (whose origins he traces to the Eastern Mediterranean and Middle East), but that short visits were also made to the site – contemporaneously – by hunters belonging to a different cultural tradition (but with similar origins) using a different (steppe-based) economic strategy and a different (Taubodrakian) lithic industry. By contrast, Manem's approach to caves might be described as cognitive-processual, being characterised both by a critique of traditional archaeology and by a rigorous use of scientific and experimental method to reveal something about how people thought and acted in the past. Reacting against traditional interpretations of French Bronze Age caves as dwelling places (as opposed to places of ritual deposition), Manem consequently sets out to distinguish domestic from ritual uses of caves with reference to the different 'operational chains' (*chaînes opératoires*) implicit in the manufacturing of pottery deposited at different types of cave sites. He identifies a restricted number of operational chains (1–5) in pottery made through homogeneous domestic production at Bronze Age sites around the English Channel, in contrast to a greater diversity of technical know-how characteristic of pottery produced and used at meeting places with a ritual function, such as the Bronze Age burial cave of Duffaits in the Charente, where 16 operational chains were identified in the pottery. Another example of a cognitive-processual approach is provided by Ordoño, who used a detailed geographical analysis to investigate changes in human territorial behaviour between the Middle and Upper Palaeolithic in Cantabrian Iberia.

By contrast again, other contributors to this volume adopt – at least in part – a more interpretive approach to caves, characterised by a critical recognition that methodological and personal biases inevitably affect archaeological research and by an interest in the perceptions and experiences of past people. So, Bjerck, for example, questions the impact of flash photography on cave archaeology: arguing that it foregrounds things never observed in dark and inaccessible caves by people in the past. He also provides a consciously subjective phenomenological description of his personal experience of entering, being in, and exiting caves containing Bronze Age paintings in North Norway. Mlekuž likewise explores how the properties 'afforded' by different caves and rockshelters (such as shelter, protection, enclosure, and passage to the underworld) were perceived, experienced, and acted upon by people whilst routinely performing practical tasks in and around these sites in the landscape.

The research methods used today in European cave archaeology are also varied. In addition to the widespread adaptation of well-established speleological and above-ground archaeological field techniques to the prospection, excavation, and recording of caves, archaeologists have incorporated many other approaches in their research designs, both before and after excavation. For example, Ordoño undertook a site location analysis where he recorded factors such as cave orientation, altitude, distance to water sources, and accessible biotopes. Gradoli and Meaden drew upon studies of place-names, folklore, and local traditions to locate caves and rockshelters of archaeological significance in the territory of Seulo in central Sardinia; while Buhagiar found it essential to combine the skills of the archaeologist and of the documentary historian to understand the human uses of caves in Malta during the later Middle Ages. Post-excavation research on the archaeological deposits of Barakaevskaya cave in the north-west Caucasus has also been multi-disciplinary: involving radiocarbon dating, use-wear analysis, physical anthropology, archaeozoology, geomorphology, sedimentology, and palynology to obtain new and detailed scientific information on the chronology, palaeoecology, and palaeoeconomy of the Neanderthal/Mousterian occupation of this site (Levkovskaya *et al.* – this volume). Several of the authors have also done extensive work on museum archives and collections in order to compile regional overviews of cave research (Bergsvik and Storvik; Orschiedt; Skeates; Weiss-Krejci – this volume).

The contextual approach of archaeology has the potential to unite these various methods and theories, particularly in the case of works of synthesis that aim to summarize and interpret knowledge about the archaeology of caves. For example, Bonsall, Pickard and Ritchie explicitly place their analysis of long-term changes in cave forms, deposits, and human uses in the area of Oban Bay in western Scotland in the context of wider geomorphological processes (such as sea level changes and talus formation) and cultural processes (such as the transition to agriculture and monument building); while Skeates outlines the contextual web of relations within which occupied caves in Sardinia were situated over the course of prehistory, particularly in relation to wider landscapes, lifeways, and beliefs. Indeed, all of the chapters in this volume contribute to the much-needed contextualization of caves, ranging from: studies of single caves or rockshelters occupied over various timescales (Arias and Ontañón; Bicho *et al.*; Levkovskaya *et al.*; Manko); to studies of single site types (painted caves, burial caves, and cave-settlements) in particular regions and periods (Buhagiar; Bjerck; Orschiedt; Weiss-Krejci); to studies of different caves in a single archaeological period (Bergsvik and Storvik; Manem); to syntheses of long-term human uses of caves in particular regions (Bonsall *et al.*; Gradoli and Meanden; Haug; Ordoño; Skeates).

We also want to emphasize the variable conditions for establishing a contextual relationship between the use of caves and the utilisation of other landscapes. For some regions and periods treated in this volume, data from caves and rockshelters are the most important sources of knowledge, either because geological or human processes have destroyed much of the archaeological evidence at open-air locations, or because surveys and excavations have focussed on these sites. According to Ordoño, 88 per cent of the known Palaeolithic sites in Cantabria are caves and rockshelters. Similarly, Mlekuž notes that the archaeological record of the karst region of North-East Italy and Western Slovenia almost exclusively consists of such sites. In other regions, such as western Norway, the situation is the other way around. Here, open-air sites predominate heavily over the number of caves and rockshelters, partly as a result of open-air sites being much more systematically and intensively surveyed than cave sites (Bergsvik and Storvik – this volume).

The thematic scope of this volume

Although not encyclopaedic in intention, the chapters in this volume do cover a wide range of geographical regions, cultural periods, and interpretive themes relating to current research on the archaeology of caves in Europe. Four major regions of Europe are used as the basis for ordering the chapters: the British Isles and Scandinavia (Bergsvik and Storvik; Bjerck; Bonsall *et al.*; Haug); Iberia and France (Arias and Ontañón; Bicho *et al.*; Manem; Ordoño; Weiss-Krejci); the Central Mediterranean (Buhagiar; Gradoli and Meaden; Mlekuž; Skeates); and Central and Eastern Europe (Levkovskaya *et al.*; Manko; Orschiedt). The majority of Europe's major archaeological periods are also covered: the Lower Palaeolithic (Arias and Ontañón); the Middle Palaeolithic (Levkovskaya *et al.*; Ordoño; Orschiedt); the Upper Palaeolithic (Arias and Ontañón; Bicho *et al.*; Bonsall *et al.*; Orschiedt; Skeates); the Mesolithic (Arias and Ontañón; Bergsvik and Storvik; Bonsall *et al.*; Haug; Manko; Mlekuž; Orschiedt; Skeates); the Neolithic and Copper Age (Arias and Ontañón; Bonsall *et al.*; Mlekuž; Orschiedt; Skeates; Weiss-Krejci); the Bronze Age (Arias and Ontañón; Bjerck; Bonsall *et al.*; Haug; Manem; Orschiedt; Skeates); the Iron Age (Bonsall *et al.*; Haug; Orschiedt; Skeates); the Roman period (Haug; Orschiedt); and the Medieval period (Arias and Ontañón; Buhagiar; Haug). But it is the themes explored by the contributors to this volume that are of greatest significance to current archaeological discourse: seven of which stand out in particular.

The first theme conerns the diversity of cave forms. The generic term 'cave', defined by Bonsall, Pickard and Ritchie, following White and Culver (2005, 81), as, 'any natural opening or cavity in bedrock large enough for a human to enter', usefully emphasizes the cultural significance of caves – aspects that may relate to dwelling or other activities which require people to enter a cave and stay for some time, but masks the variety of caves found throughout Europe – which is something archaeologists in general have tended to overlook, particularly in contrast to speleologists (e.g. Garasic 1991). In practice, it is quite difficult for archaeologists to categorize the diversity of cave forms, particularly without specialist advice, but a number of contributors to this volume do, at least, highlight a number of contrasts. Morphological contrasts, for example, encourage a basic distinction between 'caves', defined as 'generally having an opening that is deeper than it is wide', as opposed to 'rockshelters', with openings 'wider than they are deep' (Weiss-Krejci – this volume, quoting Weaver 2008, 6). More sophisticated consideration of the mode of formation of caves also leads to distinctions between structures such as karstic caves (resulting from the dissolution of limestone and similar carbonate rocks by weakly acidic groundwater, in some cases combined with the erosive action of subterranean streams), fissure caves (created by mechanical widening of fissures in bedrock), sea caves (formed primarily by wave action exploiting zones of weakness in sea cliffs), and boulder caves (created where boulders have piled up as a result of ice transport or rock-falls, typically on mountain slopes or at the bases of cliffs) (e.g. Bonsall *et al.* – this volume). Human modifications of caves add to this diversity and blur the boundaries between 'natural' and 'artificial' caves. A good example is provided by Buhagiar's descriptions of cave-settlements in late Medieval Malta: spatially extended and partitioned by the addition of artificial terraces, walls, and roofs to two or more adjoining natural caves; and camouflaged within the limestone countryside of the islands. A key question for archaeologists, then, concerns what members of past cultures made of the diversity of caves in the process of ordering the landscapes and societies that they inhabited and belonged to. As Weiss-Krejci asks, with regard to the use of large caves, rockshelters, avens (i.e. caves with vertical entrances), and rock fissures for human burials in Neolithic and Copper Age Iberia, 'did the prehistoric people perceive these natural landscape features as different types of natural place and bury different categories of people in them?'

The second theme in this volume relates to the landscape context of archaeological caves, analysed on various spatial scales (cf. Bradley 2000, 19–32; Barnatt and Edmonds 2002). An important point highlighted by Bonsall, Pickard and Ritchie is that we must take into account the long-term dynamics of the landscapes within which archaeological caves are situated. For example, they note that the sea caves around Oban Bay only became available for human utilization during the marine regression of the Early Holocene or Mesolithic (when the sea level reached a low point), and that the increasing constriction of cave entrances by talus accumulation during and after the Mid-Holocene in western Scotland contributed to a change in cave use from 'economic' to 'funerary'. On this local scale, the proximity of chosen cave sites to key subsistence resources – such as perennial water sources, good fishing rivers, hunting areas, pastures, and farmland – would also have been of significance to cave dwellers (Buhagiar; Haug – this volume). And, on a regional scale, we should take into consideration the dynamic distribution patterns of occupied caves, which can be explained, at least in part, in the context of wider strategies of settlement and subsistence. For example, in western Norway, Mesolithic rockshelters are concentrated in coastal areas, while relatively few are situated in the fjords and on the mountain plateaux. This may be explained by the strong focus on marine resources in this period (Bergsvik and Storvik – this volume). In Cantabria, the shift from a more dispersed use of caves located along the valleys of both principal rivers and tributary rivers in the Middle Palaeolithic to a more restricted use of caves mainly located along principal river valleys in the Early Upper Palaeolithic might be interpreted as reflecting the pursuit of a more specialised settlement and subsistence strategy by the first modern human groups in the region (Ordoño – this volume). And, in Sardinia, the significant concentration and growth in the human use of caves in the agriculturally marginal province of Carbonia-Iglesias in south-west Sardinia during the Copper Age – adjacent to the Campidano settlement heartland of the contemporary Monte Claro culture – might be understood in terms of a territorial expansion of herder-hunter groups belonging to the Monte Claro culture (Skeates – this volume).

These observations overlap well with the third key theme in this volume: the connectivity of cave occupants and their cave-based activities and materials. As Mlekuž puts it, a cave used by people – including the people and material things brought into it – becomes connected in a contextual web of relations, flows, and paths, particularly through the activities performed by people as they move across the landscape. This line of thought is echoed, for example, in Bicho, Cascalheira and Marreiros' chapter, which identifies the stylistic and social connections maintained between a regional Upper Palaeolithic group whose members periodically aggregated at the Vale Boi rockshelter on the Atlantic coast of Portugal and contemporary communities based on the Spanish Mediterranean coast. The movement of raw and processed materials into and, sometimes, out of caves throughout prehistory is also noted in various chapters. During the Palaeolithic, resources such as hard stones (used to produce tools) and game animals (processed for meat, bone marrow, and skins) were carried to cave sites from habitats extending from the immediate environs of caves to sources situated over 100km away (Bicho *et al.*; Levkovskaya *et al.*; Ordoño – this volume). In the Mesolithic, large quantities of seashells and other materials were transported to some coastal and inland caves, where, after processing and consumption, their remains accumulated over many years to form 'middens' (Arias and Ontañón; Bonsall *et al.*; Mlekuž – this volume). And, particularly in later prehistoric periods, increasing qualities of human remains were deposited and re-deposited in caves through primary and secondary burial rites involving the veneration and movement of human bodies and bones in, out of, and even between caves and other places in the landscape (Mlekuž; Skeates; Weiss-Krejci – this volume).

This emphasis on deposition links to our fourth theme, which concerns the precise taphonomic processes – both natural and cultural – that have led to the formation (and disturbance) of cave sediments. At the Vale Boi rockshelter, for example, in the slope area, careful recording of the orientation, size, and fragmentation of artefacts, bones, and shells, enabled the identification of a contrast between a lower area, with deposits affected by natural erosion processes (which led to the orientation of artefacts and bones generally following the trend of the slope, the size-sorting of lithic artefacts, and a relative rarity and fragmentation bones), and an upper area, containing undisturbed Upper Palaeolithic cultural deposits (characterised by artefacts oriented in all directions, the possibility of refitting bone tools and shells, and the survival of piles of limpet shells nested inside each other) (Bicho *et al.* – this volume). And, at La Garma A in Spain, it has been suggested that, during the Upper Palaeolithic, the steady accumulation of sediments in the cave led to a reduction in the inhabitable surface area, and that, for the Mesolithic, the presence of cultural deposits in the cave represents the accumulation of waste produced by activities located immediately outside (as opposed to inside) the cave (Arias and Ontañón – this volume). In some caves, such processes have led to the substantial accumulation of deposits. At Podmol pri Kastelcu in Istria, for example, more than 8m of deposits have accumulated since the Early Neolithic (Mlekuž – this volume).

A large and important fifth theme in this volume concerns the diverse human uses of caves. Essentially, four different categories of cave use can be identified:

as shelters linked to economic activities; as dwelling places; as human burial places; and as places for the performance of other rituals (cf. Bonsall and Tolan-Smith 1997). However, it is important to take into account the significant variability that has occurred within these categories of use, particularly over space and time.

Caves have often been occupied as convenient shelters linked to a variety of – sometimes specialized and seasonal – economic activities. For example, the Vale Boi rockshelter, strategically situated not far from the Atlantic coast, near to the entrance to a narrow gorge leading to a freshwater lagoon, appears to have been used in the Gravettian both as a base for the daily gathering and consumption of shellfish from the tidal sea shore and as a site for the extraction and rendering of bone grease from red deer and equids, which are thought to have been hunted in the surrounding area during the spring and summer (Bicho *et al.* – this volume). Other caves, particularly in the Trieste karst, but also in regions such as Sardinia, Norway, France, and Scotland, were arguably used by mobile herders in later periods as stock pens, characterised archaeologically by evidence of the culling, processing, and consumption of sheep and goats, and by the cyclical clearing and burning of animal dung (Bonsall *et al.*; Manem; Mlekuž; Haug, Skeates – this volume; cf. Angelucci *et al.* 2009). However, the traditional idea that such sites reflect the wider existence of nomadic pastoralist societies, particularly during the European Bronze Age, is open to question (e.g. Manem – this volume; Miracle and Forenbaher 2005, 276). During and after the Middle Ages, the economic use of caves diversified even further across Europe, to include functions such as blacksmiths' workshops, boat houses, agricultural stores, animal-driven mills, and apiaries (Buhagiar; Haug – this volume).

Caves have also been occupied widely as dwelling places, characterised by varying degrees of modification, domesticity, and permanence, and also by phases of abandonment (cf. Jacobsen 1981). For example, in the Middle Palaeolithic, Barakaevskaya cave was occupied as a base-camp by Neanderthal/Mousterian populations, but with differences in settlement intensity and mobility linked – at least in part – to oscillating climatic phases (Levkovskaya *et al.* – this volume). In the Upper Palaeolithic, an extensive and intensive Magdalenian occupation floor was formed in the entrance area of the Lower Gallery at La Garma, located near to the valley floor (Arias and Ontañón – this volume). It included at least one tent structure, 2.5 to 3m in diameter, ringed by boulders and containing a trampled floor. In western Norway, however, caves and rockshelters seem to have been used as long-term residential sites only to a minimal degree. Such sites were first of all situated in the open landscape, while the caves were used for shorter occupations (Bergsvik and Storvik – this volume). In later prehistory, notably in Sardinia, large caves were increasingly abandoned as long-term dwelling places, as open settlement sites were established more widely across the landscape (Skeates – this volume). But elsewhere, some large caves, such as Roucadour cave in the Lot region of France or Għar il-Kbir in Malta, continued to be used and structurally elaborated as dwelling places by troglodytic communities, linked to the agricultural exploitation of the adjacent countryside (Buhagiar; Manem – this volume). As dwelling contexts, then, caves have framed both everyday practices and cultural ideals, and their long-term transformation.

Burial caves represent another category of cave use characterised by diversity: in this case, a wide range of often complex mortuary practices, deposits, and meanings contextually tied to particular cultures and communities, some of which also maintained other kinds of burial places and practices. Indeed, almost every example mentioned in this volume seems somewhat unique, ranging: from the Mesolithic burial of an individual in an oak bark coffin in El Truchiro at La Garma (Arias and Ontañón); to the early Neolithic deposition of heavily selected human bones in Junfernhöhle at Tiefenellern in Bavaria (Orschiedt); to the presence of Mesolithic disarticulated bones at residential sites in western Norway (Bergsvik and Storvik); to the Neolithic deposition of around 338 individuals of all age groups at the rockshelter of San Juan ante Portam Latinam in the upper Ebro valley in Spain (Weiss-Krejci); to the Bronze Age secondary burial of skulls, long bones, and vertebrae under a pile of stones and below a spring at Su Cannisoni rockshelter in Sardinia (Skeates); to the Medieval disposal of the bodies of at least five young men, and the crushing of their skulls, in the Lower Gallery at La Garma (Arias and Ontañón), in a marginal place and rite perhaps reserved for what Weiss-Krejci describes as 'deviant social personae'. However, similarities can also be noted, particularly between some of the later prehistoric burial cave forms and practices and contemporary megalithic burial monuments and rites in various parts of Europe. For example, in Neolithic western Scotland, coastal caves might have been perceived as equivalent to the chambered tombs and passage cairns constructed by communities living further inland (Bonsall *et al.* – this volume); and, in Neolithic Westphalia, the Blätterhöhle might have been selected as a collective burial place due to the resemblance of its narrow entrance to that of a megalithic tomb (Orschiedt – this volume). Certainly, a blurring of traditional archaeological distinctions between natural and artificial monuments is called for, but the precise nature of the relationships between traditions of cave burial and of megalithic burial in Europe is a matter of enduring debate (e.g. Evans 1959, 88–92; Green 1989, 75; Barnatt and Edmonds 2002; Laporte *et al.* 2002, 75–77; Gili *et al.* 2006; Dowd 2008; Skeates 2010, 160–161; Weiss-Krejci – this volume).

Selected caves, often with distinct physical features, have also been ascribed a symbolic significance as places of natural wonder, as secret-sacred sites, and as ritual boundaries, and have consequently been used as venues for the performance of other kinds of rituals. The vulva- and womb-like appearance of some cave entrances and passages is certainly striking, as in the case of a fissure cave in North Norway known locally as Bølakointa, meaning 'the Bøla cunt' (Bjerck – this volume; cf. Gimbutas 1999, 60; Gradoli and Meaden – this volume), although whether or not such caves were also used for rites of regeneration dedicated to an Earth Mother goddess remains more doubtful. Other caves used as sacred places were often marked by visually expressive material remains deposited in less accessible, liminal, and other-worldly areas of their underground systems. For example, in the Lower Gallery of La Garma, situated 130m in from the cave entrance, a range of remarkable Magdalenian cultural features have been recorded (Arias and Ontañón – this volume). These include: painted signs in a relatively hard-to-reach space; artistic engravings (some zoomorphic) on a low ceiling; drystone walls (and speletherms) delimiting a series of small spaces; floor deposits containing a large quantity of ornaments and other portable art objects, and a relative scarcity of stone and bone artefacts; and faunal remains with an unusually high proportion of horse bones, a modified horse skull, a modified lion bone, and two nearly completed skeletons of shelduck. These features tie in with the well-known corpus of European Upper Palaeolithic cave art (e.g. Ucko and Rosenfeld 1967; Sieveking 1979; Leroi-Gourhan 1982; Bahn and Vertut 1999; Lewis-Williams 2002; Pettitt *et al.* 2007), but must also be understood in the context of local cultural practices. In later periods, certain caves continued to be used for rituals, which took old and new forms. For example, in the deepest part of Grutta I de Longu Fresu in Sardinia, small-scale Middle Neolithic rituals involved the installation of a group of anthropomorphic paintings, a circular drystone structure, secondary deposits of human bones, and a single greenstone axe head, all close to a natural spring (Gradoli and Meaden; Skeates – this volume). Bronze Age cave paintings are also known along the coast of North Norway (Bjerck – this volume). The majority comprises 'stickmen' painted in red, but representations of zoomorphic figures and of a hand is also known, as well as a few geometric motifs. These images were typically placed in inaccessible spaces, visible to only a few people, perhaps during the course of controlled ritual activities and experiences involving social transformation and communication with supernatural chthonic forces. And in late Medieval Malta, some urban and rural rock-cut churches and oratories were decorated with murals (Buhagiar – this volume). Sacred spaces and symbols such as these are likely to have acquired multiple meanings through repeated ritual practice, including the recital of myths and legends featuring human ancestors, animals, and powerful spirits, and mediated appeals to these supernatural forces (Arias and Ontañón; Bjerck; Bonsall *et al.*; Gradoli and Meaden; Haug; Mlekuž; Weiss-Krejci – this volume). As sacralised places and as ritual boundaries, then, cave contexts have offered creative materials and opportunities to imagine, express, and mediate a wide range of ideas about human relations and identities.

The sixth theme in this volume concerns the embodied, sensory, and psychological, human experiences of cave environments, particularly those of otherworldly cave environments chosen for ritual performances. The sense of sight is commonly mentioned with reference to cave environments. On the one hand, rockshelters and cave entrances can often serve as landmarks and vantage-points: being seen from afar, and affording extensive views. But, on the other hand, contrasts of light, including gradations from daylight, to twilight, to darkness, are also characteristic features of cave interiors that often restrict the sense of sight, including views of the outside world and the visibility of people and other things contained within caves (Bjerck; Bonsall *et al.*; Haug; Mlekuž; Weiss-Krejci – this volume). Other senses can also be heightened or deprived in cave environments, with cramped spaces, humidity, coolness, mouldy smells, special sound effects, or silence affecting the way people feel, even to the extent of inducing spatial and temporal disorientation and fear – feelings that can be exploited during the course of controlled ritual performances and communicating (Arias and Ontañón; Bjerck; Mlekuž – this volume; cf. Roe 2000; Lewis-Williams 2002, 214–227; Whitehouse 2001; Skeates 2007). From these observations, it is clear that caves offer a significant potential to be reconsidered in terms of human experiences of their powerful multi-sensory environments: sensed, appropriated, and modified during the course of dwelling, visiting, working, performing, and thinking.

A seventh theme is the way caves and particularly cave dwellers have been perceived in the archaeological history of cave research. During the late nineteenth and early twentieth centuries, the people who had occupied the caves were generally described as primitive and backwards – a notion which was closely connected to contemporary ideas of social evolution. In popular culture these thoughts have endured until this day; 'caveman' is still a derogatory term in many European languages. These ideas probably also relate to the fact that many of the more recent cave dwellers were groups or individuals with a low social status, such as travellers and thieves. An important part of these early interpretative frameworks was that the cave dwellers were cannibals. Human skeletons and particularly disarticulated bones found in caves were related to cannibalism, often without closer investigations of the bones themselves (see for example Reuch 1877; Fürst 1910). As Orschiedt (this volume) points out, this alternative has even been chosen recently by scholars in Germany and Poland working with Stone Age and Bronze Age human remains from caves, while other possibilities have been ignored. However, Orschiedt's detailed examinations of these bones show that cut marks and scraping on the bones were probably the results of complex funerary rituals which did not necessarily include the consumption of human flesh. This example demonstrates that archaeologists dealing with caves and rockshelters continuously have to consider deeply rooted preconceptions about how prehistoric cave dwellers behaved, not only among the general public, but also among fellow scholars.

Future research directions in European cave archaeology

This volume provides a representative snap-shot of current research on caves and rockshelters in European archaeology. But it also offers some hints for the future. Certainly, we advocate the continued production of contextualizing regional syntheses of archaeological caves and their associated ancient landscapes, to help us better understand the varied place of caves in human history. And, within these, we recommend that scholars attempt to move beyond the traditional distinction between 'economic' (or 'domestic') and 'ritual' uses of caves, to a more inclusive and sophisticated consideration of caves and their sheltered contents as culturally valued practical and symbolic resources, even though, in some cases, we must acknowledge spatial distinctions between different activities. For example, in the Lower Gallery of La Garma, 'the distribution of Magdalenian floors and paintings at this site challenges the notion that areas with cave art and habitation areas were segregated spatially', although at the same site an inaccessible gallery area was identified with nothing but painted signs (Arias and Ontañón – this volume). We also look forward to the production of new data derived from the application of established and new techniques of archaeological science to caves, including the continued replacement of relative with absolute dating to resolve multiple questions of chronology. We encourage new, interdisciplinary, research to be undertaken on the recent (historic) human uses and significance of caves in Europe. And we hope for the development of more rigorous experimental work on human experiences of cave environments, as a contribution to the wider intellectual turn towards the senses in the social sciences (cf. Hamilton *et al.* 2006). Caves and rockshelters continue to be significant and meaningful archaeological resources and we hope that this volume will serve as a stepping stone for their further study within wider archaeological agendas.

References

Ahronson, K. and Charles-Edwards, T. M. (2010) *Prehistoric Annals* and Early Medieval monasticism: Daniel Wilson, James Young Simpson and their cave sites. *The Antiquaries Journal* 90, 455–466.

Angelucci, D. E., Boschian, G., Fontanals, M., Pedrotti, A. and Vergès, J. M. (2009) Shepherds and karst: the use of caves and rock-shelters in the Mediterranean region during the Neolithic. *World Archaeology* 41/2, 191–214.

Bahn, P. G. and Vertut, J. (1999) *Journey through the Ice Age.* London, Seven Dials.

Barnatt, J. and Edmonds, M. (2002) Places apart? Caves and monuments in Neolithic and Earlier Bronze Age Britain. *Cambridge Archaeological Journal* 12, 113–129.

Barrett, J. C. (1987) Contextual archaeology. *Antiquity* 61, 468–473.

Barrowclough, D. A. and Malone, C. eds. (2007) *Cult in Context: Reconsidering Ritual in Archaeology*. Oxford, Oxbow Books.

Bergsvik, K. A. (2005) Kulturdualisme i vestnorsk jernalder. In K. A. Bergsvik and A. Engevik jr. (eds.) *Fra Funn til Samfunn. Jernalderstudier tilegnet Bergljot Solberg på 70-årsdagen.* 229–258. Bergen, Universitetet i Bergen Arkeologiske Skrifter 1.

Bonsall, C. and Tolan-Smith, C. eds. (1997) *The Human Use of Caves*. Oxford, British Archaeological Reports, International Series 667.

Bradley, R. (2000) *An Archaeology of Natural Places*. London, Routledge.

Butzer, C. W. (1982) *Archaeology as Human Ecology: Method and Theory for a Contextual Approach*. Cambridge, Cambridge University Press.

CAPRA 1999–2007. Cave archaeology and palaeontology research archive (1999–2007). <http://www.capra.group.shef.ac.uk/>, accessed 18 January 2011.

Conkey, M. W. (1997) Beyond art and between caves: thinking about context on the interpretive process. In M. W. Conkey, O. Soffer, D. Strutmann and N. G. Jablowski (eds.) *Beyond*

Art: Pleistocene Image and Symbol, 343–367. San Francisco, California Academy of Sciences.

Daniel, G. (1981) *A Short History of Archaeology*. London, Thames and Hudson.

Dowd, M. A. (2008) The use of caves for funerary and ritual practices in Neolithic Ireland. *Antiquity* 82, 305–317.

Drewett, P. L. (1999) *Field Archaeology: An Introduction*. London, Routledge.

Evans, J. D. (1959) *Malta*. London, Thames and Hudson.

Fürst, C. M. (1909) *Das Skelett von Viste auf Jäderen*. Christiania, Jacob Dybwad.

Garasic, M. (1991) Morphological and hydrological classification of speleological structures (caves and pits) in the Croatian karst area. *Geoloski Vjesnik* 44, 289–300.

Geertz, C. (1973) Thick description: toward an interpretive theory of culture. In C. Geertz (ed.) *The Interpretation of Cultures: Selected Essays*, 3–30. New York, Basic Books.

Gili, S., Lull, V., Micó, R., Rihuete, C. and Risch, R. (2006) An island decides: megalithic burial rites on Menorca. *Antiquity* 80, 829–842.

Gimbutas, M. (1999) *The Living Goddesses*. Berkeley and Los Angeles, University of California Press.

Grayson, D. K. (1983) *The Establishment of Human Antiquity*. New York, Academic Press.

Green, H. S. (1989) The Stone Age cave archaeology of Wales. In T. D. Ford (ed.), *Limestones and Caves of Wales*, 70–78. Cambridge, Cambridge University Press.

Gunn, J. ed. (2004) *Encyclopedia of Caves and Karst Science*. London, Fitzroy Dearborn.

Hamilton, S., Whitehouse, R., Brown, K., Combs, P., Herring, E. and Seager-Thomas, M. (2006) Phenomenology in Practice. *European Journal of Archaeology* 9(1), 31–71.

Harris, E. C. (1979) *Principles of Archaeological Stratigraphy*. New York, Academic Press.

Hodder, I. (1986) *Reading the Past: Current Approaches to Interpretation in Archaeology*. Cambridge, Cambridge University Press.

Hodder, I. ed. (1987) *The Archaeology of Contextual Meanings*. Cambridge, Cambridge University Press.

Holderness, H., Davies, G., Chamberlain, A., and Donahue, R. (2006) *Research Report: A Conservation Audit of Archaeological Cave Resources in the Peak District and Yorkshire Dales*. Sheffield: Archaeological Research & Consultancy at the University of Sheffield Research School of Archaeology (Research Report 743.b). <http://capra.group.shef.ac.uk/7/743Research.pdf> accessed 2 November 2010.

Inskeep, R. ed. (1979) *World Archaeology* (Caves) 10(3).

Jacobsen, T. W. (1981) Franchthi cave and the beginning of settled village life in Greece'. *Hesperia* 50(4), 303–319.

Kornfeld, M., Vasil'ev, S. and Miotti, L. eds. (2007) *On Shelter's Ledge: Histories, Theories and Methods of Rockshelter Research. Proceedings of the XV World Congress UISPP (Lisbon, 4–9 September 2006), Volume 14, Session C54*. Oxford, British Archaeological Reports, International Series 1655.

Laporte, L., Joussaume, R. and Scarre, C. (2002) The perception of space and geometry: megalithic monuments of west-central France in their relationship to the landscape. In C. Scarre (ed.) *Monuments and Landscape in Atlantic Europe: Perception and Society during the Neolithic and Early Bronze Age*, 73–83. London and New York, Routledge.

Leroi-Gourhan, A. (1982) *The Dawn of European Art: An Introduction to Palaeolithic Cave Painting*. Trans. S. Champion. Cambridge, Cambridge University Press.

Lewis-Williams, D. J. (2002) *The Mind in the Cave: Consciousness and the Origins of Art*. London, Thames and Hudson.

Miracle, P. T. and Forenbaher, S. (2005) Neolithic and Bronze-Age herders of Pupićina cave, Croatia. *Journal of Field Archaeology* 30, 255–281.

Pettitt, P., Bahn, P. and Ripoll, S. (2007) *Palaeolithic Cave Art at Creswell Crags in European Context*. Oxford, Oxford University Press.

Prijatelj, A. (2010). (Hi)stories of cave archaeology in Slovenia: politics, institutions, individuals, methods and theories. <http://www.uplandcavesnetwork.org/files%20for%20download/Abstractper cent20book%20-%20Cave%20archaeology,%20past%20present%20and%20future.pdf> accessed 2 November 2010.

Reusch H. H. (1877) Nogle norske Huler II. Sjonghelleren. *Naturen* 1(4), 49–57.

Roe, M. (2000) The brighter the light the darker the shadows: how we perceive and represent underground spaces'. *CAPRA*, 2. <http://capra.group.shef.ac.uk/2/roe.html>, accessed 7 November 2008.

Shanks, M. and Hodder, I. (1995) Processual, postprocessual and interpretive archaeologies. In I. Hodder, M. Shanks, A. Alexandri, V. Buchli, J. Carman, J. Last and G. Lucas (eds.) *Interpreting Archaeology: Finding Meaning in the Past*, 3–29. London and New York, Routledge.

Sieveking, A. (1979) *The Cave Artists*. London, Thames and Hudson.

Simek, J. F. (1994) Archaeology of caves: history. In J. Gunn (ed.), *Encyclopedia of Caves and Karst Science*, 80–82. London, Fitzroy Dearborn.

Skeates, R. (1994) Ritual, context and gender in Neolithic South-Eastern Italy. *Journal of European Archaeology* 2(2), 199–214.

Skeates, R. (2007) Religious experience in the prehistoric Maltese underworld. In D. Barrowclough, C. Malone and S. Stoddart (eds.) *Cult in Context: Reconsidering Ritual in Archaeology*, 90–96. Oxford, Oxbow Books.

Skeates, R. (2010) *An Archaeology of the Senses: Prehistoric Malta*. Oxford, Oxford University Press.

Ucko, P. J. and Rosenfeld, A. (1967) *Palaeolithic Cave Art*. London, Weidenfeld and Nicolson.

Tolan-Smith, C. and Bonsall, C. (1997) 'The human use of caves'. In C. Bonsall and C. Tolan-Smith (eds.) *The Human Use of Caves*, 217–218. Oxford, British Archaeological Reports, International Series 667.

Trigger, B. (2006) *A History of Archaeological Thought. Second Edition*. Cambridge, Cambridge University Press.

Watson, P. J. (2001) Theory in cave archaeology. *Midcontinental Journal of Archaeology*, 26(2), 139–143.

Weaver, H. D. (2008) *Missouri Caves in History and Legend*. Columbia, University of Missouri Press.

White, W. B. and Culver, D. C. (2005) Cave, definition of. In W. B. White and D. C. Culver (eds.) *Encyclopedia of Caves*, 81–85. Amsterdam, Elsevier.

Whitehouse, R. D. (2001) A tale of two caves. In P. F. Biehl and F. Bertemes (eds.) *The Archaeology of Cult and Religion*, 161–167. Budapest, Archaeolingua.

Chapter 2

From Assynt to Oban: Some Observations on Prehistoric Cave Use in Western Scotland

Clive Bonsall, Catriona Pickard and Graham A. Ritchie

This paper examines the evidence for temporal variation in the human use of caves in western Scotland. The archaeological record of cave use in this part of the British Isles extends back to at least the Late Glacial. However, there is little evidence for the regular use of caves until the later Mesolithic, around 7500 cal BC. During that period an important use of caves was as temporary shelters for food-processing activities. A significant change in the human use of caves occurred around 3900 cal BC, coincident with the appearance of farming. At that time there was a shift from economic use to funerary or ritual function. This may reflect the change in economic practices and mortuary rituals that marked the Mesolithic-Neolithic transition, but in part at least was influenced by the geomorphic evolution and changing nature of caves in Mid-Holocene Scotland.

Introduction

Caves have attributes that have given them a crucial role in understanding the past. Some caves are highly visible features in the landscape, and so were among the first sites to attract the attention of early archaeologists. Stratification is more common in caves than in many other kinds of archaeological site, facilitating the chronological ordering of cultural deposits. Moreover, the protected environment of a cave quite often leads to the preservation of animal and plant remains that would be far less likely to survive in open-air sites. In fact, in regions dominated by acid soils, like much of Northwest Europe, caves may be among the few sites that provide direct evidence of the subsistence activities of prehistoric peoples, or even their skeletal remains.

In this paper, we review the evidence for prehistoric cave use in the area of Scotland most familiar to us archaeologically – the Atlantic coast and its immediate hinterland (Fig. 2.1). Paradoxically, this is an area where some of the best-known archaeological caves were not prominent landscape features, but were revealed accidentally in the course of urban development. Yet, they offer an interesting (though often neglected) perspective on human behaviour in prehistory.

Caves in western Scotland

A cave may be defined as *any natural opening or cavity in bedrock large enough for a human to enter*. At one end of the spectrum are rockshelters, shallow undercuts in cliff faces, and at the other are large caverns that extend deep underground and receive daylight only near their entrances. Most caves are formed by some kind of weathering or erosion, even if the processes involved are not always well understood. Caves occur throughout Scotland, but are concentrated particularly in the west and north. Four main types of cave may be distinguished by their morphology and/or mode of formation: karstic caves, sea caves, fissure caves, and boulder (or talus) caves.

Karstic caves are the result of the dissolution of limestone and similar carbonate rocks by weakly acidic

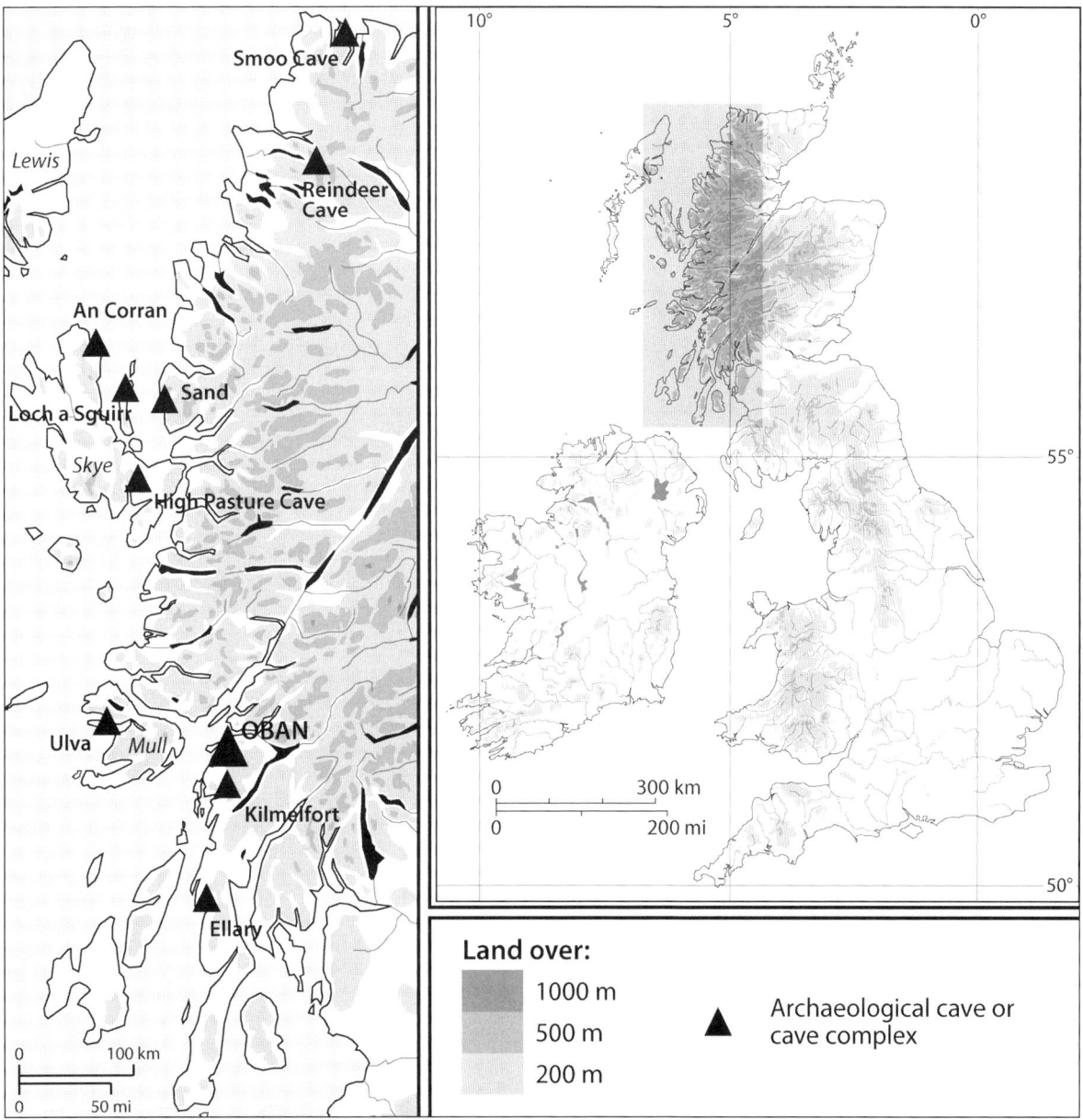

Figure 2.1: Archaeological sites and localities mentioned in the text.

groundwater, in some cases combined with the erosive action of subterranean streams. The main carbonate rock formations in Scotland are the Cambro-Ordivician dolostones and limestones of the Durness series. Their outcrop area forms a narrow band running from Durness on the north coast, south-southeast as far as the Isle of Skye. Caves are common in the 'Durness Limestone', some extending for hundreds of metres, although the entrances are often small (Fig. 2.2A). Only a few are archaeologically important; they include Reindeer Cave, in Assynt (Callander *et al.* 1927; Lawson and Bonsall 1986a, 1986b; Murray *et al.* 1993; Saville 2005), High Pasture Cave, on Skye (Birch *et al.* 2009), and Smoo Cave near Durness (Pollard 1992). The last mentioned lies at the head of a narrow sea inlet and its front part has been enlarged by marine erosion, so it combines characteristics of both a karstic cave and a sea cave.

Sea caves are formed primarily by wave action exploiting zones of weakness, such as faults or joints, in sea cliffs. Numerous relict sea caves occur along former shorelines fringing the west coast of Scotland

Figure 2.2: Examples of four types of cave found in western Scotland. A. karstic caves, Creag nan Uamh, Assynt (photo: Tim Lawson); B. sea cave, Ulva, Argyll (photo: Clive Bonsall); C. fissure cave, Carding Mill Bay, Oban (photo: Clive Bonsall); D. boulder cave, Ellary, Argyll (photo: Christopher Tolan-Smith).

(Fig. 2.2B). The most extensively developed of these ancient shorelines is the Main Rock Platform – a prominent coastal rock platform backed by a former seacliff up to 30m high – which is thought to have formed during the Younger Dryas (Sissons 1974; Stone *et al.* 1996). This relict shoreline has since been uplifted and tilted by glacio-isostatic rebound, and now lies above sea level along mainland and inshore island coasts of central-west Scotland between Skye and the Isle of Arran. The Main Rock Platform rises to a height of around 12m at the head of the Firth of Lorn. Around Oban Bay it stands at around 9m and the caves and rockshelters in its backing cliff have been the focus of archaeological interest since the discovery of Mesolithic shell middens there in the late nineteenth century (Anderson 1895, 1898; Lacaille 1954; Bonsall and Sutherland 1992). Older rock platform fragments and associated cliffs and caves occur at elevations up to around 50m above present sea level along the west coast, and several 'high-level' caves have produced important archaeological remains dating back to the Mesolithic, most notably Sand Rockshelter on the Applecross

Peninsula (Hardy and Wickham-Jones 2003, 2009) and Ulva Cave near Mull (Bonsall *et al.* 1989; Russell *et al.* 1995). On the whole, relict sea caves occur with remarkable frequency along the raised shorelines of western Scotland. During a recent survey of the coastline and islands of the sea strait between Skye and the mainland, 141 caves and rockshelters were recorded, many of which showed evidence of past human use (Hardy 2009).

Fissure caves are created by mechanical widening of thin fissures in bedrock, forming narrow, often triangular-shaped crevices (Fig. 2.2C). Though not confined to coastal areas, fissure caves are particularly common in the relict sea cliffs of western Scotland, and several of the archaeologically important caves around Oban Bay are of this type, including Carding Mill Bay I (Connock *et al.* 1993), Carding Mill Bay II (Bartosiewicz *et al.* 2010), and Raschoille Cave (Bonsall 1999).

Boulder (or talus) caves are cavities created where boulders were piled up as a result of ice transport or rockfalls. Typically, they occur on mountain slopes or at the bases of cliffs, and in rare cases they can resemble dolmens (cf. Lacaille 1954, fig. 11). Arguably, the best-known archaeological example from western Scotland is the Ellary Cave in Mid-Argyll (Fig. 2.2D), with a record of human use extending back to at least the Neolithic (Tolan-Smith 2001).

Caves and early human occupation

The dating of the earliest human occupation of Scotland is still a matter of some debate. The oldest securely dated archaeological evidence comes from open-air sites belonging to the Early Holocene, around 8400 cal BC (J. Lawson 2001; Ashmore 2004), but there are a few findspots of lithic artefacts with parallels in Late Upper Palaeolithic (LUP) assemblages from southern Britain and the North European mainland dating to the Late Glacial period between around 12,000 and 9600 cal BC (Livens 1956; Morrison and Bonsall 1989; Ballin and Saville 2003).

Evidence of pre-Holocene activity has also been suspected in two Scottish caves, Reindeer Cave in Assynt, and Kilmelfort Cave on the west coast near Oban.

At Reindeer Cave large numbers of shed reindeer antlers were recovered from Late Pleistocene deposits (Callander *et al.* 1927). Based on the data provided in the original excavation report and a single radiocarbon date on a bulk sample of antler fragments, Lawson and Bonsall (1986a, 1986b) suggested the antlers had been deposited in the cave during the Younger Dryas, with humans rather than wolves or reindeer being the likely agent of accumulation. However, further research involving single entity AMS radiocarbon dating demonstrated that the antlers were deposited over many thousands of years extending back to before the Last Glacial Maximum (LGM), and not introduced to the cave *en masse* during the Younger Dryas, which weakens the argument for human involvement (Murray *et al.* 1993).

At Kilmelfort a small limestone cave sealed by talus deposits was exposed and partially destroyed during construction work for a hydroelectric power installation. Excavation of the surviving deposits by Coles (1983) led to the recovery of over 700 flint and quartz artefacts. Some were recovered *in situ* from at least three different lithostratigraphic units, but the great majority came from disturbed contexts and so probably are a 'mixed' assemblage. Originally interpreted as 'Mesolithic' by Coles (1983, 11–12), the collection was re-examined by Ballin (2008) who noted the similarity of some of the backed tool forms to those from LUP sites on the North European mainland dated to the period between the Hamburgian/Creswellian and Federmesser technocomplexes. On this basis, Ballin suggested an age of around 12,000 cal BC for the initial occupation of Kilmelfort Cave.

The fact that there is no sound evidence of human occupation in Scotland prior to the Late Glacial raises the possibility that the archaeological record has been truncated by successive Pleistocene glaciations. Virtually the whole of Scotland was overridden by ice during the LGM (around 20,000 cal BC), which is likely to have erased the evidence of earlier human settlement — from open-air locations, at least. In parts of England covered by the last ice sheet evidence of Palaeolithic settlement is extremely rare, whereas beyond the ice sheet limit it is abundant.

If there was human settlement of Scotland before the Late Devensian, then the evidence is most likely to have survived in caves that offered some protection from glacial erosion. The potential for survival of such evidence is demonstrated by occurrences of well-preserved Pleistocene mammalian remains in several caves in the Durness Limestone in Assynt, including Reindeer Cave (see above).

Why human artefacts have not been found already is a matter for speculation. Humans may not have been present before the Late Glacial. On the other hand, very little systematic archaeological work has been undertaken in caves with known Pleistocene sequences. In fact, many caves that have attracted interest from archaeologists are relict sea caves formed *after* the LGM, like those around Oban Bay and the Inner Sound between Skye and the mainland. A possible exception is Ulva Cave near Mull (Bonsall *et al.* 1989; Russell *et al.* 1995), where Holocene deposits (the current focus of investigation) are underlain by a thick Pleistocene sequence. Whether these include deposits predating the LGM remains to be determined.

Evidence of LUP or Early Mesolithic use of caves in Scotland is still comparatively rare, and cave use on a 'regular' basis is not apparent until the later Mesolithic, after around 7500 cal BC.

The evidence for later Mesolithic cave use in western Scotland comes mainly from former sea caves along the coast. Only a small number of sites have been excavated, and radiocarbon dates are available for seven of them: Druimvargie Rockshelter, MacArthur's Cave, and Raschoille Cave (around Oban Bay), Ulva Cave (near Mull), An Corran (on Skye), Loch a Sguirr (on the Isle of Raasay), and Sand (on the Applecross Peninsula) (Fig. 2.1). A consistent feature of these sites is the presence of midden deposits composed largely of the shells of rocky shore molluscs, such as limpets (*Patella* sp.) and periwinkles (*Littorina* sp.). The shell middens usually contain other food remains, such as fish, crab, and even terrestrial plant and animal material, as well as stone and bone artefacts, although they vary somewhat in size and composition. In most cases, the small size of the sites, the preponderance of expedient over formal tools, the absence of associated architectural features, and the fact that the caves sometimes occur in relatively exposed locations without immediate access to permanent water sources or natural boat landings, suggest they were not residential sites. More likely they were food-processing loci, set up where a cave was conveniently situated near to shell beds or prime fishing spots that were some distance away from the residential base camp (Bonsall 1996). The impression gained from some recently excavated sites (e.g. Ulva Cave) is that the caves were used occasionally – perhaps once a year or every few years – over generations, centuries or millennia.

Possible exceptions to this pattern exist. At Sand Rockshelter there is evidence of a wider range of activities compared to other sites, including microlith production and skin working, and radiocarbon dating suggests the greater part of the midden could have been deposited during two periods of relatively intensive occupation around 6600 and 5500 cal BC (Hardy and Wickham-Jones 2009).

The logistical use of caves in the west Scottish Mesolithic may not always have been connected directly with the food quest. Caves can be a source of raw materials, including knappable lithic materials for the production of stone tools. 'Baked mudstone' (the result of contact metamorphism around igneous intrusions) was well represented in the Mesolithic assemblage from Sand, and was found in a number of other sites located around the sea strait between Skye and the mainland.

Currently, the only known source of this material in the region is at An Corran, where it occurs as nodules in the cliffs above the cave and as pebbles on the beaches below (Wickham-Jones 2009). It is possible that human use of the An Corran rockshelter site was in part connected with the exploitation of this raw material source.

The use of coastal caves for economic purposes seen in the Mesolithic continued into later periods, with some elements of 'Mesolithic' technology (e.g. bevel-ended tools) still found as late as the Bronze Age (Saville 1998a, 1998b). The midden at An Corran apparently continued in use into the Neolithic, while the shell middens in fissure caves at Carding Mill Bay, Oban, may not have been used before the Neolithic judging from the available radiocarbon dates and the presence of pottery and bones of domestic animals (Connock *et al*. 1993; Bartosiewicz *et al*. 2010).

Cave use and geomorphological changes

Geomorphological processes affected the way in which caves were used, as well as the chances of discovery of the evidence of past human uses. This is clearly illustrated by evidence from the Oban area.

Sea-cut caves abound in the backing cliff of the Main Rock Platform around Oban Bay, their location and morphology clearly controlled by bedrock lithology and structure (Macklin and Rumsby 1991). They are particularly well developed in outcrops of Dalradian black slates and Lower Old Red Sandstone conglomerates and intrusive andesites, which are strongly affected by faulting and jointing.

The Oban caves have attracted archaeological interest since the nineteenth century, although relatively few have been investigated systematically, and nearly all were disturbed before excavations took place (Fig. 2.3). Nevertheless, research has shown that human utilization of these sites extended back to at least the later Mesolithic (Bonsall and Sutherland 1992).

The Main Rock Platform and associated caves were likely formed during the Younger Dryas by marine erosion assisted by periglacial frost shattering (Sissons 1974; Stone *et al*. 1996). For the caves to become available for human use would have required a relative fall of sea level of perhaps several metres. Evidence suggests the Early Holocene was a time of marine regression in central-west Scotland with a low sea-level stand reached around 7500–7000 cal BC (Sissons 1983; Sutherland 1988). This was followed by a major transgression (Main Postglacial Transgression), which culminated in the Oban area around 5600 cal BC (Bonsall and Sutherland 1992) when the sea rose to above the level of the Main Rock Platform. On this evidence, the Oban caves would have stood above sea level and been available for human use for a time during the Early Holocene, although the evidence of that activity could have been removed from more exposed caves that were re-occupied by the sea around the transgression maximum.

At least three of the Oban caves contained evidence of human use around the maximum of the transgression, or earlier. MacArthur's Cave faced onto the open part of the Oban embayment, and at the maximum of the transgression the cave floor (at 11m above sea level) appears to have been within the reach of storm waves. Two Mesolithic shell midden layers were recorded, the younger one overlying a layer of rounded gravel, which was interpreted as a beach deposit, and the older one stratified within the upper part of the gravel (Anderson 1895). A bone artefact from one of the midden layers (the exact location is unknown) was radiocarbon dated by AMS to around 5600 cal BC (Table 2.1).

Even older Mesolithic deposits were recorded in Druimvargie Rockshelter and Raschoille Cave (Fig. 2.3). AMS radiocarbon dates on bone artefacts from Druimvargie range from 8340±80 to 7810±90 BP corresponding to a calendar age range of around 7400–6650 cal BC, while bone artefacts and carbonized plant material associated with shell midden deposits in Raschoille Cave produced dates between 7640±80 and 7250±55 BP corresponding to a calendar age range of around 6500–6150 cal BC (Table 2.1). That the Mesolithic deposits in these caves survived the Middle Holocene transgression may be due to the fact that both sites were located in a more protected part of the Oban embayment where wave and wind energy were lower.

An equally important environmental control on the human utilization of caves was cave infilling. Infilling of the caves around Oban Bay resulted from two sets of processes: the accumulation of refuse associated

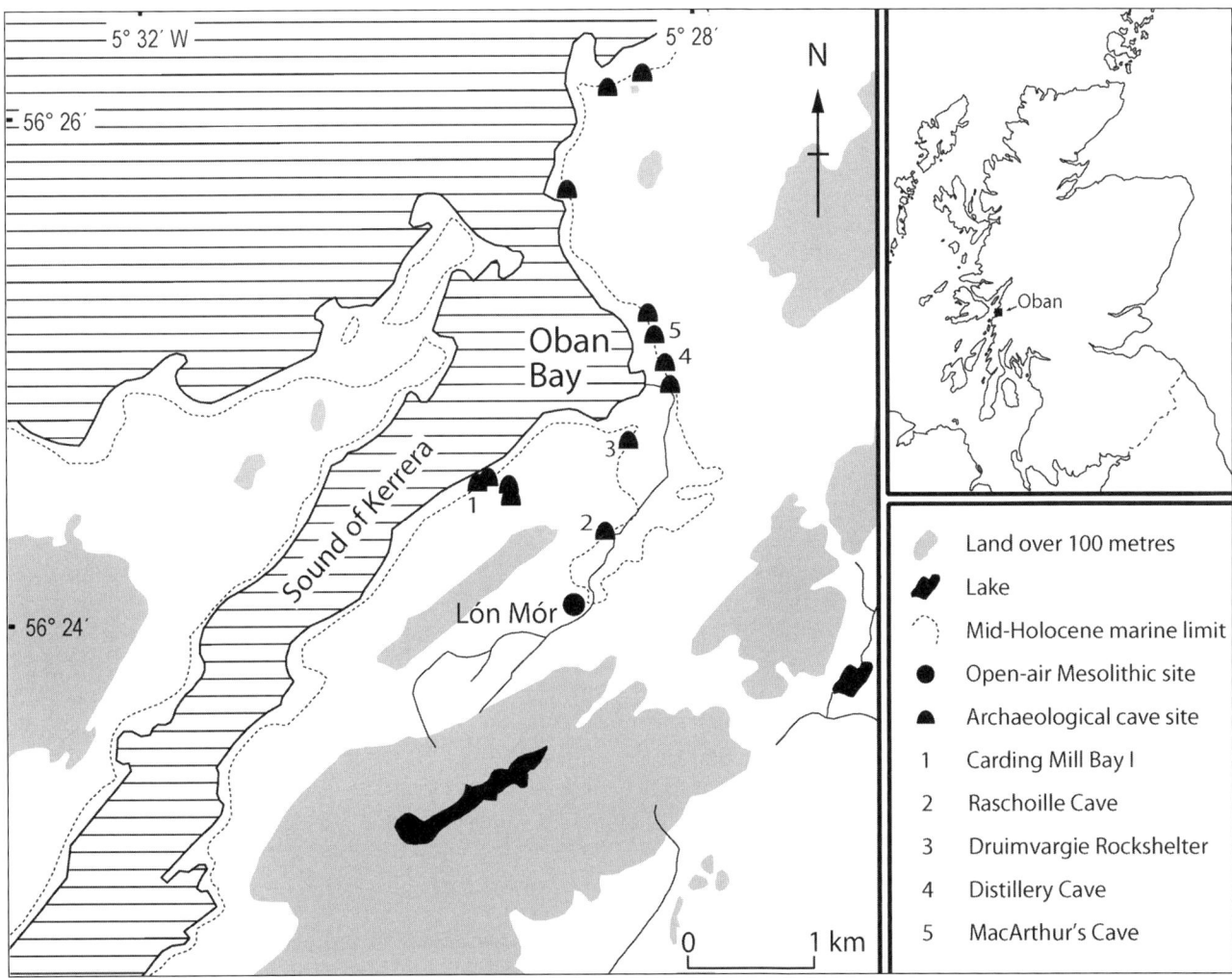

Figure 2.3: Archaeological caves around Oban Bay.

Table 2.1: Radiocarbon dates from the Oban caves. Calibrations were performed with CALIB 6.0 (Stuiver and Reimer 1993; Stuiver et al. 2005) using the IntCal09 and Marine09 datasets (Reimer et al. 2009). In applying a marine reservoir correction to the dates on mollusc shells, we have assumed a ΔR of 68±16 ^{14}C yr for the Raschoille Cave dates and a ΔR of 143±20 ^{14}C yr for the Carding Mill Bay 1 dates (following Ascough et al. 2007). The three dates from Raschoille Cave, which are all on the same sample (a single valve of the common cockle, Cerastoderma edule), should be regarded as suspect since the outer part of the shell gave a significantly younger age than the inner, suggesting recrystallization of the shell has occurred. * Weighted mean of two measurements on a single sample.

Site	Lab. Ref.	Stratigraphic context	Material	^{14}C Age BP	δ^{13}C ‰	cal BC age range 2s	Median probability (cal BC)	Reference
Carding Mill Bay 1	OxA-3740	Midden, lower part	Antler artefact	5190±85	-20.5	4236-3794	4009	Bonsall and Smith 1992
	SUERC-3953/4952	Midden, upper part	Hazelnut shell, charred	5053±29*	-23.1	3952-3783	3873	Ascough et al. 2007
	GU-2796	Midden, upper part	Charcoal (mixture of cherry, elm, hazel and oak)	5060±50	-25.6	3964-3714	3862	Connock et al. 1993; G. Cook, pers. comm.
	GU-2898	Midden, upper part	Marine shells (limpets)	5410±60	1.0	3924-3644	3777	Connock et al. 1993; G. Cook, pers. comm.
	GU-2797	Midden, lower part	Charcoal (mixture of alder, cherry, elm, hazel, and oak)	4980±50	-25.8	3942-3653	3761	Connock et al. 1993; G. Cook, pers. comm.
	GU-2899	Midden, lower part	Marine shells (limpets)	5440±50	0.6	3879-3614	3708	Connock et al. 1993; G. Cook, pers. comm.
	SUERC-4950	Midden, upper part	Marine shell (limpet)	5335±40	0.7	3694-3504	3603	Ascough et al. 2007
	SUERC-4947	Midden, upper part	Marine shell (limpet)	5330±35	1.4	3682-3507	3598	Ascough et al. 2007
	OxA-7664	Midden, lower part	Human bone	4830±45	-20.9	3703-3521	3596	Schulting and Richards 2002
	SUERC-4949	Midden, upper part	Marine shell (limpet)	5325±40	0.6	3685-3498	3592	Ascough et al. 2007
	SUERC-4948	Midden, upper part	Marine shell (limpet)	5310±40	1.3	3673-3486	3578	Ascough et al. 2007
	OxA-7663	Midden, upper part	Human bone	4800±50	-21.5	3693-3381	3571	Schulting and Richards 2002
	SUERC-3587	Midden, upper part	Hazelnut shell, charred	4775±35	-22.7	3643-3384	3568	Ascough et al. 2007
	SUERC-3588	Midden, upper part	Hazelnut shell, charred	4785±45	-25.7	3652-3381	3567	Ascough et al. 2007
	SUERC-3592/4951	Midden, upper part	Hazelnut shell, charred	4816±27*	-26.6	3653-3527	3567	Ascough et al. 2007
	OxA-3739	Midden, upper part	Bone artefact	4765±65		3654-3373	3550	Bonsall and Smith 1992
	OxA-7665	Midden	Human bone	4690±40	-21.4	3630-3368	3456	Schulting and Richards 2002
	OxA-7890	Fissure deposits	Human bone	4330±60	-22.0	3311-2765	2969	Schulting and Richards
Distillery Cave	OxA-4509	?	Antler (red deer) – artefact	3780±60		2457-2032	2211	Saville and Hallén 1994
Druimvargie Rockshelter	OxA-4608	Midden	Animal bone (large ungulate) – artefact	8340±80		7569-7177	7393	Bonsall et al. 1995
	OxA-4609	Midden	Animal bone (large ungulate) – artefact	7890±80		7043-6598	6786	Bonsall et al. 1995
	OxA-1948	Midden	Animal bone (large ungulate) – artefact	7810±90		7027-6466	6663	Bonsall and Smith 1989

Site	Lab no.	Context	Material	¹⁴C age BP	δ¹³C	Cal BC (2σ)	Cal BP	Reference
MacArthur's Cave	OxA-1949	Midden	Antler (red deer) – artefact	6700±80		5726–5488	5618	Bonsall and Smith 1989
	OxA-4487	?Lower shell layer	Human bone	2460±55	-21.9	764–409	590	Saville and Hallén 1994
	OxA-4486	?Upper shell layer	Human bone	2365±55	-21.3	753–259	471	Saville and Hallén 1994
	OxA-4488	?Lower shell layer	Human bone	2295±60	-22.3	518–196	349	Saville and Hallén 1994
	OxA-4485	Black earth	Human bone	2170±55	-21.4	380–59	235	Saville and Hallén 1994
Raschoille Cave	OxA-8396	Lower deposits	Animal bone (red deer)	7640±80	-21.8	6647–6368	6496	Bonsall 1999
	OxA-8397	Lower deposits	Animal bone (red deer)	7575±75	-21.5	6590–6253	6433	Bonsall 1999
	OxA-8395	Lower deposits	Animal bone (lynx)	7495±50	-19.9	6441–6247	6367	Bonsall 1999
	OxA-8398	Lower deposits	Animal bone (red deer) – artefact	7480±75	-21.6	6470–6214	6342	Bonsall 1999
	OxA-8535	Lower deposits	Animal bone (red deer) – artefact	7265±80	-21.4	6352–5989	6138	Bonsall 1999
	OxA-8439	Lower deposits	Hazelnut shell, charred	7250±55	-25.1	6224–6022	6126	Bonsall 1999
	OxA-8501	Lower deposits	Marine shell (common cockle)	7390±55	1.4	5966–5716	5836	Bonsall 1999
	OxA-8839	Lower deposits	Marine shell (common cockle)	7580±45	3.0	6150–5913	6022	Bonsall 1999
	OxA-8840	Lower deposits	Marine shell (common cockle)	7300±50	2.6	5866–5637	5747	Bonsall 1999
	OxA-8538	Lower deposits	Animal bone (red deer)	6460±180	-22.1	5717–5005	5403	Bonsall 1999
	OxA-8438	Lower deposits	Hazelnut shell, charred	5115±55	-26.3	4039–3780	3888	Bonsall 1999
	OxA-8440	Lower deposits	Hazelnut shell, charred	4995±45	-21.8	3943–3662	3777	Bonsall 1999
	OxA-8432	Upper deposits	Human bone	4980±50	-20.4	3942–3653	3761	Bonsall 1999
	OxA-8431	Upper deposits	Human bone	4930±50	-20.6	3904–3638	3710	Bonsall 1999
	OxA-8433	Upper deposits	Human bone	4920±50	-20.2	3893–3636	3701	Bonsall 1999
	OxA-8441	Upper deposits	Human bone	4900±45	-21.2	3783–3635	3684	Bonsall 1999
	OxA-8442	Upper deposits	Human bone	4890±45	-21.0	3775–3540	3678	Bonsall 1999
	OxA-8536	Lower deposits	Wood charcoal (hazel – single fragment)	4880±60	-27.2	3794–3524	3673	Bonsall 1999
	OxA-8404	Upper deposits	Human bone	4850±70	-21.6	3787–3381	3641	Bonsall 1999
	OxA-8443	Upper deposits	Human bone	4825±55	-20.4	3709–3383	3589	Bonsall 1999
	OxA-8434	Upper deposits	Human bone	4720±50	-21.1	3635–3373	3510	Bonsall 1999
	OxA-8444	Upper deposits	Human bone	4715±45	-21.1	3634–3372	3503	Bonsall 1999
	OxA-8435	Upper deposits	Human bone	4680±50	-22.5	3631–3362	3460	Bonsall 1999
	OxA-8400	Upper deposits	Human bone	4640±65	-20.3	3634–3117	3442	Bonsall 1999
	OxA-8399	Upper deposits	Human bone	4630±65	-21.4	3631–3108	3432	Bonsall 1999
	OxA-8401	Upper deposits	Human bone	4565±65	-21.1	3515–3030	3264	Bonsall 1999
	OxA-8537	Upper deposits	Human bone	4535±50	-21.8	3487–3034	3216	Bonsall 1999

with human activity, and natural clastic sediment input from rockfall and talus accumulation adjacent to the cliff in which the caves occur. This rockfall activity is thought to have been most intense during the earlier Holocene before 4000 cal BC, possibly linked to greater seismic activity (Ballantyne and Dawson 1997; Sutherland 1997). However, the well-developed talus slopes beside the backing cliff of the Main Rock Platform around Oban Bay must have formed largely since the maximum of the Middle Holocene marine transgression around 5600 cal BC, since older talus accumulations would likely have been eroded during that event. One consequence of the development of these talus slopes was a gradual narrowing, and in many cases the complete closure, of cave entrances (Fig. 2.4). In fact, all the known archaeological cave sites around Oban Bay were hidden behind talus deposits and only discovered when the talus was removed during building operations.

A change in cave use is evident in the Oban area after 3900 BC, when there is also evidence of a shift from a foraging-centred economy to one based on farming. Economic use of the coastal caves did not come to an abrupt halt, as the presence of post-Mesolithic shell midden deposits at Carding Mill Bay testifies, but some caves started to be used for disposal of the dead.

The earliest known burials are those from Raschoille Cave and Carding Mill Bay I (Table 2.1). Raschoille has a large series of AMS radiocarbon dates on human bone that place the burial activity there between around 3760 and 3220 cal BC. The human remains were stratified in loose, angular rock debris in the forepart of the cave.

The rock debris is likely to be the product of weathering of the steep rock face above the cave, rather than roof collapse, and this suggests the burials were emplaced when the entrance of the cave was becoming obstructed by talus accumulation.

The practice of cave burial continued into later periods in the Oban area. Nearly all the archaeological caves yielded human remains. Individual sites appear to have been used at different periods, but for a limited time in each case, with the human remains being the most recent archaeological deposits. At Raschoille Cave burial activity was confined to a few centuries during the Early Neolithic. At Carding Mill Bay I burial activity appears to have continued into the Early Bronze Age, judging from the associated pottery (Connock *et al.* 1993); while AMS radiocarbon dating places the burials from MacArthur's Cave in the Early Iron Age between around 600 and 200 cal BC (Table 2.1).

From this rather limited evidence, it is tempting to suggest that the caves around Oban Bay were utilized for disposal of the dead when their entrances (and internal space, perhaps) were becoming constricted by rockfall and they were no longer usable as temporary shelters for subsistence-related tasks. This happened at different times in different sites, but ultimately each cave was closed by talus accumulation, at which point burial activity effectively ceased – although it is not implausible that a 'burial cave' would have retained some ceremonial or symbolic significance for a time after closure.

The use of caves for burial is commonplace in the Neolithic and Bronze Age of Britain and Ireland

Figure 2.4: The Main Rock Platform and backing cliff on the island of Kerrera, photographed from Carding Mill Bay on the mainland (cf. Fig. 2.3). Caves and undercuts in the relict cliff are partially visible behind vegetated talus (photo: Clive Bonsall).

(Barnatt and Edmonds 2002; Dowd 2008). Caves may be viewed as thresholds that represent a boundary between light and dark and perhaps acted as an allegory for life and death. This liminality may have facilitated mediation with spirits and ancestors.

Caves may have been perceived as equivalent to monumental tombs with little distinction drawn between the 'made' and the natural monument in the Neolithic (Barnatt and Edmonds 2002, 125–126). The earliest use of the Oban caves as burial sites broadly coincided with the appearance of man-made burial monuments in the wider region (cf. Ritchie 1997).

From published accounts of excavations in the Oban caves it is uncertain whether the human remains represent primary burials of intact corpses, or secondary disposal of disarticulated bones or body parts, linked to the practice of excarnation. Nevertheless, similarity in ritual at man-made and cave sites can be discerned. Deliberate 'blocking' of the passages or entrances to chambered tombs may be perceived as 'metaphor' for natural closure of burial caves through talus accumulation.

Communal burial practices evident in the Early Neolithic funerary monuments were replaced in the Bronze Age with a renewed emphasis on the individual (Thomas 2000). Single grave burials, often associated with abundant grave goods, became the dominant mortuary practice. Just as burial monuments and associated funerary practices underwent change through time, so did the use of caves as places for the formal disposal of the dead. Although not as well documented for the Oban caves as for the man-made structures, similar changes in mortuary practice are suggested. Stone slabs covered the Early Bronze Age burials at Carding Mill Bay. This practice and the associated grave goods (pottery and stone tools) are reminiscent of the widespread practice of cist burial in the Scottish Early Bronze Age. Neither practice was evident with the Neolithic burials at Carding Mill Bay I or Raschoille Cave (cf. Connock et al. 1993).

Although within the British Isles caves were unlikely to have been the inspiration for the construction of burial monuments (Barnatt and Edmonds 2002) it is possible that certain coastal caves may have taken on the role of the chambered tombs and passage cairns constructed by communities living further inland where caves are less common. In the wider European context, however, it is an interesting question whether the tradition of cave burial as manifested around the Mesolithic-Neolithic transition was the forerunner of the chambered tomb phenomenon, or vice versa.

Conclusion

In western Scotland human use of caves is recorded from the Late Upper Palaeolithic onwards; however, there is little evidence for the regular use of caves until the later Mesolithic. The nature of cave use changed around 3900 BC, coincident with the appearance of agriculture in Scotland.

The change in cave use from 'economic' to 'ritual' or 'ceremonial' that is observable around Oban Bay was not unique to that area. Burial caves are common in western Scotland, though clear patterns in cave use through time are less well documented than in the Oban area. Not all burial caves were in coastal locations. Equally, some coastal caves (like Sand and Ulva) were never used for formal burials. There are also burial caves that did not have their entrances 'closed' by talus accumulation; and some caves that were used for burial in prehistory even reverted to an economic use (e.g. as stock enclosures) in later periods – an obvious example being Reindeer Cave in Assynt (Callander et al. 1927; Bonsall and Murray 1997).

But all caves had individual environmental histories, which probably influenced how they were used at different periods, or whether they were used at all. Therefore, if there is a conclusion to be drawn from this review, it is perhaps this – in interpreting *what* went on in a cave, *when*, and *why*, archaeologists need to have regard to the geomorphological history of the site!

Acknowledgements

The authors would like to thank Tim Lawson and Christopher Tolan-Smith for permission to reproduce the photographs used in Figures 2A and 2D, Gordon Cook for advice on calibration of marine shell radiocarbon dates and for additional information included in Table 2.1, and Magda Midgley for discussion of Neolithic chambered tombs.

References

Anderson, J. (1895) Notice of a cave recently discovered at Oban, containing human remains and a refuse heap of shells and bones of animals, and stone and bone implements. *Proceedings of the Society of Antiquaries of Scotland* 29, 211–230.

Anderson, J. (1898) Notes on the contents of a small cave or rock shelter at Druimvargie, Oban; and of three shell mounds on Oronsay. *Proceedings of the Society of Antiquaries of Scotland* 32, 298–313.

Ascough, P., Cook, G. T., Dugmore, A. J. and Scott, E. M. (2007) The North Atlantic marine reservoir effect in the Early Holocene: implications for defining and understanding MRE values. *Nuclear Instruments and Methods in Physics Research B* 259, 438–447.

Ashmore, P. J. (2004) Dating forager communities in Scotland. In A. Saville (ed.) *Mesolithic Scotland and its Neighbours*, 83–94. Edinburgh, Society of Antiquaries of Scotland.

Ballantyne, C. and Dawson, A. G. (1997) Geomorphology and landscape change. In K. J. Edwards and I. B. M. Ralston (eds.) *Scotland: Environment and Archaeology, 8000 BC–AD 1000*, 1st Edition, 23–44. Chichester, Wiley.

Ballin, T. B. (2008) *Quartz Technology in Scottish Prehistory*. Scottish Archaeological Internet Reports 26 (Online), <http://www.sair.org.uk/sair26/index.html>, accessed 11 November 2009.

Ballin, T. B. and Saville, A. (2003) An Ahrensburgian-type tanged point from Shieldaig, Wester Ross, Scotland and its implications. *Oxford Journal of Archaeology* 22/2, 115–131.

Barnatt, J. and Edmonds, M. (2002) Places apart? Caves and monuments in Neolithic and earlier Bronze Age Britain. *Cambridge Archaeological Journal* 12/1, 113–129.

Bartosiewicz, L., Zapata, L. and Bonsall, C. (2010) A tale of two shell middens: the natural *versus* the cultural in 'Obanian' deposits at Carding Mill Bay, Oban, western Scotland. In A. M. VanDerwarker and T. M. Peres (eds.) *Integrating Zooarchaeology and Paleoethnobotany: A Consideration of Issues, Methods, and Cases*, 205–225. New York, Springer.

Bonsall, C. (1996) The 'Obanian' problem: coastal adaptation in the Mesolithic of western Scotland. In T. Pollard and A. Morrison (eds.) *The Early Prehistory of Scotland*, 183–197. Edinburgh, Edinburgh University Press.

Bonsall, C. (1999) Raschoille Cave, Oban [radiocarbon dates]. *Discovery and Excavation in Scotland* 1999, 112.

Bonsall, C. and Murray, N. (1997) Creag nan Uamh caves. In R. E. M. Hedges, P. B. Pettitt, C. Bronk Ramsey and G. J. van Klinken, Radiocarbon dates from the Oxford AMS system: Archaeometry datelist 26. *Archaeometry* 40, 438.

Bonsall, C. and Smith, C. (1989) Late Palaeolithic and Mesolithic bone and antler artifacts from Britain: first reactions to accelerator dates. *Mesolithic Miscellany* 10, 33–38.

Bonsall, C. and Smith, C. A. (1992) New AMS ^{14}C dates for antler and bone artifacts from Great Britain. *Mesolithic Miscellany* 13(2), 28–34.

Bonsall, C. and Sutherland, D. G. (1992) The Oban caves. In W. J. C. Walker, J. M. Gray and J. J. Lowe (eds.) *The South-West Scottish Highlands: Field Guide*, 115–121. Cambridge, Quaternary Research Association.

Bonsall, C., Sutherland, D. G. and Lawson, T. J. (1989) Ulva Cave and the early settlement of northern Britain. *Cave Science* 16/3, 109–111.

Bonsall, C., Tolan-Smith, C. and Saville, A. (1995) Direct dating of antler and bone artefacts from Great Britain: new results for bevelled tools and red deer antler mattocks. *Mesolithic Miscellany* 16/1, 2–10.

Birch S., Wildgoose, M. and Kozikowski, G. (2009) *Uamh An Ard Achadh — High Pasture Cave*. <http://www.high-pasture-cave.org/>, accessed 8 November 2009.

Callander, J. G., Cree, J. E. and Ritchie, J. (1927) Preliminary report on caves containing Paleolithic relics, near Inchnadamph, Sutherland. *Proceedings of the Society of Antiquaries of Scotland* 61, 169–172.

Coles, J. M. (1983) Excavations at Kilmelfort Cave, Argyll. *Proceedings of the Society of Antiquaries of Scotland* 113, 11–21.

Connock, K. D., Finlayson, B. and Mills, C. M. (1993) Excavation of a shell midden site at Carding Mill Bay near Oban, Scotland. *Glasgow Archaeological Journal* 17, 25–38.

Dowd, M. A. (2008) The use of caves for funerary and ritual practices in Neolithic Ireland. *Antiquity* 82, 305–317.

Hardy, K. (2009) Survey and test pitting around the Inner Sound. In K. Hardy and C. Wickham-Jones (eds.) *Mesolithic and Later Sites around the Inner Sound, Scotland: The Work of the Scotland's First Settlers Project 1998–2004*. Scottish Archaeological Internet Reports 31 (Online), <http://www.sair.org.uk/sair31/section2-1.html>, accessed 20 January 2010.

Hardy, K. and Wickham-Jones, C. R. (2003) Scotland's First Settlers: an investigation into settlement, territoriality and mobility during the Mesolithic in the Inner Sound, Scotland. In L. Larsson, H. Kindgren, A. Åkerlund, K. Knutsson and D. Loeffler (eds.) *Mesolithic on the Move. Papers Presented at the Sixth International Conference on the Mesolithic in Europe, Stockholm 2000*, 369–384. Oxford, Oxbow Books.

Hardy, K. and Wickham-Jones, C. R. eds. (2009) *Mesolithic and Later Sites around the Inner Sound, Scotland: The Work of the Scotland's First Settlers Project 1998–2004*. Scottish Archaeological Internet Reports 31 (Online), <http://www.sair.org.uk/sair31.html>, accessed 20 January 2010.

Lacaille, A. D. (1954) *The Stone Age in Scotland*. Oxford, Oxford University Press.

Lawson, J. (2001) Cramond, Edinburgh [radiocarbon dates]. *Discovery and Excavation in Scotland* (new series) 2, 124.

Lawson, T. J. and Bonsall, C. (1986a) Early settlement in Scotland: the evidence from Reindeer Cave, Assynt. *Quaternary Newsletter* 49, 1–7.

Lawson, T. J. and Bonsall, C. (1986b) The Palaeolithic in Scotland: a reconsideration of evidence from Reindeer Cave, Assynt. In S. N. Collcutt (ed.) *The Palaeolithic of Britain and its Nearest Neighbours: Recent Trends*, 85–89. Sheffield, University Department of Archaeology.

Livens, R. G. (1956) Three tanged flint points from Scotland. *Proceedings of the Society of Antiquaries of Scotland* 89, 438–443.

Macklin, M. G. and Rumsby, B. (1991) *Geomorphological Survey of Caves and Rockshelters in the Oban District, Scotland. Report to Historic Scotland*. Unpublished report. Department of Geography, University of Newcastle upon Tyne.

Morrison, A. and Bonsall, C. (1989) The early post-glacial settlement of Scotland. In C. Bonsall (ed.) *The Mesolithic in Europe*, 134–142. Edinburgh, John Donald.

Murray, N., Bonsall, C., Sutherland, D. G., Lawson, T. J. and Kitchener, A. (1993) Further radiocarbon determinations on reindeer remains of Middle and Late Devensian age from the Creag nan Uamh caves, Assynt, north-west Scotland. *Quaternary Newsletter* 70, 1–10.

Pollard, T. (1992) Smoo Cave (Durness parish): shell midden. *Discovery and Excavation in Scotland* 1992, 48–49.

Reimer, P. J., Baillie, M. G. L., Bard, E., Bayliss, A., Beck, J. W., Blackwell, P. G., Bronk Ramsey, C., Buck, C. E., Burr, G. S., Edwards, R. L., Friedrich, M., Grootes, P. M., Guilderson, T. P., Hajdas, I., Heaton, T. J., Hogg, A. G., Hughen, K. A., Kaiser, K. F., Kromer, B., McCormac, F. G., Manning, S. W., Reimer, R. W., Richards, D. A., Southon, J. R., Talamo, S., Turney,

C. S. M., van der Plicht, J. and Weyhenmeyer, C. E. (2009) IntCal09 and Marine09 radiocarbon age calibration curves, 0–50,000 years cal BP. *Radiocarbon* 51, 1111–1150.

Ritchie, G. (1997) Monuments associated with burial and ritual in Argyll. In G. Ritchie (ed.) *The Archaeology of Argyll*, 67–94. Edinburgh, Edinburgh University Press.

Russell, N. J., Bonsall, C. and Sutherland, D. G. (1995) The role of shellfish-gathering in the Mesolithic of western Scotland: the evidence from Ulva Cave, Inner Hebrides. In A. Fischer (ed.) *Man and Sea in the Mesolithic. Coastal Settlement Above and Below the Present Sea Level*, 273–288. Oxford, Oxbow Books.

Saville, A. (1998a) An Corran, Staffin, Skye [radiocarbon dates]. *Discovery and Excavation in Scotland 1998*, 126–127.

Saville, A. (1998b) Comment [on bone tool from Balephuil Bay, Tiree]. In C. Bronk Ramsey, P. B. Pettitt, R. E. M. Hedges, G. W. L. Hodgins and D. C. Owen, Radiocarbon dates from the Oxford AMS system: *Archaeometry* datelist 30. *Archaeometry* 42/2, 466.

Saville, A. (2005) Archaeology and the Creag nan Uamh bone caves, Assynt, Highland. *Proceedings of the Society of Antiquaries of Scotland* 135, 343–369.

Saville, A. and Hallén, Y. (1994) The 'Obanian Iron Age': human remains from the Oban cave sites, Argyll, Scotland. *Antiquity* 68, 715–723.

Schulting, R. J. and Richards, M. P. (2002) The wet, the wild and the domesticated: the Mesolithic–Neolithic transition on the west coast of Scotland. *European Journal of Archaeology* 5/2, 147–189.

Sissons, J. B. (1974) Lateglacial marine erosion in Scotland. *Boreas* 3, 41–48.

Sissons, J. B. (1983) Shorelines and isostasy in Scotland. In D. E. Smith and A. G. Dawson (eds.) *Shorelines and Isostasy*, 209–225. London, Academic Press.

Stone, J., Lambeck, K., Fifield, L. K., Evans, J. M. and Cresswell, R. G. (1996) A Lateglacial age for the Main Rock Platform, western Scotland. *Geology* 24, 707–710.

Stuiver, M. and Reimer, P. J. (1993) Extended ^{14}C data base and revised CALIB 3.0 ^{14}C age calibration program. *Radiocarbon* 35/1, 215–230.

Stuiver, M., Reimer, P. J. and Reimer, R. (2005) *CALIB Radiocarbon Calibration (rev. 5.0.2): On-line Manual*. <http://radiocarbon.pa.qub.ac.uk/calib/manual/>, accessed 15th February 2010.

Sutherland, D. G. (1988) Glaciation and sea-level change in the south west Highlands. In A. G. Dawson, D. E. Smith and D. G. Sutherland (eds.) *INQUA Subcommision on Shorelines of NW Europe, Field Excursion and Symposium, Scotland, September 9–15th 1988*, 15–27. Coventry, Coventry Polytechnic.

Sutherland, D. G. (1997) The environment of Argyll. In G. Ritchie (ed.) *The Archaeology of Argyll*, 10–24. Edinburgh, Edinburgh University Press.

Thomas, J. (2000) Death, identity and the body in Neolithic Britain. *The Journal of the Royal Anthropological Institute* 6, 653–668.

Tolan-Smith, C. (2001) *The Caves of Mid Argyll: An Archaeology of Human Use*. Edinburgh, Society of Antiquaries of Scotland.

Wickham-Jones, C. R. (2009) Lithic raw material use around the Inner Sound. In K. Hardy and C. R. Wickham-Jones (eds.) *Mesolithic and Later Sites around the Inner Sound, Scotland: The Work of the Scotland's First Settlers Project 1998–2004*. Scottish Archaeological Internet Reports 31 (Online), <http://www.sair.org.uk/sair31.html>, accessed 20 January 2010.

Chapter 3

Mesolithic Caves and Rockshelters in Western Norway

Knut Andreas Bergsvik and Ingebjørg Storvik

This chapter presents an overview of the Mesolithic caves and rockshelters in western Norway. Altogether 32 sites are known, spanning the middle and late Mesolithic periods (8000–4000 cal BC). Most of them are situated on the coast, but there are also some in the fjords and in the mountains. There is great variability between the sites in terms of the quality and volume of data and documentation, as well conditions for preservation at the sites. When compared to the large number of Mesolithic coastal open-air sites in this region, it appears that the caves and rockshelters played relatively modest roles. The archaeological and faunal data indicate that they were primarily used as short-term field camps, transient camps and hunting stations for small family groups and crews. The burials found at some of the sites may be related to the general northern European tradition of burying people at residential sites. Alternatively they can be connected to the status of caves and rockshelters as 'natural places'.

Introduction

Prehistoric hunter-gatherer use of caves and rockshelters is one of the classic themes in archaeological research. Large numbers of investigations have been performed in caves worldwide and in many areas – where open-air sites have been destroyed because of natural or cultural processes, these sites provide almost the only information that exists about a given population. The functions of caves and rockshelters are often interpreted in utilitarian terms, and supported by information on daily activities such as repair of hunting gear, production of stone tools, or data on different types of domestic activities like preparation and consumption of food. A very common view of these sites is that they were short-term residential sites for family groups or field camps for small crews on hunting expeditions, sometimes also transient camps, lookout posts or caches (*sensu* Binford 1982). But other functions have also been discussed. Some caves are thought to have been arenas for social aggregations, in which feasting, exchange and other between-group activities took place. Such activities are indicated by particularly large variations in lithic raw-material use, the findings of special artefacts or the results of conspicuous food consumption (e.g. Conkey 1980; Miracle 2001). Furthermore, many caves are seen as important for hunter-gatherer cosmology. This is supported by evidence such as parietal art, portable art and funerals at some of these sites (e.g. Leroi-Gourhan 1968; Arias 2009).

This shows that the archaeology of hunter-gatherer cave-use is varied and the interpretative options wide. The large variety may have to do with their many different qualities: caves and rockshelters are characteristic features in the terrain that are easily spotted, easy to remember; their whereabouts are easy to explain and pass on to other people. They are also practical: cave openings and cliff overhangs provide easy shelters for small groups or individuals on the move. Some large rockshelters are well suited for social gatherings – perhaps as a contrast to everyday wooden or skin dwellings – because they can shelter many people at the same time. From a cosmological viewpoint, caves and shelters are 'natural places' (Bradley 2000). They offer natural entrances to other worlds: to the worlds of ancestors, gods or other beings. They are therefore well suited for liminal rituals and communication with these gods or beings.

In Norway, caves and rockshelters have been used by people from the Mesolithic until present times. About 150 caves and 500 rockshelters have traces of prehistoric human use. A large portion of these are found in western Norway: in the counties of Rogaland, Hordaland, Sogn og Fjordane, and Sunnmøre (Fig. 3.1). In this area, no karstic caves are present; however, there are occasional sea-caves and large numbers of rockshelters created by different geological processes such as erosion in fault-zones and fissures or as a result of rock-fall. A number of boulders have also created rockshelters which are large enough for human occupation. The early excavations of caves and rockshelters in this region were important for the establishment of archaeology as a scientific discipline in this country, and the results have contributed significantly to an understanding not only of these sites themselves, but also of the societies of which they were a part. Although many of them are well published, many are unpublished or less well known. This chapter is an attempt at compiling an overview of all caves and rockshelters that are dated to the Mesolithic period (about 9500–4000 cal BC). The main goal of the contribution is to discuss their importance to the Mesolithic populations: what were caves and rockshelters used for, and how were they related to general patterns of economic, social, and ritual life among these hunter-fishers?

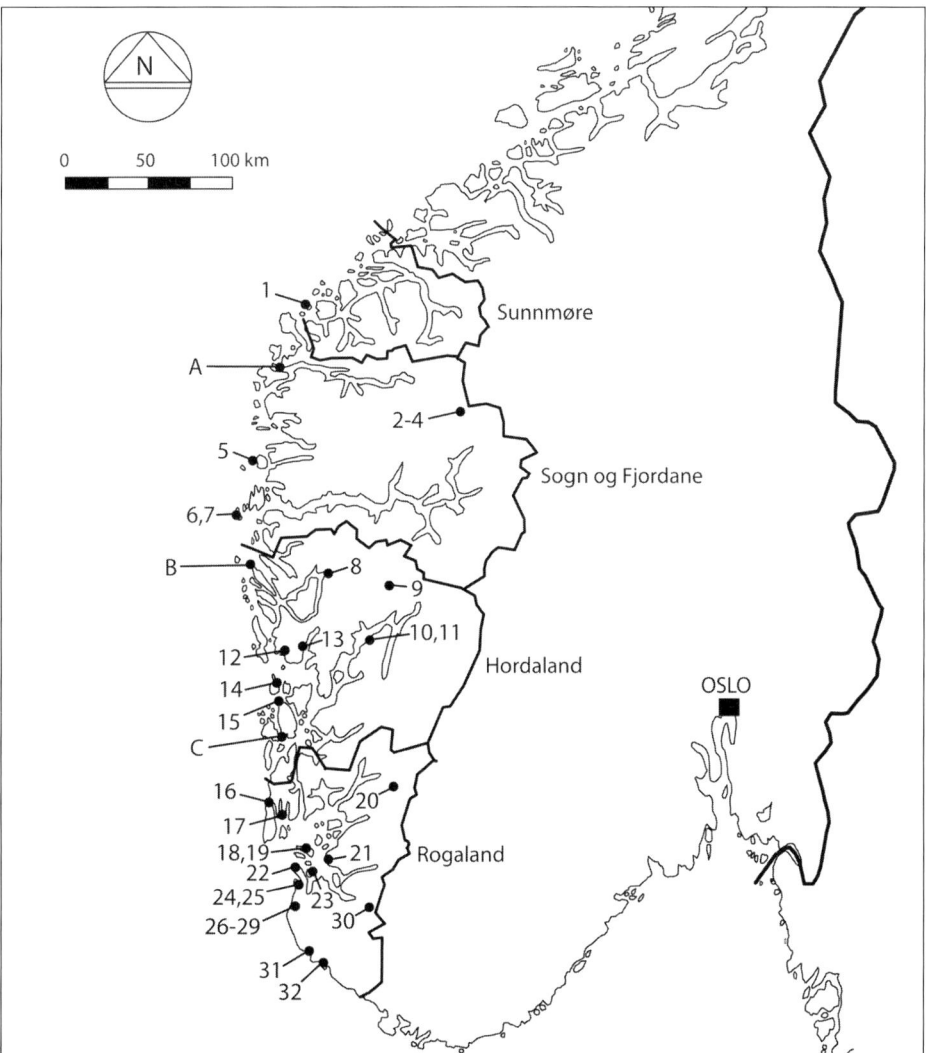

Figure 3.1: Distribution of Mesolithic caves and rockshelters in Western Norway (see Table 3.1). Other sites or areas discussed in the text are marked. A: Skatestraumen, B: Fosnstraumen, C: Bømlo.

Mesolithic cave research in Norway

As in many other European countries, antiquarians surveyed and investigated caves and rockshelters in Norway during the 1800s (e.g. Neumann 1837; Koren 1869; Bendixen 1870). The first scientific excavation of a Mesolithic cave, however, took place in Svartehålå (Viste cave) in Rogaland in 1907 (Brøgger 1908). Once published, Viste became important for the understanding of subsistence, hunting techniques, and tool industry in a North European comparative perspective. Some years later, more Mesolithic shelters were reported from Rogaland (H. Gjessing 1920, 84–163), but, due to the lack of comparative data outside the caves in this early period, little was said about the roles of the caves and rockshelters for the overall settlement pattern by Gjessing or in the overviews of the 1920s by Shetelig (1922) and Brøgger (1925). In 1932 and 1933 the rockshelter Skipshelleren in Hordaland was excavated (Bøe 1934). This fjord site also had rich faunal data and bone industry, dated to the latter part of the Mesolithic, and was an important supplement to the information from Viste, on issues such as chronology and technology. Bøe noticed that there were relatively few lithic artefacts in Skipshelleren, and that the lithic industry at open-air sites at the coast, particularly Bømlo, was much larger and more varied. Bøe therefore suggested that the coast was the 'home' of these populations, and that they utilised other areas during seasonal moves (Bøe 1938, 52, 59). A theory of caves and rockshelters as seasonal hunting stations was supported by G. Gjessing some years later (G. Gjessing 1945, 124).

During the late 1960s and 1970s Skipshelleren, and the Viste cave, were reviewed, together with Grønehelleren, which was excavated during the 1960s (Jansen 1972). Mainly on the basis of the faunal data from these sites, it was argued, contrary to Bøe's interpretation, that they had been used as base camps throughout the year (Indrelid 1978, 170; Mikkelsen 1979, 92).

Until the 1970s, most of the excavations in Norway had been organised as research projects. Caves and rockshelters had been attractive for such investigations because of the preservation conditions of faunal material and bone artefacts. From the late 1960s and onwards, the focus of Mesolithic fieldwork changed, first of all because of industrial development, and building of roads and hydroelectric power plants. Archaeological surveys related to management of the law of antiquities uncovered large numbers of open-air sites in the alpine areas as well as at the coast. A good portion of these sites were Mesolithic, and scholars specialising in this period directed their attention largely towards the study of chronology, settlement patterns, and intra-site studies based on data from these open-air settlements (see Bjerck 2008b for an updated review of this research). The results from these investigations led many to re-establish old viewpoints. Instead of being base camps, caves and rockshelters were once again interpreted as seasonal camps, as field camps or other types of short-term occupations (e.g. Bjørgo 1981, 36). They have been, as Bjerck points out, '...considerably more important for the archaeologists than to the Mesolithic people' (Bjerck 2007, 15).

No systematic survey has yet been directed specifically towards locating caves and rockshelters with prehistoric use in Norway. Such sites are usually found and reported accidentally by non-archaeologists, or as part of general cultural heritage surveys. Nevertheless, during the last few years, there has been a renewed interest in Mesolithic caves and rockshelters in Norwegian archaeology. Some of this interest has been related to their roles as funerary sites (Todnem 1999). Several excavations have been carried out (e.g. T. B. Olsen 2006; Bergsvik and Hufthammer 2009; Eilertsen 2009). The main reasons for these are the often excellent conditions for preservation of faunal material and bone tools. Open-air sites – even if they are numerous and often larger – lack this type of data because of the acid soil conditions. Furthermore, caves and rockshelters are different from the open-air sites because they offer different conditions for shelter and ritual activities. This new focus has resulted in a greater interest in the concrete functions of these sites, and also in an increased attention to how the caves and rockshelters fit into society at large.

Data

Below, a compilation of the Mesolithic caves and rockshelters in western Norway is presented. It is based on an examination of published as well as unpublished data from the university museums in Bergen and Stavanger. The sites are dated partly with reference to the general technological and typological framework for this region (Bjerck 1986, 2008b), partly on published/reported radiocarbon dates. This work has been done by Storvik for Rogaland and by Bergsvik for Hordaland, Sogn og Fjordane, and Sunnmøre. Concerning the names of the sites, note that *hule/hålå* in Norwegian means cave and that *heller/hiller* means rockshelter. The following information is included in Table 3.1 and in the list below:

– County and municipality
– Type (cave, rockshelter, boulder)
– Macro-scale topographical setting (coast, fjord, or mountain)

- Micro-scale topographical setting (variable)
- Meters above sea level
- Site information (year and type of investigation, total area, excavated area, layer thickness, sieving used)
- Dating evidence
- Information on later occupations: Neolithic (N), Bronze Age (BA), or Iron Age (IA)
- Information on artefacts, lithic raw materials, faunal data, and burial data
- Evaluation of faunal data on seasonality

In the discussion, this information will be used to perform a basic evaluation of the activities that took place at the sites, and to carry out a tentative comparative regional analysis.

1. *Dollsteinhula* (Fig. 3.2) lies on the coast of the exposed island of Sande, Sunnmøre, on a steep slope facing the North Sea, and was excavated in 1952 and 1954. Except for the faunal data (Lie 1989), the site is still unpublished. Dollsteinhula is a deep sea-cave, but the human occupations took place under a rockshelter just at its entrance. A radiocarbon dated bone indicates its use at about 4700–4300 cal BC. Fragments of stone adzes and grinding slabs were found, and also blades, and flakes of different raw materials. Some bone tools and large amounts of shells and bones were collected during the excavations, but they have not yet been securely correlated against the stratigraphy and none of it can, therefore, be securely related to the Mesolithic.

2. *Hella* rockshelter (Fig. 3.3) is situated under a large boulder on a plateau close to a small lake in the mountains of Breheimen, Stryn, Sogn og Fjordane. It was excavated in 1982 and 1983 (Randers 1986, 15–32). It was used during the period 6500–4000 cal BC based on technology and radiocarbon determinations. The lithic material consists of microblades, scrapers, cores and flakes. Quartzite is the dominant raw material. No bone tools were found. The faunal material is heavily dominated by reindeer (*Rangifer tarandus*).

3. *Styggevasshelleren* is situated underneath a boulder close to the shore of Styggevatnet in the mountains of Breheimen, Stryn, Sogn og Fjordane. It was excavated in 1982 and 1983 (Randers 1986, 46–69).

Figure 3.2: The cave Dollsteinhula during excavations in 1954 (Photo E. Hinsch, Copyright Bergen Museum).

Figure 3.3: The mountain rockshelter Hella before excavations in 1982. (Photo: K. Randers, Copyright Bergen Museum).

Map no	Site	Location	Type	Excavated year	C. total area within dripline (m²)	Area excavated (m²)	Layer thickn. Max. cm.	Sieving mesh	MASL	Appr. dating cal BC *C14-dated sites	Later occupations	Faunal data	Human bones	Shells
1	Dollsteinhula	Coast	Cave	1952, 1954	60–70	46	10–15	not sieved	69	4700–4300*	N, BA, IA	X	-	X
2	Hella	Mount.	Rockshelter	1982–83	13	30	c. 20	4mm	1157	6500–4000*	N	X	-	-
3	Styggvasshelleren	Mount.	Rockshelter	1982–83	25	47	c. 30	4mm	1156	6700–4000*	N, BA, IA	X	-	-
4	Austdalsvatn (J-2)	Mount.	Boulder	1982	2	2	-	4mm	1157	8000–4000	none	X	-	-
5	Fuglehelleren	Coast	Rockshelter	1967	-	Test-pit	-	not sieved	-	8000–4000	IA?	X	-	-
6	Grønehelleren	Coast	Rockshelter	1964, 1966	c. 60	40	?	not sieved	15	6000–4000*	N, BA, IA	X	X	X
7	Sætrevågen	Coast	Rockshelter	1974	6	Test-pit	>15	4mm	5	8000–4000	IA?	X	-	X
8	Skipshelleren	Fjord	Rockshelter	1930–31	300	94	50	not sieved	20	5300–4000*	N, BA, IA	X	X	X
9	N. Grødalsvatn (lok. 2)	Mount.	Boulder	1978	15	Test-pit	30	4mm	c. 1000	8000–4000	none	-	-	-
10	Sævarhelleren	Fjord	Rockshelter	2005–06	100	15	150	2mm	37	7000–5800*	BA, IA	X	X	X
11	Olsteinhelleren	Fjord	Rockshelter	2006	30	1.5	130	2mm	28	5600–4800*	IA	X	-	X
12	Fana	Coast	Rockshelter	1979	c. 10	4	30	4mm	c. 40	6200–5900*	N	-	-	-
13	Øvste Dåvatræhelleren	Coast	Rockshelter	1995–96	100	6	80	4mm	25	7950–4000*	N, BA, IA	X	-	-
14	Haukanes (lok. 5)	Coast	Rockshelter	2001	15	Test-pits	20	4mm	15	8000–4000	IA	-	-	-
15	Sandvik (lok. 2)	Coast	Rockshelter	2001	40	Test-pit	c. 30	4mm	20	6300–6500*	none	-	-	-
16	Fiskåhelleren	Coast	Rockshelter	2008	12	21.5	100	4mm	18	7000–5500*	N, BA, IA	X	-	X
17	Fosnaneset (lok. 2)	Coast	Rockshelter	2004–05	9	24.5	40–100	4mm	17–20	6900–6800*	BA, IA	X	-	X
18	Ertstein 1 (lok. 44)	Coast	Rockshelter	1989–90	5	1	10	4mm	20–22	8000–4000	IA	X	X	X
19	Ertstein (lok. 45)	Coast	Rockshelter	1989–90	35	2	25	4mm	15–17,5	8000–4000	IA	-	-	-
20	Storhiller	Mount.	Rockshelter	1962–63	c. 100	58	c. 20	not sieved	740	7000–6100*	N	X	-	-
21	Barka	Coast	Boulder	2009	6	17	20	4mm	34	8050–4000	N, BA, IA	-	-	-
22	Viste cave	Coast	Cave	1907, 1938–42	c. 50	125	c. 180	not sieved	16	6800–5020*	N, BA, IA	X	X	X
23	Hellesøy	Coast	Rockshelter	1920, 1925	-	>15	-	not sieved	c. 20	8000–4000	BA	X	-	X
24	Gåsehelleren	Coast	Rockshelter	1909	50	collected	-	not sieved	6	8000–4000	N, BA	X	-	X
25	Tjorahelleren	Coast	Boulder	1963–65	6.5	25	20	not sieved	14	8000–4000	IA	X	X	X
26	Kjelleberget	Coast	Boulder	1911–12	6–10	collected	-	not sieved	14	8000–4000	BA	-	-	-
27	Gåsen	Coast	Rockshelter	1914	-	collected	-	not sieved	20	8000–4000	none	X	-	-
28	Stangelandshelleren	Coast	Rockshelter	1906–07,1909	-	-	-	not sieved	45	8000–4000	N, BA, IA	X	-	X
29	Øksnevadhelleren	Coast	Boulder	1906	15	-	-	not sieved	20–23	8000–4000	none	-	-	-
30	Jensevatn	Mount.	Boulder	1980	c. 10	Test-pit	5	4mm	586	8000–4000	none	-	-	-
31	Dognesteinen	Coast	Boulder	1931	10–18	-	-	not sieved	25	8000–4000	N, IA	X	-	X
32	Skiparsteinen	Coast	Boulder	1971	10	15	50	4mm	10	8000–4000	N, IA	-	-	-

Table 3.1: Site data from Mesolithic caves and rockshelters in Western Norway.

Radiocarbon determinations indicate that it was used as early as about 6700 cal BC, and that it was revisited several times during the rest of the Mesolithic. The lithic material consists of a stone adze with a round cross-section ('trinnøks'), microblades, scrapers, borers, cores, and flakes. No bone tools were found. The faunal material was generally in poor condition (burnt). Only bones of reindeer (*Rangifer tarandus*) were determined to species.

4. *Austdalsvatnet rockshelter (J-2)* is situated underneath a boulder close to lake Austdalsvatnet in the mountains of Breheimen, Stryn, Sogn og Fjordane. It was excavated in 1982 and 1983 (Randers 1986, 33–36). A few microblades date it generally to 8000–4000 cal BC. A bipolar core and some flakes were also found. The artefacts were made of flint and quartzite.

5. *Fuglehelleren* is situated on the coast of the exposed island of Tvibyrge in Askvoll, Sogn og Fjordand. It was test-pit surveyed in 1967 (Inselset 2007). A Mesolithic stone adze with a round cross-section dates it generally to 8000–4000 cal BC. This was found together with a flint blade, a core and a few flakes of different raw materials. A soapstone bird figure was also present at the site.

6. *Grønehelleren* lies on the coast of the exposed island of Ytre Sula in Solund, Sogn og Fjordane. It is situated close to the shoreline in a protected bay. The site was excavated in 1964 and 1965 (Jansen 1972). The Mesolithic layers at the site are dated to about 6500–4000 cal BC based on the lithic industry (Indrelid 1978; Bjerck 2007). The lithic assemblage is made up of a number of late Mesolithic stone adzes and fragments of such adzes, microblades, several scrapers, and flakes of flint, quartz crystal, and other raw materials. A few bone tools were also present: a slotted point and some fishhooks. The faunal material is not very large, but relatively varied in terms of species. It consists of several mammals, among them seal (Phocidae), red deer (*Cervus elaphus*), wild boar (*Sus scofa*), and otter (*Lutra lutra*). Altogether ten marine bird species were found, among them glaucous gull (*Larus hyperboreus*), puffin (*Fratercula arctica*), and the now extinct great auk (*Alca impennis*). The fish bones were dominated by cod (*Gadhus morrua*), coalfish (*Pollachius virens*), and pollack (*Pollachius pollachius*). In addition Jansen reports that mussels (*Mytilus edulis*), common periwinkle (*Littorina littorea*), common limpet (*Patella vulgata*), and oysters (*ostreum*) were observed during excavation of the Mesolithic layers (Jansen 1972, 64). Altogether four burials were found in Grønehelleren, and also some disarticulated human bones. One of the skeletons (skeleton I) from a separate pit was radiocarbon dated to the Middle Neolithic. Skeletons II, III and IV were found close together in a pit, and may have been buried at the same time. Only one of these skeletons (II) is radiocarbon dated to 6080±16 BP (5343–4686 cal BC) (Indrelid 1996, 231) and was probably of a woman aged about 40 years (Jansen 1972, 59). There are a few seasonal indicators in the faunal data: Great auk and puffin are likely to have been caught during the spring or early summer in this area and glaucous gull probably during the winter (Jansen 1972, 62).

7. *Sætrevågen rockshelter* lies on the coast of the exposed island of Ytre Sula in Solund, Sogn og Fjordane. It is situated close to the shoreline in a protected bay. The site was test-pit surveyed in 1973 (Bjørgo 1974). A quartz microblade dates it generally to 8000–4000 cal BC. A core and several flakes of different raw materials were also found. A few fragments of fish bones were collected, and also some shells.

8. *Skipshelleren* (Fig. 3.4.) is situated in Vaksdal, Hordaland in a deep and narrow fjord, at the end of the watercourse leading down from the Voss valley. The site was excavated in 1930 and 1931 (Bøe 1934; H. Olsen 1976). The radiocarbon dated Mesolithic layers in the rockshelter span 5300–4000 cal BC (Indrelid 1978; Bjerck 2007). Despite the large excavated volume, relatively few lithic artefacts were found: only one greenstone adze, a few scrapers, cores, and flakes of different raw materials. Some conical microblade cores and microblades were also present. Several patches of red ochre were observed. Fragments of slotted bone points were found and also a number of smaller bone points. There were a number of bone fishhooks, harpoon-heads and leister prongs in the Mesolithic layers, and also awls and needles. A large amount of faunal material was collected at the site, but, unfortunately, in the unpublished report, only mammal bones are related to phase, whereas bird and fish bones are presented together (H. Olsen 1976). The Mesolithic mammal bones show a marked dominance of deer (50 per cent). Wild boar and seal represent 12 per cent each. Bøe reported that shells were relatively frequent in the Mesolithic layers, particularly mussels, and also common periwinkle and other gastropods (Bøe 1934, 13–19). Isolated human bones, from the foot, hand and finger were found at different places in the Mesolitic layer. On the basis of the faunal data, Olsen argues that the site was probably occupied during the summer months. This is supported by recent analyses of otoliths from the site (Hufthammer *et al.* 2010).

Figure 3.4: The fjord rockshelter Skipshelleren during excavations in 1931 (Photo: Johs. Bøe, Copyright Bergen Museum).

Figure 3.5: The fjord rockshelter Sævarhelleren during excavations in 2005 (Photo: K. A. Bergsvik).

9. *Nedre Grødalsvatn rockshelter (lok. 2)* is situated in the mountains of Ulvik on the eastern bank of the river Moldå. It was test-pit surveyed in 1978 (Gustafson 1982, 36). A microblade dates it generally to 8000–4000 cal BC. Some flakes of quartz/quartzite and flint were also found.

10. *Sævarhelleren* (Fig. 3.5) is situated in Herand, Jondal in the Hardanger fjord. It is located relatively close to the shoreline. It was excavated in 2005 and 2006 (Bergsvik and Hufthammer 2009). The radiocarbon dates for the Mesolithic layers at the site span 7000–5800 cal BC. Relatively few lithic artefacts were found during the excavations: some fragments of greenstone adzes and cores, microblades, scrapers and some flakes of flint, quartz and quartz crystal. The bone tool assemblage was dominated by fishhooks and fragments of such hooks. Some needles and awls were also found, and a few pendants. The conditions for preservation of the faunal material were very good at the site. Mammals such as wild boar, red deer and elk (*Alces alces*) were present, as well as otter (*Lutra lutra*). Bird bones were almost absent, only one single bone of the thrush family (Turdus sp.) was found. Concerning fish, wrasses (Labridae) and Gadids (Gadidae) dominate, particularly coalfish (*Pollachius virens*), cod (*Gadus morhua*), and pollack (*Pollachius pollachius*), but there are also Salmonids (Salmonidae), herring (*Clupea harengus*), mackerel (*Scomber scombrus*), and tusk (*Brosme brosme*). The cultural layers at the site were to a large degree made up of shells, almost without exception mussels and common periwinkle. Scallop shells (Pectinidae) were also occasionally found. One fragment of a human skull and a finger joint were identified in the Mesolithic layers (Anne Karin Hufthammer pers. comm.). The presence of salmonidae and mackerel clearly indicates that it was used during the summer, although occupations during other seasons cannot be excluded (Bergsvik and Hufthammer 2009, 444).

11. *Olsteinhelleren* lies about 50m to the west of Sævarhelleren. It was excavated in 2006 (Bergsvik and Hufthammer 2009). The radiocarbon dates show that the Mesolithic occupation of the site span 5600–4800 cal BC, thus it succeeds Sævarhelleren's occupation in time. The lithic material consists of microblades, blades, quartz crystals, grinding stones, and net-sinkers. There were also several types of cores and relatively many flakes, primarily of quartz and quartzite. The bone tools are dominated by fishhooks, and there were also preforms for the production of such hooks. Barbed points, slotted points, net-needles, and awls were also found. The conditions for preservation of faunal material were comparable to that of Sævarhelleren. The following mammals were present at the site: otter, wild boar, red deer, moose, wolf (*Canis lupus*), and red squirrel (*Sciurus vulgaris*). Very few bird bones were found; only razorbill (*Alca torda*) was identified. In contrast, fish bones were abundant. As at Sævarhelleren, these were dominated by coalfish, cod, and pollack. There were also several wrasses and mackerel, trout (*Salmo trutta*), and also (in contrast to Sævarhelleren) several deep-water species such as tusk, skate (*Hypotremata*), and ling (*Molva molva*). The shells of common mussel and common periwinkle were abundant at the site. The presence of trout and mackerel indicate summer occupations (Bergsvik and Hufthammer 2009, 444).

12. *Fana rockshelter*, Bergen, Hordaland. The site is situated on the coast in the bottom of Fanafjorden relatively high above the contemporary shoreline. It was excavated in 1979 (Randers 1979). A radiocarbon date indicates that it was occupied around 6000 cal BC. The lithic material consists of a microblade, and a few flakes of flint, quartz and quartz crystal.

13. *Øvste Dåvatræhelleren* lies on the coast in Fusafjorden at Samnøy, Fusa, Hordaland, relatively high above the contemporary shoreline. It was excavated in 1996 (Lødøen 1997). The radiocarbon determinations indicate that it was occupied already around 7950 cal BC (possibly also earlier), and that it was reoccupied several times during the rest of the Mesolithic. The lithic material consists of a stone club (without shaft-hole), a stone adze with a round cross-section, fragments of such adzes, a lanceolate point, microblades, scrapers, cores, and flakes. The lithic raw material is dominated by flint in the oldest phases but changes towards quartz, quartzite, quartz crystal and mylonites in the younger Mesolithic layers. Faunal material was collected from the site, but it has not yet been analysed.

14. *Haukanes rockshelter (lok 5)* is situated on the coast of the island Huftarøy, Austevoll, relatively close to the shoreline. It was test-pit surveyed in 2001 (Østerdal 2001). The lithic material consists of four microblades, which date it generally to 8000–4000 cal BC, and a few flakes of flint, quartzite, quartz crystal and basalt.

15. *Sandvik rockshelter (lok 2)* is situated on the coast of the island Stord, in the bay of Sandvikvågen, relatively close to the shoreline. It was test-pit surveyed in 2001 (Skrede and Baug 2001). A

radiocarbon date indicates that it was occupied at around 6400 cal BC. The lithic material consists of a few flakes of flint.

16. *Fiskåhelleren* is situated on the coast of Karmøy, Rogaland. Although its location is broadly coastal, it actually lies inland on the island Karmøy close to the shore of Fiskåvatnet (a freshwater lake). It was excavated in 2007 and 2008 (Eilertsen 2009). The Mesolithic layers were radiocarbon dated to about 7040–5560 cal BC. Stone adzes with round cross-sections were found. The lithic tool assemblage is dominated by microblades. Scalene triangles were also abundant. There were also large amounts of cores and flakes. Flint was the dominant raw material.

17. *Fosnaneset rockshelter (lok. 2)* is situated on the coast of the island of Fosen, Karmøy, Rogaland, close to the shoreline at the bottom of the bay of Vollsvika. It was excavated in 2005 (T. B. Olsen 2006). A radiocarbon date indicates that it was occupied around 6900–6800 cal BC. It was probably reoccupied several times afterwards during the Mesolithic. The lithic material consists of a stone adze with a round cross-section, two soapstone net-sinkers, and a large number of microblades. There were also relatively frequent flakes of flint, and others of quartz and quartz crystal.

18. *Ertstein rockshelter 1 (lok. 44)* is situated on the coast of the island of Rennesøy in Boknafjorden, Rogaland, relatively high above the contemporary shoreline. It was excavated in 1989 and 1990 (Høgestøl 1995, 203–206). Two microblades of flint date it generally to 8000–4000 cal BC. Several flint flakes were also found.

19. *Ertstein rockshelter (lok. 45)*, Rennesøy, Rogaland has more-or-less the same location as lok. 44 Ertstein (above) and was excavated at the same time (Høgestøl 1995, 92–94). A small soapstone net-sinker, microblades, and conical microblade cores date it generally to 8000–4000 cal BC. In addition to these artefacts, drills, scrapers, and relatively large numbers of flakes were found. Flint is the dominant raw material, supplemented by a few items of quartz crystal. Faunal material was not preserved at the site.

20. *Storhiller* is situated in the mountains in Årdalsheiene, Rogaland, close to the shore of Storhillervatnet, at an altitude of around 740m. It was excavated in 1962 and 1963 (Bang-Andersen 1987). A radiocarbon date indicates occupation(s) at around 6000 cal BC; however, the presence of triangular microliths at the site suggests that it was occupied as early as 7000 cal BC. The site may have been used repeatedly during the Mesolithic from this date. The lithic (flint)

Figure 3.6: The Viste cave during excavations in 1907 (Photo: A. W. Brøgger 1907. Copyright: Archaeological Museum, University of Stavanger).

material consists of microliths, a drill, microblades, microblade cores, and flakes. One fragment of a slotted bone point was also found. Only a small percentage of fragmented faunal material collected at the site has been determined to species, but all of the bones are from cloven-footed mammals and two of them most likely from reindeer or red deer.

21. *Barka rockshelter* is situated on the coast at Strand, Rogaland, relatively high above the shore of Idsefjorden, underneath a large boulder. It was excavated in 2009 (Eilertsen 2010). The presence of scalene triangles indicates that it may have been used as early as about 8000 cal BC. It was probably also reoccupied several times later during the Mesolithic period. The lithic material consists of a relatively large number of microblades, which were made from flint, quartz crystal, and quartzite. There were also microblade cores and flakes of various raw materials at the site.

22. *Svartehålå (Viste cave)* (Fig. 3.6) is situated on the coast in Randaberg, Rogaland, close to the shoreline in the bay of Visteviken. It was excavated in 1907 (Brøgger 1908) and in 1939–1941 (Lund 1951). The Mesolithic layers in the cave span 6800–5020 cal BC (Mikkelsen 1971; Bjerck 2007, 11). The archaeological material consists of a variety of lithic artefact types, such as ground and pecked greenstone adzes, and cores, scrapers, burins, knives, microliths, and flakes of flint and other lithic raw materials. Bone tools were also found: slotted bone points, fishhooks, harpoons, awls, and pendants (Mikkelsen 1979). The rich osteological material (70 species) is dominated by the terrestrial species wild boar and elk. Altogether 22 mammals were present. Seal was present in the deposits as well as birds (37 species) and fish (11 species) (Degerbøl 1951; Indrelid 1978; Bjerck 2007). A number of shells were also found during the excavations. The dominance of terrestrial species is puzzling, considering the location of the site. It may, however, partly be a result of coarse methods of collecting faunal material (the soil was not water sieved). During the excavations in 1907, the skeleton of a boy was found close to the inner cave wall. In addition, one joint from the foot and two finger joints of an older individual were found (Fürst 1909). The young skeleton has been directly dated to 7420±150 BP (5725–5558 cal BC) (Hufthammer and Meiklejohn 1986). According to the osteologist Degerbøl, the faunal data strongly indicate summer occupations. A few autumn-winter indicators are, however, also present, such as swan (*Cygnus cygnus*) and gannet (*Morus basemus*) (Degerbøl 1951, 77, 81). Bones from newborn grey seal (*Halichoerus grypus*) may also indicate autumn/winter occupation at this site (Indrelid 1978, 162).

23. *Hellesøy rockshelter* is situated on the coast in Idsefjorden, Stavanger, Rogaland, relatively close to the shoreline. It was excavated in 1920 and 1925 (H. Gjessing 1920, 160). A Mesolithic stone adze with a round cross-section, and microindustry, dates it generally to 8000–4000 cal BC. A hammerstone, scrapers, blades, and flint flakes were also found.

24. *Gåsehelleren* is situated on the coast at Meling, Sola, Rogaland close to the shoreline. Several artefacts were collected here in 1914 (H. Gjessing 1920, 141–143). The microindustry dates it generally to 8000–4000 cal BC. In addition, a hammerstone, blades, cores and flakes of flint were found. Bone awls and a net-needle were also present.

25. *Tjorahelleren* is situated on the coast at Sola, Rogaland, close to the shoreline, underneath a boulder facing the North Sea. It was excavated in 1963–1965 (Myhre 1967). The presence of bone fishhooks of the 'Skipshelleren type', microindustry, and conical microblade cores dates it generally to 8000–4000 cal BC. In addition to the fishhooks, bone awls were found. The lithic artefacts consisted of projectile points, scrapers, cores, and flakes, mainly of flint. Faunal material was preserved, but cannot be securely related to the Mesolithic layers at the site. Neither can a few fragments of human bone (a tooth and two finger joints) found at the site.

26. *Kjelleberget rockshelter* is situated on the coast at Hedland, Sola, Rogaland, close to the shoreline underneath a boulder. It was surface-collected in 1908 (H. Gjessing 1920, 134). A Mesolithic stone adze blank dates it generally to 8000–4000 cal BC. Flint flakes and a hammerstone were also found.

27. *Gåsen rockshelter* is situated on the coast at Sele, Klepp, Rogaland, close to a small lake, at an altitude of about 20m. It was surface-collected in 1914 (H. Gjessing 1920, 117–121). A slotted bone point dates it generally to 8000–4000 cal BC. A bone comb and the preform of a bone fishhook were also present. The lithic material consists of a grinding slab, a possible grinding stone and flint flakes. Faunal material was preserved, but cannot be securely related to the Mesolithic.

28. *Stangelandshelleren* is situated on the coast at Stangeland, Klepp, Rogaland. Although its location is generally on the coast, it actually lies inland at Jæren close to the bank of the river Figgjo. It was

excavated in 1906–1907 and in 1909 (Brøgger 1911; H. Gjessing 1920, 126–130). A Mesolithic stone adze with a round cross-section, a soapstone net-sinker, bone fishhooks of the 'Viste type', and micro-industry date it generally to 8000–4000 cal BC. In addition, the lithic material consists of fragments of blades, scrapers, drills, hammer stones, cores, and flakes of flint. Net-needles and ordinary needles of bone were also found. Faunal material was preserved, but cannot be securely related to the Mesolithic.

29. *Øksnevadhelleren* is situated on the coast at N. Øksnevad, Klepp, Rogaland, relatively close to the coast, underneath a boulder, close to the bank of the river Figgjo. It was excavated in 1906 (H. Gjessing 1920, 125,126). The presence of an adze of Mesolithic type with a round cross-section dates it generally to 8000–4000 cal BC. The lithic material consists of blades, hammerstones, and flakes of flint and quartz crystal.

30. *Jensavatn rockshelter* is situated in the mountains of Frafjord, underneath a boulder, close to a small lake at an altitude of about 586m. It was test-pit surveyed in 1980 (Hofseth 1982, 100). The presence of micro-industry dates it generally to 8000–4000 cal BC. Microblades and flakes of flint and quartz crystal were found.

31. *Dognesteinen rockshelter* is situated on the coast at Hå, Rogaland, underneath a boulder, relatively high above the contemporary shoreline. It was excavated in 1931 (Kloster 1931). The presence of a one-sided harpoon-head, and a fishhook of the 'Viste type', dates it generally to 8000–4000 cal BC. A piece of a pointed antler was found. The lithic material consists of blades, cores, hammerstones, scrapers, and flakes of flint and other rocks. Shells and faunal material was preserved, but it cannot be securely related to the Mesolithic.

32. *Skiparsteinen rockhelter* is situated on the coast at Gjedlestadvige on the island of Eigerøy, Eigersund, Rogaland, underneath a boulder, relatively close to the shoreline. It was excavated in 1971 (Bang-Andersen 1971). The presence of micro-industry dates it generally to 8000–4000 cal BC. The lithic material consisted of microblades, cores, scrapers, and flakes of flint.

Summary of the data

In sum, 32 caves and rockshelters dated to the Mesolithic period have been surveyed or excavated in western Norway. The overview shows that the first investigations took place early in the nineteenth century, and the last ones only recently. The excavation methods therefore vary significantly, from random collection of finds to detailed stratigraphic control and sieving of the deposits. The general quality of the documentation clearly also increases over time. The dating of most of the rockshelters spans most of the Mesolithic, between 8000–4000 cal BC. Although the actual use of many of the shelters may have been short and sporadic, the majority of the sites can only generally be placed within this broad frame of dating: primarily because of lack of radiocarbon dates or other more precise datable evidence. It is mainly the occurrence of micro industry (microblades) or stone adzes with round cross-sections ('trinnøkser') that indicates a broad Mesolithic date (Bjerck 1986). A number of the sites are, however, radiocarbon dated, indicating more precise time-spans for the occupation of these shelters (although several of these may also have had earlier as well as later Mesolithic occupations than those indicated by these datings). Most of the sites were then occupied during later periods. There are significant variations between the sites in terms of site extent and thickness of the cultural deposits: the largest site covers as much as about 300m^2, the smallest are less than 10m^2. One site has layers that are 1.8m thick while others only have 10–15cm thick deposits. Furthermore, the conditions for preservation of faunal material are variable. 13 of the sites have not produced faunal data. The remaining 19 have bone material preserved, but for several of these the published faunal data is not related specifically to the Mesolithic. Concerning the location of the sites, most of them are situated on the coast, some relatively close to the contemporary shorelines, and some considerably higher than these. A few rockshelters have been found in the fjords, and some are also found in the mountains.

Discussion

The earliest dates from the above sites are about 8000 cal BC (middle Mesolithic 1). Considering that western Norway was colonized about 1500 years earlier (Bjerck 2008b), this strongly indicates that the early Mesolithic populations did not occupy caves and rockshelters. We can only speculate why this was so. One possible reason may be related to the frequent residential moves of these people. Most of their sites are found along the outer coastline and some also in the mountains. Sites in both areas appear to have been the results of relatively short occupations in tents made of hide (Bang-Andersen 2003; Bjerck 2008a). These mobile homes were probably part of the standard equipment for family-groups as well as task-groups; they were quickly set up and taken down, and may have minimized the need for occupations in caves or rockshelters. Furhermore, during this period of colonization, people were constantly seeking new territories; there was frequent residential mobility by boats across long distances. The large number of small sites indicates that there was a minimal degree of reoccupation of the same camps (Bjerck 2008a; Fuglestvedt 2009, 190). This may imply that there were few connections to specific landscapes, and no detailed knowledge beforehand of where they could find convenient and accessible caves and rockshelters.

The lack of cave-use during the early Mesolithic may thus be related to a flexible or opportunistic 'fully nomadic' settlement system and mobility pattern. This could imply that the initial use of caves and rockshelters was a result of changes within these systems and patterns. Such changes came about around 8000 cal BC. A process of regionalisation was going on in western Norway during this period, which is first of all indicated by the appearance of regionally distinct ground/pecked adzes with round cross-sections made of quarried basalts (Olsen and Alsaker 1984). The regionalisation in the material culture was probably related to the emergence of sedentism, even if this is difficult to evaluate because of the general lack of sites from the middle Mesolithic period. Large coastal sites with thick cultural layers indicate, however, that coastal sedentism was a significant feature of the succeeding late Mesolithic in this area (Warren 1994; Bergsvik 2001). Simultaneously, there was also a development of socially constructed boundaries between the sedentary populations (Olsen and Alsaker 1984; Bergsvik and Olsen 2003; Skjelstad 2003). In contrast to the early Mesolithic system, the pattern of sedentism and social boundaries generally implied a reduction of the residential mobility, and in turn increased logistical mobility for generalised or specialised task groups. With the sedentary coastal sites as points of departure, these task groups repeatedly travelled within their territories to exposed islands on the coast, into the fjords and forests, and up to the mountain plateaux. These journeys led to a detailed knowledge of the landscape and of the sites it had to offer, and enabled planning ahead and a possibility to integrate caves and rockshelters into the pattern of logistical mobility.

The questions asked initially were how important the caves and rockshelters were for people, and what they were used for during the middle and late Mesolithic periods. Below, we will deal with these questions on the basis of the analyses of the above data in comparison with data from contemporary open-air sites.

One way of measuring the importance of the caves is to compare them to the number of known open-air sites in this region. A complete overview of such sites has not yet been made; however, data from some test-pit surveys may be used for such a purpose. The areas of Skatestraumen, Nordfjord (Bergsvik 2002, 101–102), Fosnstraumen, Nordhordland (A. B. Olsen 1992; Bergsvik 2001), and Bømlo, Sunnhordland (Kristoffersen and Warren 2001) are examples of areas that have been thoroughly test-pit surveyed. These surveys, which together covered no more than around 7km^2, resulted in a total of 77 sites dated to the middle and late Mesolithic period. One might argue that these areas may have been particularly attractive during the Mesolithic due to rich resources. However, considering that only 32 Mesolithic caves and rockshelters are known from the entire region (altogether around 50,000km^2), this strongly indicates that open-air sites were much more attractive. The site sizes also confirm this pattern. Generally, the caves and rockshelters in this analysis are small: the majority of them are below 50m^2. Only four shelters are larger than 100m^2. Open air sites are generally much larger. At Skatestraumen, Fosnstraumen, and Bømlo the average site sizes are 315m^2, 355m^2, and 1245m^2 respectively (Bergsvik 1991, 163; 2002:101–102; Kristoffersen and Warren 2001). Thickness of layers might also be used to quantify this difference. Most of the caves and rockshelters have Mesolithic deposits that are between 20 and 50cm thick. Four shelters have deposits which are thicker than 100cm. At Fosnstraumen the maximum average layer thickness of the 18 surveyed Mesolithic sites is 40cm (Bergsvik 1991, 163), which is about the same as the shelters. However, a problem in terms of comparison is the different conditions for preservation at the shelters compared to the open-air sites. At the latter sites, acid soils and a generally humid climate has led to a breakdown of organic components including faunal material. In many caves and rockshelters, the preservation conditions are good because of the sheltered conditions. If the faunal material had decayed also at these sites, the layers would also have been much thinner.

On this basis it is apparent that middle and late Mesolithic caves and rockshelters are greatly outnumbered by open-air sites. Furthermore, the open-air sites are generally much larger, and have much thicker cultural deposits than the shelters. This clearly shows that open-air sites were preferred for human occupation during this period. Nevertheless, caves and rockshelters were used in a variety of ways, and their study may supplement the information from the open air sites. It is not possible in this contribution to discuss in detail the functions of the individual shelters, nor to investigate changes during the middle and late Mesolithic. But it is possible to outline some trends for future research based on the collected information on the location of the sites, and on the archaeological and faunal material from the sites.

Regional differences

Concerning the location of the sites it is interesting to note that the majority of the Mesolithic caves and rockshelters are situated in the southern part of the area, first of all in Rogaland, whereas only one site lies at Sunnmøre in the north. This could theoretically be the result of there being fewer natural caves and overhangs in the south compared to further north. Such an option is difficult to evaluate, because no record of all natural caves and rockshelters exists for western Norway. However, if we look at the total number of known caves which have prehistoric traces (all periods), this shows the following numbers: Rogaland – 84 (Storvik 2011), Hordaland – 138, Sogn og Fjordane – 65, Sunnmøre – 10 (Bergsvik in prep.). These numbers indicate that natural caves and rockshelters are abundant throughout the region. If this is correct, the relatively high number of Mesolithic rockshelters in Rogaland would be the result of other factors. One such factor is that archaeologists in Rogaland actively surveyed these sites after the excavation of the Viste cave in 1907, and this resulted in the discovery of several new Mesolithic rockshelters. Still, for this factor to be significant, the same effect should be seen in Hordaland, considering that archaeologists have also been active in surveying shelters in this county. However, a similar number of shelters from this period were not recorded here. This could mean that the Mesolithic populations used caves and rockshelters more actively in Rogaland than people who lived further to the north.

Site functions and seasonality

The open-air sites from Skatestraumen, Fosnstraumen, and Bømlo referred to above are situated in favourable areas in terms of communications and access to marine and terrestrial resources. Many of them have been interpreted as sedentary base camps; archaeological and faunal data indicate that some of them were probably used over several years continuously or repetitively by the same group. Some of these large sites are also likely to have served as aggregation sites, perhaps during the wintertime. Among the known caves and rockshelters, only the Viste cave may have been used in such a manner, because of its favourable sheltered and strategic location and the relatively large and varied archaeological and faunal deposits. Egil Mikkelsen argues that this site may have sheltered three to four nuclear families at the same time, and that such a group may have exploited the northern part of Jæren (Mikkelsen 1979, 92). The functions of the remaining rockshelters were probably different from this, but they may still be linked to the same subsistence-settlement system as the large open-air base camps. Below, I will briefly discuss possible site functions of these rockshelters on the coast, in the fjords, and on the mountain plateaux. In this discussion, the connection to the shoreline is an important factor. During the relevant period 8000–4000 cal BC, the shorelines along the coastline were generally between 5 and 15m above today's levels, depending on local geological conditions. Due to increasingly stronger isostatic uplift in the eastern direction, the Mesolithic shorelines were higher in the fjords than at the coast (Prøsch-Danielsen 2006; Kaland1984; Svendsen and Mangerud 1987).

Including the Viste cave, a total of 23 rockshelters are situated on the coast. 13 of these were situated close to the shoreline, had short distances to marine as well as terrestrial resources and were favourably placed in relation to communication routes. Several of these sites – Gåsen, Kjelleberget, Helleberget, Gåsehelleren, Haukanes lok. 5, Sandvik lok. 2, Fuglehelleren and Sætrevågen – were only surface-collected or test-pitted. However, the shelters of Tjorahelleren, Skiparsteinen, Fosnaneset lok. 2, Ertstein lok. 45, and Grønehelleren have more substantial data from excavations. This indicates that they were used for a variety of purposes, such as food preparation and food consumption, scooping of wooden objects, scraping of hides or other materials, lithic tool production (on different raw materials), and maintenance activities. Among these, Grønehelleren stands out because of the faunal data. Although not very large, this shows a relatively wide subsistence base, in which both terrestrial and marine birds and mammals were consumed in addition to fish. This data may also indicate that the site was used during summer as well as winter occupations.

Some of the coastal rockshelters were situated at some distance from contemporary shorelines, probably in forests (Fana, Øvste Dåvatræhelleren, Barka, Ertstein lok. 44, Øksnevadhelleren, Dognesteinen), along a river basin (Stangelandshelleren), beside a lake (Fiskåhelleren), or high up on a steep mountain slope (Dollsteinhula). Mesolithic coastal sites are generally connected much closer to the shoreline than these sites (around 2–5m above contemporary shorelines). The relatively high location may indicate that these sites were used differently than the shorebound caves as well as open-air sites, perhaps as specialised hunting stations. This is an important result because it modifies the theory of the shoreline as the most important site location factor during this period (see also Berg-Hansen 2009 for a discussion of this problem). Although people at the coast generally seem to have preferred shorebound site locations, these data show that they also sometimes occupied sites on higher ground. These sites – in our case rockshelters – were perhaps used as specialised hunting stations. Although not securely dated to the Mesolithic, faunal material from Stangelandshelleren is dominated by bones of deer, perhaps indicating specialised big-game hunting (Myhre 1980, 30). The large number of scalene triangles at Fiskåhelleren may also support such a theory. Eilertsen relates these to terrestrial hunting by a group that used this inland shelter as a starting point for hunting expeditions (Eilertsen 2009).

From an East-West perspective, the concentration of caves and rockshelters on the coast is easy to understand, since the large majority of other sites from this period are found on the coast. A strong coastal focus is also confirmed by test-pit surveys in the fjords, which usually produce few and small sites (cf. Lødøen 1995). This is supported by the distribution maps of stray-finds of stone axes (found by non-archaeologists and delivered to the regional museums), which are mainly found at the coast (Lødøen 1995; Bergsvik and Hufthammer 2009). This strongly indicates that people mainly lived on the coast, and used the fjords and mountains much less intensively. The three fjord sites include the rockshelters Skipshelleren, Sævarhelleren, and Olsteinhelleren. These shelters are all relatively shorebound. They are also characterised much in the same way as the coastal shelters in terms of tool types, and lithic raw material variation. Several of them have rich faunal material, which includes marine as well as terrestrial species. Seen together, the data indicate that these sites were also used for a wide range of activities. Six of the rockshelters are situated in the mountains: Storhiller, Jensevatn, N. Grødalsvatn lokl. 2, Hella, Styggvasshelleren, and Austdalsvatn J2. The lithic data from these sites are very much in accordance with the data from the fjords and the coast. This indicates that the people who occupied them performed a number of activities related to hunting, maintenance and tool production. Although badly preserved at these sites, the faunal data show that reindeer was the most common hunted species, which again may indicate specialised big-game hunting.

One question is how the shelters in the fjords and in the mountains were related to the overall settlement system. Considering that there was a marked settlement concentration on the coast, and generally few sites further east, it is likely that the shelters in the fjords and in the mountains were used by groups or individuals who originated at the coast. This is supported by the similarity in technology as well as in lithic raw material use in the shelters and at other sites in the three areas. The faunal material from the three fjord sites, Skipshelleren, Sævarhelleren, and Olsteinhelleren, clearly shows that they were used during the summer. It is likely that the reindeer hunting related to the mountain shelters also happened during the summer or early autumn. On this basis one may conclude that the fjords and mountains were primarily utilized during the summer.

Concerning the more detailed functions of the sites – and also how this changed through time – much more work is needed on distinguishing between different occupational phases and on performing more thorough analyses of the archaeological and zoological data. Until such analyses are undertaken, however, one may tentatively conclude – on the basis of the relatively varied lithic and faunal data from most of these sites – that many of the caves and rockshelters were used for different tasks simultaneously, and that a number of different species were consumed in several of them during the same occupations. Most of them may, therefore, have been 'field camps' or 'transient sites'. This would imply that, on a short-term basis, they were points of departure for small hunting, fishing or gathering crews that would have utilised the resources close by.

Many of them – in particular the ones at strategically important locations – were reoccupied a large number of times over several hundred years. This accounts for shelters at the coast as well as in the fjords and in the mountains. Skipshelleren, which is situated on the boundary between the fjord system and the inland river system, is a good example of this phenomenon; as are Sævarhelleren/Olsteinhelleren which lie on the edge of a forested valley, half way into the deep Hardanger fjord. Øvste Dåvatræhelleren, Storhiller, and Grønehelleren also have favourable locations in terms of communication. The large degree of reoccupation at all of these sites probably indicates that information on the locations of these shelters was transferred between generations within social groups. There may even have been traditions for occupying the shelters, and their use may have comprised part of a planned seasonal utilisation of these particular landscapes.

The caves as funerary sites

The above discussion has largely focussed on how the caves and rockshelters were integrated in the Mesolithic settlement systems. Little attention has been paid to the ritual importance of these places. This aspect may be relevant considering that their characteristics as natural places would have made them particularly well suited for ritual communication. Still, a cave or rockshelter as such cannot automatically be ascribed such significance; there is need for additional data that can support such an interpretation. It is likely that some of the archaeological as well as the faunal data from the known caves may be relevant for a discussion of such aspects; however, as pointed out above on site functions, this requires a much more detailed comparative analysis than that which is possible in the current contribution. Here, only the burials will be discussed briefly. Mesolithic skeletons have been found in the Viste cave and in Grønehelleren, indicating primary burials, and fragments of human bones have been found in the cultural deposits of Viste, Skipshelleren, Sævarhelleren, (and possibly Tjorahelleren), indicating secondary burials or perhaps the loss of amulets. These burials may either have taken place at the same time that the sites were used as residential sites, or they may have happened during separate events – this cannot be determined on the basis of the site reports.

An important question concerns the relationship between these cave burials and other mortuary practices among the Mesolithic populations. Unfortunately, despite the large number of excavations at open-air residential sites, little is known about these practices due to the poor conditions for the preservation of bones. Only two other burial sites are known from Norway from this period. One was found in an old beach gravel deposit at Bleivik close to Haugesund in Rogaland (Indrelid 1996, 53). Skeletal material from two, possibly three, other individuals was found at a submerged site in Søgne, close to Kristiansand in Vest-Agder, Southern Norway (Sellevold and Skar 1999). The skeletons at Bleivik and Søgne were both found at open-air locations, but they cannot be securely related to settlement sites. When looking at the Danish and Swedish Mesolithic burial evidence, it indicates a relatively large diversity in terms of what kinds of sites were used (on, or close to, dwelling sites, in the sea, individual, cemeteries) and how the bodies had been treated (inhumation, cremation, secondary burial) (e.g. Jensen 2001, 221–231). It is perhaps reasonable to believe that diversity in burial practice also characterised the Norwegian Mesolithic. One possible interpretation of the burials in the current analysis is that they were first of all a variant of dwelling site burial (Todnem 1999). Such burials would perhaps normally have taken place on or close to the large open-air base camps, but they could also have been performed when people lived in caves and rockshelters – even if these were occupied only briefly. If this is correct, it would undermine somewhat the importance of caves and rockshelters in Mesolithic Norway as natural places of particular ritual significance; a rockshelter may have been 'just another' site for dwelling as well as for burial. Alternatively, the known burials were explicitly connected to the caves or rockshelters as fixed places of stone. These sites – perhaps mainly the ones that already had been used for dwelling – were sought out in order to use them as primary or secondary burial sites.

Conclusions

From these discussions it seems clear that caves and rockshelters in western Norway played a modest role for the Mesolithic populations – and no role whatsoever during the first 1500 years. They are few compared to the sites that are found in the open, and they are also generally smaller and have thinner deposits. Still, it would be a mistake to disregard them for these reasons. What makes them particularly interesting is that they provide invaluable faunal data. Contrary to the open-air sites, the conditions for preservation of bone material are often very good. When excavated carefully, they are excellent 'laboratories' for the study of: strategies for hunting, fishing and shell collecting; the processes of treating and consuming food, use, and discard of bone tools; and Mesolithic funerary practices. Data on these issues are almost impossible to come by elsewhere. Another important aspect of the caves and rockshelters is their quality as natural places. As such they belong to a category of their own compared to open-air sites, and this particular quality should be taken into consideration when discussing their functions or roles within Mesolithic patterns of settlement, mobility, and utilisation of the landscape.

The main goal of this chapter was to discuss the importance of the caves and rockshelters; to investigate what they were used for and how they related to general patterns of settlement, subsistence and funerary rituals during the Mesolithic. It is argued here that they were probably integrated in the same settlement system as the large base camps on the coast. These large open-air camps were occupied during most of the year, and seem to have been attractive for winter occupations, perhaps by several households at the same time. Areas outside of these settlement areas were utilised by small families or crews on shorter or longer expeditions. Sometimes caves and rockshelters were used during such forays, and they are generally interpreted here as different kinds of short-term sites, such as field camps, transient camps, and possibly hunting stations. The rockshelters show that expeditions extended to remote islands on the outer coast, to forests, rivers or lakes closer to the coast, and also to the fjords, and to the mountain plateaux. It appears from the faunal data from the caves that these expeditions – at least the ones that went to the fjords and mountain plateaux – mainly took place during the summer. One important feature is that several of the caves and rockshelters were reoccupied a large number of times. This may indicate that they were a part of a planned utilisation of the landscape, perhaps by one social group and by the descendants of this group, to which the information about these places was transferred. It seems likely that the caves and rockshelters can be related to patterns of subsistence, settlement and mobility. But they were probably also integrated in the ritual life of these people, as is shown by the burials and the single human bones.

References

Arias, P. (2009) Rites in the dark? An evaluation of the current evidence for ritual areas at Magdalenien cave sites. *World Archaeology* 41(2), 262–294.

Bang-Andersen, S. (1971) Rapport fra Utgravning av Lokaliteten Skiparsteinen, Eigerøy, Egersund k. Rogaland. Unpublished Report on File in Topographical Archive, Archaeological Museum, University of Stavanger.

Bang-Andersen, S. (1987) Storhiller - en 8000 år gammel boplass i Årdalsheiene. *Fra Haug ok Heidni* 1987(1), 166–171.

Bang-Andersen, S. (2003) Southwest Norway at the Pleistocene/Holocene transition: landscape development, colonization, site types, settlement patterns. *Norwegian Archaeological Review* 36(1), 5–26.

Bendixen, B. E. (1870) Adjunkt B. E. Bendixens beretning om undersøgelsen af Stensvikshulen og Bremsnæshulen. *Foreningen til Norske Fortidsminnesmerkers Bevaring. Aarsberetning* 1869, 170–178.

Berg-Hansen, I. M. (2009) *Steinalderregistrering. Metodologi og Forskningshistorie 1900–2000 med en Feltstudie fra Lista i Vest-Agder*. Oslo, Kulturhistorisk museum, Universitetet i Oslo.

Bergsvik, K. A. (1991) Ervervs- og Bosetningsmønstre langs Kysten av Nordhordland i Steinalder belyst ved Funn fra Fosnstraumen. En arkeologisk og geografisk Analyse. Unpublished Cand. Philol. Thesis, University of Bergen.

Bergsvik, K. A. (2001) Sedentary and mobile hunter-fishers in Stone Age Western Norway. *Arctic Anthropology* 38(1), 2–26.

Bergsvik, K. A. (2002) *Arkeologiske Undersøkelser ved Skatestraumen. Bind 1*. Bergen, University of Bergen.

Bergsvik, K. A. and Hufthammer, A. K. (2009) Stability and change among marine hunter-fishers in Western Norway 7000–4500 BC. Results from the excavations of two rockshelters. In P. Crombé, M. Van Strydonck, J. Sergant, M. Boudin and M. Bats (eds.) *Chronology and Evolution in the Mesolithic of N(W) Europe. Proceedings of and International Meeting, Brussels, May 30th–June 1st 2007*, 435–450. Cambridge, Cambridge Scholars Publishing.

Bergsvik, K. A. and Olsen, A. B. (2003) Traffic in stone adzes in Mesolithic Western Norway. In L. Larsson, H. Kindgren, K. Knutsson, D. Loeffler and A. Åkerlund (eds.) *Mesolithic on the Move*, 395–404. Oxford, Oxbow Books.

Binford, L. R. (1982) The archaeology of place. *Journal of Anthropological Archaeology* 1(1), 5–31.

Bjerck, H. B. (1986) The Fosna-Nøstvedt problem. A consideration of archaeological units and chronozones in the South Norwegian Mesolithic period. *Norwegian Archaeological Review* 19(2), 103–121.

Bjerck, H. B. (2007) Mesolithic coastal settlements and shell middens (?) in Norway. In N. Milner, O. E. Craig and G. N. Bailey (eds.) *Shell Middens in Atlantic Europe*, 5–30. Oxford, Oxbow Books.

Bjerck, H. B. (2008a) Lokalitet 48 Nordre Steghaugen – tidligmesolittiske boplasser med ildsteder og telttufter. In H. B. Bjerck, L. I. Åstveit, T. Meling, J. Gundersen, G. Jørgensen and S. Normann (eds.) *NTNU Vitenskapsmuseets undersøkelser. Ormen Lange Nyhamna*, 217–256. Trondheim, Tapir Akademisk Forlag.

Bjerck, H. B. (2008b) Norwegian Mesolithic trends. A review. In G. Bailey and P. Spikins (eds.) *Mesolithic Europe*, 60–106. New York, Cambridge University Press.

Bjørgo, T. (1974) Befaring av Fylkesveianlegget Storøy-Ytrøy-Hjønnevåg. Unpublished Report on File in Topographical Archive, Bergen Museum,University of Bergen.

Bjørgo, T. (1981) Flatøy. Et Eksempel på Steinalderens Kronologi og Livbergingsmåte i Nordhordland. Unpublished Mag. Art. Thesis, University of Bergen.

Bøe, J. (1934) *Boplassen i Skipshelleren på Straume i Nordhordland*. Bergen, A. S. John Griegs Boktrykkeri.

Bøe, J. (1938) Fangstmann og bumann. In A. Bugge and S. Steen (eds.) *Norsk Kulturhistorie. Bind 1*, 29–90. Oslo, J. W. Cappelens forlag.

Brøgger, A. W. (1908) *Vistefundet, en Ældre Stenalders Kjøkkenmødding fra Jæderen*. Stavanger, Stavanger Museum.

Brøgger, A. W. (1911) Jæderske stenalderbosteder. *Naturen* 35, 313–334.

Brøgger, A. W. (1925) *Det Norske Folk i Oldtiden*. Oslo, Instituttet for Sammenliknende Kulturforskning.

Conkey, M. W. (1980) The identification of prehistoric hunter-gatherer aggregation sites: The case of Altamira. *Current Anthropology* 21(5), 609–630.

Degerbøl, M. (1951) Det osteologiske materiale. In H. E. Lund (ed.) *Fangst-boplassen i Vistehulen på Viste, Randaberg, Nord-Jæren*, 52–84. Stavanger, Stavanger Museum.

Eilertsen, K. S. (2009) Skjevtrekantene fra Fiskåvatnet – en preliminær presentasjon av mikrolittmaterialet fra en heller ved Fiskåvatnet, Karmøy kommune, Rogaland. In M. Nitter and E. S. Pedersen (eds.) *Tverrfaglige Perspektiver*, 67–74. Stavanger, Archaeological Museum, University of Stavanger.

Eilertsen, K. S. (2010) Helleren på Barka i Strand – oppholdssted og naturformasjon i årtusener. *Fra Haug ok Heidni*, 2010(1), 3–10.

Fuglestvedt, I. (2009) *Phenomenology and the Pioneer Settlement on the Western Scandinavian Peninsula*. Lindome, Bricoleur Press.

Fürst, C. M. (1909) *Das Skelett von Viste auf Jäderen*. Christiania, Jacob Dybwad.

Gjessing, G. (1945) *Norges Stenalder*. Oslo, Norsk Arkeologisk Selskap.

Gjessing, H. (1920) *Rogalands Stenalder*. Stavanger, Stavanger Museum.

Gustafson, L. (1982) *Arkeologiske Registreringer i Flåms- og Undredalsvassdraget*. Bergen, Historical Museum, University of Bergen.

Hofseth, E. H. (1982) *Kulturminner i Bjerkreimsvassdraget*. Stavanger, Archaeological Museum in Stavanger.

Hufthammer, A. K., Høie, H., Folkvord, A., Geffen, A. J., Andersson, C. and Ninneman, U. S. (2010) Seasonality of human site occupation based on stable oxygen isotope ratios of cod otolihs. *Journal of Archaeological Science* 37(2010), 78–83.

Hufthammer, A. K. and Meiklejohn, C. (1986). Direct dating of the Viste burial, coastal Norway. *Mesolithic Miscellany* 7(2), 7–8.

Høgestøl, M. (1995) Arkeologiske Undersøkelser i Rennesøy Kommune, Rogaland, Sørvest-Norge. Stavanger, Archaeological Museum in Stavanger.

Indrelid, S. (1978) Mesolithic economy and settlement patterns in Norway. In P. Mellars (ed.) *The Early Postglacial Settlement of Northern Europe. An Ecological Perspective*, 147–176. London, Duckworth.

Indrelid, S. (1996) *Frå Steinalder til Vikingtid. Strilesoga Bind 1*. Bergen, Eide.

Inselset, S. (2007) Notat Angående B13129. Unpublished Report on File in Topographical Archive, Bergen Museum, University of Bergen.

Jansen, K. (1972) Grønehelleren, en Kystboplass. Unpublished Mag. Art. thesis, University of Bergen.

Jensen, J. (2001) *Danmarks Oldtid. Stenalder 13.000–2.000 f.Kr*. København, Gyldendal.

Kaland, P. E. (1984) Holocene shore displacement and shorelines in Hordaland, Western Norway. *Boreas* 13, 203–242.

Kloster, R. (1931) Innberetning fra Utgravningene av Dognesteinen, Hå, Rogaland. Unpublished report on file in topographical archive, Archaeological Museum, University of Stavanger.

Koren, V. (1869) Personel kapellan V. Korens indberetning. *Foreningen til Norske Fortidsminnesmerkers Bevaring. Aarsberetning* 1869, 6–12.

Kristoffersen, K. and Warren, E. J. (2001) *Kulturminner i trekant-traséen. De arkeologiske Undersøkelsene i forbindelse med Utbygging av Trekantsambandet i Kommunene Bømlo, Sveio og Stord i Sunnhordland*. Bergen, Department of archaeology, University of Bergen.

Leroi-Gourhan, A. (1968) *The Art of Prehistoric Man in Western Europe*. London, Thames & Hudson.

Lie, R. W. (1989) Animal remains from the post-glacial warm period in Norway. *Fauna Norvegia. Serie A* 10, 45–56.

Lund, H. E. (1951) *Fangstboplassen i Vistehulen*. Stavanger, Stavanger Museum.

Lødøen, T. K. (1997) Arkeologiske undersøkelser i Øvste Dåvatræhelleren. Unpublished Report on File in Topographical Archive, Bergen Museum, University of Bergen.

Lødøen, T. K. (1995) Landskapet som Rituell Sfære i Steinalder. En Kontekstuell Studie av Bergartsøkser fra Sogn. Unpublished Cand. Philol. Thesis, University of Bergen.

Mikkelsen, E. (1971) Vistefunnets kronologiske stilling. Trekk av Rogalands eldre steinalder. *Stavanger Museums Årbok* 1970, 5–38.

Mikkelsen, E. (1979) Seasonality and Mesolithic adaptation in Norway. In K. Kristiansen and C. Paludan-Müller (eds.) *New Directions in Scandinavian Archaeology*, 79–119. Copenhagen, National Museum of Denmark.

Miracle, P. (2001) Feast of famine? Epipaleolithic subsistence in the Northern Adriatric Basin. *Documenta Praehistorica* 38, 177–197.

Myhre, B. (1967) Tjorahelleren. Et bidrag til Rogalands tidlige steinalder. *Stavanger Museums Årbok* 1967, 7–40.

Myhre, B. (1980) *Sola og Madla i Førhistorisk Tid*. Stavanger, Archaeological Musum in Stavanger.

Neumann, J. (1837) Bjerghulerne i Bergens Stift. *Urda* 1, 201–229.

Olsen, A. B. (1992) *Kotedalen – en Boplass gjennom 5000 år. Bind 1: Fangstbosetning og tidlig Jordbruk i Vestnorsk Steinalder: nye Funn og nye Perspektiver*. Bergen, Historical museum, University of Bergen.

Olsen, A. B. and Alsaker, S. (1984) Greenstone and diabase utilization in the Stone Age of Western Norway: technological and socio-cultural aspects of axe and adze production and distribution. *Norwegian Archaeological Review* 17(2), 71–103.

Olsen, H. (1976) Skipshelleren, Osteologisk Materiale. Unpublished thesis, University of Bergen.

Olsen, T. B. (2006) Et lite, men lunt oppholdssted i 6000 år. *Fra Haug ok Heidni* 2006(1), 26–34

Østerdal, A. (2001). Forslag til Reguleringsplan for Haukanes gnr. 46 bnr. 6 og 7 Austevoll kommune. Unpublished Report on File in Topographical Archive, Bergen Museum, University of Bergen.

Prøsch-Danielsen, L. (2006) *Sea-level studies along the coast of Southwestern Norway*. Stavanger, Archaeological Museum in Stavanger.

Randers, K. (1979) Utgravning av Heller i Fana. Unpublished Report on File in Topographical Archive, Bergen Museum, University of Bergen.

Randers, K. (1986) *Breheimenundersøkelsene 1982–1984 I: Høyfjellet*. Bergen, Historical museum, University of Bergen.

Sellevold, B. and Skar, B. (1999) The first lady of Norway. In G. Gundhus, E. Seip and E. Ulriksen (eds.) *NIKU 1994 –1999. Kulturminneforsknings Mangfold*, 6–11. Oslo, Norsk Institutt for Kulturminneforskning (NIKU).

Shetelig, H. (1922) *Primitive Tider i Norge*. Bergen, John Griegs Forlag.

Skjelstad, G. (2003) Regionalitet i Vestnorsk Mesolitikum. Råstoffbruk og Sosiale Grenser på Vestlandskysten i Mellom- og Senmesolitikum. Unpublished Cand. Philol. Thesis, University of Bergen.

Skrede, M. A. and Baug, I. (2001) Ev 39, Strekninga Jektevik – Sandvikvåg Stord og Fitjar kommune. Unpublished Report on File in Topographical Archive, Bergen Museum, University of Bergen.

Storvik, I. (2011) Bruken av Huler og Hellere i Rogaland fra Steinalder til Middelalder. Unpublished MA Thesis, University of Bergen.

Svendsen, J. I. and Mangerud, J. (1987) Late Weichselian and Holocene Sea-level History for a Cross-section of Western Norway. *Journal of Quarternary Science* 2, 113–132.

Todnem, R., 1999, Holer og hellere, for de levende eller for de døde? In I. Fuglestvedt, T. Gansum and A. Opedal (eds.) *Et Hus med Mange Rom: Vennebok til Bjørn Myhre på 60-årsdagen*, 103–120. Arkeologisk museum i Stavanger, Stavanger.

Warren, J. E. (1994) Coastal Sedentism during the Atlantic Period in Nordhordland, Western Norway? The Middle and Late Mesolithic Components at Kotedalen. Unpublished M.A. Thesis, Memorial University of Newfoundland, St. John's.

Chapter 4

Rockshelters in Central Norway: Long-Term Changes in Use, Social Organization and Production

Anne Haug

From central Norway there is an extensive collection of finds from caves and rockshelters; material which, relatively speaking, has not been utilized by archaeologists. Little has been published and there has been no attempt to compile material from cave and rockshelters found in this region. Based on the archaeological excavation of a rockshelter carried out in 2005 at Monge in Rauma, Møre og Romsdal County, this paper outlines a project that will examine the use of rockshelters in central Norway from a long-term perspective. In particular, this project will focus on social organization, production and changes in production. It aims to investigate whether these rockshelters were in continuous use or tied to specialized, seasonal or ritual activities, and whether or not their use changed through time and space.

Introduction

Most people cannot deny the feeling of being fascinated upon entering a cave or a deep rockshelter. You walk from the daylight into a gradually darker room, and the feeling of being in another place is strong. Did the past people who took these rockshelters and caves into use feel the same way – and did this influence how the caves were used – or were people unaffected by these surroundings and used the caves for practical purposes only?

The large number of archaeological finds from caves and rockshelters in Norway generally indicate a great variation in the use of these places (e.g. Brøgger 1908, 1910; Brinkmann and Shetelig 1920; Bøe 1934; Gjessing 1943; Lund 1951; Jansen 1972; Bakka 1973; Prescott 1991; Østebø 2008; Bommen 2009; Bjerck this volume). This is also the case in central Norway (Fig. 4.1). The caves and shelters in this region seem to have been used for both sacred and domestic purposes. Some of them show only traces of short visits, while others show long term occupations with thick cultural deposits. Most of the excavations in central Norway took place between 1900 and 1950, and they demonstrate that such sites were used from the Neolithic to the Iron Age and also in the medieval period (e.g. Bendixen 1870; Petersen 1910, 1917, 1920; Nummedal 1911, 1913, 1920a, 1920b; Hougen 1929; Hultgren *et al.* 1985; Haug and Sauvage 2005; Lie 2007, 2008).

Recently, there have been several large-scale archaeological excavation projects in central Norway focusing on production and economic development from the Mesolithic period to the Iron Age, but most of these projects have been carried out on open-air sites (e.g. Haug 2007; Bjerck *et al.* 2008). The preservation of bones at these sites is poor because of the acidic soils, and so this is one of the aspects that make the caves and rockshelters interesting. The often well preserved bone material from the caves and rockshelters is important for studying changes in the hunting and trapping methods and in seasonality, both on land and at sea, but it is also vital for studying livestock husbandry and the introduction of agriculture. Traces of funerary rituals have also been found at several of the rockshelters. Fortunately, large numbers of bones from wild and domesticated animals and also some human bones from the old excavations of caves and rockshelters have been analyzed by osteologists. However, very little of this data has been the subject of comparative studies and even less has been published. It is a challenge to organize and understand the archaeological as well as faunal material and human skeletons from these excavations, especially because this was a period when methods of documentation, stratigraphical analyses and dating methods were less developed than today.

Below, a new project will be outlined, in which some of the existing material from the caves and rockshelters from central Norway will be subjected to new research and analysis. The project has a main focus on social organization; it aims to investigate whether these places were long-term settlement sites or tied to specialized or seasonal activities, and how their use changed through time. An important goal is also to discuss the social identities of the people who stayed in these shelters. In order to demonstrate the potential of this data I shall present as a case-study the excavation of a rockshelter carried out in 2005 at Monge in Rauma, Møre and Romsdal County (Fig. 4.1). On the basis of this example, I will outline some interesting themes for further discussion and study.

Caves and rockshelters in central Norway

Caves and rockshelters are often treated as a uniform group, but a rockshelter differs from a cave in many ways. Caves are enclosed on three sides and extend into the mountain. Two main types exist in central Norway: karstic caves and caves formed by sea-waves. A rockshelter is mainly a cliff overhang of stone or big blocks of stone that form a shelter or chamber. Rockshelters in this area have been formed by a variety of geological processes, such as rockslides, wind blasting, and dissolution by water or temperature changes. During the current project, archive studies uncovered that a total of 113 caves and 160 rockshelters in this area are known to have traces of prehistoric use County (Fig. 4.1). Among these, 27 rockshelters and 12 caves have been excavated in the area extending from Møre and Romsdal (Nordmøre and Romsdal) to the municipality of Rana in Nordland. The remaining have only been registered or surveyed and mapped, but several of them have a potential for further studies and excavations. Most of the registered caves and rockshelters are situated in the coastal areas and fjords.

The Rockshelter at Monge, Rauma: a case-study

During 2005, the Museum of Natural History and Archaeology at the Norwegian University of Science and Technology (NTNU) carried out an excavation of the rockshelter 'Smiehelleren' in Rauma, in Møre and Romsdal County (Sauvage 2005). The rockshelter lies in a valley where there have been several rock avalanches, close to the River Rauma, approximately 75m above present-day sea level. The rockshelter was first documented in 1847, when it was sold to a farmer who wanted to use it for storing his river boats. In the contract, the rockshelter is mentioned under the name of 'Smiehelleren' meaning 'the smith's shelter' (Fig. 4.2). The rockshelter is made out of large stone blocks that form a chamber of approximately 35m^2. The opening is about 3.5m high, and is oriented to the south-east.

The excavation revealed a thick cultural layer inside the shelter that was approximately 90cm thick. This cultural layer was divided into four different phases, relating to different activities in the shelter. The top layer was 30–50cm thick and contained a lot of charcoal, iron slag and small pieces of iron, demonstrating that the rockshelter had indeed been used by a blacksmith for iron production, as indicated by the 1847 document. The three other layers contained sand, small pieces of charcoal, osteological material and different types of bone artefacts. Most of the bones were found in the bright sandy layer and were not burnt. As the terrain just outside the shelter formed a relatively flat surface, it was decided to examine this area too, in order to shed further light on the use of the shelter. Close to the opening of the shelter we found five cooking pits and traces of a small house or cabin (Fig. 4.3).

The osteological material

During the excavation of the rockshelter, 3117 fragments of bones were discovered (Hufthammer 2007). Analyses of this osteological material revealed a variety of wild and domestic animal bones: moose (*Alces alces*), red deer (*Cervus elaphus*), otter (*Lutra lutra*), beaver (*Castor fiber*) wild boar/domestic pig (*Sus scofa*) sheep (*Ovis*), goat (*Capra*), cattle (*Bos taurus*), horse (*Equus caballus*), different types of fish, such as salmon (*Salmo salar*), trout (*Salmo trutta*) and cod (*Gadus morhua*), and also different kinds of birds and game (Fig. 4.4). As a whole, the osteological material from the rockshelter at Rauma exhibits several similarities with other excavated rockshelters in the region, especially regarding the variety of different species.

During the pre-Roman Iron Age, around 300 cal BC, it seems that the inhabitants of the shelter started to keep sheep and goat in addition to traditional hunting and trapping. Before this period, domestic bones are absent from the faunal assemblage. This is

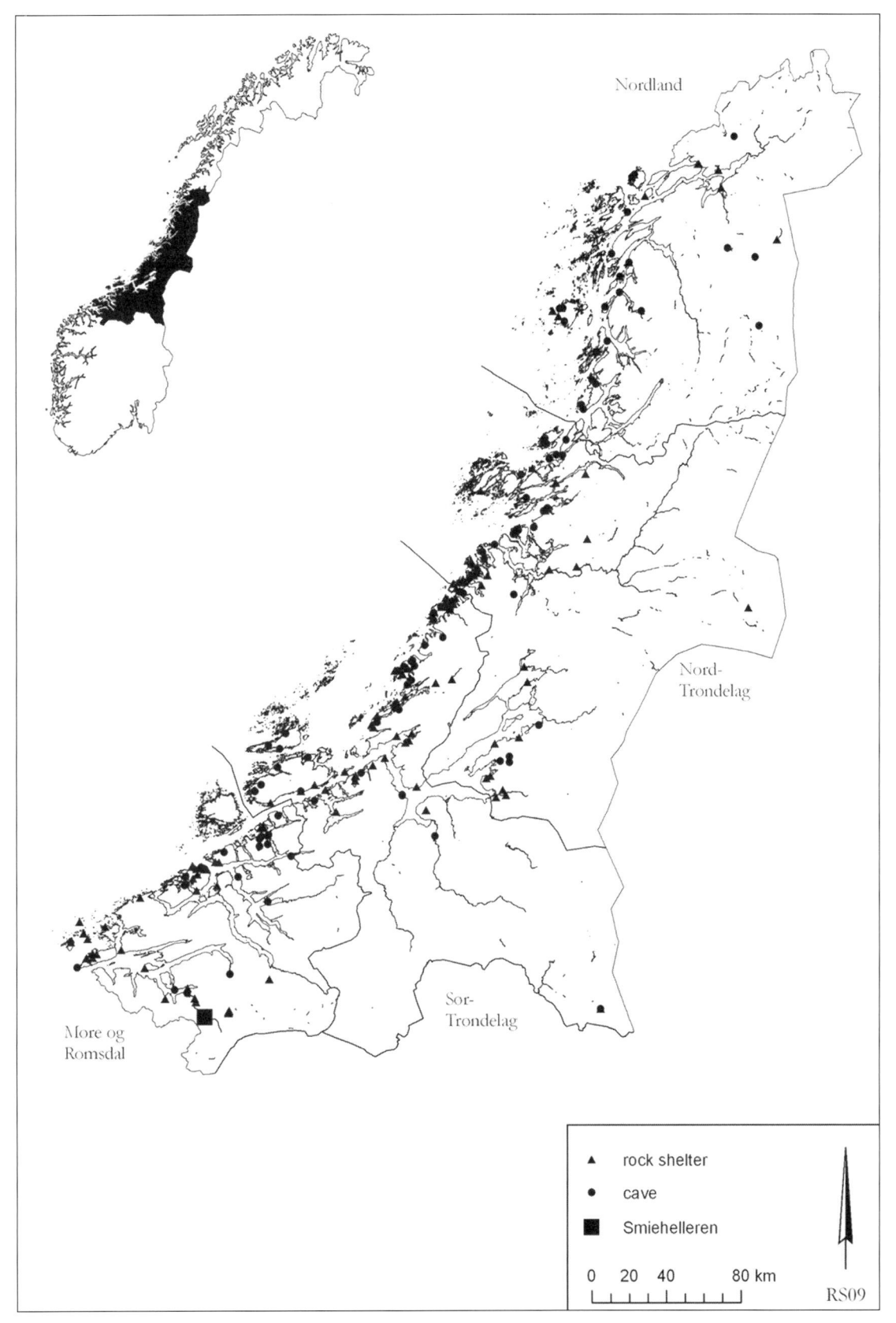

4. Rockshelters in Central Norway: Long-Term Changes in Use, Social Organization and Production

Figure 4.1: Caves and rockshelters in the counties Møre og Romsdal, Sør-Trøndelag, Nord-Trøndelag and Nordland. From Sauvage (2005). Reproduced with permission.

Figure 4.2: The rockshelter 'Smiehelleren' at Monge, Møre og Romsdal. Photo: Anne Haug.

interesting because it represents a change in their food consumption, which, in turn, might indicate a change in their economic and social life. Why did this happen? The domestic bones represent approximately three per cent of the total osteological material. Sheep and goat were mainly used for food consumption, but of course they also provided the inhabitants with wool and milk. If we look at contemporary settlements at open-air sites in central Norway during the period 300 cal BC–AD 200, the excavations demonstrate that there is an increase in farming and livestock husbandry in several areas. There is undoubtedly an interesting link between the open-air sites and the domestic bones found in the rockshelters, particularly concerning when these processes started and developed, but further studies are needed. The sheep and goat bones from the rockshelter at Rauma are dated to between 395 and 370 cal BC. Pig bones from the 'Smiehelleren' have been dated to between AD 0 and 60. Further analyses are needed to decide whether they are of wild or domestic species.

Altogether 296 fish bones were found. They make up approximately nine per cent of the total osteological material. One might have expected more fish bones, since the rockshelter is situated close to the River Rauma. One possible explanation is that fish bones in general do not preserve as well as other osteological material. Considering the fact that, today, the River Rauma is good for catching salmon and trout; it would be strange if fish were not an important part of their diet. From the fish bones, several fragments of cod were identified. This is interesting because the River Rauma is a freshwater river whereas the cod lives in saltwater. Did the people who inhabited the rockshelter bring the cod with them, or is there another explanation? One cod bone from Layer 5 dates to between 765 and 710 cal BC.

Human bones from two individuals represented another intriguing discovery from the rockshelter (Hufthammer 2007). These bone fragments were found in the layers dated to the Bronze Age and early Iron Age the central part of the rockshelter over an area of $2m^2$. They have not been directly dated. The osteological analyses identified one young adult and one infant. The material is fragile and badly preserved, but included one tooth and some small pieces of a cranium. Preliminary analysis indicates that the bones from the infant were not burnt while those from the young adult were burnt. Whether or not this difference is of significance is difficult to say, but it could lead to some interesting questions concerning ritual and burial practices in the Bronze and Iron Ages. Further studies of the human bones are necessary, including DNA, isotope and morphological studies. From an archaeological perspective, the human bones are interesting as they were found inside a rockshelter together with much household waste, including wild and domestic bones. It is

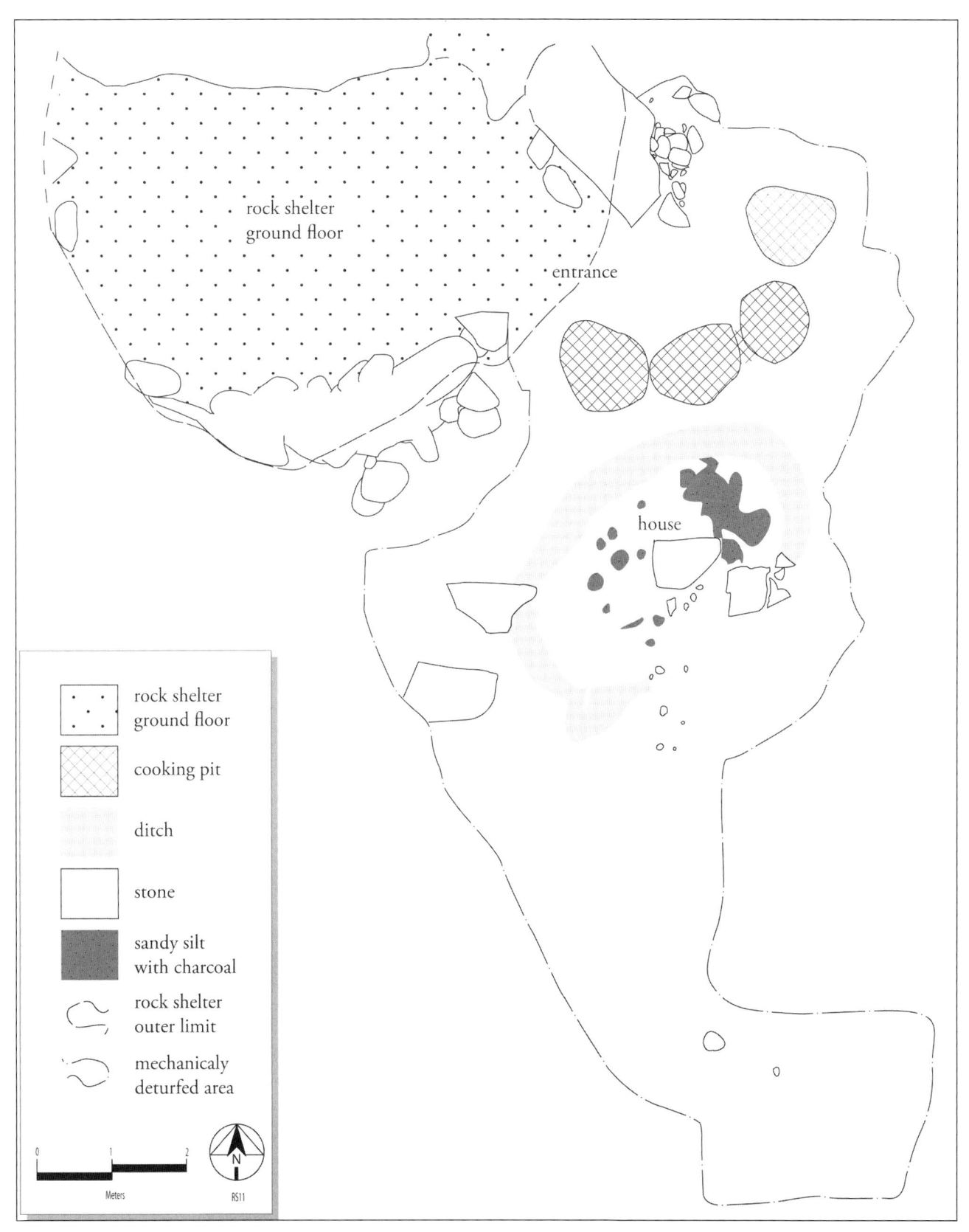

Figure 4.3: Plan drawing of the rockshelter at Monge. Based on Sauvage (2005). Reproduced with permission.

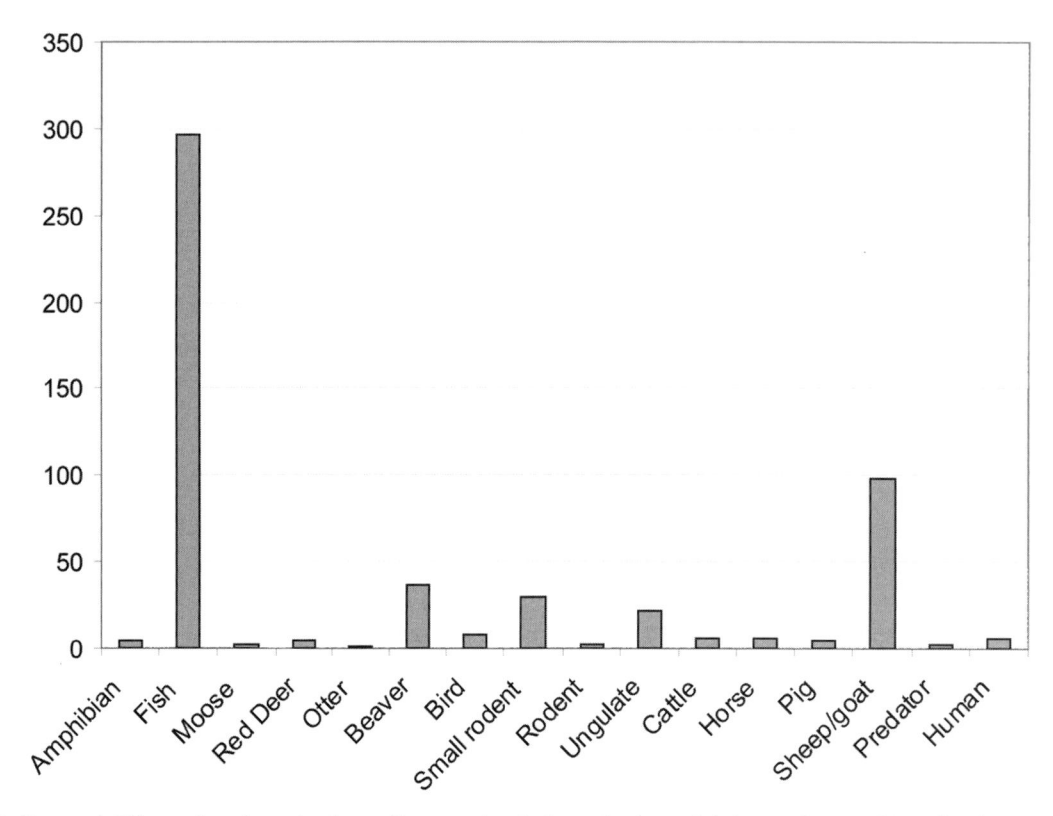

Figure 4.4: Bones of different faunal species from all occupational phases in the rockshelter at Rauma. Note that bones not identified to species are not shown here. From Sauvage (2005). Reproduced with permission.

interesting to note that several of the rockshelters in central Norway have similar material. So far, eight rockshelters and seven caves containing human bones have been discovered in the region (e.g. Todnem 1999; Lie 2007).

The artefacts

The number of artefacts from the rockshelter was quite moderate compared to the osteological material. Except for some flint flakes mainly found outside the shelter, there were six tools of bone, comprising four fish hooks (Fig. 4.5) and two awls. These tools where found in the sandy layers in the inner chamber of the rockshelter, where preservation conditions were good. Scrapers of flint and quartz and one grindstone were also recovered. In addition, fragments of ceramics from three different vessels, some of them with decoration, were found. The vessels are of Bronze Age and early Iron Age type. Some fragments of a soapstone vessel and an iron fish hook were also discovered. Finally, there were found, in connection with the iron working activity area, different tools, fragments of five melting pots, more than 3300 pieces of iron and slag, and some small pieces of leather.

Figure 4.5: Bone fishhook found in the sandy layers in the inner room of the shelter. Length: 2.6cm. Photo: Vitenskapsmuseet.

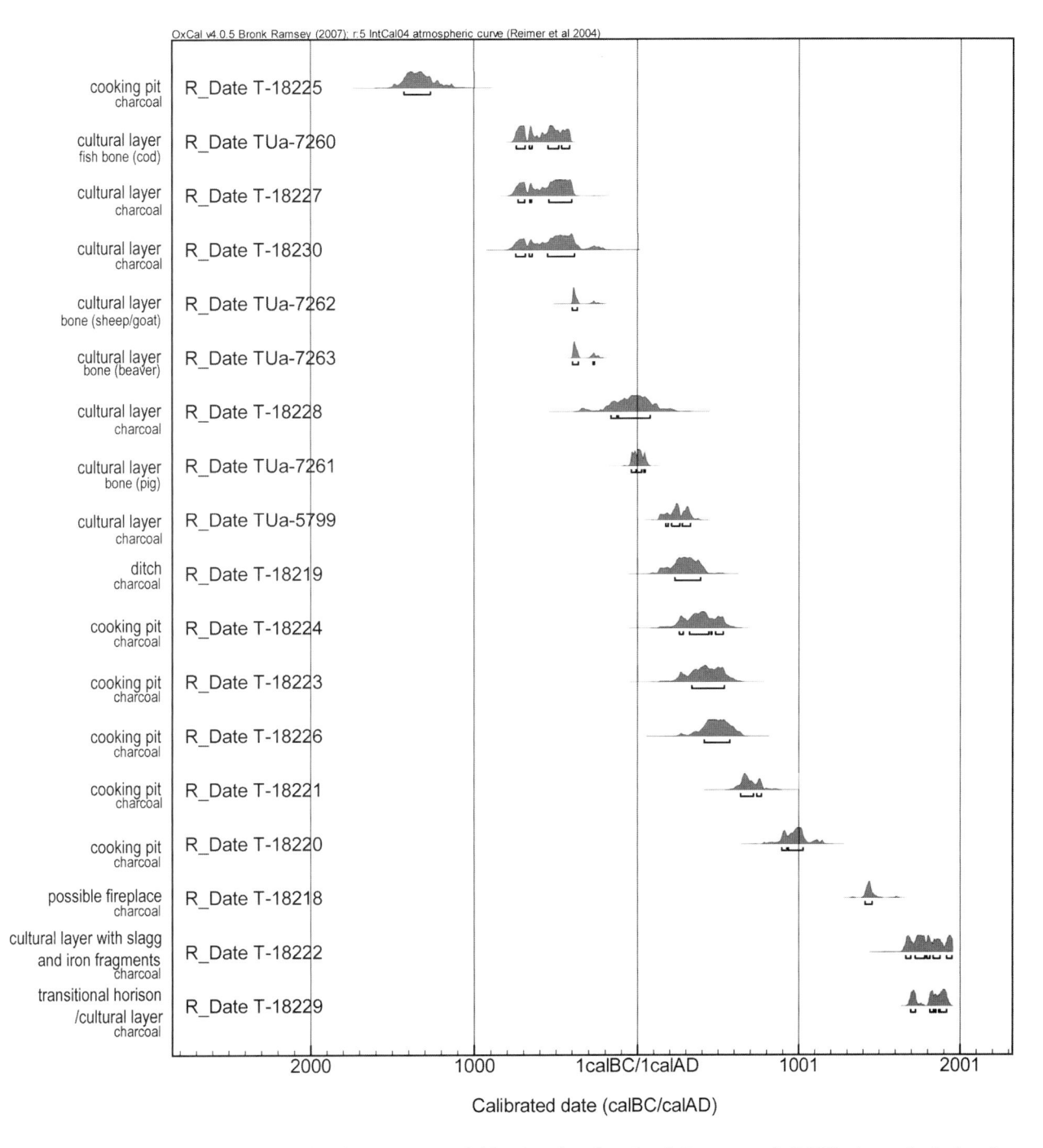

4. Rockshelters in Central Norway: Long-Term Changes in Use, Social Organization and Production 45

Figure 4.6: Radiocarbon dates from the rockshelter at Rauma. Calibrations based on Bronk Ramsey et al. (2007). Atmospheric data from Reimer et al. (2004). From Sauvage (2005). Reproduced with permission.

The age of the rockshelter

The artefacts and the osteological material show that the rockshelter at Rauma has been used for different purposes in different time periods. Much work remains to be done on the material from the site, but some patterns are worth noting. The rockshelter at Rauma seems to have been used as a specialized hunting and fishing station in the Bronze and Iron Ages, for sacred or burial purposes probably at one time in the pre-Roman period, and as a smithy in the late Middle Ages.

So far we have 18 radiocarbon dates from the rockshelter (Fig. 4.6). 14 of them are from charcoal taken from layers and structures situated inside and outside the rockshelter. Four are from bones found in the rockshelter. From these dates it is possible to offer the preliminary interpretation that there are at least four phases represented in the archaeological material. The first phase represents the oldest layers, with bones and artefacts dated to the early Bronze Age. One of the cooking pits outside can be dated to the same period, to 1415–1230 cal BC. The second phase is represented by layers containing asbestos (used as temper in Bronze Age and Early Iron Age pottery), ceramics and animal bones, both wild and domestic. This phase can be dated to the late Bronze Age – early Iron Age, to 700–100 cal BC. The third phase is represented by the smithy, whose oldest layers can be dated to the late medieval period around AD 1350–1400. The fourth phase is represented by the present-day use of the rockshelter.

As this case-study demonstrates, there are a lot of interesting questions to be asked of rockshelters in Norway. This example is just one of many in central Norway. The rockshelter in Rauma also highlights the complex stratigraphy that is a feature of many of the rockshelters. Preliminarily investigations from the region show that there are approximately 40 to 50 rockshelters with similar material, but further studies are needed to confirm their age and use.

The rockshelter project: a project outline

As the rockshelter project in central Norway is in its initial phase, it is important to outline some of the main subjects and focuses for the future work. An important part of the work must be considered as basic research, as no one before has attempted to synthesize the cave and rockshelter material from central Norway. As pointed out above, the data is large; altogether 273 caves and rockshelters with traces of human use are known from the counties Møre og Romsdal, Sør-Trøndelag, Nord-Trøndelag and Nordland (Fig. 4.1). The diversity and complexity of this material provides an opportunity to study the rockshelters in detail, especially concerning cultural and social change. Five key subjects have been chosen for further work:

1. *Length of occupation and seasonality*. Were the caves and rockshelters occupied for shorter or longer periods, and is it possible to say something about seasonality? The composition of archaeological data may be used for this purpose, and particularly the faunal data, which is often abundant at these places.

2. *Activities*. What type of activities took place in the rockshelters? Were they used for ordinary settlement from by people who hunted, trapped, fished and herded animals close to the shelters, were they sites for specialised hunting parties or production, or were they particular sites for funerals or other rituals? In order to answer these questions, interdisciplinary approaches are necessary, archaeological and palynological material are analysed together.

3. *Landscape*. Do site locations reflect these activities, and do they change over time? How are their relationships to important resources, such as good fishing rivers, hunting areas or pastures? Are they situated far from or close to contemporary open air agricultural settlement sites? During the last few years, site location analyses as well as various kinds of methods based visual and other criteria have been applied in landscape archaeology. These methods may also be relevant for the caves and rockshelters in central Norway.

4. *Social status and identities*. What were the social statuses, identities and ethnicities of the people who occupied the caves and rockshelters? In which way were they integrated in the general territorial, social and economic organisation in the different time periods? These questions have been important for the rockshelter research in Norway for about 150 years (see (Bergsvik 2005) for a review of this debate on the Iron Age), and are still relevant. In order to investigate these problems, it is necessary to develop a broad interdisciplinary approach, in which data from open-air sites should be drawn into the discussion in a comparative perspective.

5. *Regional variation*. Is it possible to trace regional differences in the use of caves and rockshelters; were they used in different ways at the coast compared to in the inland and mountains and in how they were used by the various populations along the coast of Norway? Recently, several investigations have been performed on caves and rockshelters in western Norway (e.g. Østebø 2008; Barndon 2009; Bommen 2009;). Comparisons with the results from these investigations are likely to produce interesting results.

Conclusion

A large number of caves and rockshelters in central Norway have been surveyed and excavated, and the data indicate that they have served many purposes during the different periods of prehistory. An example of this is the rockshelter 'Smiehelleren', which was excavated in 2005. During the Bronze Age, this site was first used by crews of hunter-fishers, later by groups that also consumed domesticated animals at the site.

Funerary rituals have also been performed in the shelter in this period. During late medieval times the site was utilised as a smithy – an activity that may have provided modern name of the place. Despite of the large and varied data from the caves and rockshelters from central Norway, very little research has been performed on these important sites. In an effort to improve this situation, a new project is planned by this author. In a long-term comparative and interdisciplinary perspective this will deal with what these places were used for, how their uses were organised, and with the social statuses and identities of the people who used them.

References

Bakka, E. (1973) Omkring problemet om kulturdualisme i Sør – Noreg. In P. Simonsen and G. S. Munch (eds.) *Bonde – Veidemann, Bofast – Ikke Bofast i Nordisk Forhistorie*, 109–127. Tromsø, Universitetsforlaget.

Barndon, R. (2009) Caves and rockshelters – sacred crafts and smith's graves in a long-term perspective. In I. Holm, K. Stene and E. Swensson (eds.) *Liminal Landscapes. Beyond the Concepts of 'Marginality' and 'Periphery'* Oslo Archaeological Series 11, 47–65. Oslo, Unipub.

Bendixen, B. E. (1870) Adjunkt B. E. Bendixens Beretning om Undersøgelsen af Stensvikshulen og Bremsnæshulen. *Foreningen til Norske Fortidsminnesmerkers Bevaring. Aarsberetning* 1869, 170–178.

Bergsvik, K. A. (2005) Kulturdualisme i vestnorsk jernalder. In K. A. Bergsvik and A. Engevik jr. (eds.) *Fra Funn til Samfunn. Jernalderstudier Tilegnet Bergljot Solberg på 70-Årsdagen.* Universitetet i Bergen Arkeologiske Skrifter Nordisk 1, 229–258. Bergen, Department of Archaeology, University of Bergen.

Bjerck, H. B., Åstveit, L. I., Meling, T., Gundersen, J., Jørgensen, G. and Normann, S. eds. (2008) *NTNU Vitenskapsmuseets Undersøkelser. Ormen Lange Nyhamna*. Trondheim, Tapir Akademisk Forlag.

Bommen, C. (2009) *Bruken av Hellarar i Eldre Jernalder i Sunnhordland*. Unpublished Masters thesis, Bergen, University of Bergen.

Bøe, J. (1934) *Boplassen i Skipshelleren på Straume i Nordhordland*. Bergens Museums Skrifter 17. Bergen, A. S. John Griegs Boktrykkeri.

Brinkmann, A. and Shetelig, H. (1920) *Ruskenesset. En Stenalders Jagtplass*. Norske Oldfunn 3. Kristiania, Universitetets Oldsaksamling.

Brøgger, A. W. (1908) *Vistefundet, en Ældre Stenalders Kjøkkenmødding fra Jæderen*. Stavanger Museums Årshefte 1907. Stavanger, Stavanger Museum.

Brøgger, A. W. (1910) Vestnorske hulefund fra ældre jernalder. *Bergens Museums Aarbok* 1910, no. 16, 1–22.

Bronk Ramsey, C. Brenninkmeijer, C. A. M., Jöckel, P., Kjeldsen, H. and Masarik. J. (2007) Direct measurement of the radiocarbon production at altitude. *Nuclear Instruments and Methods in Physics Research B*, 359, 558–654.

Gjessing, G. (1943) *Træn-Funnene*. Instituttet for Sammenlignende Kulturforskning, Serie B, 41. Oslo, Aschehoug & Co.

Haug, A. (2007) I Nummedals fotspor. *SPOR* 22 (2), 28–32.

Haug, A. and Sauvage, R. (2005) Huler og hellere og deres mange hemmeligheter. *SPOR* 20(2), 4–9.

Hougen, B. (1929) En stenalders boplass på Hegge i Skatval. *Det Kongelige Norske Vitenskapers Selskabs Skrifter* 1929, no. 7, 1–12.

Hufthammer, A. K. (2007) *Osteologiske Analyser fra Smiehelleren, Rauma, Møre og Romsdal*. Unpublished report, Bergen, University of Bergen.

Hultgren, T., Johansen, O. S. and Lie, R. W. (1985) Stiurhelleren i Rana *Viking* 48, 83–102.

Jansen, K. (1972) *Grønehelleren, en Kystboplass*. Unpublished Magister Artium Thesis, Bergen, University of Bergen.

Lie, R. O. (2004) Huler og hellere på Averøy. *Minner fra Averøy, Averøy Historielag Årbok*, 5–18.

Lie, R. O. (2007) Huler og hellere på Nordmøre. Del 1. *Årbok for Nordmøre* 2007, 95–131.

Lie, R. O. (2007) Huler og hellere på Nordmøre. Del 2. *Årbok for Nordmøre* 2008, 5–25.

Lund, H. E. (1951) *Fangstboplassen i Vistehulen*. Stavanger Museums skrifter 6. Stavanger, Stavanger Museum.

Nummedal, A. (1911) *Dalehelleren og Valsethulen*. Det Kgl. Norske Videnskapers Selskaps Skrifter 1910, no. 11. Trondhjem, Aktietrykkeriet i Trondhjem.

Nummedal, A. (1913) *Bjørneremsfundet*. Det Kgl. Norske Videnskapers Selskaps Skrifter 1912, no. 12. Trondhjem, Aktietrykkeriet i Trondhjem.

Nummedal, A. (1920a) *Hellerne ved Laksevaagen, Kristiansund*. Det Kgl. Norske Videnskapers Selskaps Skrifter 1919, no. 4. Trondhjem, Aktietrykkeriet i Trondhjem.

Nummedal, A. (1920b) *Bopladsfund paa Hjalmøy og Dønna*. Det Kgl. Norske Vitenskapers Selskaps Skrifter 1919, no. 5. Trondhjem, Aktietrykkeriet i Trondhjem.

Østebø, K. (2008) *Hellerbruk i Vestnorsk eldre Jernalder belyst ved Lokalisering*. Unpublised Masters thesis, University of Bergen, Bergen.

Petersen, T. (1910) *Hestneshulen*. Det Kgl. Norske Videnskapers Selskaps Skrifter 1910, no. 2. Trondhjem, Aktietrykkeriet i Trondhjem.

Petersen, T. (1917) *Haugshulen paa Leka*. Det Kgl. Norske Videnskapers Selskaps Skrifter 1916, no. 4. Trondhjem, Aktietrykkeriet i Trondhjem.

Petersen, T. (1920) *Bopladsfund paa Halmøy og Dønna*. Det Kgl. Norske Videnskapers Selskaps Skrifter 1919, no. 5. Trondhjem, Aktietrykkeriet i Trondhjem.

Prescott, C. (1991) *Kulturhistoriske Undersøkelser i Skrivarhelleren*. Arkeologiske rapporter 14. Bergen, Historical Museum, University of Bergen.

Reimer, R. W., Remmele, S., Southon, J. R., Stuiver, M., Talamo, S., Taylor, F. W., van der Plicht, J. and Weyhenmeyer, C. E. (2004) IntCal04 terrestrial radiocarbon age calibration, 0–26 cal kyr BP. *Radiocarbon* 46(3), 1029–1058.

Sauvage, R. (2005) *Utgravning av Smiehelleren på Monge gnr/bnr 69/8 i Rauma, Møre og Romsdal, i forbindelse med Reguleringsplan for ny E-136 Strekningen Monge – Marstein*. Unpublished report, Trondheim, Museum of Natural History and Archaeology, Norwegian University of Science and Technology (NTNU).

Todnem, R. (1999) Holer og hellere, for de levende eller for de døde? In I. Fuglestvedt, T. Gansum and A. Opedal (eds.) *Et Hus med mange Rom: Vennebok til Bjørn Myhre på 60-Årsdagen*, AmS-Rapport 11A, 103–120. Stavanger, Arkeologisk Museum i Stavanger.

Chapter 5

On the Outer Fringe of the Human World: Phenomenological Perspectives on Anthropomorphic Cave Paintings in Norway

Hein Bjartmann Bjerck

Norwegian cave paintings are confined to eleven localities, all located in North Norway, and seem to date to the Bronze Age. The majority of the paintings are strikingly similar red stickmen, suggesting a common tradition. This paper focuses on the context of these images, in making explicit the conspicuous sensory world of the caves, and discusses the phenomenological implications of the relation between paintings and caves. Paradoxically, scientific conformity seems to have overpowered even the most forceful caves: traditional recording (plans, sections, flash photos) bring forth things never observed by the peoples of the past, and omit the subjective impressions that are essential to understand the embodied experience of caves. Very likely, embodied experiences were imperative in the images' cultural context, and in the cosmological and ritual meanings of caves. Most things that characterize the world of humans (life, light, motion, colours, change, sounds and odours) do not exist in caves. This dichotomy, and the fact that most caves do not have a definite termination (only points where the human body cannot pass), nourish a notion of caves as connections to realms beyond the reach of humans.

Introduction

Along the rugged coast of Norway there is a large number of coastal caves – impressive formations which have probably provoked interest, enticed and frightened inhabitants of the region as well as travellers, through all times. The most famous of these is the gigantic and spectacular tunnel-shaped hole right through the prominent mountain of Torghatten near Brønnøysund (Johansen 2008; Møller and Fredriksen 2009). Some of the caves are several hundreds of metres deep, with roof heights up to 50m (Fig. 5.1). In a few of these caves, probably during the Bronze Age, simple line drawings of people (stickmen) were painted in red. All of the known Norwegian localities (and as far as I know in Fennoscandia) are found in North Norway, between Lofoten and Namdalen. In total, eleven localities have been documented, with a combined total of around 100 figures (Fig. 5.2). The uniformity of the figures, their connection with and placing within the caves, and also the fact that the cave painting phenomenon appears to be limited to a specific region indicates that the paintings belong to a common tradition of rituals and world view.

My first encounter with the red stickmen, in 1992, was by chance. I had been invited to Røst by geologist Jacob Møller at Tromsø Museum. Møller had been researching coastal caves in North Norway for many years (Møller 1985; Møller *et al.* 1992), and had exciting results from a cave named Helvete ('Hell') on Trenyken, on the outer edge of the Lofoten archipelago. Shell samples from beach sediments covering the interior floor had been dated to as early as 33,000 BP (uncalibrated), which meant that the cave was a unique archive with great potential for shedding light on late Pleistocene trends and events which in almost all other places had been lost during ice movements and fluctuations in relative sea level. The aim of our visit was to take advantage of this rare opportunity to track evidence of possible Late Glacial hunters.

However, this was to be yet another example of field studies that take a different direction to that originally planned. Within a few minutes of entering the cave the light from our headlamps fell on red figures which had not been seen by anyone for a very long time. The discovery was like a shooting star – unexpected, undeserved and magnificent – and was accompanied by a desperate longing to experience such a discovery one more time. I subsequently visited many caves, discovered more new localities with red stickmen, and also visited those already known about. The urge for new discoveries moved from longing to addiction, and I allowed myself to be willingly drawn into the caves' multiple sensory world. These impressions played an important role in an article with strong phenomenological undertones that I wrote some years later (Bjerck 1995). The article thus became a part of a research tradition that has a more developed theoretical foundation in the work of Christopher Tilley (e.g. 1994; 2004) and Julian Thomas (e.g. 1993; 1996) (see also Brück 2005; Olsen 2010, 26–32).

'Phenomenology' aims to describe the character of human experience, specifically the ways in which we apprehend the material world through our intentional intervention in our surroundings (Brück 2005, 46). Joanna Brück (2005) presents a critical retrospective view of this research tradition in her article 'Experiencing the past? The development of a phenomenological archaeology in British prehistory'. Brück emphasizes the main criticism by raising the question of how we can know whether the phenomenological relations we focus on today were regarded as meaningful in the distant past. A further objection is that most studies point out phenomenological relations between archaeological traces and physical environments, but rarely research in depth their significance for people who lived a long time ago – people with a different structure of intelligibility through which they understood the world.

This criticism is important, but nevertheless it should not obscure the fact that the sensory element in contextual relations is an important opening to a deeper understanding of the archaeological record. I believe that part of the problem is that most of the works which have gained status as theoretical references within this field have empirical starting points in British landscapes where phenomenological relations seldom extend beyond the visual impressions surrounding Neolithic monuments. The Scandinavian landscapes are well known for their grandeur: deep fiords, roaring waterfalls, coastal cliffs with pounding waves, mountain peaks, glaciers and snow patches, rock falls and gigantic boulders, and everywhere – steep rock faces with cracks, crevasses, holes, ledges, and bedrock structures. These formations will shape images of everything you can imagine – mythological creatures turned to stone, faces, entrances, doors, paths, and remains from dramatic incidents. An example of the latter is the large tunnel through Torghatten – that is part of the legend about the trolls' rivalry over the beautiful 'Lekamøya' (mythological princess, today a rock formation), and an arrow that was shot right trough the hat of one of the fighting trolls. In fact, these shapes constitute witnesses that prove these legends to be 'real'. See for yourself: here is the hat, and there is the hole, and over there – the princess herself turned to stone. The majority of these shapes and phenomena are constants through time, and hence were part of past peoples' surroundings. How these elements were integrated in past world views we do not know. My point here is that the grandeur of the Scandinavian landscapes and land shapes, and all things and phenomena herein constitute a wide and varied basis for strong sensory impressions and an exceptionally large repertoire of phenomenological lines to follow. Joakim Goldhahn's (2002) work on audio-visual perspectives related to rivers and rock art is thought provoking, as are Knut Helskog's (1999) analysis of 'the shore connection' and the works of Tore Slinning (2005) and Antti Lahelma (2005) on rock paintings, together with my own work on cave paintings (Bjerck 1995). The latter brings me directly to the main criticism of the phenomenological approach. As I shall elaborate in the following, I think that sensory impressions related to entering, being in, and departing from caves are so profound and diverse that it is possible to advance beyond Brück's core criticism – and explore how sensory impressions may have evoked meaning in past peoples' notion of 'being in the world'.

Figure 5.1: Coastal caves at Sanden, Værøy in Lofoten. The cave with paintings is situated to the left, outside the frame of the photograph (see Fig. 5.8). Photo H. Bjerck.

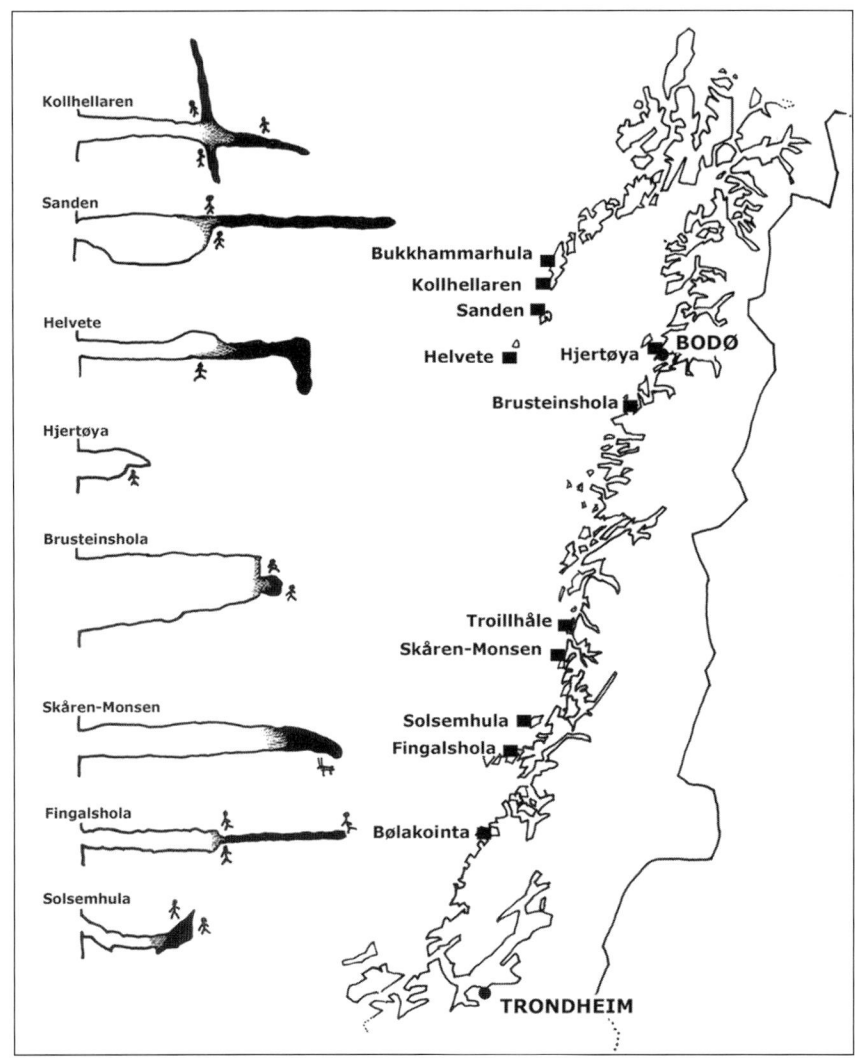

Figure 5.2: Map of the eleven cave painting localities in Norway, all situated in along the c. 500km long coast between Namdalen and Lofoten, Nord-Trøndelag and Nordland counties. To the left, a schematic presentation of the layout of selected caves (plan), including placement of paintings and the light–dark division.

Facts about Norwegian cave paintings: caves, figures, artefacts, and dating

In the following, I will refer to the caves by their Norwegian names. The ending *hula* (or *hola*, *håle*) means 'the cave', (i.e. Solsemhula is 'the Solsem cave'). *Heller* (or *hammer*) means 'rockshelter', but the distinction between caves and rockshelters is not consistent in local names (e.g. Kollhellaren is a proper cave). Some caves have separate names (e.g. Fingalshola), while others are related to place names (e.g. Bølakointa, literally 'the Bøla cunt'). Names also refer to supernatural beings (e.g. 'trolls': *troill*, *rise*), or include labels such as 'church' (*kirke*) and 'hell' (*helvete*), which underline long traditions in perceiving caves as powerful and mystic places.

The paintings

The cave paintings are part of a far larger group of painted rock art known from the whole of the Fennoscandian area (Gjessing 1936; Simonsen 1958; Slinning 2005; Taskinen 2006; Andreassen 2008). Apart from technique, the relation between rock carvings and painted rock art is unclear. The latter have survived the ravages of time on more or less protected rock surfaces, and have probably existed in far greater numbers than are known today. The rock paintings comprise a wide range of motifs, including human figures, animals, boats, and geometric figures, and appear to date to the Neolithic and Bronze Age (as discussed below). There are many examples of painted rock art on or near prominent rock formations which apparently support the notion of links between the depictions, landscapes and world view (Hesjedal 1994; Lahelma 2005; Slinning 2005).

In addition to their special context, the cave paintings have a strikingly restricted repertoire of motifs: more than 90 per cent are characteristic red 'stickmen', 25–40cm tall, with a simple line for the body connected to a round head and splayed arms and legs (Figs 5.3–5.4). However, there are also other motifs. For example, in Helvete there is also a considerably larger figure with antenna-like horns, outstretched arms with life-size hands, legs with knees, and feet with soles. Nearby is a painting of a single hand related to field of red colour that may represent a similar but more weathered figure. There are also zoomorphic figures (Skåren Monsen, Fingalshola), and a few geometric figures, e.g. the 2.6m and 3.3m long crossing lines in Solsemhula (Fig. 5.3), and the around 20m long horizontal line in Helvete reported by Norsted (2006, 32–33). There are no stickmen in Skåren Monsen, where a long-horned animal is accompanied by four indeterminate figures. This cave also contains the only carved rock art recorded in the Norwegian caves – a 'halo', without traces of paint, on a large rock in the cave's outer part exposed to natural light (for details, see Sognnes 1983; 2009; Bjerck 1995; Norsted 2006, 2010). The only paintings found in a karst cave (Resholå) (Berglund 1993) have characteristics that indicate a different tradition to that of the depictions in the coastal caves.

Associated archaeological finds

Very few archaeological artefacts have been found in the caves with paintings. However, there are interesting exceptions. Beneath the innermost paintings in Fingalshola 'palette stones' and lumps of pigment

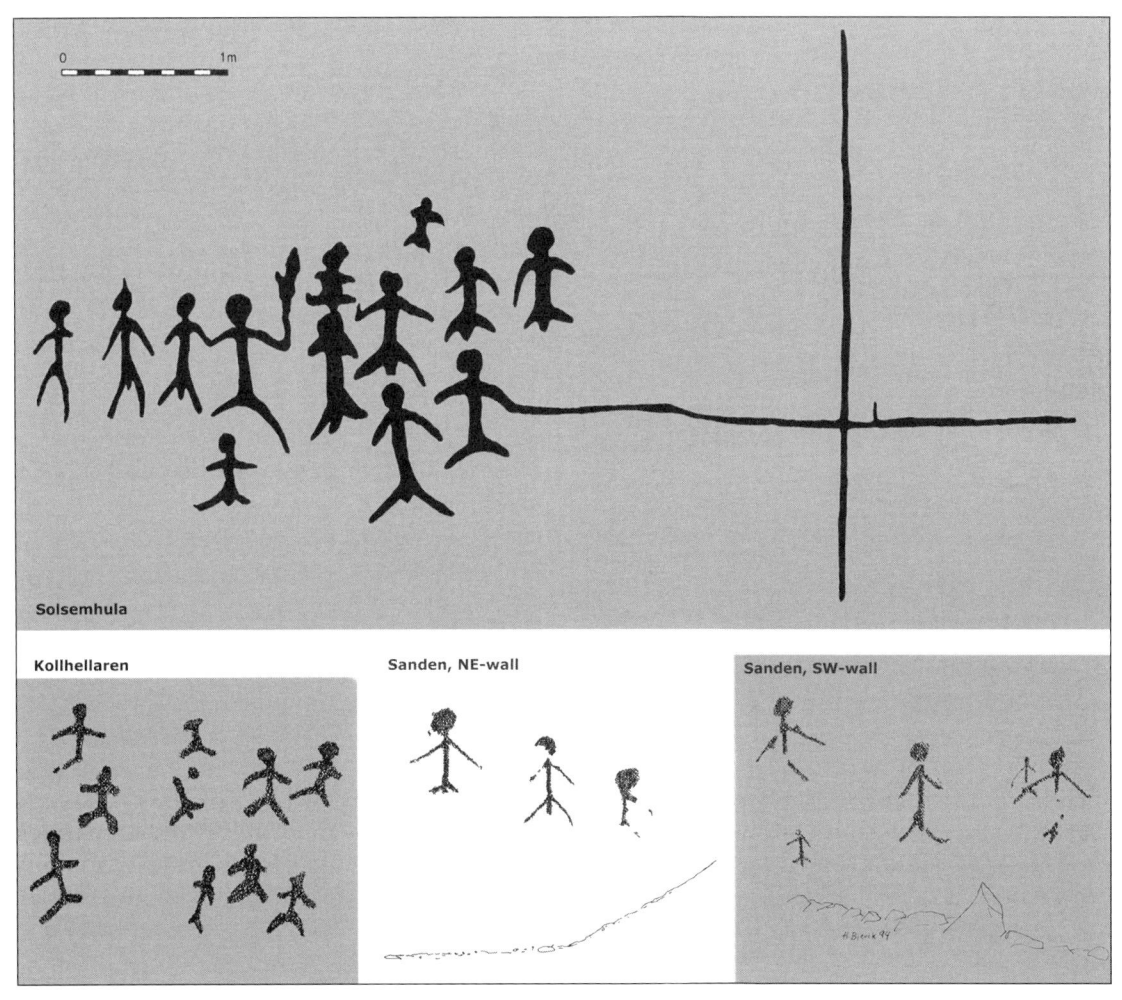

Figure 5.3: A selection of stickmen from three different caves adjusted to the same scale, demonstrating both regional similarities and local styles in size and proportions. More than 90 per cent of the Norwegian cave paintings are stickmen similar to these. Drawing from Solsemhula from Sognnes (1983), Kollhellaren after Hauglid et al. 1991, Sanden. Reproduced from Bjerck (1995).

of the same type as used in the depictions were found (Marstrander 1965, 159). A collection of grey seal (*Halichoerus gryphus*) bones – comprising skull fragments, teeth, claws, and small joints from the flippers, with frequent cut marks indicating that a whole animal was butchered at the spot – has been reported from the innermost dark zone in Helvete (Bjerck 1995).

In Solsemhula (Petersen 1914; Sognnes 1983; 2009) a *c*. 20cm deep cultural layer containing around 1500 bones from fish, birds and mammals (seal, goat, sheep, ox, horse) has been excavated. A few human bones were also recovered, of both adults and children. Theodor Petersen interpreted the cultural layer as remnants of a dwelling site – which, when considering the cave's inaccessibility and the lack of lithic waste, is not very likely (cf. Gjessing 1936, 14; Sognnes 2009). From the innermost part of the cave (with the paintings), Petersen collected a beautiful arrowhead of polished slate, a seal bone with cut notches, a seagull long bone with a drilled hole and an obliquely cut end, and a great auk amulet made from an ox bone (Sognnes 1983, fig. 8). It has been speculated that the notched seal bone and the long bone are sound instruments, a type of guiro and fipple flute respectively (Lund 1987; Bjerck 1995). Given the cave's special acoustics, this interpretation is worth consideration (cf. Waller 1993, 2002; Lahelma 2007, 131). Even the weakest sounds become amplified in such a space, and the fact that sounds might have been strange and incomprehensible may have been more important than clarity and volume.

Age

Some of the objects from Solsemhula have been radiocarbon dated: a shell (*Patella vulgata*) (1870–1560 cal BC), a whalebone artefact (790–410 cal BC), and a sheep (or goat) bone with cut marks (390–120 cal BC) (Sognnes 1983, 2009). A seal vertebra from Helvete is dated to 1600–1400 cal BC (Bjerck 1995). The radiocarbon dates and general agreement on the stylistic dating of stickmen suggests that the cave paintings date to the Bronze Age (e.g. Gjessing 1936; Marstrander 1965; Sognnes 1983, 2009; Hesjedal 1994). One should, however, be cautious about connecting the radiocarbon dated artefacts to the paintings. The many-sided functional and ritual potential of caves may have resulted in varied activities that were not related in time or tradition. It is very unlikely that the tradition of painting red stickmen in caves lasted as long as the 1500 year period suggested by the dates. One indication of this is the presence of local styles. The stickmen from Sanden have large round heads and are generally larger and slimmer than the figures from the nearby caves Kollhellaren and Helvete, which are smaller and have comparatively broader strokes (Fig. 5.3). Thus, the different localities seem to reveal individual expressions within a common practise, indicating that this cave painting tradition was rather short lived. In my opinion, only future direct dating of the paintings may clarify their exact age.

Caves and the placement of paintings

All known cave paintings are from coastal caves which have been formed at weak zones in the rock (faults, folds, intrusions) by marine erosion during several ice ages in the Pleistocene (Møller *et al.* 1992). This origin is clearly evident from how the caves lie in relation to the Late Pleistocene marine limit, and also from the fact that many have rounded and smooth polished walls, as well as remains of beach sediments on the floor in the interior (Fig. 5.10).

The topographical location of figures within the caves varies (cf. Bjerck 1995 for details). The paintings in the southernmost caves (Bølakointa, Fingalshola and Solsemhula) lie in the innermost dark spaces – in common with the animal figures in Skåren-Monsen.

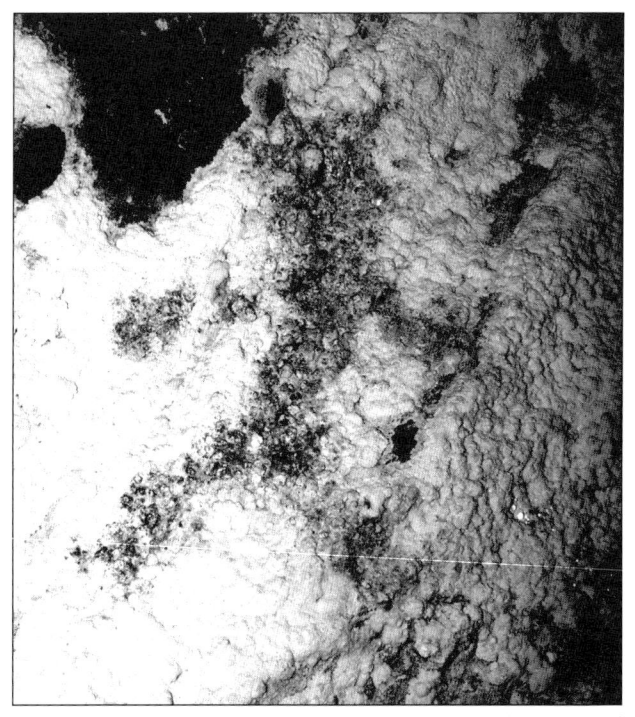

Figure 5.4: Figure in Brusteinshola, painted on white mineral precipitate on the ceiling of the inner chamber (see Figs. 5.5 and 5.6). With very few exceptions, all the paintings seem to be made by six rough strokes: a round head, body, splayed arms, and legs. Photo H. Bjerck.

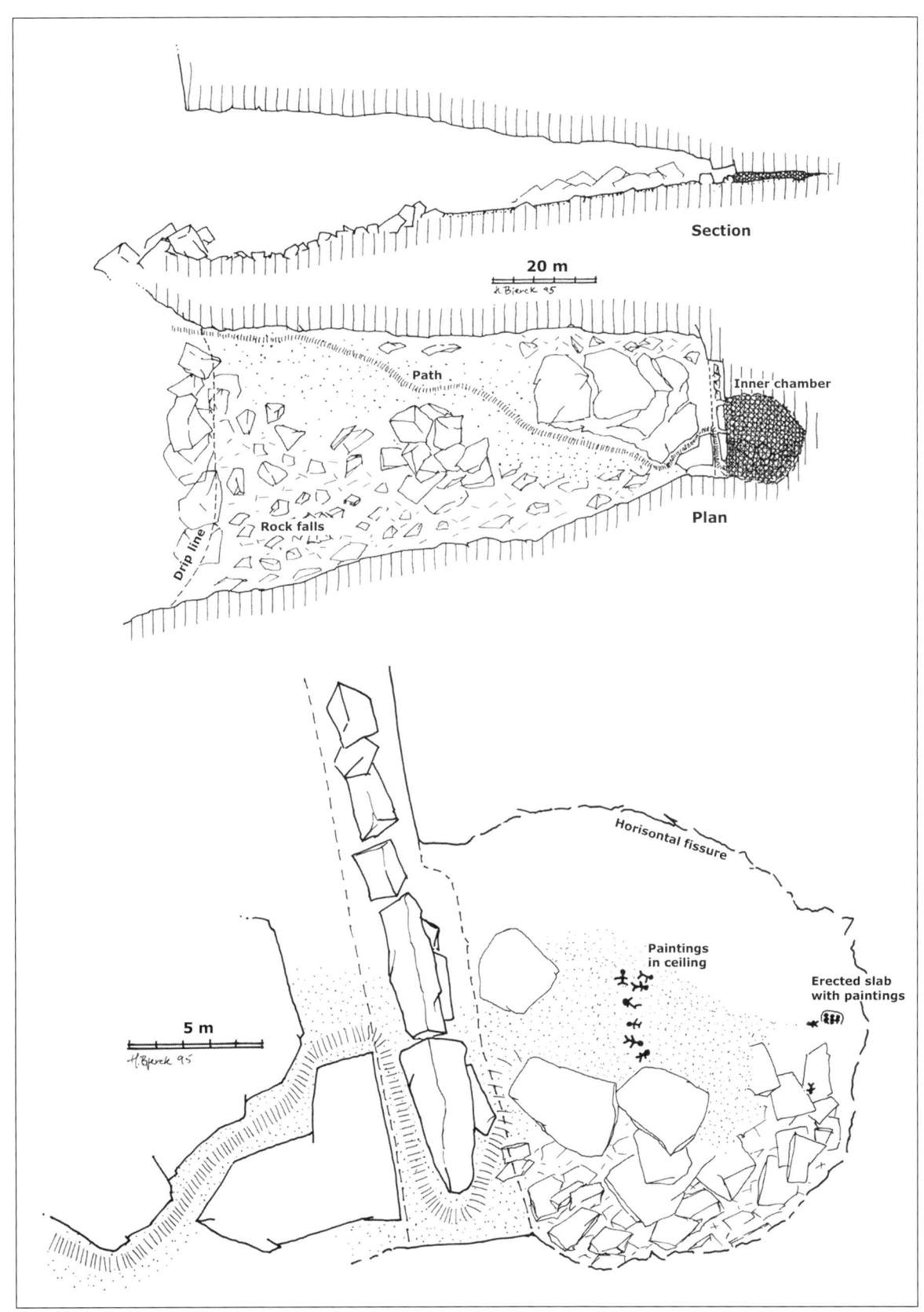

Figure 5.5: Section and plan of Brusteinshola, including a close-up of the inner dark chamber with stickmen. All the paintings are made on the ceiling of the cave (see Fig. 5.6), except the three stickmen on the erected stone slab by the horizontal fissure surrounding the chamber.

Figure 5.6: Martinus Hauglid (who discovered the paintings here) and myself studying two of the stickmen on the ceiling of Brusteinshola. Photo T. Ueland.

In Fingalshola there are also figures linked to a pronounced narrowing of the cave that also marks the transition zone between light and dark. This is similar to Refsvikhola, Sanden and Helvete, where paintings are linked to a combination of narrowing in the cave and the twilight zone. Hjertøya is a small cave without dark zones, and is the only site where cave paintings are exposed to daylight.

The placing of the figures in Brusteinshola is of particular interest (Fig. 5.5). The large and tunnel-shaped outer part of the cave is open and light. The innermost chamber (around 20 by 20m) is dark and low, below head height. At the highest part of the roof (1.6m), there is a somewhat disorderly group of seven human figures (Fig. 5.6). The cave 'ends' in a horizontal fissure, where the roof height decreases to 20–30cm. Here, three stickmen are painted on a flagstone, and two more on the roof just above. The flagstone is supported by other rocks, and it appears that the people who painted the three figures also placed the stone precisely in this position. That someone actively arranged the setting in order to place the figures exactly at this spot emphasizes the significance of this particular place. It is clear that the cave continues beyond this point, but human beings are unable to venture any further than this.

Caves without paintings

Apart from the recent discoveries in Bølakointa and Bukkhammarhula, I have visited all known Norwegian cave painting localities (Fig. 5.2). I have also examined 20 other caves in the same area, all with satisfactory conditions for preservation (indicated by smooth polished walls and floors with beach debris), without being able to detect paintings. Many of these are even bigger and more impressive than caves where paintings are recorded, and it is difficult to understand the code for selecting caves for this practise. This would be interesting to explore in more detail. However, how much emphasis one can attribute to my set of negative data is uncertain: it is reported that the painting in Bølakointa was impossible to see in the light from a normal headlamp (such as that which I have used) – it is only visible in strong bluish light, which enhances faint red colour (Ragnar Vennatrø pers. comm.).

Research history: cave paintings without caves

The red stickmen have very likely been observed, forgotten and rediscovered many times since they were painted some 2000–3000 years ago. The discovery of the paintings in Solsemhula in 1912 was the first to enter the present archaeological record (Petersen 1914). Thereafter, almost 50 years passed until the next find was reported – from Fingalshola (Marstrander 1965). Since 1980 yet more localities have been reported: Skåren-Monsen near Brønnøysund (Sognnes 1983; 2009), Kollhellaren on the outer edge of the Lofoten archipelago (Hauglid *et al.* 1991), Troillhåle in Vevelstad (Johansen 1988; Berglund 1993), Helvete on Trenyken, the caves at Sanden on Værøy and Nordvika on Hjertøya near Bodø, Brusteinshola (Bjerck 1995), and most recently Bølakointa in Flatanger (Vennatrø 2005, 93) and Bukkhammarhula in Lofoten (Norsted 2006).

The studies of the Norwegian cave paintings represent a journey through almost 100 years of research history. It is astonishing to see how the shifting research focus and theoretical paradigms have directed what has been documented and discussed. The paintings in Solsemhula were without parallel in Norway when they were found in 1912. Nevertheless, the paintings are not the main theme in Theodor Petersen's article on the discovery, but rather the finds' significance for the discussion of 'the relationship between the Arctic and Scandinavian Stone Age, the slate culture and the flint culture' (Petersen 1914, 25). The paintings are first mentioned towards the end of Petersen's article, mainly as a curious discovery of a religious annex to what he interpreted as a dwelling area. Sverre Marstrander's (1965) work on Fingalshola highlights the paintings and discusses similar images in a cross-regional context, their dating, and the cave's culture-historical context. However, the ritual meaning

Figure 5.7: Photo from the mouth of Troillhåle. The paintings are made in a narrow chamber in the side of the main cave. To enter the chamber, one has to crawl through the lower horizontal opening, which is around 1m high. Photo: H. Bjerck.

of the site is first touched upon only in the very last paragraph of the almost 20 pages long article:

> 'Everything indicates that it is the wish, or perhaps more correctly the need, to come into contact with nature's immense powers, which has been the cause of the visits to the cave. We can also imagine with what awe and horror the shaman, or whatever we will label him, has approached the cave's innermost and darkest part, where he, through ceremonies and paintings of magical pictures, intended to secure the powers' help for success in hunting and trapping.' (Marstrander 1965, 163, my translation)

The works of Petersen and Marstrander fall within the 'cultural archaeology' research tradition, whereas Kalle Sognnes' work in the wake of the discovery in Skåren-Monsen in the early 1980s bears the hallmark of the contemporary 'processual', functionalist research tradition (Sognnes 1983). Caves and depictions are thoroughly described, focussing on parallels, age, and the culture-historical context of the caves. In fact, the discussion is more focussed on subsistence patterns *outside* the caves than the meaning of the paintings *within*. The first studies that actually pinpoint the peculiarities of the paintings' context – the cave – were not developed before the 'post-processual' reorientation; 80 years after the first cave paintings entered the archaeological record. In his article on the new discovery in Troillhåle, Arne B. Johansen (1988) pointed out the exclusivity in the placing of the paintings – inaccessible, concealed, in dark confined spaces, and only visible to a few (Fig. 5.7). Anders Hesjedal (1990, 208–213; 1994, 13–14) has elaborated similar thoughts, emphasizing that the cave's characteristics may have been important for both rites of passage and rituals related to communication with supernatural forces.

Against the backdrop of my own impressions from the caves (see below), I am left surprised by the enormous mental forces embedded in scientific conformity. All researchers have connected the depictions to ritual actions, but always within the rigid frames of what contemporary academic discourse has defined as interesting and acceptable. The strong sensory impressions of the paintings' context are more or less absent – not because they have not been experienced, but because they were not deemed valid within a research tradition built on objective and quantifiable data. With our backs to the dark we have measured cave floors and roof heights, drawn plans and sections, and recorded positions of paintings. Yet all must have noticed that there is more. All have groped for erasers or tape measures on dark cave floors, but none seem to have seen the darkness itself. For how could Theodor Petersen have measured the density of darkness in Solsemhula? Would not Sverre Marstrander have been regarded as frivolous and weak if he had elaborated on how he imagined the shaman's 'awe and horror' in the innermost chamber in Fingalshola? Why (and how) should Kalle Sognnes document the goose pimples on his forearms, or the frequency of raised hairs on his neck when standing before the paintings in Skåren-Monsen? On what scale could I have measured the extent of my own feeling of being far from home when in the depths of Helvete?

Archaeology has well-established traditions for recording the physical dimensions of caves, but apart from the mandatory stippled line marking the dripline at a cave's mouth; the wide repertoire of sensory dimensions is seldom included. Information on light (and dark), scents and temperature, and how – in combination – they affect people who submit themselves to a cave's embrace is difficult to document, and are normally omitted. Paradoxically, the meticulous recordings of plans and sections reproduce something that was never observed by people in the past, and fail to reproduce what people actually saw in here: the cave opening filled with glaring daylight, darkness, and fragments of the cave's walls, ceiling and floor in the flickering light of a flaming torch (see Figs 5.9–5.10). One may agree that these sensory phenomena rarely

Figure 5.8: The mouth of the impressive cave at Sanden is nearly blocked by a large talus formation. The present entrance (marked) is less than 2m high. Photo H. Bjerck.

Figure 5.9: The mouth of Kollhellaren, seen from inside. The dimensions of the cave are breathtaking – note the person standing between the two boulders on top of the rockfall formation at the mouth of the cave. Photo H. Bjerck.

constitute entities with defined limits that may be recorded with a solid line on a piece of paper. The position of a cave's twilight zone may vary with the position of the sun (cf. Sognnes 2009, 85), but this does not mean that the light–dark dichotomy or other sensory impressions are irrelevant (see Lewis-Williams 2002, 214–227; Waller 2002; Clottes 2003).

But how do you do this? How do you detect highly subjective phenomenological relations in an objective manner? Of course, all this may be objectified and broken down to things that may be measured by light meters, thermometers, humidity detectors, sonar signals, and echo sounders (e.g. Waller 1993; Hamilton *et al.* 2006), but would this multitude of objective recordings ensure a true and complete picture of the sensory surroundings of cave paintings? And would this comprehensive set of data be an all-encompassing basis for the understanding of how sensory impressions are perceived, mediated, combined, and processed through the filter of cultural references and past experiences in the human brain?

Without claiming any general validity for the outcome, I would like to explore another line of reasoning, namely the conglomerate of impressions from my own visits to the realm of cave paintings. Like Tilley (1994), I believe that the size and shape of my body are similar to that of the people who made the paintings, as are our senses and basic physical surroundings. I will therefore elaborate on my own embodied experiences and references in order to provide peepholes into the phenomenology of cavescapes.

Figure 5.10: From the central part of Helvete, looking towards the c. 30m high formation of rock falls in the mouth of the cave (note the person standing here). In the inner part, the cave floor is covered by beach sediments that are dated to around 30,000 BP. The paintings are located on the rock face to the left. Photo H. Bjerck.

Sensing caves

From a distance, most caves are conspicuous and unfamiliar formations – big holes in the normal world of mountains and rock faces. On coming closer, the breathtaking dimensions of large caves evoke a feeling of being small and inferior. Many of the coastal caves are more or less blocked by rock debris at their mouths (Figs. 5.8–5.9). From there, you may look down and into the eerie cavescape. To advance, you often have to overcome enormous rockfalls which make many caves difficult to reach without ropes or ladders. The full 'effect' of the cave is not experienced before you have reached the point where the light shifts to darkness. You have moved from broad daylight to muted reflections from matt rock walls. The mouth of the cave is almost unnatural green with lush ferns, wide-bladed grass and wet moss. Inside there is a dim interplay of colour, with alternating stripes of rock mineral precipitates – white, yellow, red, brown, black, and grey. Further in, there are only the smoothest rock surfaces, which shine faintly (Fig. 5.10). Between the faint reflections there is the darkest black, with endless space for anything that cannot tolerate daylight. You breathe in a slight odour of earth, mould and dust – damp and cool, like a refrigerator which has not been opened for a (very) long time or a newly opened sack of (unwashed) potatoes, which triggers a childhood memory of the depths of your grandmother's cellar. About now, you also notice the cold and the fine hairs rising on your forearms. Both breath and speech become visible in fleeting mists, your damp shirt feels unpleasant. When you listen, it is the lack of sound that strikes you. The cave's own sounds are only distant dripping and the distorted directionless echoes of daily life outside the cave. There is a silence that amplifies and adds an unfamiliar reverberation to sounds from your own body and movements.

All in all, it is the *absence* of everything that hits you – the absence of movement, colours, smells, and sounds which we are used to in the life and day outside. The cave is monotonous, silent, unmoving, and unseen – the opposite of the living world. There is no day, no winter, no summer, and nothing that grows. The cave extends beyond life, and beyond time. On the floor there are remains of dead animals and birds, cadavers with hair and feathers, but without any swarms of flies or crawling maggots. Leaning against the wall is a young bird, with its upward pointing beak optimistically open, contrasting with its matt black eyes without the brightness of life (Fig. 5.11). Life has left it, but it is still sitting there. How long has it sat there? It is not easy to tell, because time does not seem to exist here in the conspicuous absence of motions and life. What *is* present is a kind of baseline for being, including what is beyond life, before and after.

As you penetrate the darkness of the cave, you encounter branches leading upwards, downwards, to the side, or beneath gigantic rocks. Regardless of which route you choose to follow you eventually reach a point where you cannot go further. Yet at the same time you observe that these points are not the cave's

Figure 5.11: Dead bird, leaning against the wall in Helvete. Photo: H. Bjerck.

end. It is just too cramped for us humans to be able to press further in. A child could squeeze a little further, but even the smallest child would (probably) shortly reach a point where the physical embrace of the cave would set a limit – at least for humans. Thus, caves are not confined spaces; it is the human body which restricts how far we can go, not the cave. Within lies the never-ending dark, and all that belongs to it. Here, the night hides when the day rules, and from here the darkness and stillness seep out when the sun sets.

Sometimes I have entered this world of darkness alone, or remained behind alone to finish taking notes or measurements. It is hard to admit, but such occasions are always accompanied by flashes of indefinable apprehension. The comfort you may find in the ray of light from the headlamp is disturbed by the pressure of the darkness that is always behind you – shadows seem to advance from all parts that are not lit up. The shaky well-being you may find in the beam of light is disrupted by the fact that it also makes you very visible, vulnerable, a highlighted eye-catcher for nameless things you are not able to see. It is ten times the uneasy feeling of being the 'target' in the child's game where all the others try to get closer each time you turn your back on them, and you wait for the uneasy moment when you are touched by the winner. It is tempting to slip away from this by switching off the light, and entering a state of a peaceful nothingness that normally is not reached by living persons without the use of strong mind altering drugs. It is like being dead or unborn.

On several occasions I have experienced the paralysing anxiety which today is known as 'claustrophobia'. For me it is like a sudden horrible stench – one that is not noticeable until it is too late, after you have inhaled it and it is inside you. This may occur when you reach the deep and narrow parts of the cave, where it is not possible to go any further forward and the way back is blocked by the others who are following. It is like being seven years old, head first in a sleeping bag, only to realize that your older playmate is not going to let you out. At times you have to crawl, and you suddenly realize that you cannot get further and there is no space to turn around. Even if these waves of anxiety may be accompanied by the conviction that there is really nothing to fear, this suffocating feeling of dread inexplicably remains, seemingly coming from outside and inside simultaneously. Perhaps humans long ago also recognized, named, and had their own explanation for this phenomenon? Did they sense this as something coming from within themselves – or as entities in the cave itself?

However, the most profound embodied experience is perhaps when you re-enter life and the light outside: your reset senses are suddenly bombarded by all the ordinary things that filled them and your brain to a degree that borders on invisibility prior to the time spent in the dark void. It is a strange feeling to be surprised by all normal things: the lazy sound of bumble bees, the smell of the sea and bird excrement, grass and flowers, the warmth of a summer day, the abundance of movement and colour, swarms of insects against the bright, red and yellow flowers that nod in the wind, and the blinking reflection of sunlight from the surface of the deep blue sea – the living everyday things we always see but hardly notice.

Caves as ritual landscapes

Compared to the general picture of Norwegian rock art traditions (e.g. Lødøen and Mandt 2010), and despite some variations, one may conclude that no other group of rock art is so uniform in technique, design and format as the cave paintings. Similarly striking is the characteristic placing of the paintings themselves, linked to dark caves, in a specific region. This implies that the paintings are related to a common practice, rooted in a common tradition in rituals and worldview. Hence, it is reasonable to believe that the interface between the placement of paintings and the phenomenology of caves has the potential to reveal aspects of the ontological backdrop for this ritual practice.

In general terms, caves are silent and static – the chaotic complexity of movements and sensory impressions in the day life outside is *absent*. In this lies the possibility to arrange, control and manipulate sensory impressions. Séances can be directed with endless combinations of light, voices, noise, stench and scent. One may show or conceal the location and origin of sensory impressions. In other words, a cave is a 'black box', an ideal work platform for priests, shamans, and necromancers (and also illusionists and theatre directors), where participants may be directed towards intended sentiments and moods that will enable them to experience and understand the complexity of symbols and cultural references that constitute a ritual (Turner 1968, 2; Doty 1986, 81).

A telling example is the séance described by Knud Rasmussen (1921, 15), where the shaman's control over what the participants see and hear is decisive. The séance itself is a bombardment of sound and movement in absolute darkness (within a winter house), where different spirits are recognized through certain movements and acoustic images. Equally important is the relation between what the participants saw *prior* to the ceremony and what they see when the lights are lit *afterwards*: the shaman, with his arms tied behind his back, in a helpless state, with drumsticks lying between his feet and the drum, and the dry, stiff hide across the entrance which will make a noise when someone (or something) passes through.

The ability to control sensory impressions is equally widely practised in present-day rituals. Many of the caves' characteristics have parallels in churches today – impressive, separated from everyday life, with heavy doors, door handles at face height, keys which need both hands to turn them – all elements intended to impose a feeling of the visitor being 'small' and overpowered. Coloured glass directs light to form particular images, but also obstructs the view outwards. Controlled acoustics give richness and colour to the sound image linked to the ritual, and spoken words acquire special meanings. At the same time, irregularities, the creak of the floor from someone arriving late, or inattentive chattering are amplified. The odour of stuffy air which has been played through organ pipes, sung with, cried into, prayed with, coughed and yawned into. The placing of items, people and functions, and all that is sensory are integrated in a liturgical whole which goes far beyond the formal programme of a religious service. Thus, caves are suitable *arenas for ritual activities*, and possess a range of phenomenological qualities with strong symbolic value for a variety of rituals (cf. More and Myerhoff 1977, 4).

Being new in the world: 'rites de passage'

Anders Hesjedal (1990, 208–213) has suggested that the Norwegian cave paintings may have been linked to 'rites of passage' – ritual markings of major transitions in a person's life, both social and biological. Such rituals are known from most human societies in one form or another (Van Gennep 1960 [1909]), such as the well-known sequence in the Western world: christening (birth), baptism (puberty), marriage (reproduction), and burial (death). Arnold Van Gennep characterized three stages that are commonly found in rites of passage: separation, transformation and incorporation. The middle phase, the change itself – the liminal phase – is the core of the ritual. The liminal phase is a threshold, a tunnel, a mixture of both and neither/nor, a state of oppositions – living and dead, adult and child, animal and human. The liminal phase may last for days or weeks, or be over in a matter of minutes. Humiliations and ordeals are often part of this phase, a downfall in parallel with the construction of the post-liminal state. It is also common that the liminal phase is marked by a separation from the remainder of society (Turner 1977, 37).

It is not surprising that caves are often reported to be central in such rites, and puberty rites in particular (Turner 1977; Eliade 1964, 41). Caves are a physical separation, a place to be away from everything else. They have qualities which can symbolize contrast-filled stages of rites of passage – places out of life and time, without movements, colours, weather, scents, and sounds. These are places where it is easy to imagine physical and mental ordeals. With simple arrangements, ordeals may easily be increased to levels of sheer horror – bodily and spiritual stress which are likely to enhance the importance of the post-liminal status and provide a sweet feeling of self-esteem and pride. I will also stress the emotional and symbolic value of the return to the outside world after a lengthy stay in a cave's embrace – with 'reset' and highly receptive senses bombarded by movements, scents, colours, contours, sounds – a cacophony of life that can enhance a feeling of being new in the world in the finale of the separation–transformation–incorporation sequence. Thus, caves are a perfect place to emerge from, to re-enter the world with a new status.

There is yet another aspect that may hint at a relation to rites of passage, especially rituals related to the child–adult transition. The liminal phase often implies deprivation – stripping people who undergo such rituals of property or social positions. In short, they are all alike, naked, regardless of their former roles or status (Turner 1969, 95). Thus, the strikingly simple and uniform appearance of stickmen in the caves – from Leka to Lofoten – may well symbolize a ritual 'undressing' and homogenizing of participants in initiation rites.

Osmotic membrane: connections to other worlds

Characteristic stone formations, stone blocks, cliffs, mountains, islets, taluses, fissures, and caves are all often part of cosmologies, myths, and rites. Characteristics of such places can symbolize important aspects of myths and rituals: the connections and contrasts between people and nature, humans and supernatural powers, individuals and society, past and future, and social and ethnic groups (Eliade 1964, 51). This is well illustrated in Sami cosmology, where there is a clear relation between shamanistic rituals and characteristic shapes in the landscape (e.g. Manker 1957; Anisimov 1963; Schanche 2000; Sveen 2003; Äikäs *et al.* 2009). In fact, these shapes constitute important elements in the notion of the three-tier world and the *axis mundi* that seems to be central in Eurasian shamanistic cosmology. The location and motifs of rock art hint that this tradition has deep roots (e.g. Helskog 1999; Goldhahn 2002; Lahelma 2005; Slinning 2005; Zvelebil 2008).

David S. Withley's (1992) study of rock art in California and Nevada is an excellent illustration of how rituals, motifs, and rock art locations may interrelate. There, rock art was a living tradition well into historical times, and extensive ethnographic material informs about the cultural context of sites and depictions. Rock art is linked to differing rituals, but places and images all appear to be related to shamans – in fact, the names of the sites are commonly variations of 'Shaman's place'. Some localities have been important in puberty rites, and ethnographic sources point to male and female symbols in geometric patterns. Nevertheless, most sites are linked to places where shamans in 'altered states of consciousness' (trance) were able to connect with supernatural worlds and beings. These powers are found 'in the rock', and it is not surprising that many rock art sites are found near caves and fissures which symbolize lines of communication (Withley 1992, 91–92). Many of the motifs are linked to the connection with the supernatural. Lizards, snakes and bears – animals which frequently venture into fissures, cracks and caves – are all part of our world and also the realms beneath. They were seen as important spiritual helpers or guardians in passages between the worlds, or as assistants in shamans' efforts to make contact with supernatural powers (Withley 1992, 101).

In other parts of the world too, researchers have linked rock art and caves to shamanism (e.g. Lewis-Williams 1982; 2002; Lewis-Williams and Dowson 1988; Skeates 1991). However, we do not need to leave our own Scandinavian tradition in order to find notions of (mostly evil) supernatural powers that inhabit rocks. There is no reason to doubt that these beliefs have deep roots in history. Very often the caves have names which indicate connections to the underworld, such as Trollhola, Risehula, and Helvete. In many cases caves are linked to stories which emphasize the unknown. These are stories which some of us actually wish to believe in or at least which we love to pass on to new generations:

> '... and one day, a boy about your age heard faint barking from a cave on the distant island over there. Inside, he found a dog, thin as a cat, naked as a pig, almost see-through, shivering, with a haunted look in his eyes. Had it not been for the collar, nobody would believe this, but it was, beyond any doubt, the very dog that had disappeared three weeks earlier, in the very cave we are in now ...'.

Departing from my own embodied engagement with cavescapes, it is easy to understand that these formations have the capacity to nourish beliefs of caves as connections to other worlds. In fact, they may be seen as tangible evidence of the very existence of a dark world beneath. One is the fact that caves are highly visible and impressive openings in the world of humans, leading to eerie chambers that impose a feeling of being out of place. Another characteristic is the fact that such chambers do not have a definite, observable termination. On the contrary, further connections may be observed in the form of innumerable openings which eventually become so narrow that it is impossible for humans to proceed – an osmotic membrane which allows some beings to pass through while others (humans) cannot. This adds further support to the sense of connection to places that people cannot normally reach.

This is enhanced by phenomenological aspects of the chambers with paintings – the lack of life and time, powerful metaphors underlining that you are on the outer fringe of the human world. The paintings are located at different places within the cavescapes – in the twilight zones, by a marked narrowing in the passages, where passages divide, or in the inner chambers. Morphologically speaking, departing from plans and sections of caves, the detailed locations of paintings seem chaotic and without any pattern that may hint at meaning. Phenomenologically speaking, however, the location of paintings seem to come together in points of change, marking borders that seem to underline the importance of tangible transition zones within the caves – where you *sense* that you are leaving the world of humans – or that you are entering the outer chambers of a world beyond.

Concluding remarks

The subjective exploration of my experiences of caves and stickmen is neither aimed at revealing the full meaning of past beliefs and rituals nor finding the exact limits of the validity of phenomenological approaches. I do not know for sure that the caves were arenas for rites of passage or ritual markings of human presence at the fringes of the world of humans, i.e. whether the stickmen are guardians or helpers – or, for that matter, representations of human foetuses in the womb of an animated Mother Earth (see Figs. 5.9–5.10). However, I believe that exploring how caves evoke human senses, and how the human brain processes and mixes sentiments against a backdrop of highly subjective personal memories and cultural references, has the potential to reach beyond the more pessimistic criticisms of the phenomenological approach, as represented, for example, by the conclusion to Joanna Brück's review of this research tradition:

> 'Some writers have suggested that embodied engagement with the landscape in the present provides an insight into past experiences and interpretations of place. For the reasons outlined in this paper, however, I would argue that this cannot be the case. Nonetheless, phenomenology can encourage us to think imaginatively about the social and political layout and landscape setting and in this it has been very successful.' (Brück 2005, 65)

The ontological depths of past phenomenological relations have probably evaporated for good, as are the filters of cultural references and individual memories that all humans use to make meaning of sensory impressions. Nonetheless, I share Tilley's (2004) basic assumption that we and the peoples of the past share 'carnal bodies', size, shape, and sensory organs – that we engage with the same physical world and the same repertoire of phenomena – and that these factors may serve as 'constants' that constitute a relevant backdrop for the interpretation of past actions, praxis and beliefs. I also believe that phenomenological relations are imperative in forming the infinite assembly of dichotomies that constitute past and present world views. With respect to the engagement with caves, the cave painters are likely to have experienced and ascribed meaning to the following:

(1) the lack of life, colours, sounds, smells, movement, and time;
(2) the fact that human senses fail to produce the information we need to orient ourselves, move and detect danger;
(3) the elusive termination of chambers and passages, which evokes a feeling that there is more – which is beyond the reach of humans; and
(4) the dichotomy between the caves' realm of absence and the 'reinforced awareness' of all 'normal' things in the world on returning from caves.

In addition to fascination, I believe that the engagement with caves induces a certain level of apprehension on a scale ranging from uneasiness to shear fear. The loss of sensory control any human being will experience here is per definition frightening. Your eyes fail to provide the information you need to move (do not even think of running) and to sense things (and threats) that surround you. Your ears fail to detect the directions and provenience of sounds, and you are not able to see what is making sounds. Your nose is filled with an all-encompassing odour of chilled decay – death, nothingness, and eternity. The phenomenon 'claustrophobia' – the strange and paralysing feeling that seems to percolate from narrow spaces, simultaneously coming from within and outside – may also have evoked a feeling of being overpowered and out of place. The level of uneasiness may vary, however enough people are likely to have experienced this to define and mediate the phenomenon.

Very likely, some of these features contributed to the much wider array of rationales that constructed the rituals that included painting red stickmen in Norwegian caves. As all humans share carnal bodies in a similar physical world, one should not be surprised that fundamental divisions similar to the *axis mundi* in Eurasian shamanism are found in the majority of known cosmologies around the world. In the most basic and simplistic form this is the World of Humans – the surface of the Earth that we trod and build our houses on, and hunt, collect and harvest from. And then there is a World Above and a World Beneath – both out of reach for living humans – 'Heaven' and 'Hell'. This is why some trees are holy; as they are a very tangible part of the human world, but are also rooted in and nourished by the underground, and stretch towards sky, sun, and heaven, above the human world. This is why churches have towers and crypts, why a variety of pyramids occur at different times and places. This is why trolls are believed to inhabit mountains and why they are petrified by the sunlight, why rituals and offerings are connected to summits ... and to caves.

With or without paintings, caves may very well have been an important factor in this cosmological construction. I think the very tangible sensuous world of caves may have nourished the notion of a world axis and a number of associated dichotomies (light/dark, warm/cold, good/bad, life/death). They are a place one may see for oneself – step by step the living world vanishes, to become replaced by nothingness – neither time nor life and movement – a dead world that may actually be seen and sensed by the living. And furthermore, the elusive termination of caves hint at their continuum to realms beyond the human world.

Today, the natural sciences have, to a large extent, blocked off the mysteries of the underworld to many of us. We may still be enticed by the scary thrills from the deep, dark and the evil in feature films such as The 'Abyss' and 'Descent'. However, the big mysteries in current cosmology – the realms beyond human reach and understanding, where beliefs and science are equals – lie at the other end of the world axis: the universe. Yet do not worry: as you read these lines, our lone and brave messenger is under way. Since September 5th 1977, the unmanned space probe Voyager has travelled in the same direction towards the big and dark unknown, carrying 'The Golden Record', a valuable and varied cargo of information about humans, humanity, our position and physical surroundings – for whom it may concern (NASA 2010). This is the current outer fringe of the human realm – and NASA's 'cave painting' on behalf of the USA and planet Earth as a whole.

Reference

Äikäs. T., Puputti, A.-K., Núñez, M., Aspi, J. and Okkonen, J. (2009) Sacred and profane livelihood: animal bones from Seidi sites in Northern Finland. *Norwegian Archaeological Review* 42(2), 109–122.

Andreassen, R. L. (2008) Malerier og ristninger. Bergkunst i Fennoskandia. *Viking* 71, 39–60.

Anisimov, A. (1963) Cosmological concept of the people in the North. In H. Michael (ed.) *Studies in Siberian Shamanism*, 157–229. Toronto, University of Toronto Press.

Berglund, B. (1993) Nyfunne hulemalerier på kysten av Sør-Helgeland. *Spor* 1993(2), 36–37.

Bjerck, H. B. (1995) Malte menneskebilder i 'Helvete'. Betraktninger om en nyoppdaget hulemaling på Trenyken, Røst, Nordland. *Universitetets Oldsaksamling Årbok* 1993–4, 121–150.

Brück, J. (2005) Experiencing the Past? The development of a phenomenological archaeology in British Prehistory. *Archaeological Dialogues* 12(1), 45–67.

Clottes, J. (2003) Caves as landscapes. In K. Sognnes (ed.) *Rock Art in Landscapes – Landscapes in Rock Art*, 9–30. Det Kongelige Norske Videnskabers Selskab Skrifter 4. Trondheim, Tapir Akademic Press.

Doty, W. G. (1986) *Mythography: The Study of Myths and Rituals*. Alabama, University of Alabama Press.

Eliade, M. (1964) *Shamanism: Archaic Techniques and Ecstasy*. London, Routledge and Kegan Paul.

Gjessing, G. (1936) *Nordenfjelske Ristninger og Malinger av den Arktiske Gruppe*. Instituttet for Sammenlignende Kulturforskning Serie B, Skrifter Bd. 30. Oslo, Aschehoug.

Goldhahn, J. (2002) Roaring rocks: an audio-visual perspective on hunter-gatherer engravings in Northern Sweden and Scandinavia. *Norwegian Archaeological Review* 35(2), 29–61.

Hamilton, S., Whitehouse, R., Brown, K., Combes, P., Herring, E. and Thomas, M. S. (2006) Phenomenology in practice: towards a methodology for a 'subjective approach'. *European Journal of Archaeology* 9(1), 31–71.

Hauglid, M. A., Helberg, B. and Hesjedal, A. (1991) Strekfigurene i Kollhellaren på Moskenesøya, Vest-Lofoten. *Tromura, Kulturhistorie* 19, 157–169.

Helskog, K. (1999) The shore connection. Cognitive landscape and communication with rock carvings in Northernmost Europe. *Norwegian Archaeological Review* 32(2), 73–95.

Hesjedal, A. (1990) *Helleristninger som Tegn og Tekst. En Analyse av Veideristningene i Nordland og Troms*. Unpublished thesis, University of Tromsø.

Hesjedal, A. (1994) The hunters' rock art in Northern Norway. Problems of chronology and interpretation. *Norwegian Archaeological Review* 27(1), 1–28.

Johansen, A. B. (1988) Bildene i Troillhåle. *Årbok for Helgeland* 1988, 29–32.

Johansen, A. B. (2008) Landemerket Torghatten. Naturfenomen, bosetningen, betydningen – en undersøkelse av langvarighet. *Viking* 71, 7–38.

Lahelma, A. (2005) Between the worlds. Rock art, landscape and shamanism in Subneolithic Finland. *Norwegian Archaeological Review* 38(1), 29–47.

Lahelma, A. (2007) 'On the back of a blue elk': ethnohistorical sources and 'ambiguous' Stone Age rock art at Pyhänpää, Central Finland. *Norwegian Archaeological Review* 40(2), 113–137.

Lewis-Williams, J. D. (1982) The economic and social contexts of southern San rock art. *Current Anthropology* 23, 429–449.

Lewis-Williams, J. D. (2002) *The Mind in the Cave: Consciousness and the Origins of Art*. New York, Thames and Hudson.

Lewis-Williams, J. D. and Dowson, T. (1988) The signs of all times: entopic phenomena in Upper Palaeolithic art. *Current Anthropology* 29, 201–245.

Lund, C. (1987) *Fornnordiska Klanger*. LP recording, Musica Sveciae 101.

Lødøen, T. and Mandt, G. (2010) *The Rock Art of Norway*. Oxford, Windgather Press.

Manker, E. (1957) *Lapparnas heliga Ställen*. Acta Lapponica 13. Uppsala, Gebers.

Marstrander, S. (1965) Fingalshulen i Gravvik, Nord-Trøndelag. *Viking* 29, 147–165.

Moore, S. and Myerhoff, B. G. (1977) *Secular Ritual*. Assen, Van Gorcum.

Møller, J. (1985) Coastal caves and their relation to early postglacial shore levels in Lofoten and Vesterålen, North Norway. *Norges Geografiske Undersøkelser, Bulletin* 400, 51–65.

Møller, J., Danielsen, T. K. and Fjalstad, A. (1992) Late Weichselian glacial maximum on Andøya, North Norway. *Boreas* 21(1), 1–15.

Møller, J. and Fredriksen, P. T. (2009) The tunnel through Torghatten, northern Norway: landforms and discussion of formation. *Norsk Geografisk Tidsskrift* 63, 268–271.

NASA (2010) The Voyager: the Interstellar Mission. <http://voyager.jpl.nasa.gov/> (accessed January 2010).

Norsted, T. (2006) Hulemaleriene i Norge. Egenart, kontekst, mening og konservering. In I. M. Egenberg, B. Skar and G. Swensen (eds.) *Kultur – Minner og Miljøer: Strategiske Instituttprogrammer 2001–2005*, 11–46. NIKU-Tema 18. Oslo, The Norwegian Institute for Cultural Heritage Research (NIKU).

Norsted, T. (2010) Safeguarding the cave paintings in Lofoten, Northern Norway. *Rock Art Research* 27(2), 239–250.

Olsen, B. (2010) *In Defense of Things: Archaeology and the Ontology of Objects*. Lanham, New York, Toronto and Plymouth, AltaMira Press.

Petersen, T. (1914) Solsemhulen paa Leka. En boplads fra arktisk stenalder. Foreløpig meddelelse. *Oldtiden* 4, 25–41.

Rasmussen, K. (1921) *Myter og Sagn fra Grønland. Bd. I: Østgrønlændere*. København/Kristiania, Gyldendalske Boghandel/Nordisk Forlag.

Schanche, A. (2000) *Graver i ur og Berg: Samisk Gravskikk og Religion fra Forhistorisk til Nyere Tid*. Karasjok, Davvi girji.

Simonsen, P. (1958) *Arktiske Helleristninger i Nord-Norge II*. Instituttet for Sammenlignende Kulturforskning, Serie B, Bind 49. Oslo, Aschehoug.

Skeates, R. (1991) Caves, cult and children in Neolithic Abruzzo, Central Italy. In P. Garwood, D. Jennings, R. Skeates and J. Toms (eds.) *Sacred and Profane*, 122–134. Oxford University Committee for Archaeology Monograph 32. Oxford, Oxford University Committee for Archaeology.

Slinning, T. (2005) Antropomorfe klippeformasjoner og fangstfolks kultsted. In J. Goldhahn (ed.) *Mellan Sten och Järn*, 489–502. Gotarc Serie C. Arkeologiska Skrifter No 59. Göteborg, Göteborg University.

Sognnes, K. (1983) Prehistoric cave paintings in Norway. *Acta Archaeologica* 53(1982), 101–118.

Sognnes, K. (2009) Art and humans in confined space: reconsidering Solsem Cave, Norway. *Rock Art Research* 26(1), 83–94.

Sveen, A. (2003) *Mytisk Landskap: Ved Dansende Skog og Susende Fjell*. Stamsund, Orkana.

Taskinen, H. (2006) Rock painting sites in Finland. *Adoranten, Scandinavian Society for Prehistoric Art*, 19–27. Underslös, Tanums Hällristningsmuseum.

Thomas, J. (1993) The hermeneutics of megalithic space. In C. Tilley (ed.) *Interpretative Archaeology*, 73–97. Providence, Berg.

Thomas, J. (1996) *Time, Culture and Identity: An Interpretative Archaeology*. London/New York, Routledge.

Tilley, C. (1994) *A Phenomenology of Landscape: Places, Paths and Monuments*. Oxford, Berg.

Tilley, C. (2004) *The Materiality of Stone: Explorations in Landscape Phenomenology*. Oxford, Berg.

Turner, V. (1968) *The Drums of Affliction: A Study of Ndembu Ritual*. Ithaca, Cornell University Press.

Turner, V. (1969) *The Ritual Process: Structure and Anti-structure*. London, Routledge and Kegan Paul.

Turner, V. (1977) Variations on a theme of liminality. In S. Moore and B. G. Myerhoff (eds.) *Secular Ritual*, 36–54. Assen, Van Gorcum.

Van Gennep, A. (1960 [1909]) *The Rites of Passage*. London, Routledge and Kegan Paul.

Vennatrø, R. (2005) *Numedals Reffne – en Sprekk i Virkeligheten*. Unpublished MA thesis, Norwegian University of Science and Technology (NTNU), Trondheim.

Waller, S. J. (1993) Sound reflection as an explanation for the content and context of rock art. *Rock Art Research* 10(2), 91–95.

Waller, S. J. (2002) Rock art acoustics in the past, present and future. In P. Whitehead and L. Loendorf (eds.), *1999 IRAC Proceedings, Volume 2*, 11–20. Ripon, Wisconsin, American Rock Art Research Association.

Withley, D. S. (1992) Shamanism and rock art in Far Western North America. *Cambridge Archaeological Journal* 2(1), 89–113.

Zvelebil, M. (2008) Innovating hunter-gatherers: the Mesolithic of the Baltic. In G. Bailey and P. Spikins (eds.) *Mesolithic Europe*, 18–59. Cambridge, Cambridge University Press.

Chapter 6

On the (L)edge: The Case of Vale Boi Rockshelter (Algarve, Southern Portugal)

Nuno Bicho, João Cascalheira and João Marreiros

This chapter focuses on rockshelter research in Southwestern Portugal, at the Palaeolithic site of Vale Boi (Vila do Bispo, Algarve). It describes each phase of human occupation at the site, starting some 28,000 years ago (radiocarbon years) with one of the earliest Gravettian occupations in Iberia, followed by Proto-Solutrean, Solutrean and Magdalenian horizons. During each phase, hunter-gatherer occupations took place at different areas of the site, suggesting differences in the functions of each site area. Subsistence, technology, and stylistic attributes related to the artefacts retrieved during excavation indicate the presence of a strong social network connecting the coastal human communities during the Gravettian and Solutrean from Valencia, in de Spanish Levante, to the Algarve. Finally, the chapter also describes technological aspects related to subsistence, such as grease rendering, and lithic reduction sequences for each Upper Palaeolithic Phase.

Introduction

Rockshelters have been most important in the study of prehistoric archaeology, particularly in what concerns Palaeolithic times. In Portugal, however, very few rockshelters have been found, tested and excavated that shed light on the Palaeolithic occupation of the Western edge of Europe.

This chapter will focus on the case of the rockshelter of Vale Boi (Algarve, Southern Portugal), a site with a long Upper Palaeolithic sequence. Vale Boi rockshelter was discovered in 1998 and excavations started two years later (Bicho *et al.* 2010). The results provided a new perspective on the Upper Palaeolithic occupation of Southwestern Europe, with a large volume of new data on the Gravettian, Solutrean and Magdalenian. Subsistence information shows a very wide spectrum of resources from very early on, including marine and terrestrial prey. Probably the most important aspect of subsistence technology is the use of grease rendering starting some 28,000 years ago (radiocarbon calendar). Stylistic information of body adornments and lithic technology in the Gravettian and Solutrean horizons shows an interesting set of characteristics. These support the idea of a strong social network with the Spanish Mediterranean world. We will also describe the cultural characteristics of each of these Palaeolithic phases and provide some interpretation on the site and on the diversity of human occupations, focusing on differences across time based on various aspects, such as site formation processes, slope deposits and intra-site spatial organization.

Cave and rockshelter research in Portugal: a short overview

During the mid nineteenth century Palaeolithic became a focus of archaeological research in Portugal. Archaeological excavations were carried out by a group of geologists from the Serviços Geológicos de Portugal (the government agency responsible for the geological survey and mapping of Portugal), of which the most important names were Nery Delgado and Carlos Ribeiro. During that period, a series of caves with Palaeolithic occupations were completely excavated, confirming that there was early human occupation in central Portugal (Fig. 6.1), at least since Middle Palaeolithic times (Zilhão 1997). Then, for more than five decades there was no research on caves or rockshelters in the region, except for the excavation of Abrigo das Bocas in the late 1930s by Manuel Heleno, director of the National Museum of Archaeology (Bicho 1992). In 1964 another important cave was discovered by accident, under the supervision of Manuel Farinha dos Santos, and, together with a very important Neolithic necropolis, the first Portuguese Palaeolithic rock art was brought to light. This took place at the cave of Santiago do Escoural, in Alentejo, south of the Tagus Valley (Zilhão 1997). This same cave was the subject of more recent research by a Belgiam-Portuguese team, coordinated by Marcel Otte and António Carlos Silva that exposed a new Middle Palaeolithic occupation near of the old cave entrance (Bicho 2004). It was the work of Jean Roche, in the 1970s, which brought cave research to life again in Portugal. In fact, two important caves were excavated in the Vale Roto, north of Lisbon. These were Gruta Nova da Columbeira, a site discovered in 1962, with Middle Palaeolithic occupations and a few Neanderthal remains, excavated by O. Veiga Ferreira but supervised by Roche (Raposo and Cardoso 1998) and Lapa do Suão excavated by Roche and a team from the Universidade do Porto, with Upper Palaeolithic occupations and human remains (Roche 1982). Recently, João Zilhão has been responsible for a more structured and long-lasting line of Palaeolithic research in cave contexts (Zilhão 1997). In the early 1980s, Zilhão noticed that Palaeolithic research needed to be anchored to a diversity of environmental data and to an absolute chronology that could be of particular quality if they came from caves and rockshelters. Thus, his work in two caves, Caldeirão and Pego do Diabo, were the trigger for a fruitful and long sequence of projects, that helped to mould, either directly or indirectly, the work of other archaeologists: Helena Moura, Thierry Aubry, and Francisco Almeida have

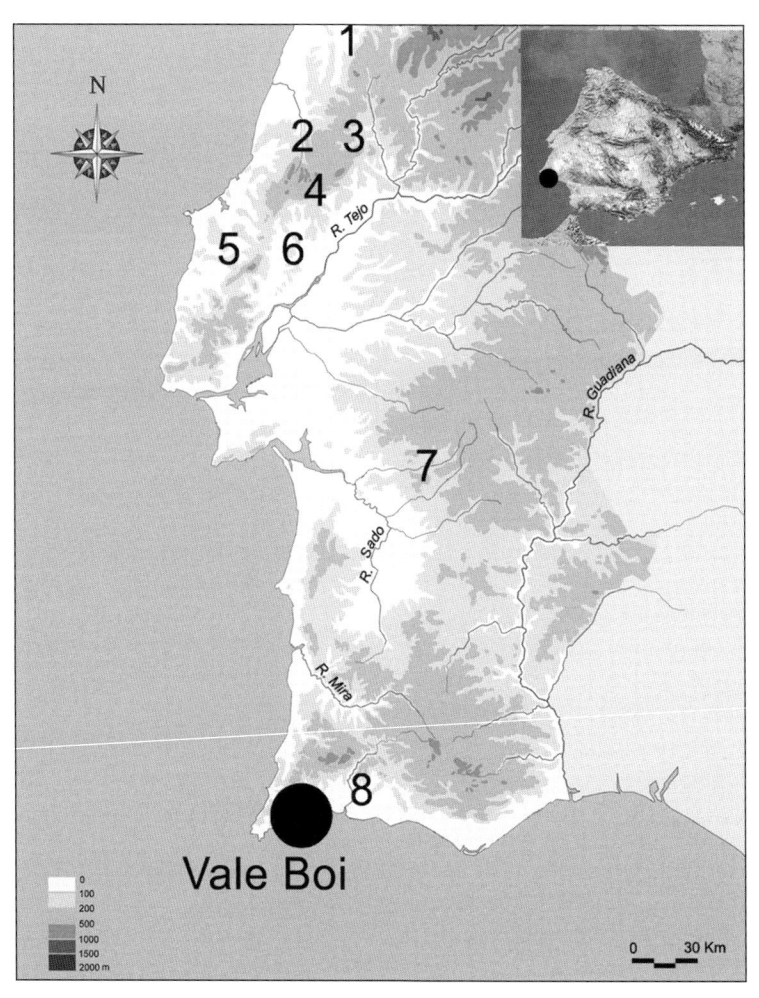

Figure: 6.1: 1. *Buraca Grande and Buraca Escura caves;* 2. *Lagar velho and Alecrim rockshelters;* 3. *Caldeirão cave;* 4. *Picareiro and Almonda caves (including Coelhos and Galeria Pesada);* 5. *Suão and Columbeira caves;* 6. *Bocas rockshelter;* 7. *Escoural;* 8. *Ibn-Ahmmar.*

since excavated caves and rockshelters in Central Portugal. These include Buraca Grande and Buraca Escura, south of Coimbra, excavated by Moura and Aubry; Coelhos in the Almonda karsic complex, Lagar Velho and Alecrim rockshelters in the Lapedo Valley, excavated by Almeida; and Almonda (various caves) excavated by Zilhão. Anthony Marks, who, since the 1980s, had worked in cooperation with Zilhão in both Almonda and in the region of Rio Maior, started a specific project in the Almonda complex with the excavation of the Galeria Pesada where he found a late Middle Pleistocene human occupation (Marks 2005). This is still the only cave in Portugal with Lower Palaeolithic material. Meanwhile, Marks developed another research group, based on members of graduate programs in the Southern Methodist University and the University of Wisconsin, Madison. Nuno Bicho and Jonathan Haws are carrying out research in both central and southern Portugal with the excavation in Portuguese Estremadura of Picareiro cave (with a long sequence including Middle Palaeolithic, Upper Palaeolithic, Mesolithic, Neolithic and Bronze Age); and in the Algarve with the excavation of Ibn Ammar cave (Middle Palaeolithic, Early Neolithic and Bronze age) and Vale Boi rockshelter, where there is a long sequence including all of the phases of the Upper Palaeolithic, Mesolithic and early Neolithic occupations. The Upper Palaeolithic in Vale Boi is the focus of this chapter.

The site of Vale Boi

Between 1996 and 2001 a large, extended survey of the Algarve for Palaeolithic and Mesolithic sites was carried out as a project funded by the Portuguese National Science Foundation (Bicho 2003). Some 60 sites were found, with chronologies varying from the Middle Palaeolithic (Bicho 2004) up to the Chalcolithic. Mesolithic and Early Neolithic sites were used in two Ph.D. dissertations (Carvalho 2007; Valente 2008) due to their data and absolute dating results. Middle Palaeolithic sites have been the subject of various publications (e.g. Bicho 2004; Ferring *et al.* 2000), although no extensive excavations have been carried out.

Upper Palaeolithic remains have been found at various surface sites in the western Algarve. At the same time, a couple of other Upper Palaeolithic sites have been excavated or tested, confirming the existence of the Solutrean and Magdalenian in southern Portugal (Zambujo and Pires 1998). Nevertheless, the data and the site preservation were either exiguous or problematic. It was the discovery of Vale Boi in 1998

that allowed a much better knowledge of the Upper Palaeolithic in Southern Portugal (Bicho *et al.* 2010).

The site of Vale Boi is located on the eastern side of a small valley that starts somewhere in the schist region some 3km further north and has its estuary on the Atlantic coast about 2km south of the site. The archaeology extends over a slope of at least 10,000m^2 running from a limestone cliff with a 10m face down to a terrace, above the alluvial plain of the small water course. Testing of the site started in 2000 and lasted two years, with augering and two 1 × 1m test pits located on the slope. The incredible amount of materials, including thousands of stone tools, marine shells, ornaments, and terrestrial fauna proved that Vale Boi was a very important site with an extensive chronology covering the whole of the Portuguese Upper Palaeolithic. Excavations started in 2002 with two areas contiguous to the test pits, of respectively with 8 and 5m^2. New testing was carried out in the following years, resulting in two new areas of excavation, one in the lower part of the site, called the Terrace, and one near the limestone face, named the Rockshelter, that are still the subject of fieldwork (Fig. 6.2).

The Terrace

Work on the lower area of Vale Boi started with a single test pit, about 1.5m deep. Since it was clear that there were various levels there, it was decided to open an excavation area of 3 by 4m that was later extended 3 more meters. The stratigraphy is composed of five geological layers (Fig. 6.3) and at least eight archaeological horizons: one early Neolithic, three Solutrean, one Proto-Solutrean and, at the bottom, three Gravettian levels. The early Neolithic occupation is in Layer 2, and it is dated to around 6000 BP (all dates in this chapter are Radiocarbon years) (Table 6.1). It includes a wide range of knapped lithic materials, ceramics (some decorated but no cardial is present), bones and shell. In addition there are various habitation features, such as hearths and a stone pavement extending over most of the excavation area (Carvalho 2007). Faunal analysis confirms the presence of mammalian species such as rabbit (*Oryctolagus cuniculus*), bovids (*Bos* sp.), red deer (*Cervus elaphus*), ovicaprids (*Capra/Ovis* sp.), and pig (*Sus scrofa*). While there are definitely wild species (aurochs, red deer, goat and rabbit), others may be of domestic origin; though due to the small sample size no confirmation of domesticates was possible. Together with the mammalian fauna, there are also some remains of marine shells as well as fish and birds, including some kind of large raptor (Carvalho 2007; Dean and Carvalho in press).

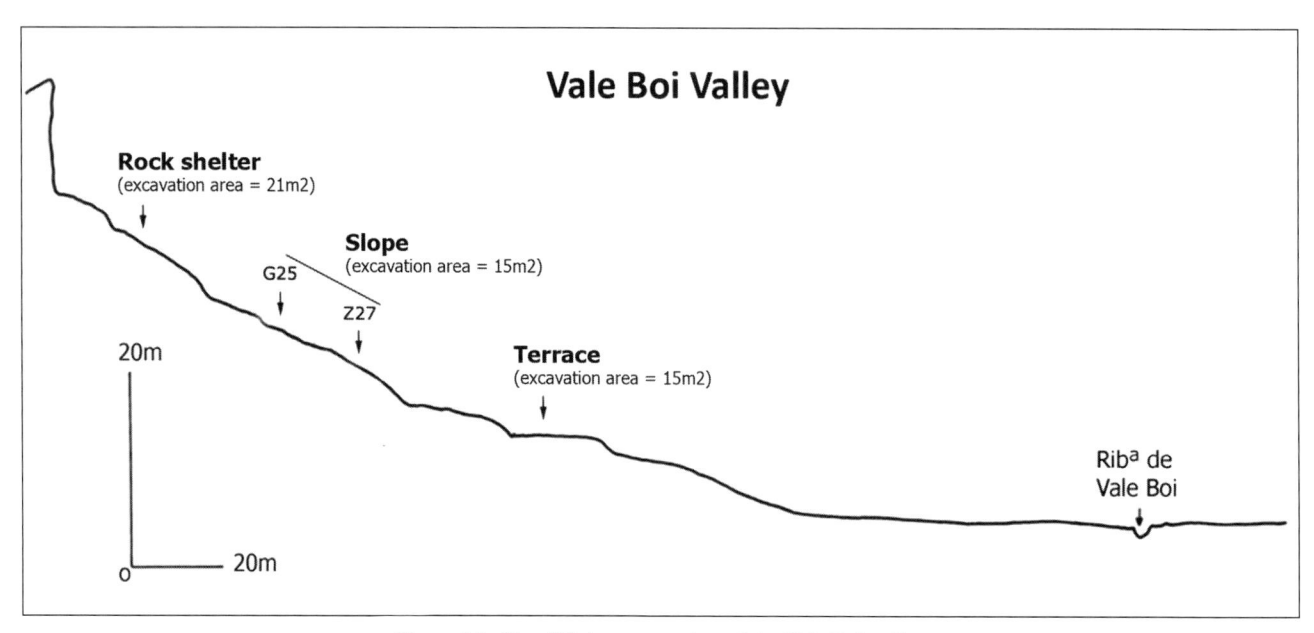

Figure 6.2: Simplified cross-section of the Vale Boi valley.

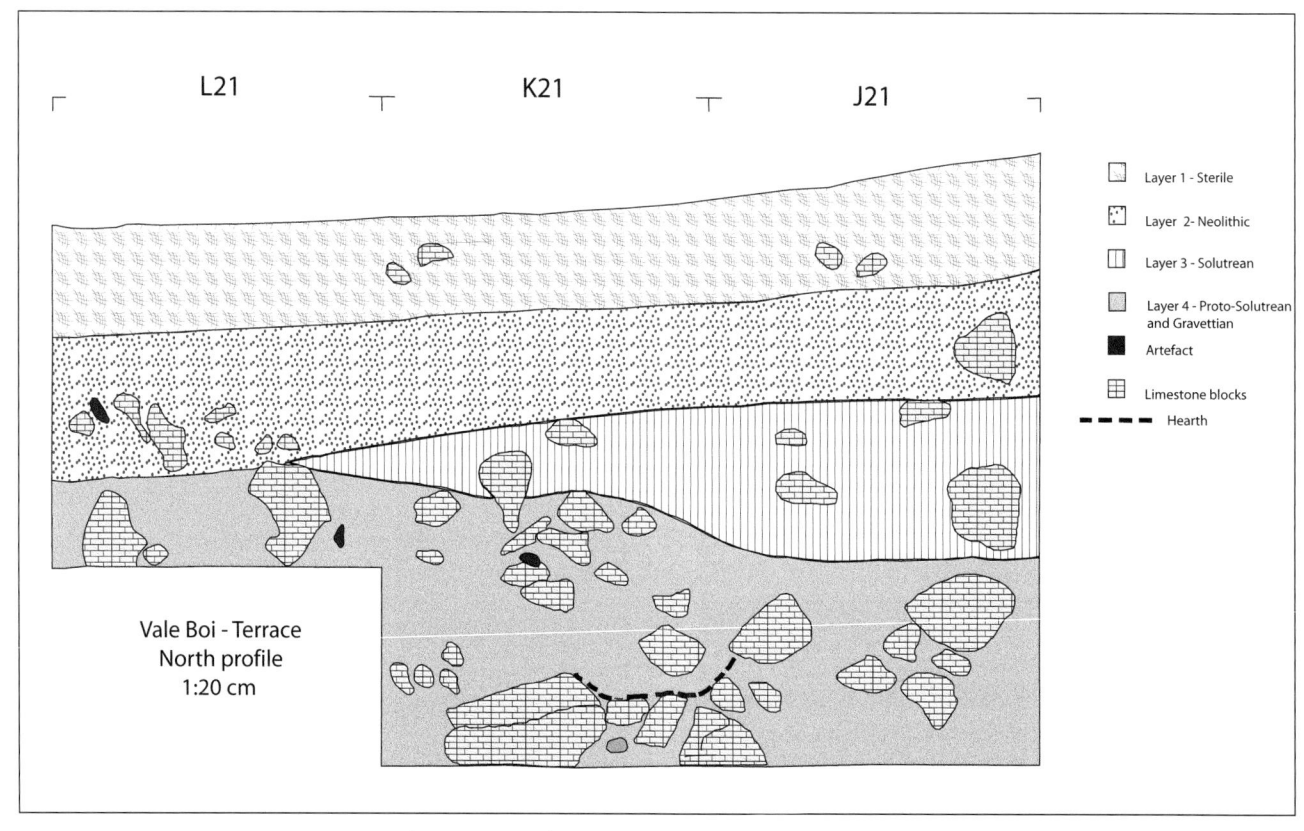

Figure 6.3: North section of the Terrace excavation area.

A single human tooth was unearthed together with the Neolithic material (Carvalho 2007). It was radiocarbon dated and proved to date to the Mesolithic, to around 7500 BP, suggesting that a Mesolithic occupation took place at the site and either was eroded away or is located in an area not yet identified (Table 6.1). However, a clear disconformity between the Holocene and Pleistocene layers (respectively 2 and 3), and the set of various radiocarbon samples from Layer 3 (containing Solutrean material) with slightly older results than the human tooth, suggest that there was likely a moment of erosion when Holocene charcoal infiltrated vertically through the upper Solutrean levels.

The Solutrean levels in Layer 3 show some evidence of mixing, as they incorporated some of the materials that rolled down the slope (possibly from both Solutrean and Gravettian contexts), and are marked by a very high degree of natural damage to the faunal remains and very rare shells. By contrast, the Solutrean

Cultural context	Area	Layer	Lab. reference	Sample	Age BP	References
Early Neolithic	Terrace	2	Wk-17030	Bone	6036±39	Bicho 2009b; Carvalho 2007
Early Neolithic	Terrace	2	OxA-13445	Bone	6042±34	Bicho 2009b; Carvalho 2007
Early Neolithic	Terrace	2	Wk-17842	Bone	6095±40	Carvalho 2009
Early Neolithic	Terrçce	2	Wk-17843	Bone	6018±34	Carvalho 2009
Mesolithic	Terrace	2	TO-12197	Human tooth	7500±90	Carvalho, 2007
Solutrean	Terrace	3	Wk-13685	Charcoal	8749±58*	This chapter
Solutrean	Terrace	3	Wk-24761	Charcoal	8886±30*	This chapter
Solutrean	Slope	2	Wk-12131	Bone	17,634±110	Bicho 2009a
Solutrean	Shelter	B6	Wk-24765	Shell	18,859±90	Bicho 2009a
Solutrean	Shelter	B6	Wk-24763	Charcoal	19,533±92	Bicho 2009a
Solutrean	Shelter	B1	Wk-17840	Charcoal	20,340±160	Bicho 2009a
Solutrean	Shelter	C4	Wk-26800	Charcoal	20,620±160	Bicho et al. 2010
Solutrean	Shelter	D2	Wk-26802	Charcoal	20,570±158	Bicho et al. 2010
Proto-Solutrean	Slope	2	Wk-12130	Bone	18,410±165**	Bicho 2009a
Late Gravettian	Shelter	D4	Wk-26803	Shell	21,896±186	Bicho et al. 2010
Late Gravettian	Slope	3	Wk-16415	Shell	21,830±195	Bicho 2009a
Late Gravettian	Slope	3	Wk-13686	Bone	22,470±235	Bicho 2009a
Early Gravettian	Terrace	4	Wk-24762	Charcoal	24,769±180	Bicho 2009a
Early Gravettian	Slope	3	Wk-12132	Charcoal	24,300±205	Bicho 2009a
Early Gravettian	Slope	3	Wk-16414	Shell	23,995±230	Bicho 2009a
Early Gravettian	Slope	3	Wk-17841	Shell	24,560±570	Bicho 2009a
Early Gravettian	Terrace	5	Wk-26801	Charcoal	27,720±370	Bicho et al. 2010

*Erroneous dates, probably due to vertical migration of surface charcoal during erosion in the Tardiglacial or early Holocene.
** Since the per cent of Nitrogen values in this sample is very low (.18), the sample has probably been partly eroded and the result should be considered as a minimum date.

Table 6.1: Uncalibrated AMS dates from Vale Boi.

and Proto-Solutrean bones from Layer 4 show very little damage and there is no evidence of mixing with the slope materials.

Two Gravettian levels are located in Layer 4, while the third and lowest Gravettian level is in Layer 5. While the lowest Gravetian horizon in level 4 is dated to around 25.000 BP, we have recently obtained a new date for level 5, around 28,000 BP indicating that this is one of the earliest Gravettian dates for Iberia.

The Slope area

The slope area is the most extensive area of the site, covering over two thirds of the total site. It is marked at certain points by flatter areas that correspond with the major concentrations of artefacts. There are no known habitation features from the two main excavation areas (which extend from the first test pits excavated in Vale Boi) and the deposits are middens, likely representing rubbish dumps that accumulated over at least 15 millennia.

The stratigraphy is marked by three main layers. The upper one is the surface of the site, with some organic materials and very few bones, but plenty of lithic artefacts: one area (around the test pit G25) can be assigned to the Magdalenian, the other to the Late Solutrean. The second layer corresponds to most of the Solutrean, while the lowest layer encompasses the Gravettian. There are radiocarbon dates for both the Gravettian and the Solutrean, indicating that, respectively, they date here to between around 24,500–21,500 BP and around 20,500–17,000 BP. Between these two phases, the Proto-Solutrean was identified in the upper area of the slope.

The upper area, around the G25 test pit, has an undisturbed deposit with artefacts oriented in all directions and with random inclinations (frequently against the slope) and no size sorting. Bone tools and shells also re-fit within the same spit or, more rarely, between two continuous spits. In the lowest Gravettian level it is common to find piles of limpets inside each other, clearly resulting from human daily consumption. Thus, the deposit seems to be in a pristine condition without any major natural erosional processes. In contrast, in the lower area (around the Z27 test pit) the deposit clearly shows evidence for some washing and erosion, at least in the Solutrean section: there was artefact sorting, with far fewer small lithic artefacts (smaller than 1cm) than in the upper area; no refitting has so far been achieved; and bones are rarer and tend to be more fragmented. It was also common to find artefacts or bones with vertical or oblique inclinations, generally following the trend of the slope.

The Rockshelter

The Rockshelter component of the site represents a collapsed ledge in front of what we think is a cave entrance. On the southern limit of the excavation area there is, behind sediments but exposed due to a 2009 animal burrow, a vertical, smooth limestone wall, curving outwards to the south and inwards to the north, in the direction of the centre of the archaeological horizons. Excavations have extended over an area of 21m^2 in the Rockshelter and have descended to a depth of at least 4.5m in the south-eastern corner, where, in 2009, a trench was dug to test the archaeological potential below the Gravettian level (Fig. 6.4).

There is one Magdalenian level (Level Z), located in the south-western corner of the area, above the collapsed ledge. It represents a short-term occupation, since artefacts are not abundant and there are rare bones or shells. The Solutrean is characterized by three important levels, with slightly different spatial

Figure: 6.4: East section of the Rockshelter excavation area.

distributions, but extending over all of the excavation area. These levels are known as A, B, and C and have radiocarbon dates with results extending between 17,000 and 20,000 BP. A hearth was excavated at the base of layer B. The Solutrean levels have a high frequency of artefacts, bones and some shells of edible marine molluscs. They are also characterized by the presence of bone tools, ornaments and portable art. The materials from the Solutrean levels (like those from the Magdalenian levels) are in an excellent condition, showing no signs of disturbance, with no size sorting and with some refits, mainly of the bone tools and shells, for which it is easier to carry out refitting. Also, the orientation of the artefacts is not according to the prevalent slope and is slightly different from level to level.

The Gravettian occupation is known as Level D. It is not nearly as well preserved as the overlying levels, since it is not a continuous horizon over the whole excavation area. While in those areas where it is present the materials are in excellent condition with no size sorting and with random orientation and horizontal positions of the artefacts, it is likely that some areas were partially eroded, since no or very few artefacts are found in certain excavation squares.

Below the Gravettian horizon there are no artefacts for up to 1.5m, as shown by the results of the 2009 trench. However, on the last day of the fieldwork, a chert flake and a fragment of bone from a medium to large mammal were found at the bottom of the trench. Thus, it seems that there will be another archaeological horizon, either very early Upper Palaeolithic or Middle Palaeolithic.

The earliest human occupation of Vale Boi: the Gravettian

At present, there are no absolute dates for the Mousterian occupation in the southern Algarve, although there are various sites with Middle Palaeolithic materials (Bicho 2004). The type of site location and the diversity of faunal remains present, including marine resources at Ibn Ammar Cave (Bicho and Haws 2008) suggest that Neandertals used the Algarve landscape until fairly late. Of course, this idea is also confirmed by the very late dates known from other Portuguese sites in Estremadura as well as from Gibraltar (Finlayson et al. 2006). In addition, the earliest dates for Upper Palaeolithic deposits in southern and western Iberia seem to indicate that modern humans were present in the region only some time after 28,000 BP (Bicho et al. 2010), and possibly even later than that in some specific areas. Thus, the few radiocarbon dates from various Gravettian levels at Vale Boi of between 25,000 and 24,500 BP are clearly in agreement with the earliest presence in the Algarve of modern humans. Nonetheless, the presence of Gravettian deposits below the artificial spit in the G25 test pit, though similar in content to the overlying levels, indicate that modern humans were at the site sometime earlier than 25,000 BP. This is also corroborated by the presence of one Gravettian horizon, at the Terrace, below the hearth dated to 24,700 BP. The interesting aspect to note is that this new level, uncovered in 2009 and apparently identical to the new Gravettian level just found in the Rockshelter, is markedly different from those known previously from the Terrace and the Slope: the faunal remains seem less fragmented and larger, and the lithic materials are marked by the presence of double-backed bi-pointed bladelets, similar to those from the Magdalenian of Estremadura (Bicho *et al.* 2006) and in the Northern European Mesolithic, where they are usually known as Sauveterre points. Thus, it is likely that there is some chronological difference between these new Gravettian horizons and the overlying ones, the former being earlier than 25,000 BP.

Unfortunately, we still do not have detailed information on faunal patterns for each archaeological horizon, but only for each broad phase of the Upper Palaeolithic (Manne and Bicho 2009). Still, there are very interesting patterns present in each phase. The Gravettian faunal list is fairly long, clearly marking a pattern of diversity attesting to both intensification and diversification with the presence of a wide dietary breadth. There are marine resources, including shellfish, crustaceans and rare mammalian and fish remains; the terrestrial specimens include carnivores, herbivores, lagomorphs (rabbits), and reptiles; and finally some medium sized birds are present.

The marine remains include a variety of shells (Table 6.2), of which the most important are various species of limpets (Manne and Bicho in press). In fact, there are thousands of fragments of limpets in the Gravettian deposits, forming, in the case of the G25 midden, a fairly solid level dated to around 25,000 BP. Other common species are *Mytilus* and *Ruditapes decussates*. It is also interesting to note the presence of a crustacean – a barnacle (*Pollicipes pollicipes*) – which is edible, was very common in the regional Mesolithic (Carvlho 2007; Valente 2008), and is much appreciated as a gourmet food today in Portugal. The presence of a few large, broken, *Pecten maximus* shells may suggest that these were also gathered as food. However, the difference in the state of preservation of the outer shell, when compared to that of other species, and the fact that they are from fairly deep marine habitats, suggest that the scallop shells were gathered empty

| | GRAVETTIAN ||| SOLUTREAN ||| MAGDALENIAN |||
SPECIES	NISP	MNI	%MNI	NISP	MNI	%MNI	NISP	MNI	%MNI
Mytilis sp.	76	12	1.1	259	25	5.1			
Pecten maximus	22	2	0.2	32	3	0.6			
Acanthocardia sp.				11	3	0.6			
Cerastoderma edule	1	1	0.1	228	11	2.2	3	1	25
Callista chione	1	1	0.1						
Ruditapes decussatus	37	5	0.5	30	5	1.0			
Veneridae	4	1	0.1	9	2	0.4			
Patella sp.	8134	1025	97.2	2875	437	89.2	142	3	75
Monodonta lineata				1	1	0.2			
Nucella lapillus	1	1	0.1						
Thais haemastoma	2	1	0.1						
Cerithiidae	2	1	0.1						
Naticidae	3	1	0.1						
Policipes policipes	3	2	0.2						
TOTALS	8286	1053		3445	487		145	4	

Table 6.2: Vale Boi marine shells and crustaceans, excluding species used for ornaments. (MNI: Minimum Number of Individuals, NISP: Number of Identified Specimens.) (Adapted from Manne and Bicho in press.)

on the shore, perhaps to be used as tools, ornaments or for some other kind of decorative function. In any case, the diversity of shells coming from a wide array of marine environments, including rocky and sandy bottoms, indicates that shellfish gathering not only took place with a clear knowledge of the resources' distribution, but also of variations in tidal timetables and of the continuous lunar month (spring and neap tides), and of the lunar day (low and high tides). This knowledge would have given hunter-gatherers, presumably women, a safe and daily scheduled means to increase their economic resources.

The terrestrial fauna also have some interesting aspects (Table 6.3). The first is the wide diversity of species, both herbivores and carnivores. The most important are red deer and horse, followed by aurochs and wild ass, although wild boar and goat are also present. However, the most common species is the rabbit, known by the thousands in the Gravettian levels of Vale Boi. The carnivores include rare bones of fox, wolf and lion, though the most important species is the Iberian lynx. While there are a few ethnographic and historical reports of the use of some of these species as food, it is likely that they were hunted for their skins and not as food.

Another important aspect of these faunal assemblages is the presence of high numbers of foetal and neonate red deer and equids (Stiner 2003). Their presence suggests a spring to summer occupation of the location, but also a hunting technique that caused a high mortality rate and random death, without any choice over the age or type of animal, including pregnant females.

In the Gravettian of Vale Boi there is a very early presence of bone grease rendering (Stiner 2003). This technique can be identified by the very high numbers of green bone fractures, resulting in the opening of the medullary cavity of virtually every major element of all of the large and medium ungulate species; the high frequency of fire-cracked non-local quartz; and, finally, the presence of hundreds of large and heavy anvils of greywacke, heavily pitted and with deep, straight indentations. In addition, the most common retouched tool is the scaled piece, of which one was found embedded in a red deer phalanx. This technique allowed for a greater extraction of all of the energy, grease and proteins out of the bones. However, it does not necessarily reflect a strategy to counter the depletion of resources. In fact, it may reflect a preference for some particular type of grease (unsaturated as opposed to saturated fatty acids) derived from different types and parts of the bones (Morin 2007; Manne and Bicho 2009), or a system

	Gravettian		Solutrean		Magdalenian	
Mammals	NISP	% NISP	NISP	% NISP	NISP	% NISP
Bos primigenius	20	0.58	74	1.54	4	0.55
Equus caballus	115	3.33	574	11.97	42	5.78
Equus sp.	15	0.43	47	0.98		
Cervus elaphus	472	13.65	1533	31.96	186	25.58
Capra/Ovis	4	0.12	7	0.15		
Sus scrofa	1	0.03	2	0.04		
Vulpes vulpes	9	0.26	4	0.08	6	0.83
Canis lupus	2	0.06	4	0.08		
Panthera leo	3	0.09	1	0.02		
Lynx pardina	11	0.32	5	0.10	2	0.28
Oryctolagus cuniculus	2803	81.06	2540	52.96	487	66.99
Cetacea	1	0.03				
Aves						
Aquila chrysaetos			1	0.02		
small sized bird			1	0.02		
medium sized bird	2	0.06	3	0.06		
large sized bird			1	0.02		
Totals	3458		4797		727	

Table 6.3: Vale Boi mammalian and avian remains (from Manne and Bicho 2009).

of risk avoidance (Manne and Bicho 2010). Another possible explanation is the fact that such a technique allowed a longer preservation of cooked meat inside the solid lard derived from the boiling of the bones. These blocks of white lard with meat inside could also be easily transported in moving between settlements (either residential or logistic); and, thus, they could have helped to increase mobility.

With regard to technology, it is clear that hunting strategies and techniques employed both lithic and bone weaponry. Bone tools are frequent in Vale Boi; in fact, there are more bone tools at the site than in all other Portuguese Upper Palaeolithic sites (Bicho and Stiner 2006; Évora 2007; Bicho 2009a). There seems to be a diversity of bone points, including a type that is frequently thought to be a kind of fish hook for large fish (Aura and Pérez Herrero 1998).

Lithic materials are made mostly on local or regional rocks. Quartz and greywacke cobbles probably came from the river bed, while large greywacke slabs for anvils were picked up slightly further away, but within a radius of less than 5km. Though there is a local chert source, less than 1km from the site, most chert materials seem to have been acquired in the Sagres promontory (Veríssimo 2005), both from primary and secondary sources in nodule or tabular form. Quartz was used in two ways: one for grease rendering, with low quality large cobbles broken into smaller pieces; and small good quality cobbles for knapping. This was used in a simple, expedient way leading to the production of mostly flakes and scarce bladelets. Quartz was reserved for simple morphological tool types such as scaled pieces, notches and denticulates.

Chert was also used in a simple, almost expedient manner. Both flakes and bladelets were produced as blanks (Fig. 6.5), and were used at different stages in the reduction sequence. Each stage produced specific types of retouched tools such as endscrapers, burins and backed tools, including backed bladelets. In the case of the earliest Gravettian occupations, there are some doubled-backed bi-ponted bladelets (Fig. 6.6), and even though their shape suggests some kind of projectile tool, preliminary use wear does seem to corroborate this cutting function. Another possibility is their use as perforators. An experimental program might clarify this question.

Lithic weaponry is fairly scarce, based on backed microlithic elements. The low numbers of these types of tools might be a consequence of different hunting techniques: using nets and traps to catch large numbers of animals in one hunting episode. This would be consistent with the presence of foetuses and neonatal animals in the age distribution. The location of the site near to the entrance of a narrow gorge leading to a fresh water lagoon at the time of Gravettian occupation may also support this hypothesis, since either the river crossing or the canyon passage would have been perfect for this type of hunting. Also, bone tools might have been used as replacements for the lithic projectiles.

Body adornment and art are also an important part of the Gravettian lifestyle in Vale Boi. There are a few tens of perforated shells (*Littorina obstusata/ mariae*, *Trivia* sp., *Dentalium* sp., and a fluvial species, *Theodoxus fluviatilis*) as well as a perforated red deer canine (Fig. 6.7). The perforated shells exhibit a tendency for small-sized specimens, suggesting that there was only the use of a single color (apparently, smaller sized *Littorina* have different colours than larger sized ones – Vanharen and D'Errico 2002). On the other hand, the small size might be a consequence of the use of only one of the *Littorina* species (e.g. *L. mariae*), which are in general smaller. In either case, it was likely to have been a stylistic marker since the body decoration would have been very different with the use of either one of the *Littorina* species (Bicho *et al.* 2003; Bicho 2009a).

In addition to the perforated shells, some indication of body or clothing decoration is provided by the presence of many iron oxide nodules, and ochre in various forms of preparation, as well as many anvils with traces of coloring on the edges. Finally, there are a few engraved plaquettes and a couple of engraved bones.

The presence of mobile art, specifically engraved small plaquettes, the existence of a fairly large bone tool assemblage and the fact that the body adornments (Bicho 2009a) and the backed tools are stylistically different from those found in Portuguese Estremadura seems to indicate that the cultural affiliation of the Gravettian from Vale Boi is strongly related to the coastal Spanish Levante.

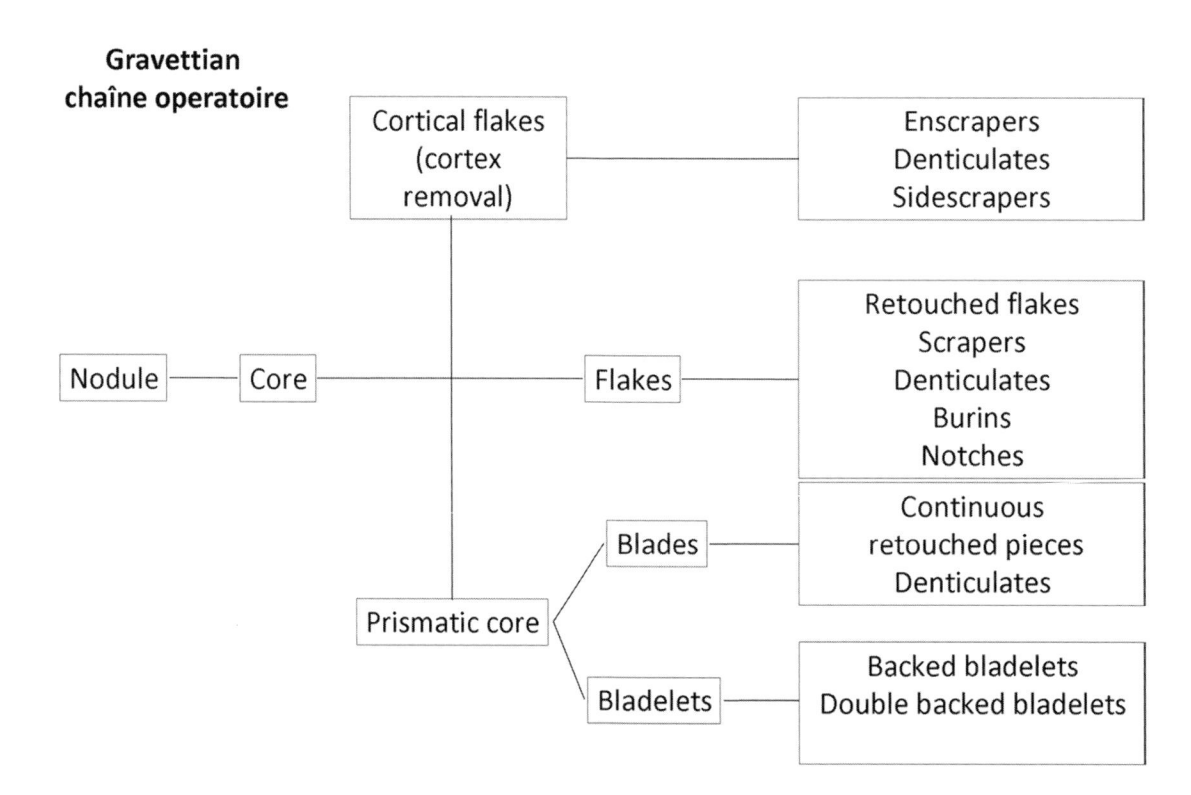

Figure 6.5: Gravettian chert reduction sequence.

The Last Glacial Maximum in Vale Boi: the Proto-Solutrean and the Solutrean

With the arrival of the Last Glacial Maximum, Portugal saw important human technological and settlement changes. While in the Algarve the environmental changes and based on limited faunal data (Manne and Bicho in press) were not as strongly marked as further north in Portuguese Estremadura – at least from the technological point of view – evolution and change were identical to the cultural phases in the North. Unfortunately, there is insufficient information to confirm any alterations in the use of space and of the landscape.

It is true, however, that shellfish remains were reduced to less than half when compared to the earlier phase. This apparent significant modification in human dietary habit (and thus in the settlement system) was probably related to changes in the distance from the site to the coastline: in fact, with the progressive regression of the sea, the coastline moved further away from the site, and so the transportation of live shellfish declined through time (Bicho and Haws 2008; Dias *et al.* 2000). The continuous presence of shellfish, as well as the diversity of species (though not as wide as in the Gravettian), confirm that this was still an important element in the human diet and a key factor in site location. In any case, it seems apparent that what was an almost specialized gathering of limpets, and thus of tidal rocky shores, became during the Last Glacial Maximum a more balanced use of rocky and sandy bottoms, marked by an increase in the exploitation of various types of cockles and clams. Although no information is available at present, it will be important to know if there is any evidence of over-exploitation of limpets through time, reflected in the diminution of shell size.

Another dietary change seems to be represented by the alterations in the relative frequency of the ungulate species. Certainly, the most important change is the striking reduction in the use of rabbit, from over 80 per cent to slightly more than 50 per cent. This change, as well as others, is probably related to modifications in the environment, from more wooden, covered vegetation to a more open landscape preferred by species such as the equids (whose relative importance rose from around 20 per cent to close to 30 per cent) and the aurochs. Nevertheless, the primary prey species remained (not counting with rabbit bones) the red deer with 68.5 per cent of the faunal assemblage (Table 6.4). Both ibex and wild boar remained secondary prey compared to the ungulates.

Art and body decoration were still important social elements. In fact, there seems to be an increase in artistic

Figure 6.6: Examples of Early Gravettian bi-ponted double-backed bladelets from Rockshelter Layer D and Terrace Layer 5.

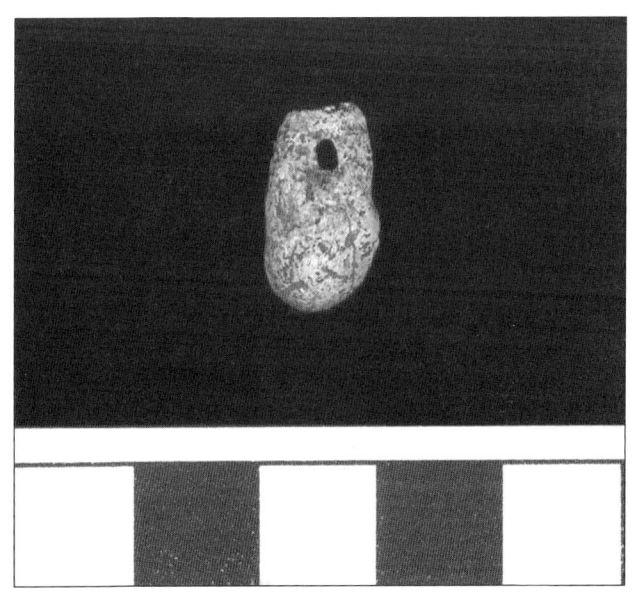

Figure 6.7: Gravettian perforated red deer teeth.

production with a few more engraved plaquettes with animal representations (Fig. 6.8). The characteristics of body adornments are very much the same, with the same species of marine shells being used. However, there seems to be a change towards larger *Littorina* shells, away from the bimodal size distribution seen during the Gravettian (Bicho 2009a).

Bone tools do not seem to have changed, although there seems to have been a slight reduction in their frequency. By contrast, there is a clear modification in lithic technology between the end of the Gravettian and the Solutrean. Proto-Solutrean technological

	GRAVETTIAN		SOLUTREAN		MAGDALENIAN	
	NISP	%NISP	NISP	%NISP	NISP	%NISP
Aurochs	20	3.2	74	3.3	4	1.7
Horse	115	18.3	574	25.7	42	18.2
Wild ass	15	2.4	47	2.1		
Red deer	472	75.3	1533	68.5	186	80.2
Ibex	4	0.6	7	0.3		
Wild boar	1	0.2	2	0.1		
TOTAL	627	100.0	2237	100.0	232	100.0

Table 6.4: Number of Identified Specimens (NISP) for ungulate remains in the Upper Palaeolithic of Vale Boi.

and typological characteristics appear clearly in two areas of Vale Boi – in the G25 and in Level 4 of the Terrace (Marreiros 2009). These are very much the same transitional elements seen in the Portuguese Estremadura as described by Zilhão (1997) and by Almeida (2000) in their Ph.D. dissertations, namely: a frequent use of quartz, and a clear presence of both blades and bladelets, carinated endscrapers and Vale Comprido points. The following phase already includes *point à face plan* and, thus, was by then a fully Solutrean tehcnology.

In Vale Boi the Solutrean lithic technology is highly efficient. On the one hand it is very economic, using local raw materials, mostly local and regional chert from Sagres (but also quartz and greywacke), producing small artefacts, and showing a tendency for the miniaturization (and microlithization through time – see Cascalheira 2009) of weaponry. On the other hand, this same technological system involved a reduction sequence that comprised a simple, but long enough chain (Fig. 6.9), that produced in different phases both blanks for unifacial reduction (e.g. bladelets for backing to make *à cran* points and backed bladelets, flakes for endscrapers, burins, etc) and for bifacial reduction (flakes, blades and bladelets for *point à face plan*, laurel leafs, *Parpalló* points). The diversity of typological characteristics of the projectiles (Fig. 6.10), including the traditional Solutrean leaf points together with the Mediterranean types of *Parpalló* and small *à cran* backed points and the larger (also rarer – in Vale Boi) Atlantic *à cran* points with invasive retouch, show that Vale Boi was certainly an important site receiving many influences, although those from the Levantine Mediterranean Spanish coast were probably the most important, suggesting that Vale Boi was one of the aggregation sites (Conkey 1980) in south-western Iberia.

Like in the Gravettian, the presence of small engraved plaquettes and of a large bone tool assemblage together

Figure 6.8: Engraved schist plaquette with aurochs.

with the highly characteristic *Parpalló* points and the other types of Solutrean projectile points (as well as traits such as the diversity and very small size found both in Algarve and in Mediterranean coast of Spain) seem to indicate that the cultural affiliation of the Gravettian and Solutrean from Vale Boi are directly related to the Spanish Levantine Upper Palaeolithic. At least, the material culture has very few traits related to those found in Portuguese Estremadura (Zilhão 1997). Thus, most likely the coastal region of Southwestern Iberia formed a territory that included a large social network (Bicho 2009a), providing a large gene pool (as well as technological and stylistic aspects) that allowed the presence of small hunter-gatherer bands that used the land- and seascapes from Algarve to Valencia that would meet periodically in large partially ritualized sites such as Vale Boi or Parpalló.

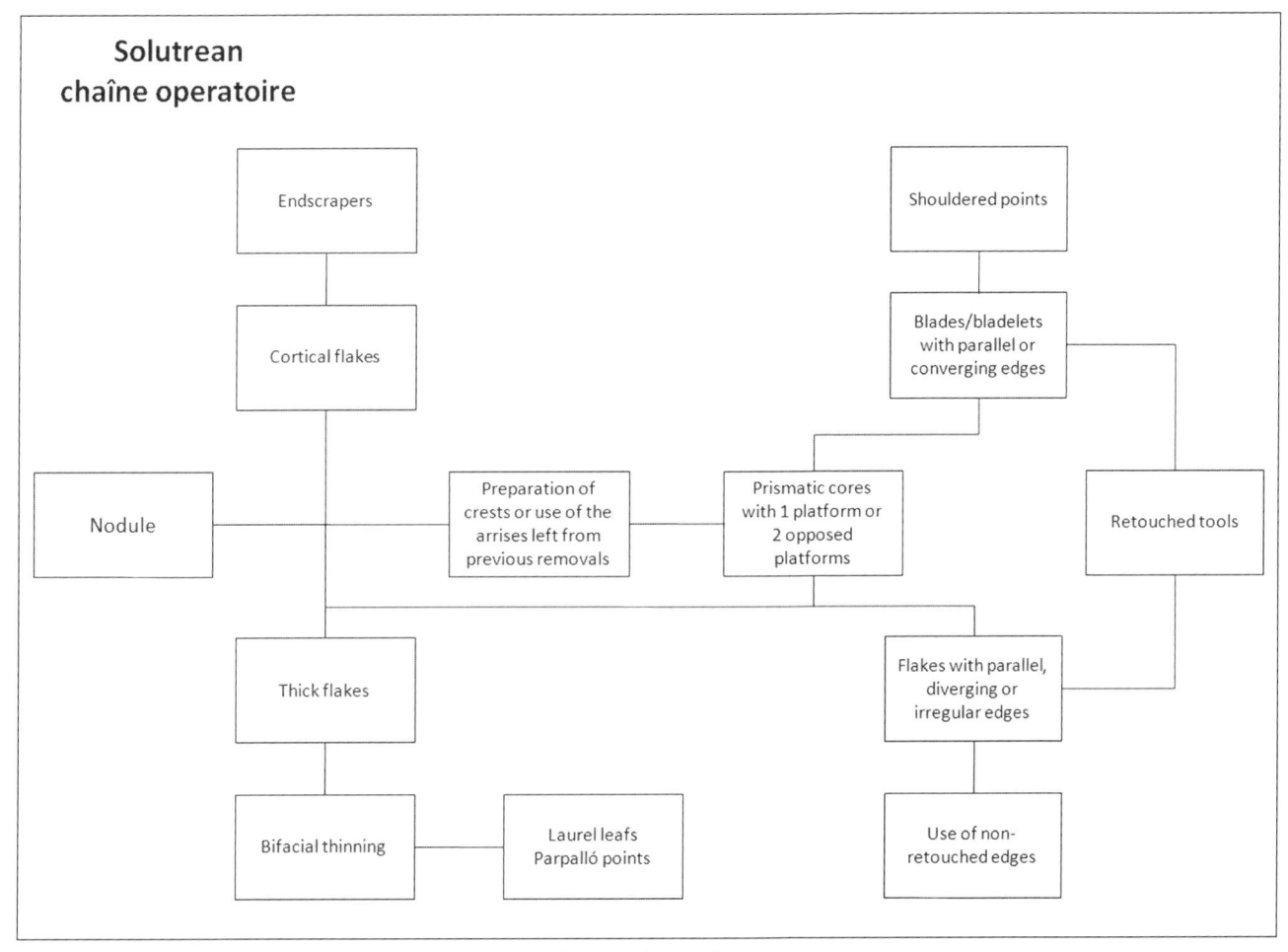

Figure 6.9: Solutrean chert reduction sequence.

Magdalenian times

Even though Zilhão has argued in favour of the presence of a Solutreo-Gravettian phase for central and southern Portugal (Zilhão 1997; Zambujo and Pires 1998), the reality is that no such transitional phase is documented in Vale Boi. It is true and relevant that there is an obvious trend towards microlithization and a replacement of bifacial weaponry by small backed *à cran* bladelet points in the late Soutrean at the site. These aspects are also marked by an increase in simple backed bladelets, but these are, in any case, very scarce (Cascalheira 2009) when compared to the traditional focus of the Solutreo-Gravettian in the Valencia region.

Very little data exists at present for the Algarve, and specifically for Vale Boi, but this trend of decrease in the use of bifacial technology and the increase in bladelets and in backing could hint at the appearance of the Magdalenian technology at the site. In fact, these two characteristics are present in the more recent levels of Vale Boi, thought to date to the Tardiglacial (although no absolute dates have been obtained so far). But the preliminary technological study carried out by Mendonça (2009) in a series of sites from Western Algarve indicate that the late lithic assemblages found at the Rockshelter and at the Slope excavations are in fact Magdalenian, probably covering the Middle and Late phases of the regional Magdalenian.

The Magdalenian of Vale Boi is characterized by the presence of quartz, greywacke and cherts, very much like the earlier Upper Palaeolithic phases. Chert seems to be most important raw material, although small nodules were used to produce small blanks: either flakes or bladelets. Small and thin endscrapers are frequently present, as well as burins. Simple retouched bladelets and backed bladelets are also present, and clearly mark the retouched stone tool assemblages, although they are not very expressive numerically. As a distinctive element from the Spanish and Portuguese Estremadura sites,

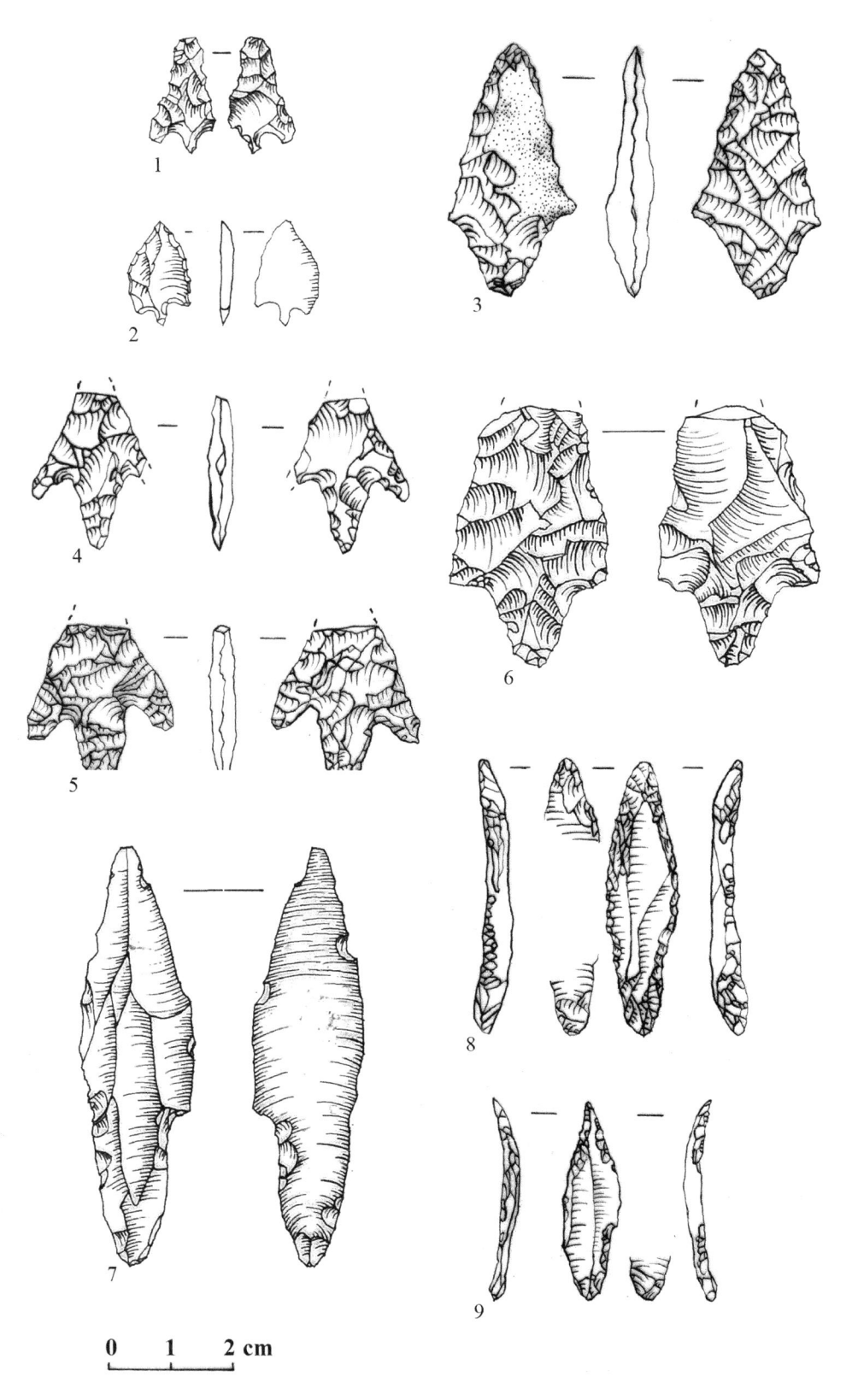

Figure 6.10: Solutrean points. 1–6, Parpalló points; 7, Atlantic à cran point; 8 and 9, Mediterranean à cran bladelet backed points.

the Magdalenian of the Algarve has a strikingly high number of scaled pieces, both in chert and in quartz. The presence of this tool type, closely comparable to that of the Gravettian and the Solutrean, was probably related to the grease rendering technique.

The lack of radiocarbon dates is due to the erosional processes that have affected the Magdalenian levels, destroying most organic materials from these levels, since they are close to the present surface. Naturally, this same process impacted negatively the preservation of other elements, such as ornaments and portable art, though they are occasionally present. Faunal remains also suffered with surface erosion and, thus, both bones and shell dating to this period are scant. Still, the few marine shells do seem to suggest a similar tend to that of the previous Solutrean phase, with the presence of both rock and sandy bottom species (limpets and cockles), while the mammalian fauna indicates again a more closed and vegetated landscape, with red deer and rabbit frequencies rising again to new maximums, as one would expect after the Last Glacial Maximum and on into the climatic amelioration of the Tardiglacial in southern Iberia. In any case, the pattern of wide dietary breadth is the same, marked by diversification and intensification of resources, very much like the scenario in central Portugal.

Final words

The work at Vale Boi shows how important a cave or rockshelter can be in providing a complete picture of a period such as the Upper Palaeolithic, particularly in the case of the Algarve where *in situ* occupations are extremely rare. The calcareous context of the limestone provided the right chemical environment to preserve the faunal remains, including bones and marine shells from a wide variety of species. This preservation allowed a unique view of the past, revealing the introduction of the new technology of grease rendering, with its clear advantages to the economy, to mobility and to the general health of the hunter-gatherer population of Southwestern Iberia.

Vale Boi also brings a new perspective on the strength of social networks across coastal Iberia, from the Spanish Levante to the Atlantic coast of Portuguese Estremadura. There are elements – stylistic and others – that suggest a movement of people, ideas, raw materials and identities through space, leading to the formation of contacts and to the establishment of social territories during the Upper Palaeolithic.

Many of the results referred to above are clearly preliminary and still need much research to be better understood. Nevertheless, they allow a series of interpretations that, in same cases, are without any doubt cognitive in nature and, thus, new to Upper Palaeolithic research in Portugal: Palaeolithic social networking across Iberia, or the mapping and calendrical (monthly and daily) knowledge of the tidal system, both aiding the more efficient use of resources and the safety of the gathering group (most likely women and children).

Finally, there are various aspects that have only been touched lightly in this chapter which will deserve more attention as more data are obtained through new field work. This is the case with the earliest Gravettian occupation and its characteristic doubled-backed bi-pointed bladelets. Clearly, the presence of such retouched tools, unique in nature and possibly also in function, needs a plausible explanation; this will be possible with more research – both excavation to enlarge the sample and an experimental program of use-wear analysis. Another important aspect that needs more research is the possible Neandertal occupation at Vale Boi. There are various lithic artefacts that are technologically and typologically Mousterian at the site. These are present in the lower levels of the Rockshelter and Slope areas. However, since they were collected within the Gravettian contexts, it is not possible to establish unequivocally the presence of a Middle Palaeolithic occupation at Vale Boi. Perhaps with the development of the work in the lower section of the Rockshelter this will be confirmed once and for all.

Acknowledgements

We wish to thank Knut Andreas Bergsvik for the invitation to participate in the session on Caves and Rockshelters at the European Association of Archaeologists' annual meeting in Malta and to contribute to this volume. Research funding was provided by the Fundação para a Ciência e Tecnologia (Grant PTDC/HAH/64184/2006), the National Geographic Society and the Archaeological Institute of America.

References

Almeida, F. (2000) *The Terminal Gravettian of the Portuguese Estremadura: Technological Variability of the Lithic Industries*. Unpublished Ph.D. dissertation, Southern Methodist University.

Aura, E. and Pérez Herrero, C. (1998) Micropuntas dobles o anzuelos? Una propuesta de estudio a partir de los materiales de la Cueva de Nerja (Málaga). In J. Sanchidrián Torti and M. Simón Vallejo (eds.) *Las Culturas del Pleistoceno Superior en Andalucia*, 339–348. Nerja, Patronato de la Cueva de Nerja.

Bicho, N. (1992) *Technological Change during the Pleistocene-Holocene Boundary in Rio Maior, Portugal*. Unpublished Ph.D. dissertation, Southern Methodist University, Dallas.

Bicho, N. (2003) A importância dos recursos aquáticos na economia dos caçadores-recolectores do Paleolítico e Epipaleolítico do Algarve. *Xelb* 4, 11–26.

Bicho, N. (2004) The Middle Palaeolithic occupation of Southern Portugal. In N. Connard (ed.) *Settlement Dynamics of the Middle Palaeolithic and Middle Stone Age, Volume II*, 513–531. Tübigen, Kerns Verlag.

Bicho, N. (2009a) Fashion and glamour: weaponry and beads as territorial markers in Southern Iberia. In F. Djindjian, J. Kozlowski and N. Bicho (eds.) *Le Concept de Territoires dans le Paléolithique Supérieur Européen. Proceedings of the XV Congress of the International Union for Prehistoric and Protohistoric Sciences*, 243–252. British Archaeological Reports International Series 1938. Oxford, Archeopress.

Bicho, N. (2009b) On the edge: Early Holocene adaptations in Southwestern Iberia. *Journal of Anthropological Research*, 65(2),185–206.

Bicho, N., Gibaja, J. and Stiner, M. (2010) Le Paléolithique supérieur au sud du Portugal; le site de Vale Boi. *L'Anthropologie*, 114(1), 48–67.

Bicho, N, and Haws, J. (2008) At the land's end: marine resources and the importance of fluctuations in the coast line in the prehistoric hunter-gatherer economy of Portugal. *Quaternary Science Review* 27, 2166–2175.

Bicho, N., Haws, J. and Hockett, B. (2006) Two sides of the same coin – rocks, bones and site function of Picareiro Cave, Central Portugal. *Journal of Anthropological Archaeology* 25, 485–499.

Bicho, N., Manne, T., Cascalheira, J., Mendonça, C., Évora, M., Gibaja, J. and Pereira, T. (2010) O Paleolitico superior do sudoeste da Península Ibérica: o caso do Algarve. In X. Mangado (ed.) *El Paleolítico Superior Peninsular. Novedades del Siglo XXI*. Barcelona, Universidad de Barcelona.

Bicho, N. and Stiner, M. (2006) Gravettian coastal adaptations from Vale Boi, Algarve (Portugal). In J. Sanchadrian, A. Belén Marquez and J. Fullola y Pericot (eds.) *La Cuenca Mediterránea durante el Paleolítico Superior. Reunión de la VIII Comisión del Paleolítico Superior*, 92–107. Nerja, Fundación Cueva de Nerja.

Bicho, N., Stiner, M., Lindly, J., Ferring, C. R. and Correia, J. (2003) Preliminary results from the Upper Palaeolithic site of Vale Boi, southwestern Portugal. *Journal of Iberian Archaeology* 5, 51–66.

Carvalho, A. (2007) *A Neolitização do Portugal Meridional: Os Exemplos do Maciço Calcário Estremenho e do Algarve Ocidental*. Unpublished Ph.D. dissertation, Universidade do Algarve.

Cascalheira, J. (2009) *Tecnologia Solutrense do Abrigo de Vale Boi*. Unpublished M.A. thesis, Universidade do Algarve.

Conkey, M. (1980) The identification of prehistoric hunter-gatherer aggregation sites: the case of Altamira. *Current Anthropology* 21(5), 609–630.

Dean, R. and Carvalho, A. F. (in press) Surf and turf: the use of marine and terrestrial resources in the Early Neolithic of coastal southern Portugal. In N. Bicho, J. Haws and L. Davis (eds.) *Trekking the Shore: Changing Coastlines and the Antiquity of Coastal Settlement*. New York, Springer.

Dias, J. A., Boski, T. Rodrigues, A. and Magalhães, F. (2000) Coast line evolution in Portugal since the Last Glacial Maximum until present – a synthesis. *Marine Geology* 170, 177–186.

Évora, M. (2007) *Utensilagem Óssea do Paleolítico Superior Português*. Unpublished M.A. thesis, Universidade do Algarve.

Ferring, R., Lindly, J., Bicho, N. and Stiner, M. 2000. The Middle Palaeolithic of Algarve. In R. E. Balbin, N. Bicho, B. Carbonell, A. Hockett, L. Moure, L. Raposo, M. Santonja and G. Vega Toscano (eds.) *Paleolítico da Peninsula Ibérica. Actas do 3º Congresso de Arqueologia Peninsular*, 271–276. Porto, ADECAP.

Finlayson, C., Giles Pacheco, F., Rodríguez-Vidal, J., Fa, D., Guiterrez López, J. M., Santiago Pérez, A., Finlayson, G., Allue, E., Baena Preysler, J., Cáceres, I., Carrión, J. S., Fernández- Jalvo, Y., Gleed-Owen, C. P., Jiménez Espejo, F. J., López, P., López Sáez, J. A., Riquelme Cantal, J. A., Sánchez Marco, A., Giles Guzman, F., Brown, K., Fuentez, N., Valarino, C. A., Villalpando, A., Stringer, C., Martinez Ruiz, F. and Sakamoto, T. (2006) Late survival of Neanderthal at the southernmost extreme of Europe. *Nature* 443, 850–853.

Manne, T. and Bicho, N. (2009) Vale Boi: rendering new understandings of resource intensification and diversification in southwestern Iberia. *Before Farming* 2009/2,.article 1. < http://www.waspjournals.com/journals/beforefarming/journal_20092/abstracts/download.php?filename=2009_2_01.pdf>

Manne, T. and Bicho, N. (2010) Never too much of a good thing: continuation of subsistence patterns during the Upper Palaeolithic in Southwestern Portugal. Unpublished paper presented at the 75th Annual Meeting of the Society for American Archaeology, St. Louis, USA.

Manne, T. and Bicho, N. (in press) Prying new meaning from limpet harvesting at Vale Boi during the Upper Palaeolithic. In N. Bicho, J. Haws and L. Davis (eds.) *Trekking the Shore: Changing Coastlines and the Antiquity of Coastal Settlement*. New York, Springer.

Marks, A. (2005) Micoquian elements in the Portuguese Middle Pleistocene: assemblages from the Galeria Pesada. In N. Bicho. (ed.), *Paleolítico. Actas do IV Congresso de Arqueologia Peninsular*, 195–206. Faro, Universidade do Algarve.

Mendonça, C. (2009) *A Tecnologia Lítica no Tardiglaciar do Algarve*. Unpublished M.A. thesis, Universidade do Algarve.

Marreiros, J. (2009) *As Primeiras Comunidades do Homem Moderno no Algarve Ocidental: Caracterização Paleotecnológica e Paleoetnográfica das Comunidades Gravetenses e Proto-Solutrenses da Vale Boi (Algarve, Portugal)*. Unpublished M.A. thesis, Universidade do Algarve.

Morin, E. (2007) Fat composition and Nunamiut decision-making: a new look at the marrow and bone grease indices. *Journal of Archaeological Science* 34(1), 69–82.

Raposo, L., and Cardoso, J. (1998) Las industrias líticas de la Gruta Nova da Columbeira (Bombarral, Portugal) en el contexto del Mustierense Final de la Península Ibérica. *Trabajos de Prehistoria* 55 (1), 39–62.

Roche, J. (1982) A gruta chamada Lapa do Suão (Bombarral). *Arqueologia* 5, 5–18.

Stiner, M. (2003) Zooarchaeological evidence for resource intensification in Algarve, southern Portugal. *Promontoria* 1, 27–61.

Valente, J. (2008) *As Últimas Sociedades de Caçadores-Recolectores no Centro e Sul de Portugal (10,000–6,000 Años BP): Aproveitamento dos Recursos Animais*. Unpublished Ph.D. dissertation, Universidade do Algarve.

Vanhaeren, M. and D'Errico, F. (2002) The body ornaments associated with the burial. In J. Zilhão and E. Trinkaus (eds.), *Portrait of the Artist as a Child. The Gravettian Human Skeleton from the Abrigo do Lagar Velho and its Archaeological Context,* 154–186. Lisboa, IPA.

Veríssimo, H. (2005) Aprovisionamento de materiais líticos na Pré-história do Concelho de Vila do Bispo (Algarve). In N. Bicho (ed.) *O Paleolítico. Actas do IV Congresso de Arqueologia Peninsular,* 509–523. Faro, Universidade do Algarve.

Zambujo. G. and Pires, A. (1998) O sítio arqueológico da Vala, Silves: Paleolítico Superior e Neolítico Antigo. *Revista Portuguesa de Arqueologia* 2.1, 5–24.

Zilhão, J. (1997) *O Paleolítico Superior da Estremadura Portuguesa.* Lisbon, Edições Colibri.

Chapter 7

The Use of Caves and Rockshelters by the Last Neandertal and First Modern Human Societies in Cantabrian Iberia: Similarities, Differences, and Territorial Implications

Javier Ordoño

The Middle to Upper Palaeolithic Transition has stimulated much debate over the last decades within archaeological research. Several factors have been suggested to explain the different changes that happened in this controversial period (human biology, technological evolution, language, symbolism). However, there are still many unknown areas to analyze. One of them is the territorial behaviour of the populations involved in the process. Until now there has been very little insight in the spatial distribution of the sites and even less about the decisions and conditionings that led them to settle and move back and forth across the territory. This contribution analyzes the case of the Cantabrian Region, an excellent setting for an archaeo-geographical study because of its ample archaeological record (approximately 125 sites) and its specific geography lie. The main caves and rockshelters (88 per cent of the total) are examined in order to gather information on their spatial distribution, their geographical and topographical position with regard to their natural environment, their orientation, their relationship with the main economic resources (especially water sources, accessible biotopes, and lithic raw material supplies) and, where possible, on their quality and functionality. The interpretation of these variables allows us to detect differences and similarities in the territorial behaviour of the 'transitional populations'.

Introduction

The effort made by prehistoric archaeology to define the transition from the Middle to Upper Palaeolithic has been important in recent years, especially since discussion of it was reopened following the famous discovery of Saint-Césaire in the Charente in 1979 (Lévêque and Vandermeersch 1980). Beyond the controversies that arose regarding its nature and development, research has mainly been directed towards the study of the anthropological, technological, economic, and symbolic aspects concerning the two main groups: the last Neandertals and the Early Modern Humans. However, little is known about the territorial behaviour of these people, a point which must surely provide new arguments for the characterization of this complex historical process.

In order to plug this gap, a new archaeo-geographical analysis, supported by a broader concept of territoriality than those proposed previously, is presented here for the first time. It is based on the study of very diverse variables (e.g. geographical or palaeo-economic ones), that is applied – still in its initial phase – to the outstanding archaeological record identified for that period in the Cantabrian Region, where caves and rockshelters are of paramount importance. This analysis has yielded some preliminary results, which are outlined and discussed in this essay.

The Middle to Upper Palaeolithic Transition and the analysis of territorial behaviour: a brief introduction

The pioneering chrono-cultural and typological proposals made by Breuil, Peyrony, Garrod, Bordes, Delporte or Pradel are now distant. It was mainly in the 1980s that the foundations of what we understand today as the Middle to Upper Palaeolithic Transition were laid, thanks especially to the growth of fieldwork and the development of new theoretical and methodological tools that allowed scholars to construct many explanatory theories, which would be quite difficult to summarize in these pages. Drawing on one of the first systematic efforts to define the nature of this period in the European continent, made by P. Mellars in the early 90s (Mellars 1991), it could be said that the Transition was characterized by a series of fundamental changes that would break with the preceding situation (the Neandertal one) and would form the germ of the new Upper Palaeolithic universe: in the biological context, the replacement of Neandertals by Modern Humans; from a technological point of view, the disappearance of flake manufacture in favour of diverse standardized blade assemblages and the incorporation of bone technology; with regard to symbolism, the appearance of artistic representations and the production and diffusion of personal ornaments; in the social field, the increase of population density and the size of human groups; finally, in the economy: the specialization of animal exploitation, the development of new hunting techniques, and long-distance contacts.

This disruptive view, outlined by Mellars in earlier works (1973; 1989a; 1989b) and followed by others (Klein 1994; Bar-Yosef 2002), understands that in this supposed *revolution* biological and cultural change went hand-in-hand (the Chatelperronian being the result of the acculturation of Neandertals by Modern Humans). Contrary to this, we find other interpretative lines which call attention to the increasing similarities observed between the last Middle Palaeolithic and the first Upper Palaeolithic cultures, lending a gradual (and sometimes even continuous) character to the process, but without denying the possible influence of Modern Humans over Neandertals. Thus, we can cite, among others, the pioneering works of Laplace (1962) and Brace (1962), or, from the 80s, those of White (1982), Clark and Lindly (1989), Otte (1990), Clark (1992), Rigaud (1996), Straus (1996), Cabrera *et al.* (2000), or Harrold and Otte (2001). Other authors, however, advocate a Transition without acculturation, arguing for an independent Neandertal ability to evolve in order to explain the emergence of *modern* nuances in the final stages of the Middle Palaeolithic (d'Errico *et al.* 1998; Zilhão and d'Errico 2000). Of course, the large number of meetings, symposiums, and publications exclusively devoted to the problem of the Transition in Europe in the last 20 years has not escaped these controversies (Camps 2009).

Whatever the correct answer is to the big question of the Transition, it is true that in recent years some preliminary evidences have been detected that allow us to glimpse a gradual and complex process (especially in its chronological development and its different geographical impact): an ample re-occupation or settlement continuity of many Neandertal sites in the Early Upper Palaeolithic; the existence of Middle Palaeolithic assemblages with a presence of incipient bone and blade technologies; the uncertain authorship of some transitional assemblages, like the early Aurignacian and/or Protoaurignacian, the Uluzzian, the Szeletian, etc. (and even the Chatelperronian?); the observation of similar patterns in faunal exploitation and raw material procurement; the matter of possible Neandertal-made symbolic practices (e.g. burials) and ornamentation.

Future research must check whether this apparent gradual change is evident in other areas of the archaeological record. One of them is the *territorial behaviour* of human groups. The study of this particularly important aspect must become a new source to describe the transitional process in terms of continuity or rupture (it is further assumed that it is in transitional times when the contrasts in all behavioural spheres are best appreciated), taking into account the differences or similarities observed in the territorial behaviour of the last Neandertals and early Modern Humans. This will require the analysis of more specific issues, such as settlement patterns, the exploitation of biotic and abiotic resources, the annual mobility of groups, communication routes, the hierarchical organization of settlements, the size and characteristics of the controlled territory, intergroup contacts, diffusion of technical-stylistic and symbolic conventions, etc.

Despite the need to use a regional or even transregional perspective that transcends the usual individual study of each site and its immediate surroundings, territorial behaviour has been insufficiently analyzed by research, and although we find a significant number of works concerning some of these issues within the European Palaeolithic research, few of them deal specifically with the period of the Transition. Among them, we must mention the ones related to settlement patterns (Ashton 1983; Bahn 1984; White 1985; Blades 1999; Cabrera *et al.*

2004; Svoboda 2006), faunal resources (Altuna and Mariezkurrena 1988; Chase 1989; Pike-Tay and Knecht 1993; Grayson and Delpech 1998; Richards *et al.* 2001; Yravedra 2002; Dari 2003; Coward 2004), mobility and the procurement of lithic raw materials (Geneste 1988; Turq 1993; Féblot-Augustins 1993; Demars 1998; Lieberman 1998; Tarriño 2001; Bordes *et al.* 2005), and even to the definition of technocultural (Arrizabalaga *et al.* 2007) and artistic territories (*territorios gráficos*) (Fortea 1994; García and Eguizábal 2008).

We will not detail the contribution of these works to the discussion surrounding the nature of the Transition, since it would exceed the aim of this essay. But we must say that these studies usually analyze only specific aspects of the many dimensions that make up what we understand by territorial behaviour. So, it will be necessary to develop more eclectic approaches that integrate the most useful points of this amalgam of studies (Djindjian 2009), based on a reformulation of the concept of territory with reference to the characteristic Palaeolithic archaeological record. This is the key to achieve a more complete knowledge of prehistoric territoriality and, in general, of the behaviour of human groups, understood as social beings with the capacity to apprehend, assimilate, and manage the space surrounding them.

Towards a new approach to territorial behaviour

Redefining the concept of Palaeolithic territoriality

The definition of a notion of *territory* (in our opinion a concept less vague than *landscape*) that fits better the nature of Palaeolithic hunter-gatherers has been under consideration by research in recent years, as reflected by the emergence of more and more published monographs devoted to the matter (Jaubert and Barbaza 2005; Vialou *et al.* 2005; Bressy *et al.* 2006; Djindjian *et al.* 2009). Working from positions not dependent upon scholastic orthodoxies (mainly processual or pos-processual ones) or methodologies advocated by them, we can begin to overcome the traditional multi-faceted and fragmented view of territoriality in which one aspect (e.g. ecological-economic, cultural, or symbolic) takes precedence over the others when defining the characteristics of territories (Ordoño 2007, 98). These circumstances have promoted the appearance of new proposals that deal with the knowledge of Palaeolithic territories from a more holistic perspective, involving the study of the various spheres composing territoriality.

Thus, Merlet (1996) distinguishes three scales to be analyzed: the *subsistence territory* (designed to meet the immediate needs of a human group), the *covered space* (including intergroup exchange, alliance, or social relations), and the *cultural space* (in which a spiritual or cultural communion or convergence reflected in the archaeological record would be evident).

From a similar position, Julien and Connet (2005) suggest the differentiation of various places and spaces reflected on Palaeolithic territoriality (Table 7.1), based on the ethnographic observations made by Bonnemaison (1996) in the Vanuatu islands.

From a different perspective, Bracco (2005) – following the definition of Bourgeot (1991, 698) which understands the territory as an objectively organized and culturally invented space (by human beings) – proposes a territoriality composed of two interrelated areas: *physical space*, in which man would meet his material and cultural needs; and *symbolic space*, this one being a product of the ideological and spiritual conception of environment by man, and manifested in the main representation systems (e.g. language, art, or mortuary practices).

Far from suggesting a compartmentalization of space, and working from the premise that space is appropriated once it is perceived by men (Thomas 2001, 73), we believe that it is more relevant to define territoriality as all of the interactions that take place when the human being appropriates or socializes his natural environment (Ordoño 2007, 105). The chronological and spatial coincidence of man and the environment involves the necessary existence of a range of interactions (whether direct or indirect) that are the reflection of the coexistence between them: man, as a rational animal, uses the resources provided by nature to meet his material and spiritual needs; and for its part, the natural environment determines the availability of resources and the suitability for human habitation. Thus, a *quid pro quo* relationship arises, in which human beings and the environment plays a role in the functioning of a territory.

Such interactions, which constitute an archaeological document (as a product of human action), can be naturally different, linked to, for example, the material and mental appropriation and delimitation of space, the occupation and organization of territory, economic activities, or even the symbolic representation of the environment. The analysis of them should become the basis of a modern study of territorial behaviour. But, how can we analyze them? What aspects of the archaeological record could contribute information on this matter?

A new analytical approach

To understand these interactions, we have designed an archaeo-geographical approach, whose principal foundations are for the first time presented here. It has been necessarily adapted to the characteristics of the human groups involved (hunter-gatherer societies) and to the particular limitations of the archaeological record under examination. Thus, combining interesting perspectives (especially concerning the methodology) included in relatively recent works (White 1985; Duchadeau-Kervazo 1986; Márquez and Morales 1986; Ramil 1990; Fano 1998; Cruz 2004; Julien and Connet 2005; Djindjian 2009) with others arising from our reflections, we intend to study and discuss a series of key variables on three analytical levels:

The site and its geographical and physical position
On this level are variables that provide information on the choice of the location (mainly caves and rockshelters, but also some open air places) and its suitability for habitation: geomorphological unit and sub-unit; physiographical location within the geomorphological sub-unit; main river basin; stretch within river basin; ranking within river basin; bank within river basin; absolute altitude; relative altitude; slope; orientation of the entrance; potential sunshine in the entrance.

The site and its role in the immediate environment
This level is divided into three different functions:

ECONOMIC FUNCTION

Variables that tell us about the economic resources provided by the natural environment, to be compared with the real (archaeological) use of them: proximity to water resources; potential visibility from the site; site exploitation territory (SET); accessible biotopes within the SET; proximity to main communication routes or strategic enclaves; distance to lithic raw material sources; distance to singular element sources (molluscs, amber, etc.); distance to estimated Palaeolithic sea shore.

SOCIAL FUNCTION

Variables for determining the role and hierarchy of sites within the social life of human groups: intensity of the occupation/s; length of the occupation/s; human activities carried out at the site; distance to closest contemporary sites; inter-visibility with closest contemporary sites.

\multicolumn{3}{c}{Contemporary societies of Vanuatu (Bonnemaison, 1996)}	Palaeolithic societies
Types of places	Types of spaces	Function and uses	Adaptation to archaeological events
Cultural places: sacred and mythic places.	Magic space.	Religious function: origins of lineages, receptacle of magic powers.	Space apparently not evident until the Chatelperronian, but probably emergent from the Aurignacian onwards.
Social places: places for human gathering. Habitation and household places.	Household space.	Political and residential function.	Nature and chronological length of the habitat (seasonal or permanent) and site function.
Places of use: interior plots and gardens.	Economic space.	Farming territory.	Procurement spaces.
Exterior frontiers: forest zones and natural boundaries.	Political space.	Territorial function.	Usual frequentation or not?
Itineraries: marine and terrestrial routes.	Alliance space.	Exchange function: transterritorial and transregional relations.	What kind of abroad relations?

Table 7.1: Theoretical application of the territorial system observed by Bonnemaison in contemporary societies of the Vanuatu islands to Palaeolithic territories (modified from Julien and Connet 2005, 134).

SYMBOLIC FUNCTION
Variables that tell us about the special meaning of the site regarding the surrounding landscape: monumental character of the site; visibility of the site from the surrounding landscape.

The site in its regional context
Variables on this third level are related to the role of the site beyond its immediate environment: territorial organization; settlement networks; probable communication routes.

In order to obtain information on these different variables and to analyze their contribution to our knowledge of the territorial behaviour of the last Neandertals and early Modern Humans of the Cantabrian Region (in the end the main goal of our project), a research design was also formulated, consisting of several essential stages. First, the data compilation through: a bibliographic review of the history of research and the archaeological record of each site; data compilation at the sites (*in situ*) through: a standardized Geographic Data Form designed by us (see an example in Fig. 7.1), a photographic report, and a preliminary examination of the environment, something that needed our personal visit to each cave, rockshelter and open air site; obtaining geographical data through topographic and thematic maps. Second, the creation of a database that compiles data related to the value of variables in each site. Third, data analysis, divided into three sub-phases: the statistical and cartographic (GIS-based) analysis of each variable from a diachronic point of view to observe long-term changes, and from a regional perspective, to observe geographic variability; the statistical and cartographic analysis of all the variables together, also observing diachronic and regional changes, in order to finally produce a general model for the territoriality of human groups; and the comparison of this model with the paleoenvironmental data already obtained from the archaeological record, with the existing publications relating to the territorial expression of cultural diversity (e.g. technocultural or artistic territories), and with the models of territoriality proposed by previous research.

The Cantabrian Region: an ideal geographical setting

Having defined the methodology, we must talk about the selected geographical setting. The Cantabrian Region can be defined as a biogeographic unit, located at the north-eastern part of the Iberian Peninsula, and extending along approximately 600km from the Aquitainean Plain and the Western Pyrenees in the East, to the Galician Massif in the West (Fig. 7.2). Its geographic configuration is very characteristic since, despite its length, its width almost never exceeds 70km, due to the limitation imposed by the Cantabrian Sea to the North and by the limestone barrier formed by the Cantabrian Mountain Range and the Basque Mountains to the South, with heights that commonly rise to 1,500m and sometimes even to 2,000m. Thus, it is relatively easy to go from the coast to high lands in some tens of kilometres. This gradient has enabled the rivers of the zone to form valleys with steep hillsides. These are almost always oriented from North to South and open onto the coastal plain, an axis that has been traditionally used as a way of communication (a natural corridor) between them. This layout, together with the oceanic climate characteristic of the region, has made the existence of different ecosystems (coastal, valley or mountain) possible, and in very confined spaces that human beings have inhabited and exploited for their benefit throughout history. The richness of this landscape in this exceptional space certainly provides a good basis on which to apply a territorial analysis like the one proposed above.

But if, from a geographical perspective, the region is interesting, from the archaeological one it is no less so. Thus, the great analytical possibilities of this region are due, on the one hand, to the important local research tradition into the Palaeolithic (especially on the Transition), reflected in a large body of internationally known works; and on the other, to the outstanding role of the Cantabrian Region in the Transitional process, principally focused on two aspects. First, its ample archaeological record, comprising more than 125 sites assessed from the Final Middle Palaeolithic to the Gravettian. Among these, the caves and rockshelters are of utmost importance. Their presence is favoured by the lithology of Cantabrian lands (with abundant limestone, at least in its eastern and central part). It enabled the preservation of a noteworthy human record in such well-known sites as Gatzarria, Isturitz Lezetxiki, Axlor, Covalejos, El Castillo, Morín, El Pendo, Esquilleu, Sopeña, El Sidrón, La Viña, and El Conde. This is the reason why caves and rockshelters constitute an essential part in our study. Second, the Cantabrian region is important as it is one of the two main communication routes between Europe and the Iberian Peninsula (apart from Catalonia), with possible relations to the famous records of the French Pyrenees and Aquitaine.

7. The Use of Caves and Rockshelters by the Last Neandertal and First Modern Human Societies

IDENTIFICACIÓN DEL SITIO					
NOMBRE ACTUAL	CUEVA DEL CONDE	OTRAS DENOMINACIONES	C. DEL FORNO, DE LOS GITANOS o LOS AMANTES	SIGLA ASIGNADA	CON

LOCALIZACIÓN ADMINISTRATIVA					
TERRITORIO / PROVINCIA	ASTURIAS	COMARCA	OVIEDO	MUNICIPIO	SANTO ADRIANO
TÉRMINO MUNICIPAL	TUÑON		TOPÓNIMO		
DESCRIPCIÓN ACCESO		Desde la localidad de Tuñón, tomamos la carretera As-228 en dirección a Villanueva. Alcanzada la última casa de Tuñón, a mano izquierda, asciende un sendero sobre la carretera que en pocos metros nos deja frente a la boca del abrigo.

LOCALIZACIÓN ESPACIAL					
	COORDENADAS UTM				
Huso	Banda	x	y	Datum	ED50
30	T	258 073	4797294		

ALTITUD REAL (m. s. n. m.)	ALTITUD RELATIVA 1 (sobre altura mínima entorno)	ALTITUD RELATIVA 2 (bajo altura máxima entorno)
+180	+40	

NATURALEZA Y CARACTERÍSTICAS FÍSICAS DEL SITIO					
AIRE LIBRE (A)		ABRIGO (B)	X	CUEVA (C)	
Extensión (m²)		Longitud de visera	18 m.	Nº bocas	
Desnivel total (%)		Profundidad de Visera	15 m.	Dimensiones de la/s boca/s	
Orientación ladera		Orientación	NW (320°)	Orientación de la/s boca/s	
				Longitud total (m.)	
				Nº galerías y salas	
POSICIÓN (prim. / sec.)	Primaria	EXTENSIÓN OCUPACIÓN	300 m²	POTENCIA OCUPACIÓN	1.7 m.
OTRAS DESCRIPCIONES		Posee 2 pequeñas concavidades. Conserva testigos y grabados lineales.

EMPLAZAMIENTO GEOGRÁFICO		
UNIDAD DE RELIEVE	SUBUNIDAD	DESCRIPCIÓN EMPLAZAMIENTO FISIOGRÁFICO
Planicie	Llanura / Meseta	Emplazamiento en ladera de valle. Posición privilegiada para el control del valle y del tránsito de las manadas de ungulados.
Eminencia	Loma / Colina / Cerro / Montaña / Sierra	
Depresión	Cuenca / Valle X / Garganta / Barranco / Hondonada / cubeta / dolina	

CUENCA HIDROGRÁFICA	Río Nalón	TRAMO	Medio	MARGEN	Izq.
CURSO DE AGUA MÁS PRÓXIMO	Río Trubia	DISTANCIA (m)	100 m.	RANGO	2

CARACTERÍSTICAS GENERALES DEL ENTORNO	
LITOLOGÍA Y MORFOLOGÍA	HIDROLOGÍA
Caliza de Montaña (Carbonífero inferior)	Escorrentía favorecida por la pendiente y las formaciones kársticas
FAUNA ACTUAL	FLORA ACTUAL
Especies de bosque y media montaña	Encinar Cantábrico / robledal mixto

Figure 7.1: Example of a completed Geographic Data Form (in this case applied to El Conde's Asturian cave).

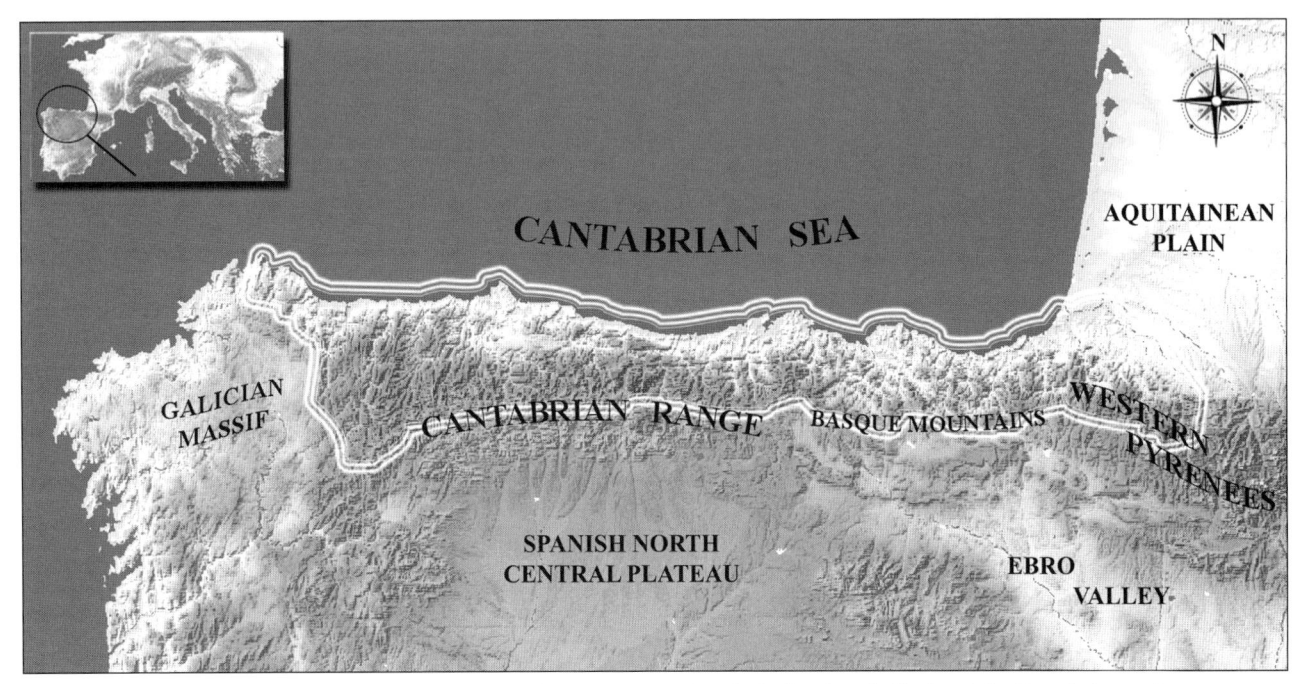

Figure 7.2: Map of the north-western part of the Iberian Peninsula, indicating the limits of the Cantabrian region.

Turning data into knowledge: some preliminary results

Below, a brief outline of results is presented. After defining the sample, we will show the individual contribution of some of the variables listed above, most of them related to the geographical position of the sites, including the general spatial distribution of the sites, their rank according to their situation within the river basins, their altitude, and the orientation of their entrances (in the case of caves and rockshelters), the rest related to economic resources, such as the proximity to water resources, accessible biotopes, and raw material supplies.

The sample

The number of archaeological sites selected for our project amounts to 127, including all those sites found within the physical boundaries of the Cantabrian Region in which human presence or action has been mentioned for the period between the Final Middle Palaeolithic and Early Upper Palaeolithic (including the Gravettian). In the case of sites whose only preserved record is rock art, those linked to an archaic style (styles I to III of Leroi-Gourhan 1965) and/or to a pre-Magdalenian chronology (both Aurignacian-Gravettian and Gravettian-Solutrean cycles) are included, despite their ever-difficult chronological attribution. We have also incorporated a small number of sites of the same period whose location in principle transgress the boundaries of the region, but which would have an outstanding role inside it, due to their location in areas with easier access to other regions and/or to the mineral and hunting wealth of their surrounding landscapes (e.g. the Upper and Middle Ebro Valley), as demonstrated in recent studies (Tarriño 2001; Sáenz de Buruaga 2004; Barandiarán and Cava 2008).

The chosen sample is quantitatively important, but at the same time is heterogeneous from both a qualitative and a chronological-geographical point of view. Concerning the quality of the sample, the information contributed by each site is very different, due to factors such as the site's own nature, the different preservation and inherent variability of the archaeological record, or the variable quality of research. This last factor, in particular the differential impact of survey in the region, could be one of the reasons for the apparent disproportion observed in the representation of the sites according to their nature: 95 of them are caves (75 per cent), 17 rockshelters (13 per cent) and 15, open air sites (12 per cent). With regard to the chronology, a very different representation of the studied periods is also found (Table 7.2). Thus, we have very well represented periods (e.g. the Aurignacian and Gravettian, with 84 and 79 sites respectively) but others with poor representation, such as the Chatelperronian (14 sites, 9 of them doubtfully attributed), a fact that could be related to the aforementioned research problems or, perhaps more likely, with an actual limited impact of these cultural *facies* in the Cantabrian Region. The same is the case with some internal phases of other periods (e.g. the final stages of the Aurignacian and Gravettian), poorly represented in our area compared to other regions (e.g. the Perigord or French Pyrenees). We must also add the traditional difficulty of placing sites within the time scale of the Middle Palaeolithic (in our case 61), and especially within the final stages of this period, and the much-discussed chrono-cultural attribution of rock art, particularly due to the lack of absolute dating (whose authority is, however, questionable). Finally, there is an important geographical variation in the distribution of sites, highlighted by the lack of sites to the West of the Nalón Valley. This is perhaps due to the absence of limestone lithologies allowing the existence of habitable caves and rockshelters, although one should not reject the idea of a possible lower population density in this area. However, there are other outstanding gaps across the Cantabrian Region, probably related to the lower intensity of survey projects (mainly reflected in the scarcity of open air sites, whose discovery is, nevertheless, difficult due to the poor archaeological visibility that is characteristic of the Cantabrian landscapes), since there is no reason to deny its occupation during the analyzed periods.

We can not ignore these biases, so it will be necessary to value the relevance and real contribution of each archaeological site. On the positive side, this heterogeneity enables us to analyze the great variety of human expressions reflected in the territory.

Geographical patterning of the sites
Spatial distribution

Regarding the spatial distribution of the sites, shown in Fig. 7.3, we have a great spread of sites from the Adour aquitainean river to the Nalón river in Asturies (note again the absence of sites to the West of it, as mentioned previously), of which the majority are concentrated in the main valleys of the Cantabrian region, to the North of the Cantabrian mountain range.

Period	Caves	Rockshelters	Open air	Total
Middle Palaeolithic	43	10	8	61
Chatelperronian	11	1	2	14
Aurignacian	67	11	6	84
Gravettian	62	10	7	79

Table 7.2: Distribution of sites by period and by nature. (Note that in a significant number of sites several periods are represented.)

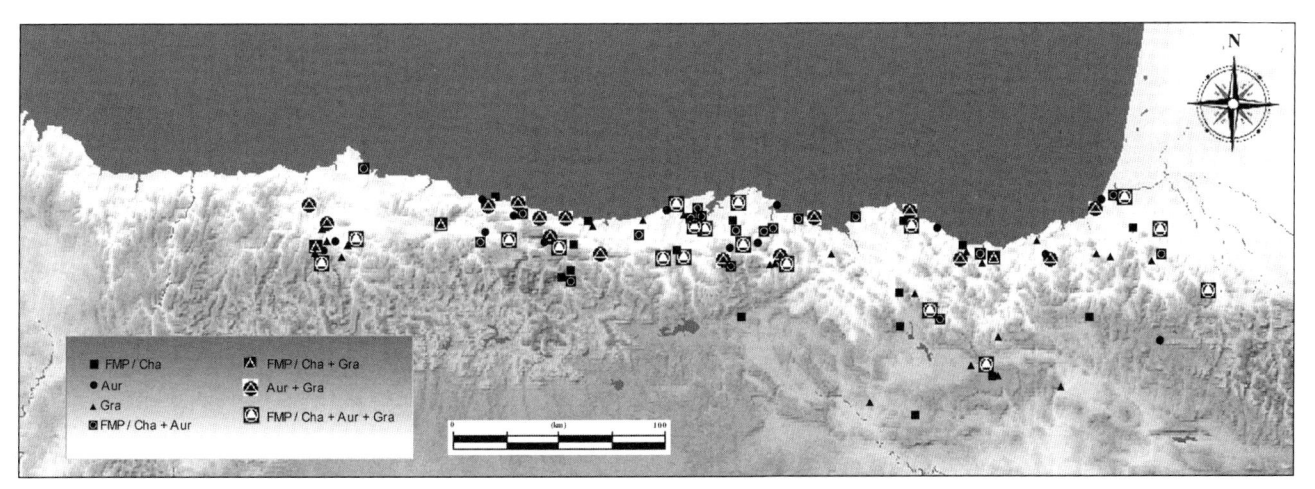

Figure 7.3: Map of the Cantabrian region with the sites analyzed in this project. They are differently represented depending on the periods contained in their archaeological record.

Within this overall pattern, we can observe an apparent continuity between the Final Middle and Early Upper Palaeolithic settlement distribution, especially in the central part of the Cantabrian Region (where there are many clearly defined sites with evidence of recurrent habitation or use, like Sopeña and Llonín in eastern Asturies; or Hornos de la Peña, El Castillo, El Pendo, Morín, and La Garma in Cantabria). On its western and easternmost sides this continuity is less common (excepting cases like La Viña and El Conde in Asturies, Lezetxiki in the Basque Country, and, perhaps, Coscobilo in Navarre). This tendency may be related to a more intensive settlement of the central area, being more diffuse at both ends, though we cannot overlook the importance of areas like the Nalón Valley or the Basque coast. Only to the East, in the northwestern foothills of the Pyrenees, do we find other important sites exhibiting long sequences (such as Isturitz, Le Basté, or Gatzarria), but these were possibly more influenced by the population streams coming from the Aquitainean plain (Perigord) and the Pyrenees. On the other hand, there are few sites containing Middle Palaeolithic/Chatelperronian and Gravettian levels without intermediate Aurignacian ones (perhaps the basque cave of Amalda is the clearest case). More common is the successive pairing of Middle Palaeolithic/Chatelperronian followed by Aurignacian (at sites such as La Güelga, Arnero, Covalejos, El Ruso, Otero, or Labeko koba), and Aurignacian followed by Gravettian (as at Salitre, El Cuco, Antoliña, or Aitzbitarte), which seems to endorse the reoccupation tendency observed between periods. Likewise, we find a low percentage (around 40 per cent of the total, including doubtful sites) with a single period represented, of which 15 per cent are short-term open air occupations.

As for more concrete differences in the spatial distribution, it seems that there are many more Middle Palaeolithic and Gravettian sites in the uplands (to the South), a fact that could be related to better climatic conditions or, as in the case of the Basque Country, to the minor barrier effect of the southern mountains and the exploitation of inland raw material sources. By contrast, no clear Aurignacian sites, excepting the strange case of Zatoya with a level (IIbam) dated to 28,870 +760 -690 BP (Barandiarán and Cava 2001, 68) and located at almost 1,000m above sea level in the southern foothills of the Pyrenees, appear in the mountainous and supposed steppe inlands, even in the Basque case (since doubt remains about the disappeared site of Coscobilo, which was related to the exploitation of Urbasa's flint).

Despite these interesting data, we must note that this general view is conditioned by the existence of a great lack in the research of some areas, as previously mentioned. We especially need to take into account the consequences of Quaternary sea level fluctuations: major studies (Pujos 1976; Cearreta et al. 1992; Siddall et al. 2003) estimate the shore line of the interval 50,000–20,000 BP to be between -50 and -100m below sea level, so we must acknowledge that the majority of the sites of the period (both caves and rockshelters and especially open air sites), together with the Palaeolithic coastal plain, are now surely under water.

Ranking of the sites within the river basins
This variable is in our opinion a very interesting approach for the study of settlement patterns, but has been rarely used in research, although pioneering studies (White 1985; Duchadeau-Kervazo 1986) and more systematic works (Cruz 2004) can be found. It takes into account the position of the sites with regard to the main and secondary river basins, since settlement choice depends on the importance of the rivers (as natural ways of communication and prey movement). Thus, several categories of sites are established. In our case these are: Category 1 – sites located on the banks of the main river or less than 1.5km away from it, and those placed by large tributary rivers (those contributing more than 33 per cent of the flow to the main river basin); Category 2 – sites located by tributary rivers (except the large ones) or less than 1km away from them and those situated in the sub-tributaries of these large tributary rivers; Category 3 – sites located by sub-tributary rivers or less than 0.5km from them and those placed by the sub-sub-tributary rivers of the large tributary ones; Category 4 – sites located by sub-sub-tributary rivers or within their immediate environment.

The results of applying these categories to the studied record are interesting. By period, there is a different distribution: Middle Palaeolithic and Chatelperronian sites are predominantly ascribed to ranks 2 (47 per cent) and 1 (38 per cent), much more than the third (13 per cent) and fourth (2 per cent), while for Aurignacian and Gravettian sites rank 1 predominates (54 and 48 per cent respectively) over the second (38 and 41 per cent), with similar values regarding the third and fourth (less than 10 per cent). Interestingly, comparison of these results with those obtained by R. White (1985, 97–103) for the Early Upper Palaeolithic in the Perigord allows us to observe great similarities: with Chatelperronian (Lower Perigordian) sites located mostly on the tributary rivers (66 per cent), while from the Aurignacian onwards, a preference for principal valleys is very evident (more than 50 per cent).

It is difficult to explain the observed differences if we do not pay attention to the analysis of other variables, such as palaeoeconomic ones, although in principle the results could indicate perhaps more specialized settlement patterns in the early stages of the Upper Palaeolithic, directed to a greater and more effective control of communication routes with surely economic or, possibly, other aims. By contrast, Middle Palaeolithic and Chatelperronian sites appear to exhibit a less marked pattern, even favouring the occupation of secondary rivers, although this could be due to different (not 'worse') economic strategies compared to the Upper Palaeolithic ones.

Altitude of the sites

Regarding the altitude of the sites, we can say that generally the majority (83 per cent) of the caves and rockshelters are located under 300m above sea level, but observing the chart by period (Fig. 7.4), we can also recognize some differences. Under 300m above sea level, we observe a similar tendency in the position of Final Middle Palaeolithic/Chatelperronian and Aurignacian sites: with a great number of them under 100m (near the coast), an outstanding decrease between 100 and 150m, an increase between 150–200m, and then a normal progressive decrease of sites at higher altitudes. This pattern would be logical were it not for the first decrease, and it could perhaps relate to the dichotomy between winter camps (under 100m) and summer camps (up to 150m) traditionally argued by researchers to explain the annual mobility of hunter-gatherers within the Cantabrian Region. By contrast, Gravettian sites have a significantly smaller percentage of sites under 100m, with an increase (instead of a decrease) up to 150–200m, and then the same progressive decrease. Then, up to 300m the situation changes again, with an almost total absence of Aurignacian sites (a few clear examples being Sopeña, Salitre, Lezetxiki, and perhaps Coscobilo and Zatoya), but a more important number of Final Middle Palaeolithic (but not Chatelperronian) and Gravettian ones, especially up to 500m, which practically join the inland sites of the Basque Country and Navarre.

It is especially interesting to note the Gravettian preference for higher altitudes (an average of 197m against, e.g. 148m for Aurignacian sites). This could easily be related to favourable climatic conditions (a factor that could equally explain the presence of Final Middle Palaeolithic sites at higher altitudes) which would lead Gravettian groups to settle inland to the detriment of the coastlands. However, in my opinion this view seems simplistic. This better climate is not evidenced at all in the paleoecological record (especially when the majority of the dated occupations coincide with the Kesselt-Tursac stadial, placed between 26/25,000 and 23,000 BP, and their final stages with

ALTITUDE OF THE SITES

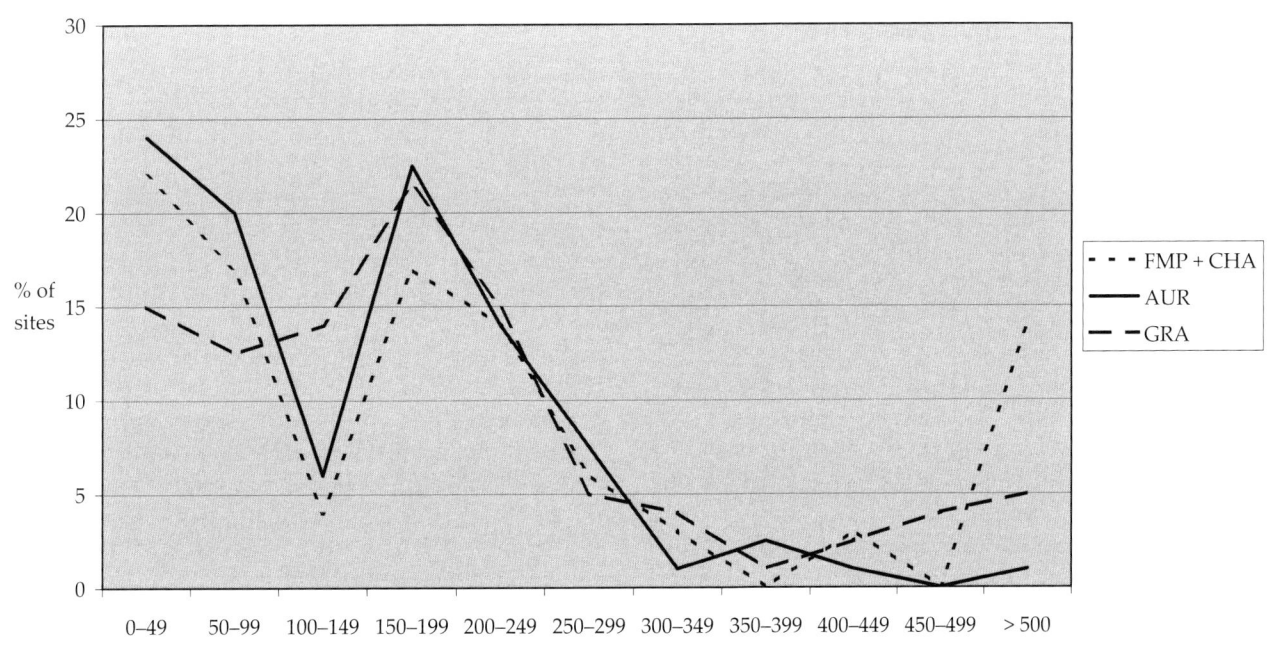

Figure 7.4: Altitude (meters above sea level) of the sites compared by period.

the Last Glacial Maximum, after 22,000/21,500 BP). Furthermore, we have to take into account the bias created by sites being covered by the Holocene sea transgression, that surely had an outstanding importance for Gravettian societies, as proved by the great increase in the production of ornaments made of marine molluscs (Álvarez 2007). That is why Gravettian settlement could be perhaps more closely related to a better and wider occupation and control of the uplands and their resources (e.g. flint supplies) and/or even to an increase in population density. Aurignacian sites, however, appear to be more spatially concentrated in low-middle lands, something that can also be said of the few Chatelperronian sites (in contrast to the greater dispersal of Final Middle Palaeolithic sites), but we must consider whether or not it was really related to a more restricted territoriality.

The comparison of these results with well-known studies (e.g. Demars 2000) does not seem that effective, since the geographies studied by them and, therefore, the significance of the altitude of the sites is different from the Cantabrian Region, an exceptional setting.

Orientation of the entrances of the caves and rockshelters
Concerning the orientation of the caves and rockshelters (Figs 7.5–7.7), in the Final Middle Palaeolithic and Chatelperronian no well-defined pattern can be identified, though caves are commonly oriented South/Southeast. The substantial chronological duration of the Middle Palaeolithic involves the presence of numerous climatically stable periods (one of them curiously between 60,000 and 40,000 BP) in contrast to what occurs in the Upper Palaeolithic: something that could explain the habitation of a large percentage of less favourably oriented (W, NW, N and NE) sites, since it is illogical to think that Neandertal groups did not worry about being protected from a harsher climate.

By contrast, we find more similar patterns for the Gravettian and Aurignacian sites, which are mainly oriented Southwest and South/South-Southeast. For sure these results are not unique, since one must suppose that Early Upper Palaeolithic groups had the same requirement of protection in such unstable climatic conditions, and therefore looking for the best protected locations. But in exploring the reason for the two main different preferences (SW *vs.* S/SSE), it is possible to compare the results with the altitude of the sites, and to observe that Southwest oriented ones are more common in the lowlands (at an average of 155m above sea level) while caves situated inland in more inhospitable areas are south/southeast oriented and better protected from the coldest winds.

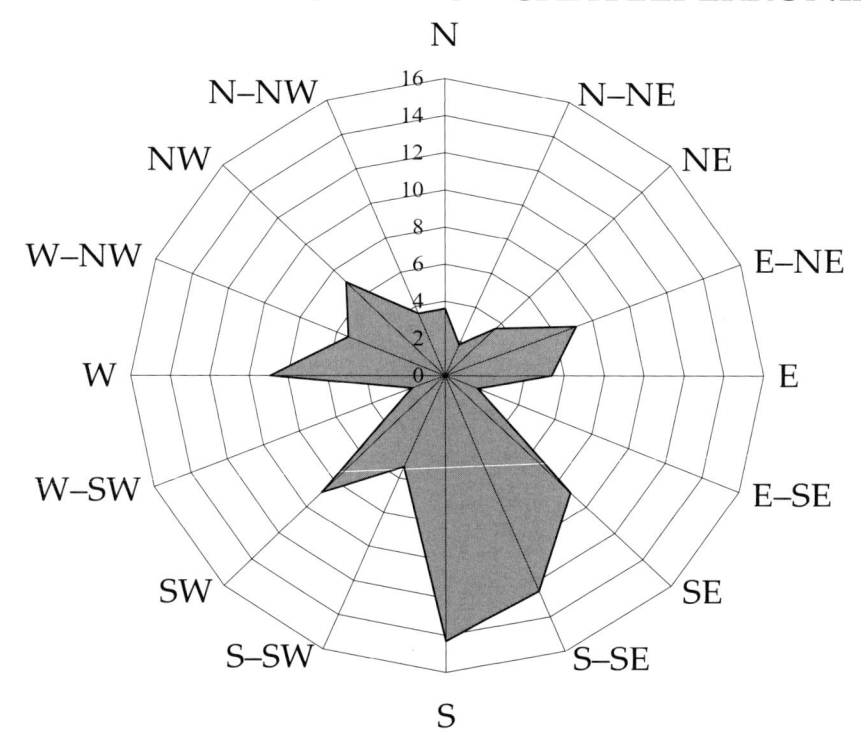

Figure 7.5: Orientation of Middle Palaeolithic and Chatelperronian caves and rockshelters.

AURIGNACIAN

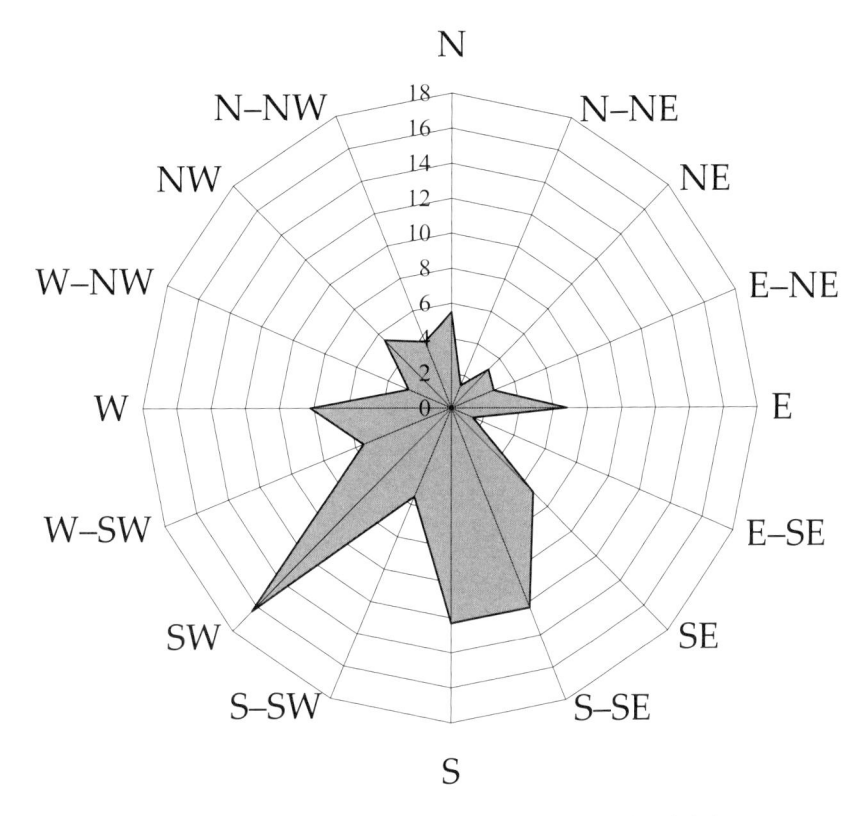

Figure 7.6: Orientation of Aurignacian caves and rockshelters.

GRAVETTIAN

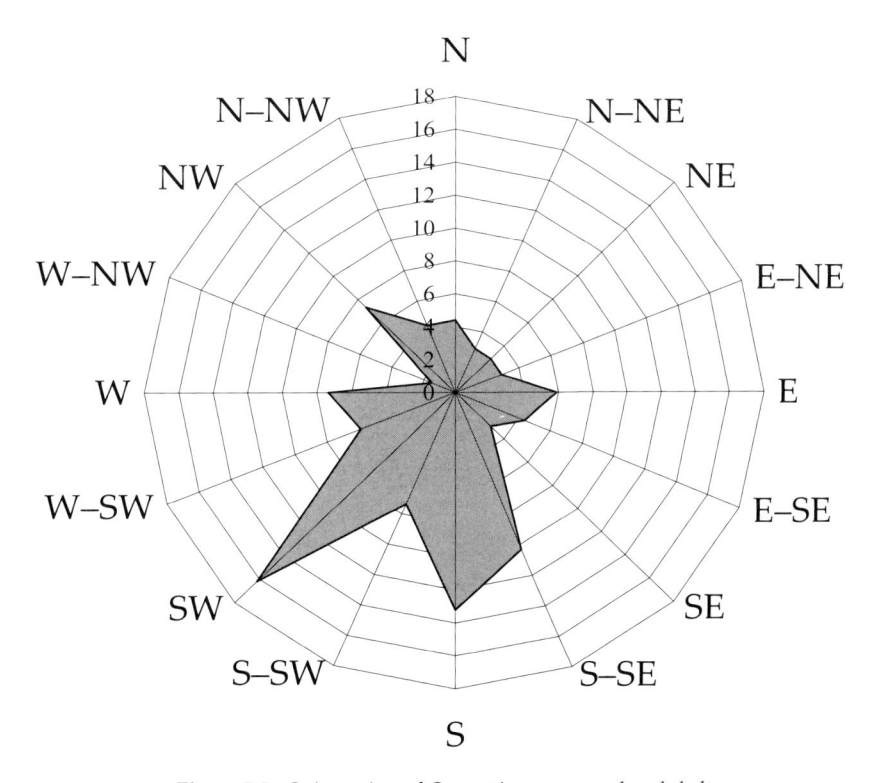

Figure 7.7: Orientation of Gravettian caves and rockshelters.

On the other hand, almost all of the unfavourably oriented caves and rockshelters are located at low altitudes (e.g. Cueva Oscura, Arnero, Morín) or are usually protected at the front by higher hillsides (Chufín, El Ruso, Otero, Axlor, Aitzbitarte, Olha), which helped them resist the rigours of the climate. The rest are related to sanctuaries, if we consider the archaeological record (mainly rock art) found inside them (Candamo, Salitre, Covalanas, Pondra, Arenaza), with few signs of habitation. But we cannot discard the possible use of protective structures (e.g. windbreaks) at other sites, which unfortunately are not usually evident in the archaeological record. Finally, these results should be compared with those obtained from the study of potential sunshine, since this is a factor that also influenced the choice of sites.

Palaeoeconomic resources

Distance to nearest water sources

Analyzing the distance of the sites to their nearest rivers, it can be stated that they are usually not far from water sources (always less than 800m distance), but usually less than 200m (68 per cent). This seems logical, considering that water is surely the first necessary resource for biological subsistence. For instance, we note a similar tendency when comparing periods (Fig. 7.8), apart from some variations due to the topographical position of some sites at higher altitudes, which consequently lie further from water courses. Similar results have been provided by other studies (White 1985; Duchadeau-Kervazo 1986).

Accessible biotopes

Although we are not yet in a position to present the results of the analysis concerning site exploitation territory (SET) and its real use manifested in archaeological faunas, we can draw some general interpretations of the biotopes economically useful surrounding each site, mainly in relation to their geographical position. Establishing a radius of 10km around the sites, we find no significant differences by period concerning the possible accessible biotopes, which can be divided into three basic categories (coast, plain-valley, and mountain). From each site there are at least two biotopes within a few kilometres (a fact favoured by the singular Cantabrian mountain geography), which provided great ecological and

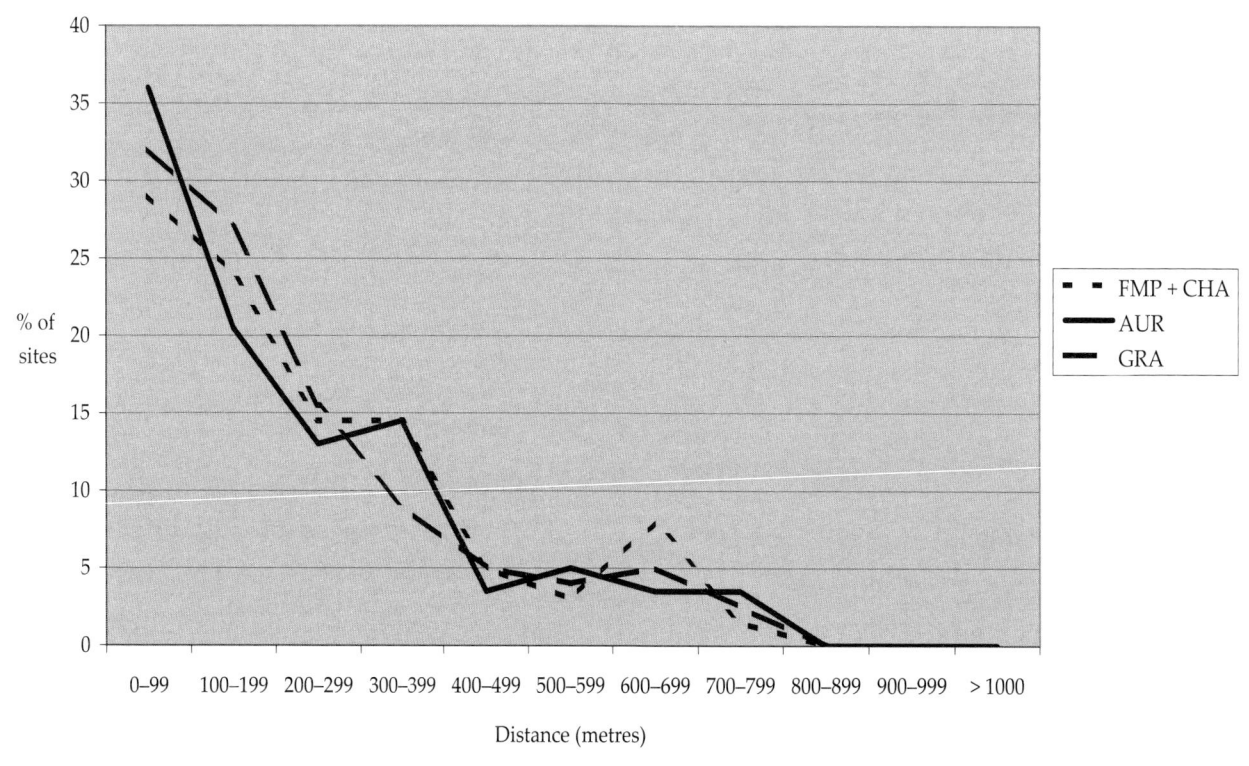

Figure 7.8: Distance from the sites to the nearest watercourses compared by period.

economic diversity for human groups. The percentage of sites with access to the valley-mountain pairing is important, because we know better those sites located inland than those of the original coast, unfortunately currently submerged under water. We also find a significant percentage (more than 20 per cent) of sites with three accessible biotopes (coast-valley-mountain), which reinforces the idea of an intentional choice of habitats situated not far from the coast, but which always looked to the protection of the valleys.

Of course, the real use of biotopes at each site would not comply with this general overview, since faunal exploitation depends more on human preferences and different economic strategies, the study of which is an essential part of this research project.

Lithic raw material supply
In recent years new studies have promoted a qualitative change in our knowledge of the mobility involved in the acquisition of lithic raw materials, especially flint, breaking with the traditional idea of Palaeolithic groups depending on local resources (Geneste 1985; Demars 1994; Féblot-Augustins 1998; Terradas 2002).

The study of the eastern part of the Cantabrian Region (approximately eastern Cantabria, the Basque Country, Navarre, and the Atlantic French Pyrenees), where high quality flint is found, has revealed the main sources offered by this territory (Sarabia 1999; Tarriño 2001) and some interesting patterns: in the Final Middle Palaeolithic it was more usual to exploit local flint (without excluding occasional long-distance exploitation), although other raw materials were also of some importance, sometimes greater than that of flint (Carrión *et al.* 1998, 90). Only in the final stages of the Middle Palaeolithic can we observe an increase in the percentages of flint (e.g. at sites such as Morín, El Ruso, Axlor, or Lezetxiki) (Sarabia 1990, 444), whereas in the Early Upper Palaeolithic a progressively greater preference and mobility for flint and a much lower importance of other raw materials (usually less than 5–10 per cent) has been recognized (Cabrera and Bernaldo de Quiros 1993, 177; Arrizabalaga 1998, 98).

A very preliminary observation on the distance of these eastern sites to their main flint sources (the most strongly represented in the archaeological record of each site) tells us that the main sources are usually located less than 20km from the sites, indicating a preference for local or semi-local flint. Nonetheless, there are low percentages of high quality flint (normally used in the production of specific tools) whose sources lie over 100km away (Fig. 7.9). This is not strange if we consider that in some Aquitainean Aurignacian (Tarriño and Normand 2002; Bordes *et al.* 2005) and Central French Pyrenean Gravettian sites (Foucher and San Juan 2005) flint was carried over distance of as much as 200–250km. The question of whether this exploitation of flint sources is based on long journeys linked to a wide mobility of human groups or to repeated intergroup exchanges (Bordes *et al.* 2005, 195) will be one of the main ones to be addressed by future research.

However, we still lack information on the sources of other lithic raw materials (e.g. quartzite, quartz, ophite, sandstone, or limestone), which were of great importance in the Middle Palaeolithic, especially at the sites of Asturies and western Cantabria (even Upper Palaeolithic ones) where flint is scarce. Only in these cases can we say that the exploitation patterns depend on the lithological richness of the environment. The usual local character of these raw materials disguises the real value of human mobility in these areas (better represented by flint procurement), but they provide an excellent source of information about the exploitation of the territory and, above all, about technological behaviour.

Conclusions

The analysis of human territoriality is a top priority issue within Palaeolithic research. Nowadays it is essential to promote research projects that take into account all the dimensions composing the concept of *territoriality*, defined by us as the interactions that take place when the human being appropriates or socializes his natural environment. For this reason, we have presented a project that tries to include different useful perspectives and methods contributed by some well-known publications and others arising from our reflection, in an attempt to achieve a more complete archaeo-geographical view of Palaeolithic territories. The study is essentially focused on the analysis of the value of several variables whose nature is very different (e.g. geo-topographic, economic, social or even symbolic). The statistical and cartographical treatment of them will allow us to produce territorial models to be compared with others proposed by previous research.

The Cantabrian region offers an excellent setting to apply our project to the knowledge of the territorial behaviour of Palaeolithic human groups, especially in the Middle-to-Upper Palaeolithic Transition, because of its ample archaeological record and its geographical character. The data analyzed to date enable us to outline some provisional statements regarding territorial behaviour in the Transitional period.

First, the use of caves and rockshelters seems to be more common than the occupation of open air

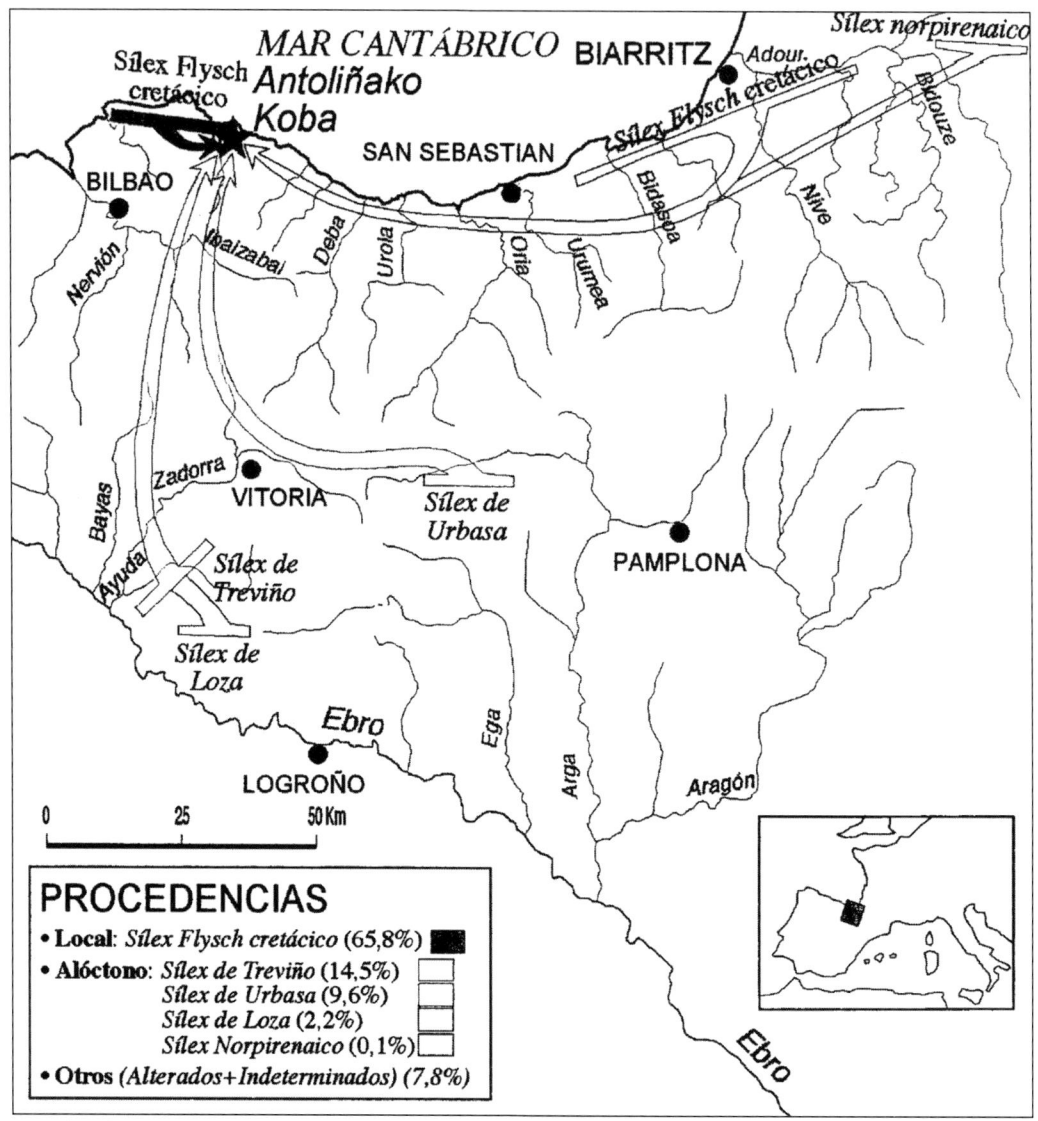

Figure 7.9: Example of a modern study of raw material procurement: main flint sources exploited by the Palaeolithic inhabitants of Antoliña's Basque archaeological cave (after Tarriño and Aguirre 2002, 16 – reproduced with permission).

sites in both the Final Middle Palaeolithic and the Early Upper Palaeolithic periods. The increase in the number of Gravettian open air sites could be a sign of different settlement and territorial patterns, more than an indication of real climatic amelioration. But this overview suffers from gaps related to the different impact of survey projects and the effects of the Holocene sea transgression.

Second, we find a quite important continuity in the distribution of caves (at least in the middle part of the Cantabrian Region), amongst which some well-known sites with recurrent evidence of habitation stand out. Those sites representing just one period are fewer and are usually represented by short-term open air occupations.

Third, with regard to the topographical position of the sites, we find some differences between periods. Concerning the ranking of the sites within the river basins, we note that Middle Palaeolithic sites are mainly located in secondary valleys while Early Upper Palaeolithic ones prefer principal rivers, a fact that could be related to a progressive specialization in settlement patterns perhaps intended for a better control of human and faunal communication routes. Also, significant variations regarding the altitude of the sites are evident, although the general tendency is for them to lie under 300m above sea level. Thus, we note the presence of more Middle Palaeolithic and Gravettian sites at higher altitudes – something that in the second case could be due to a wider territorial occupation or even to an increase in population density, if we dismiss the climatic amelioration as the main cause.

Fourth, the differences between Middle Palaeolithic and Early Upper Palaeolithic in the orientation of the caves and rockshelters could indicate better climate conditions in the Middle Palaeolithic (something which will be analyzed in relation to the palaeoenvironmental record) rather than different patterns in the selection of habitats. Early Upper Palaeolithic sites follow a similar trend, with a predominance of Southwest and Southeast oriented ones offering the best protection against the rigours of the climate.

Fifth, we do not observe too many differences in the selection of the sites in relation to distance to water sources and accessible biotopes, which, without forgetting the important role of plant resources, actually constituted the basic subsistence elements for those hunter-gatherer groups and, for this reason, had to be close to the sites. The future comparison of these data with the real use of faunal resources (represented in the archaeological record) will be essential.

Sixth, it seems that Palaeolithic groups (mainly Upper Palaeolithic ones) had an ample mobility for searching for and transporting of lithic raw materials, especially flint. However, a progressive specialization on the use of flint (as opposed to other raw materials) can be identified from the first stages of the Upper Palaeolithic. In the western part of the Cantabrian Region, the local geology, which is scarce in flint sources, disguises the mobility of human groups.

To conclude, it is hoped that the future development of this research project, analyzing many other variables and comparing their contribution to other data (e.g. technology, art, ornaments) and theoretical models concerning Palaeolithic human territoriality, will enable us to better characterize the changes that took place in the territorial behaviour of both Final Middle and Early Upper Palaeolithic societies.

Acknowledgements

This article was produced as a consequence of the author's receipt of a Pre-Doctorate Grant for Researchers (BFI05.431) provided by the Basque Government during the years 2005–9. I also want to thank Dr. Álvaro Arrizabalaga for his support, advice, and corrections, and, at the same time, all the other members of the *Grupo de Investigación Consolidado y de Alto Rendimiento de Prehistoria* (9/UPV 155.130-14570/2002) of the University of the Basque Country for their constantly useful suggestions.

References

Altuna, J. and Mariezkurrena, K. (1988) Les macromammifères du Paléolithique moyen et supérieur ancien dans la région Cantabrique. *Archaeozoologia* 1(2), 179–196.

Álvarez, E. (2007) La explotación de los moluscos marinos en la Cornisa Cantábrica durante el Gravetiense: primeros datos de los niveles E y F de la Garma A (Omoño, Cantabria). *Zephyrus* 60, 43–58.

Arrizabalaga, A. (1998) El aprovisionamiento en materias primas líticas durante la génesis del Leptolítico: el Cantábrico oriental. *Rubricatum* 2, 97–104.

Arrizabalaga, A., Bon, F., Maíllo, J. M., Normand, C. and Ortega, I. (2007) Territoires et frontières de l'Aurignacien dans les Pyrénées occidentales et les Cantabres. In N. Cazals, J. E. González Urquijo and X. Terradas (eds.) *Frontières Naturelles et Frontières Culturelles dans les Pyrénées Préhistoriques*. Monografías del Instituto Internacional de Investigaciones Prehistóricas de Cantabria 2, 301–318. Santander, PubliCAN-Ediciones de la Universidad de Cantabria.

Ashton, N. M. (1983) Spatial patterning in the Middle-Upper Palaeolithic transition. *World Archaeology* 15, 224–235.

Bahn, P. G. (1984) *Pyrenean Prehistory: A Palaeoeconomic Survey of the French Sites*. Warminster, Aris and Phillips.

Barandiarán, I. and Cava, A. (2001) El Paleolítico superior de la Cueva de Zatoya (Navarra): actualización de datos en 1997. *Trabajos de Arqueología de Navarra* 15, 5–99.

Barandiarán, I. and Cava, A. (2008) Identificaciones del Gravetiense en las estribaciones occidentales del Pirineo: modelos de ocupación y uso. *Trabajos de Prehistoria* 65, 13–28.

Bar-Yosef, O. (2002) The Upper Palaeolithic revolution. *Annual Review of Anthropology* 31, 363–393.

Blades, B. S. (1999) Aurignacian settlement patterns in the Vézère Valley. *Current Anthropology* 40(5), 712–719.

Bonnemaison, J. (1996) *Gens de Pirogue et Gens de la Terre. Les Fondements Géographiques d'une Identité. L'Archipel du Vanuatu. Essai de Géographie Culturelle. Livre 1*. Paris, Éditions de l'Orstrom.

Bordes, J-G., Bon, F. and Le Brun-Ricalens, F. (2005) Le transport des matieres premieres lithiques à l'Aurignacien entre le Nord et le Sud de l'Aquitaine: faits attendus, faits nouveaux. In J. Jaubert and M. Barbaza (eds.) *Territoires, Déplacements, Mobilité, Échanges durant la Préhistoire. Actes des 126 Congrés Nationaux des Societés Historiques et Scientifiques (Toulouse, 2001)*, 185–198. Paris, Éditions du Comité des Travaux Historiques et Scientifiques.

Bourgeot, A. (1991) Territorio. In P. Bonte and M. Izard (eds.) *Diccionario de Etnología y Antropología*, 698–699. Madrid, Akal.

Bracco, J-P. (2005) De quoi parlons-nous? Réflexions sur l'appréhension des territoires en Préhistoire Paléolithique. In J. Jaubert and M. Barbaza (eds.) *Territoires, Déplacements, Mobilité, Échanges durant la Préhistoire. Actes des 126 Congrés Nationaux des Societés Historiques et Scientifiques (Toulouse, 2001)*, 13–16. Paris, Éditions du Comité des Travaux Historiques et Scientifiques.

Brace, C. L. (1962) Refocusing on the Neanderthal Problem. *American Anthropologist*, new series 64(4), 729–741.

Bressy, C., Burke, A., Chalard, P. and Martin, H. eds. (2006) *Notions de Territoire et de Mobilité. Exemples de l'Europe et des Premières Nations en Amérique du Nord avant le Contact Européen*. Études et Recherches Archéologiques de l'Université de Liège (ERAUL) 116. Liège, Université de Liège.

Cabrera, V. and Bernaldo de Quirós, F. (1993) L'Aurignacien de la région Cantabrique Espagnole. In L. Bánesz and J. K. Kozlowski (eds.) *Aurignacien en Europe et au Proche Orient. Actes du XIIe Congrés International de l'Union International des Sciences Préhistoriques et Protohistoriques (Bratislava), vol. 2*, 173–181. Bratislava, Institut Archéologique de l'Académie Slovaque des Sciences.

Cabrera, V., Pike-Tay, A. and Bernaldo de Quirós, F. (2004) Trends in Middle Palaeolithic settlement in Cantabrian Spain: the Late Mousterian at Castillo Cave. In N. J. Conard (ed.) *Settlement Dynamics of the Middle Palaeolithic and Middle Stone Age*, 437–460. Tubingen, Kerns Verlag.

Cabrera, V., Pike-Tay, A., Lloret, M. and Bernaldo de Quirós, F. (2000) Continuity patterns in the Middle-Upper Palaeolithic transition in Cantabrian Spain. In C. B. Stringer, R. N. E. Barton and C. Finlayson (eds.) *Neanderthals on the Edge*, 85–93. Oxford, Oxbow Books.

Camps, M. (2009) Where there's a will there's a way? 30 years of debate on the Mid-Upper Palaeolithic transition in Western Europe. In M. Camps and C. Szmidt (eds.) *The Mediterranean from 50 000 to 25 000 BP: Turning Points and New Directions*, 1–10. Oxford, Oxbow Books.

Carrión, E., Baena, J. and Conde, C. (1998) Aprovisionamiento de materias primas en el Paleolítico medio de Cantabria. *Rubricatum* 2, 89–96.

Cearreta, A.; Edeso, J. M.; Ugarte, F. M. (1992) Cambios del nivel del mar durante el Cuaternario reciente en el Golfo de Bizkaia. In A. Cearreta and F. M. Ugarte (eds.) *The Late Quaternary in the Western Pyrenean Region*, 57–94. Bilbao, Universidad del País Vasco.

Chase, P. C. (1989) How different was Middle Palaeolithic subsistence? A zoological perspective on the Middle and Upper Palaeolithic transition. In P. Mellars and C. Stringer (eds.) *The Human Revolution: Behavioural and Biological Perspectives in the Origins of Modern Humans*, 321–337. Edinburgh, Edinburgh University Press.

Clark, G. A. (1992) Continuity or replacement? Putting modern human origins in an evolutionary context. In H. Dibble and P. Mellars (eds.) *The Middle Palaeolithic: Adaptation, Behaviour and Variability*, 183–205. Philadelphia, University of Pennsylvania.

Clark, G. A. and Lindly, M. (1989) The case of continuity: observations on the biocultural transition in Europe and Western Asia. In P. Mellars and C. Stringer (eds.) *The Human Revolution: Behavioural and Biological Perspectives in the Origins of Modern Humans*, 626–676. Edinburgh, Edinburgh University Press.

Coward, F. (2004) *Transitions, Change and Identity: The Middle and Upper Palaeolithic of Vasco-Cantabrian Spain*. Unpublished Doctoral Thesis, Southampton, University of Southampton.

Cruz, M. (2004) *Paisaje y Arte Rupestre: Ensayo de Contextualización Arqueológica y Geográfica de la Pintura Levantina*. Unpublished Doctoral Thesis, Madrid, Universidad Complutense.

Dari, A. (2003) *Comportements de Subsistance pendant la Transition Paléolithique Moyen- Paléolithique Supérieur en Cantabrie à partir de l'Étude Archéozoologique des Restes Osseux des Genes Mammiferes de la Grotte d'El Castillo*. Unpublished Doctoral Thesis, Paris, Muséum National d'Histoire Naturelle.

Demars, P.-Y. (1994) *L'economie du silex au Paléolithique superieur dans le Nord de l'Aquitaine*. Thèse de Doctorat d'Etat, Bordeaux, Université de Bordeaux.

Demars, P.-Y. (1998) Circulation des silex dans le Nord de l'Aquitaine au Paleolithique superieur. L'occupation de l'espace par les derniers chasseurs-cueilleurs. *Gallia Préhistoire* 40, 1–28.

Demars, P.-Y. (2000) Altitude des sites suivant le climat au Paleolithique superieur et au Mesolithique en France. *Praehistoria* 1, 47–54.

Djindjian, F. (2009) Le concept de territoires pour les chasseurs cueilleurs du Paléolithique supérieur européen. In F. Djindjian, J. K. Kozlowski and N. Bicho (eds.) *Le Concept de Territoires dans le Paléolithique Supérieur Européen. Proceedings of the XV World Congress of the UISPP (Lisbon, 4–9 September 2006), Volume 3*. British Archaeological Reports International Series 1938, 3–25. Oxford, Archaeopress.

Djindjian, F., Kozlowski, J. K. and Bicho, N. eds. (2009) *Le Concept de Territoires dans le Paléolithique Supérieur Européen. Proceedings of the XV World Congress of the UISPP (Lisbon, 4–9 September 2006), Volume 3*. British Archaeological Reports International Series 1938. Oxford, Archaeopress.

Duchadeau-Kervazo, C. (1986) Les sites paléolithiques du bassin de la Dronne (nord de l'Aquitaine). Observations sur les modes et emplacements. *Bulletin de la Société Préhistorique Française* 83(2), 56–64.

d'Errico, F., Zilhão, J., Julien, M., Baffier, D. and Pelegrin, J. (1998) Neandertal acculturation in Western Europe? A critical review of the evidence and its interpretation. In R. G. Fox (ed.) *The Neanderthal Problem and the Evolution of Human Behavior. Current Anthropology* 39(3) (sup.), 1–44.

Fano M. (1998) *El Hábitat Mesolítico en el Cantábrico Occidental. Transformaciones Ambientales y Medio Físico durante el Holoceno Antiguo*. British Archaeological Reports International Series 732. Oxford, Archaeopress.

Féblot-Augustins, J. (1993) Mobility strategies in the Late Middle Palaeolithic of Central Europe and Western Europe: elements of stability and variability. *Journal of Anthropological Archaeology* 12, 211–265.

Féblot-Augustins, J. ed. (1997) *La Circulation des Matieres Premières au Paléolithique: Synthèse des Données, Perspectives Comportementales*. Études et Recherches Archéologiques de l'Université de Liège (ERAUL), 75. Liège, Université de Liège.

Fortea, J. (1994) Los 'santuarios' exteriores en el Paleolítico cantábrico. *Complutum* 5, 203–220.

Foucher, P. and San Juan, C. (2005) La circulation des matières siliceuses dans le Gravettien pyrénéen. In J. Jaubert and M. Barbaza (eds.) *Territoires, Déplacements, Mobilité, Échanges durant la Préhistoire. Actes des 126 Congrés Nationaux des Sociétés Historiques et Scientifiques (Toulouse, 2001)*, 199–217. Paris, Éditions du Comité des Travaux Historiques et Scientifiques.

García, M. and Eguizábal, J. (2008) *La Cueva de Venta Laperra. El Grafismo Parietal Paleolítico y la Definición de Territorios Gráficos en la Región Cantábrica*. Bilbao, Ayuntamiento de Carranza.

Geneste, J-M. (1985) *Analyse Lithique d'Industries Mousteriennes du Périgord: Une Approche Technologique du Comportement des Groupes Humaines au Paléolithique Moyen*. Thése de Doctorat, Bordeaux, Université de Bordeaux.

Geneste, J.-M. (1988) Systemes d'approvisionnement en matières premières au Paléolithique moyen et au Paléolithique superieur en Aquitaine. In J. K. Kozlowski (ed.) *L'Homme de Néandertal. Volume 8: La Mutation*. Études et Recherches Archéologiques de l'Université de Liège (ERAUL) 35, 61–71. Liège, Université de Liège.

Grayson, D. K. and Delpech, F. (1998) Changing diet breadth in the Early Upper Palaeolithic of Southwestern France. *Journal of Archaeological Science* 25, 1119–1129.

Harrold, F. B. and Otte, M. (2001) Time, space and cultural process in the European Middle-Upper Palaolithic transition. In M. A. Hays and P. T. Thacker (eds.) *Questioning the Answers: Re-solving the Fundamental Problems for the Early Upper Palaeolithic*. British Archaeological Reports International Series 1005, 3–11. Oxford, Archaeopress.

Jaubert, J. and Barbaza, M. (eds.) (2005) *Territoires, Déplacements, Mobilité, Échanges durant la Préhistoire. Actes des 126 Congrés Nationaux des Societés Historiques et Scientifiques (Toulouse, 2001)*. Paris, Éditions du Comité des Travaux Historiques et Scientifiques.

Julien, M. and Connet, N. (2005) Espaces, territoires et comportements des Châtelperroniens et Aurignaciens de la Grotte du Renne à Arcy-Sur-Cure (Yonne). In D. Vialou, J. Renault-Miskovsky and M. Patou-Mathis (eds.) *Comportements des Hommes du Paléolithique Moyen et Supérieur en Europe: Territoires et Milieux. Actes du Colloque du GDR 1945 du Centre National de la Recherche Scientifique (Paris, 2003)*. Études et Recherches Archéologiques de l'Université de Liège (ERAUL) 111, 133–146. Liège, Université de Liège.

Klein, R. G. (1994) The problem of modern human origins. In M. Nitecki and D. Nitecki (eds.) *Origins of Anatomically Modern Humans*, 3–17. New York, Plenum Press.

Laplace, G. (1962) Recherches sur l'origine et l'evolution des complexes leptolithiques. Le problème des Perigordiens I et II et l'hypothèse du synthetotype Aurignaco-Gravettien, essai de typologie analytique. *Quaternaria* 5, 153–240.

Leroi-Gourhan, A. (1965) *Préhistoire de l'Art Occidental*. Paris, Mazenod.

Lévêque, F. and Vandermeersch, B. (1980) Découverte de restes humains dans un niveau Castelperronien à Saint-Césaire (Charente). *Comptes Rendues de l'Academie des Sciences de Paris* 291, 187–189.

Lieberman, D. E. (1998) Neandertal and early modern human mobility patterns. In K. Akazawa, K. Aoki and O. Bar-Yosef (eds.) *Neanderthals and Modern Humans in Western Asia*, 263–275. New York, Plenum Press.

Márquez, J. E. and Morales, A. (1986) La habitabilidad de las cuevas: análisis morfológico. *Arqueología Espacial* 7, 169–181.

Mellars, P. (1973) The character of the middle-upper Palaeolithic transition in South-West France. In Renfrew, C. (ed.) *The Explanation of Cultural Change: Models in Prehistory*, 255–276. London, Duckworth.

Mellars, P. (1989a) Technological changes across the Middle-Upper Palaeolithic transition: economic, social and cognitive perspectives. In P. Mellars and C. Stringer (eds.) *The Human Revolution: Behavioural and Biological Perspectives in the Origins of Modern Humans*, 338–365. Cambridge, McDonald Institute.

Mellars, P. (1989b) Major issues in the emergence of modern humans. *Current Anthropology* 30, 349–385.

Mellars, P. (1991) Cognitive changes and the emergence of modern humans in Europe. *Cambridge Archaeological Journal* 1(1), 63–76.

Merlet, J.-C. (1996) Les Magdaléniens dans le bassin de l'Adour: territoires de subsistance et espaces parcourus. In H. Delporte and J. Clottes (eds.) *Pyrénées Préhistoriques. Arts et Sociétés.*

Actes du 118e Congrés National des Sociétés Historiques et Scientifiques, 225–230. Paris, Éditions du Comité des Travaux Historiques et Scientifiques.

Ordoño, J. (2007) *La Noción de Territorio en la Historiografía de la Transición del Paleolítico Medio al Superior en la Región Cantábrica. Génesis, Evolución y Estado Actual*. Unpublished dissertation, Vitoria-Gasteiz, Universidad del País Vasco.

Otte, M. (1990) From the Middle to the Upper Palaeolithic: the nature of the transition. In P. Mellars (ed.) *Origins and Dispersal of Modern Humans*, 438–456. Cambridge, Cornell University Press.

Pike-Tay, A. and Knecht, H. (1993) La caza y la transición del Paleolítico Superior. In V. Cabrera (ed.) *El Origen del Hombre Moderno en el Suroeste de Europa*, 287–314. Madrid, Universidad Nacional de Educación a Distancia.

Pujos, M. (1976) *Écologie des Foraminifères Benthiques et des Thecamoebiens de la Gironde et du Plateau Continental Sud-Gascogne. Application à la Connaissance du Quaternaire Terminal de la Région Oust-Gironde*. Thése de Doctorat d'Etat. Mémoires de l'Institut de Géologie du Bassin d'Aquitaine 8. Bourdeaux, Université de Bourdeaux.

Ramil, E. (1990) Habitabilidad cavernícola: elección de asentamientos. *Brigantium* 6, 191–197.

Richards, M. P., Pettit, P. B., Stiner, M. C. and Trinkaus, E. (2001) Stable isotope evidence for increasing dietary breadth in the European Mid-Upper Palaeolithic. *Proceedings of the National Academy of Sciences of the United States of America* 98(11), 6528–6532.

Rigaud, J.-P. (1996) L'emergence du Paléolithique supérieur en Europe occidentale: le role du Castelperronien. In O. Bar-Yosef, L. Cavalli-Sforza, R. March and M. Piperno (eds.) *Proceedings of the XIII International Congress of the Prehistoric and Protohistoric Sciences (Forlì)*, 219–223. Forlì, Abaco Edizioni.

Sáenz de Buruaga, A. (2004) Las primeras manifestaciones del Paleolítico Superior Antiguo en Araba y la explotación de las materias primas silíceas: algunas reflexiones. *Estudios de Arqueología Alavesa* 21, 1–16.

Sarabia, P. M. (1990) L'utilisation du silex dans les industries du Paléolithique de Cantabria (Espagne du Nord). In M. R. Seronie-Vivien and M. Lenoir (eds.) *Le Silex de sa Gènese a l'Outil*, 443–447. Cahiers du Quaternaire 17.

Sarabia, P. M. (1999) *Aprovechamiento y Utilización de Materias Primas en los Tecnocomplejos del Paleolítico en Cantabria*. Unpublished Doctoral Thesis, Santander, Universidad de Cantabria.

Siddall, M., Rohling, E. J., Almogi-Labin, A., Hembelen, C., Meischner, D. Schmelzer, I. and Smeed, D. A. (2003) Sea-level fluctuations during the last glacial cycle. *Nature* 423, 853–858.

Straus, L. G. (1996) Continuity or rupture; convergence or invasion; adaptation or catastrophe; mosaic or monolith: views on the Middle to Upper Palaeolithic transition in Iberia. In E. Carbonell and M. Vaquero (eds.) *The Last Neanderthals, the First Anatomically Modern Humans*, 203–218. Tarragona, Universitat Rovira i Virgili.

Tarriño, A. (2001) *El Sílex en la Cuenca Vasco-Cantábrica y Pirineo Navarro: Caracterización y su Aprovechamiento en la Prehistoria*. Unpublished Doctoral Thesis, Leioa, Universidad del País Vasco.

Tarriño, A. and Aguirre, M. (2002) Datos preliminares sobre la procedencia de los sílex recuperados en el yacimiento de Antoliñako Koba. In N. Cazals (ed.) *Comportements Techniques et Économiques des Sociétés du Paléolithique Supérieur dans le Contexte Pyrénéen. Project Collectif de Recherche 2002*, 6–25. Toulouse, Service Régional de l'Archéologie de Midi-Pyrénées.

Tarriño, A. and Normand, C. (2002) Estudio preliminar de la procedencia de los silex del Auriñaciense antiguo (C4b1) de Isturitz (Pyrénées Atlantiques). *Espacio, Tiempo y Forma* 15, 135–143.

Terradas, X. (2004) *La Gestión de los Recursos Minerales en las Sociedades Cazadoras-Recolectoras*. Treballs d'Etnoarqueología 4. Madrid, Centro Superior de Investigaciones Científicas.

Thomas, J. (2001) Archaeologies of place and landscape. In I. Hodder (ed.) *Archaeological Theory Today*, 165–186. Cambridge, Polity.

Turq, A. (1993) L'approvisionnement en matieres premières lithiques au Mousterien et au debut du Paleolithique superieur dans le Nord-Est du bassin Aquitain. In V. Cabrera (ed.) *El Origen del Hombre Moderno en el Suroeste de Europa*, 315–325. Madrid, Universidad Nacional de Educación a Distancia.

Svoboda, J. A. (2006) The Danubian gate to Europe: patterns of chronology, settlement archaeology, and demography of Late Neandertals and early modern humans on the Middle Danube. In N. J. Conard (ed.) *When Neanderthals and Modern Humans Met*, 233–268. Tübingen, Kerns Verlag.

Vialou, D., Renault-Miskovsky, J. and Patou-Mathis, M. eds. (2005) *Comportements des Hommes du Paléolithique Moyen et Supérieur en Europe: Territoires et Milieux. Actes du Colloque du GDR 1945 du Centre National de la Recherche Scientifique (Paris, 2003)*. Études et Recherches Archéologiques de l'Université de Liège (ERAUL) 111. Liège, Université de Liège.

White, R. (1982) Rethinking the Middle/Upper Palaeolithic transition. *Current Anthropology* 23(2), 169–192.

White, R. (1985) *Upper Palaeolithic Land Use in the Perigord. A Topographic Approach to Subsistence and Settlement*. British Archaeological Reports International Series 253. Oxford, Archaeopress.

Yravedra, J. (2002) Subsistencia en la transición del Paleolítico medio al Paleolítico superior de la Península Ibérica. *Trabajos de Prehistoria* 59, 9–28.

Zilhão, J. and d'Errico, F. (2000) La nouvelle bataille Aurignacienne. Une révision critique de la chronologie du Châtelperronien et de l'Aurignacien ancient. *L'Anthropologie* 104(1), 17–50.

Chapter 8

La Garma (Spain): Long-Term Human Activity in a Karst System

Pablo Arias and Roberto Ontañón

Human activities in caves have changed during the history. The karst system of La Garma, in northern Spain, provides most detailed information about the evolution of the relationship between caves and human societies. Ten archaeological sites, nine of them in caves, have been studied in this hill, dating from at least 175,000 years ago to the Middle Ages, with most intermediate periods represented. These sites were used in the most variable fashion during the Upper Palaeolithic, when the entrances of some caves were used as residential camps, whereas the inner part of the karst was dedicated to graphic expression and probably ritual activities. Another period of intense human activity was the third and second millennia cal BC, when several galleries were used as burial places. It is emphasized here that human presence in the karst cannot only be explained by utilitarian factors. On the contrary, the symbolic significance of such a particular environment should be considered.

Introduction

Although karst environments have played a vital role in the research into certain periods of prehistory (particularly the European Palaeolithic, see for example Straus 1979), caves are not necessarily the best options for human activity. It is quite likely that their relevance within archaeological research is connected with the better conditions for the preservation and discovery of remains in this geological setting, rather than with their true importance as the location of the activities of past human groups. Without a doubt, caves do have certain advantages, such as providing a ready-made shelter from inclement weather, which could well have been a significant factor in colder phases in the Pleistocene. However, these advantages are counteracted by the disadvantages, such as the darkness, humidity or spatial constrictions that characterise the inner parts of caves. Neither can certain cultural factors be ignored. Caves are frequently associated with a feeling of mystery, which often makes them the setting for myths and legends. In many cultures, caves are the home of worrisome beings, like fairies and goblins, or simply terrifying creatures such as dragons and other monsters (think of the terrible Cyclops Polyphemus in the Odyssey or Fafnir in Germanic mythology). It is likely that this is connected with a characteristic trait of caves: the darkness. Our species, a diurnal primate, defenceless against its enemies in the night, may associate caves with terrible ancestral fears, probably born in the very dawn of the hominization process. In fact, in contrast with the image of prehistoric man as a cave-man, so widespread in popular culture, we might wonder about the reasons why they chose to enter these kinds of places, and occasionally to visit them repeatedly.

From this perspective, La Garma is an ideal observatory. This hill, 185m high, located some 4km from the present-day shores of the Bay of Biscay (Fig. 8.1), contains ten archaeological sites, most of them located inside caves, which have revealed evidence of human presence from the Lower Palaeolithic to the Middle Ages, with most of the periods in between represented (Fig. 8.2).

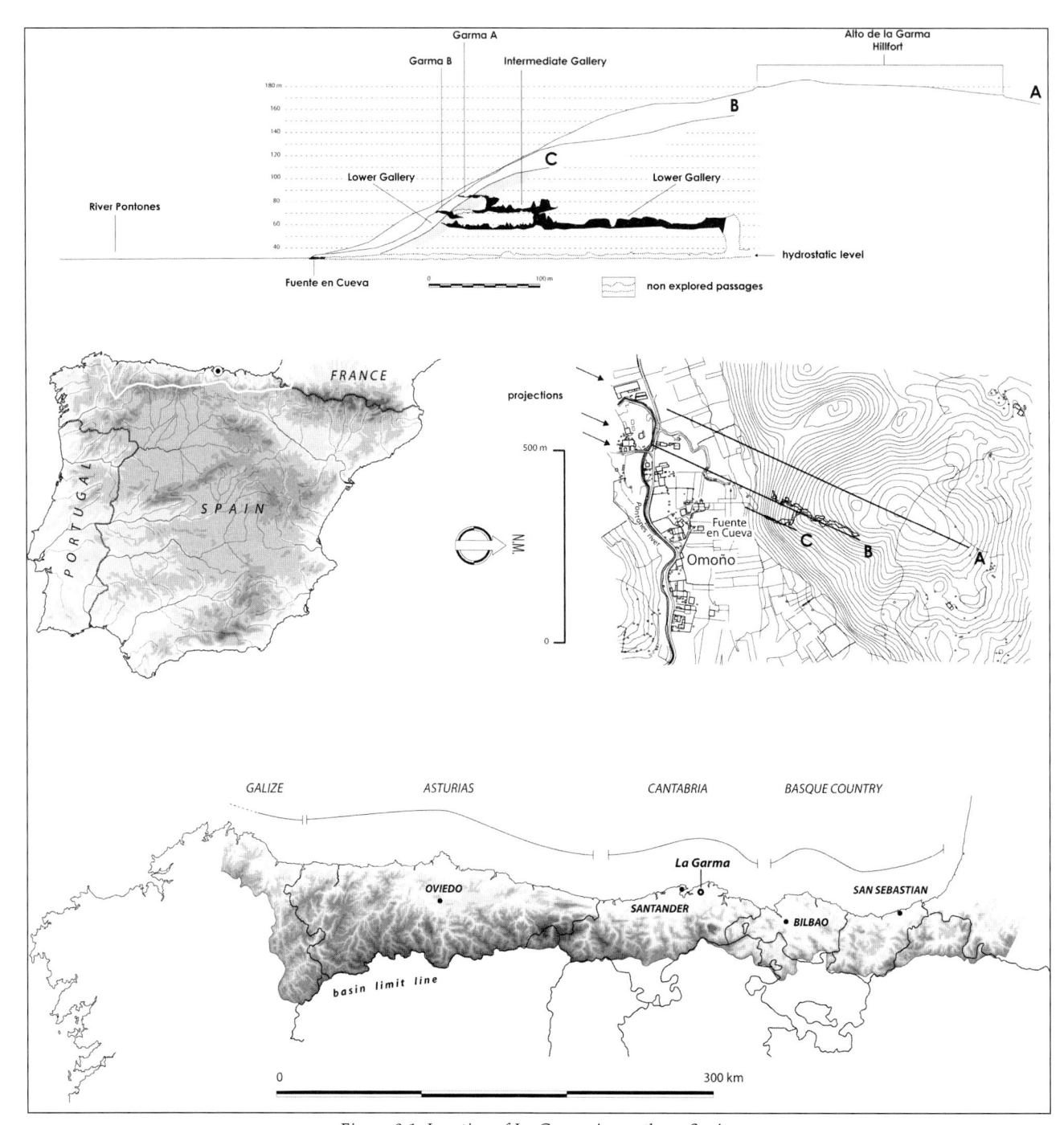

Figure 8.1: Location of La Garma in northern Spain.

These sites are concentrated on the southern side of the hill, in different levels of a complex cave system, where seven levels have been identified so far. The most important archaeological sites are located in the levels at 59m and 80m above sea level: the Lower Gallery and La Garma A. The former is a passage 300m long, whose original entrance became blocked in the Pleistocene, where several areas with concentrations of Magdalenian remains have been found on the surface of the cave floor, and which has an important ensemble of Palaeolithic cave art, of varying chronology. It was also used as a burial cave, and further evidence of the exploration of the cave in the Middle Ages has been found.

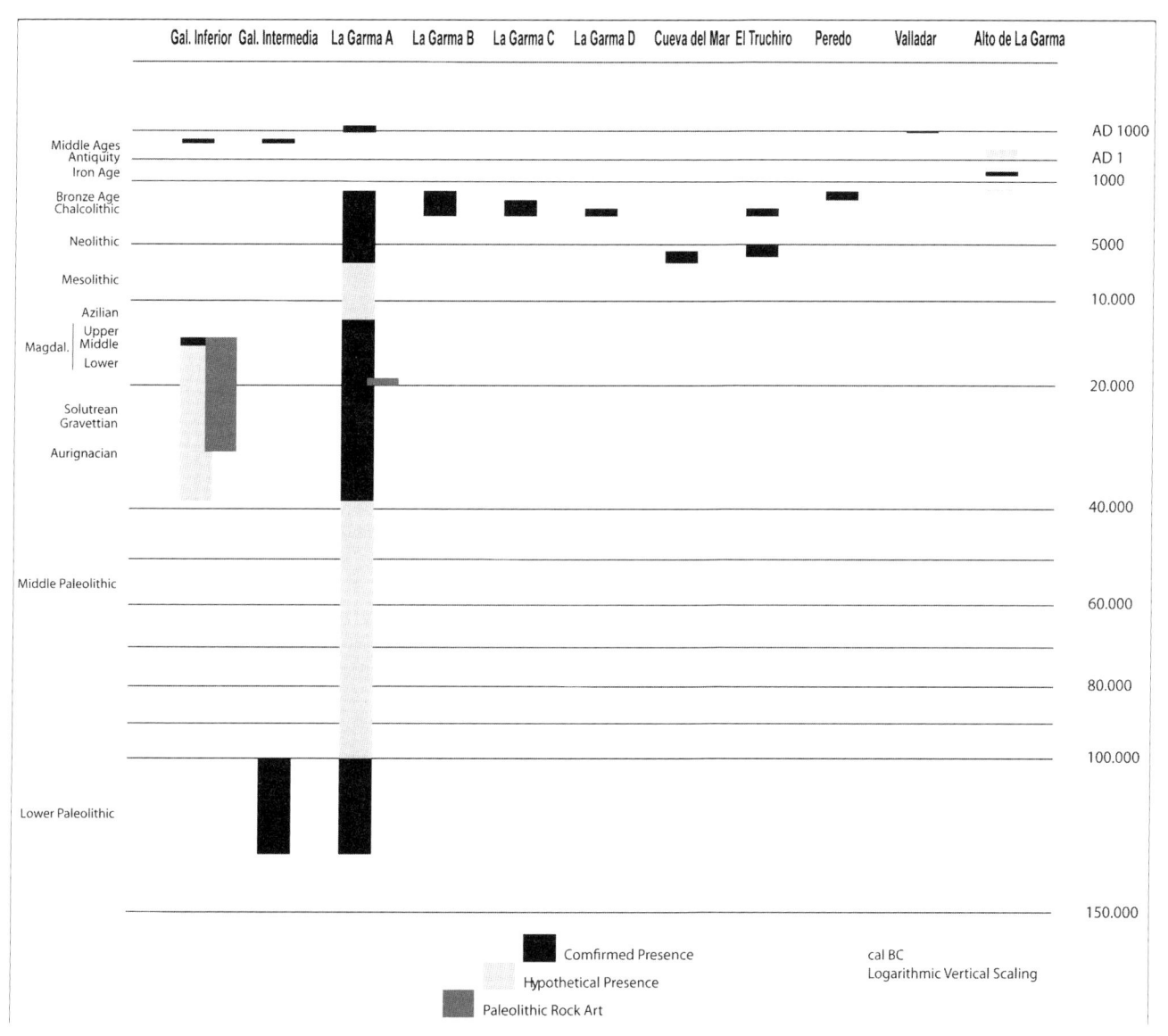

Figure 8.2: Chronological schema of the archaeological sites in La Garma hill.

In turn, La Garma A, which is now the only entrance into La Garma cave system, has yielded a long stratigraphic sequence, which begins in the Lower Palaeolithic, and includes strata of all the main phases of the Upper Palaeolithic (Aurignacian, Gravettian, Solutrean, Magdalenian), and of the Mesolithic, Neolithic, Chalcolithic and Bronze Age, in addition to some medieval objects on the surface.

Between these two sites, there is another level at 70m above sea level, in whose entrance (La Garma B) a Chalcolithic and Bronze Age burial deposit has been excavated, and whose interior (Intermediate Gallery) contains some Palaeolithic cave paintings and a Lower Palaeolithic deposit, as well as palaeontological remains (bears).

Two sections of cave passage on a higher level than La Garma A in the same karst system, called La Garma C and La Garma D, were the location of Chalcolithic burial deposits.

Finally, in the lower level, situated on the present river terrace, at about 35m above sea level, El Truchiro Cave has Mesolithic and Chalcolithic human burials.

Archaeological sites have been studied in other parts of the hill, apparently unconnected with the main cave system. These are the caves named Peredo, with Bronze Age burials, El Valladar, with medieval remains, and Cueva del Mar, with a badly deteriorated deposit that includes Mesolithic levels. Finally, the only open-air site *sensu stricto* is found on the top of La Garma: an Iron Age hillfort, dated to the seventh-sixth centuries BC.

Before *Homo sapiens*: Lower Palaeolithic evidence at La Garma

The earliest evidence of humans at La Garma is located in two places that are near each other and which were probably connected originally; the area outside La Garma A, and the Intermediate Gallery.

In the former of these two areas, a thick deposit has been studied (Fig. 8.3), made up of a succession of layers of silt and flowstone, with a characteristic Lower Palaeolithic industry (including bifaces and cleavers), several molars of narrow-nosed rhinoceros (*Stephanorhinus hemitoechus*), and numerous remains (deciduous teeth and a tusk fragment) of straight-tusked elephant (*Palaeoloxodon antiquus*) and Deninger's bear (*Ursus spelaeus deningeri*). The study currently being made by Jesús Tapia suggests that the chronology of this deposit can be situated at least in the MIS 5e (Marine Isotope Substage 5e – a temperate phase at the beginning of the Upper Pleistocene). Recent amino acid racemization dates of rhino molars appear to confirm this, since an age around 126,500 years has been found for the top of this sequence. This stratigraphy has filled a cave passage which, according to the preliminary data, would have connected the level of La Garma A (80m above sea level) with the Intermediate Gallery (at 70m above sea level). To date, the deposit has been dug to a depth of 3.5m reaching an altitude of 77m above sea level, where dates of around 175,000 years (MIS 6, a cold phase at the end of Middle Pleistocene) have been obtained. The sedimentological study shows that these strata formed in a cave environment, which suggests that at the time of their formation the area was covered by a large overhang, which has retreated with the erosion of the hillside. The information obtained during the excavation of the contiguous Upper Palaeolithic deposit in La Garma A shows that the erosion of the overhang to approximately its present situation was completed between the Solutrean and the lower Magdalenian.

The Intermediate Gallery is a section of the 70m above sea level, now isolated from its entrance (La Garma B) by the growth of stalagmites. When this Gallery was discovered in 1991, the cave floor was marked with numerous bear dens, associated with bones of brown bear (*Ursus arctos*) and other carnivores, as well as certain Lower Palaeolithic

Figure 8.3: Earliest section of the Garma A stratigraphy, showing the Lower Palaeolithic layers.

implements (including cleavers). The excavation carried out in 1998 showed that this association of fauna and industry was the result of post-depositional processes. The Lower Palaeolithic industry came from a silty layer, which in places had been disturbed by the bears when they dug their dens.

The Lower Palaeolithic remains at La Garma pose some complex questions. What is the origin of these deposits? How and why did the archaeological remains reach these two very different sites? What relationship is there between the implements recovered in these contexts and the remains of fauna and flora found in association with them? In short, what activities of the human groups are connected with these deposits? At the moment we are unable to give definite answers to these questions, as the deposits are still being excavated. Furthermore, if a detailed taphonomic study is necessary at any archaeological site to determine its formation, in the case of such old levels – which have undergone a complex geological history and which are associated with hominids different from our species – even greater precautions should be taken. With the available data, it seems possible to postulate the existence of some type of human activity in the former entrance of La Garma A. We have clear evidence of the transport to the site of raw materials brought from different locations in the surrounding area (flint, ophite) and possibly the remains of fauna were processed in some way there. An analogous site, from this point of view, would be Levels 24–26 at El Castillo (Cabrera 1984), about 28km southwest of La Garma.

The case of the Intermediate Gallery is more enigmatic. The area where the Lower Palaeolithic remains have been found is situated inside the cave, in an area which is not too far from the surface, but certainly beyond the limits of daylight and half-shade zones (Rouzaud 1997). This means that we should consider very seriously whether this is a secondary deposit coming from La Garma A, with which it was connected in the past. However, so far we have been unable to define any convincing geological processes to explain this movement. The sediments containing the Lower Palaeolithic artefacts appear to have been deposited in a low energy environment (probably a flooded area), which is unlikely to have been responsible for moving large, heavy objects, such as ophite cleavers, over any great distance.

There is, therefore, evidence of an early human presence in the karst at La Garma, corresponding to hominids prior to our species (probably Neanderthals, like those documented at El Castillo (Garralda 2005–6). The exact meaning of these occupations should be studied in detail in future fieldwork seasons.

The Golden Age at La Garma: the Upper Palaeolithic

The period that has provided the most significant information at La Garma is the Upper Palaeolithic. As we saw above, the occupations in that phase of prehistory are mainly located in two of the passages that have their entrances on the southern side of the hill: La Garma A and the Lower Gallery. The former site underwent some major changes in the late Pleistocene, which altered its shape. Thus, at the start of the Upper Palaeolithic, it was a large rockshelter, some 6m high, much deeper than the small vestibule which exists now. There, a large area (at least some 30m^2) was occupied by the groups of the Aurignacian. Later, the cave was abandoned for a long period, during which a layer of sterile silty-clay was deposited. The cave was re-occupied at about 24,500 cal BC, during the Gravettian. Some notable evidence has been documented belonging to that period, such as an ibex (*Capra pyrenaica*) metapode with a large perforation near the distal epiphysis and series of short engraved lines (Arias and Ontañón 2005) and a large rock with engravings. During the Last Glacial Maximum, in the Solutrean, the cave was occupied frequently, and this continued during the lower, middle and upper Magdalenian, until the end of the Pleistocene. The accumulation of sediment began to cause a reduction in the inhabitable surface area, as the cave was steadily filled. Eventually, the deposit reached the roof in some places, dividing the former large rockshelter into two chambers connected by a narrow corridor. In the middle Magdalenian, the areas of these two chambers were used differently, and a roughly semi-circular stone structure was built against the rear wall of the inner chamber (Arias *et al.* 2005). The archaeological remains that have been recovered suggest that the outer vestibule would basically have been a passageway, although some evidence has been found of butchery work and knapping, whereas most of the tasks would have been carried out in the inner chamber. The space separated off by the semi-circular structure could have been a kind of rest area (Arias *et al.* 2005).

As stated above, the Lower Gallery is a level in the karst where the cave entrance became blocked at the end of the Pleistocene. It has preserved an important ensemble of Palaeolithic cave art and an outstanding sample of the last occupational phase of the cave, in the middle Magdalenian (Arias, Ontañón *et al.* in press). It is possible to observe directly, without excavation, a surface area of about 800m^2 of an extraordinarily rich deposit, including several stone structures (Ontañón 2003).

Zone I is the name of the sector nearest the original entrance of the cave (Figs. 8.4 and 8.5). It includes the original vestibule and the first section of the underground passage. It is the largest part of the cave and possesses the form of a large chamber 70m long with an average width of about 7m. Here can be seen, in full view, an impressive, large Magdalenian floor, in a unique state of conservation. Thousands of bones, flint, and antler implements (including *batons percés*, spear-throwers and decorated ribs, bone points and spatulas), portable art objects, sea-shells, and other remains of human activity are spread over an area of about 500m².

Although the spatial analysis of this occupation is still in its early stages, the first observations have revealed some interesting patterns. Thus, dense concentrations of bones can be seen against the walls, which could partly be due to the sedimentary matrix being washed away and removed by the water running down the walls, but also to an accumulation of food remains and worked bones in the peripheral areas of the chamber.

But perhaps the most outstanding feature of this extraordinarily well preserved Magdalenian deposit is the presence of artificial structures built with stones and pieces of stalagmites. The structure for which we have the fullest spatial data (Fig. 8.5) is located about 40m from the entrance, next to the western wall of the passage and underneath an overhanging shelf of rock which projects out 1.5m above the floor. It has the form of a sub-circular ring of boulders and pieces of limestone, occasionally juxtaposed but mostly separated by gaps of several centimetres. The maximum diameter of this structure is 3m and the minimum is 2.5m. On its southeast side it is made of small stones, considerably smaller in size than the boulders that make up most of the circle, which might suggest that an opening existed on that side. The nature of the occupation floor is clearly affected by the building of this structure: the floor inside the stone circle has been trodden more than the surrounding space. This could identify the space inside the structure as a living area. The stones that remains visible today would have formed the base of a light structure made from

Figure 8.4: Middle Magdalenian floors near the original entrance of the Lower Gallery (Zone I).

perishable materials (possibly sticks and branches, bark or skins), making a more-or-less conical frame which would probably have leant against the rock shelf that overhangs the structure. They were probably used to hold this kind of tent to the floor, resembling some examples of hut foundations documented by ethnoarchaeology (Oliver 1997).

This impressive habitation floor suggests that this sector of the cave was occupied intensely in the Middle Magdalenian. The studies underway should allow us to determine whether its formation corresponds to a continuous, prolonged use or to the repetition of shorter visits. Evidence such as the existence of remains beneath and on top of the constructions described above might be an indicator of more than one occupation episode in the cave which, because of its excellent conditions for habitation, would have been a good shelter for human groups in the region during the Late Glacial. The very presence of substantial structures and other traits of habitation have often been used to argue in favour of a long occupation. These criteria are valid in general terms and in the case of the Lower Gallery they are met fully. Finally, another argument in favour of the long duration of the Magdalenian occupation at La Garma is the consideration of the structural differences in comparison with contemporary open-air camps, like those in the Paris basin (Julien *et al.* 1988; Olive *et al.* 2000). Those sites, the result of brief occupations, display a more clearly arranged structure (organised into clearly differentiated functional areas), whereas at La Garma the boundaries between the areas of activity are less defined. Thus, it may be inferred that the archaeological remains that can be seen on the surface of Zone I are the result of a dilated occupation during a longer period of time. Assuming in both cases a limited amount (but never absence) of post-depositional alterations, it can be inferred that the apparently disorderly surface deposit in the Lower Gallery is the consequence of complicated formation processes, which can be mainly attributed to the effects of prolonged, multiform human visits. This resulted in along series of actions that produced superimpositions, overlaps and movements of the remains.

Figure 8.5. Middle Magdalenian floor in the rear part of the vestibule of the Lower Gallery (Zone I). A circular stone structure can be observed in the upper left part of the picture.

Zone III, located about 90m from the entrance, and therefore in a deeper area, in total darkness, is an elliptical chamber, about 15m long and 8m wide, whose roof slopes down steeply on its western side, creating a smaller area where it is difficult to move about. The whole floor in this chamber is covered with hundreds of remains of implements and fauna. These include several fragments of partially cut reindeer (*Rangifer tarandus*) antler and other evidence of bone-working, as well as some magnificent examples of portable art. It is also worth noting the large number of bovine bones (*Bison priscus* or *Bos primigenius*). In several parts of this zone, both constructed and latent structures have been recognised, such as a number of stones piled up around a hollow, as a kind of footing for a post, and, above them, a space in a natural ledge with a decorated rib inside it. On the northern side, the presence of an unusual concentration of antler points is also striking.

Here we can find further remains of Palaeolithic constructions. The most noticeable is a sub-circular area, next to the west wall, beneath the low roof mentioned above, and formed by large limestone boulders and some pieces of calcite set upright. The floor inside this enclosure, about 5m^2, appears to have been lowered in comparison with the surrounding area, and the density of objects inside it is considerably less than in the area outside.

As the characteristics of this area are not too different from those seen in Zone I, its position in an interior part of the cave, away from the normal areas for habitation – those reached by daylight or at least in an area of half-shade – means that an alternative function needs to be found.

Finally, Zone IV is a sector of the Lower Gallery some 130m from the entrance of the cave that was used by Palaeolithic groups. As in Zone III, the Magdalenian remains are concentrated in an area with a low ceiling, which drops from 1.7m to 0.5m. There, archaeological objects (mainly mammal bones, but also lithic artefacts, shells, charcoal, red ochre and portable art items) are distributed over an area of about 55m^2.

Three stone structures are found next to the western wall of the chamber. Two of them (adjacent structures IV-A and IV-B) (Fig. 8.6) are roughly rectangular areas demarcated by speleothems, where IV-B is much larger than IV-A (3.18m^2 compared with 1.5m^2), whereas IV-C (5.35m^2), located 1.4m to the northeast of IV-B, was built by digging out the stones (mostly fragments of speleothems) from the floor and piling them up around its edges to form two low drystone walls, as well as by placing large slabs vertically.

The provisional results of the archaeological research carried out in this area so far (Arias, Ontañón *et al*.

in press) show several unusual features for a middle Magdalenian site. The distribution of archaeological items is strongly biased towards faunal remains. Bones of horse (*Equus caballus*) dominate the assemblage, amounting to approximately 61 per cent of the material recorded (and 70 per cent of the material corresponding to consumed mammals), whereas in any other Magdalenian site in northern Spain red deer (*Cervus elaphus*) or ibex (*Capra pyrenaica*) are usually the dominant species. Moreover, some unusual features have been recorded, such as the presence inside structure IV-B of the skull of an equid whose dome had been removed. We must also highlight the presence of bones of cave lion (*Panthera (Leo) spelaea*) with human cut-marks. In a limited area at the southern end of structure IV-C, two nearly complete skeletons of shelduck (*Tadorna tadorna* L.) whose bones do not present butchery marks or burnt surfaces, suggest that the birds were left as whole bodies on the cave floor.

In contrast to the large human impact on the cave's interior, especially regarding the abundant faunal material, the scarcity of the lithic assemblage is quite striking: so far, fewer than nine objects per square metre have been recorded. The bone and antler industry density is also very low compared with other Magdalenian sites.

Other outstanding traits are the high indices of objects of adornment (most of them perforated marine mollusc shells or teeth) and portable art items (bone or antler sculptures, engraved bones, and engraved slabs of limestone or calcite). Some evidence suggests that at least some of these were made in that part of the cave: there are two unfinished objects, a partially perforated bear canine and a bone cutout, and most of the engraved *plaquettes* appear to have been made from calcite slabs gathered in the cave.

Several engravings have been observed on the ceiling above the southern part of structure IV-B. Magdalenian in style, these include representations of horses, bison and hinds.

Although Zone IV is still under study, the evidence gathered so far, and the comparison with other Middle Magdalenian sites in the Pyrenean area, suggest that this might be an area where ritual activity was predominant (Arias 2009).

The Lower Gallery also contains an important ensemble of Palaeolithic cave art. Some 500 graphic units have been documented, of which 92 are animal figures, 109 are signs, and 40 stencilled hand- or finger-prints. The paintings and engravings are spread throughout the whole cave, from the original entrance to the end of the passage. However, according to C. González's study (González Sainz 2003), clear diachronic differences exist in the spatial distribution of the images. Thus, while the paintings that can be attributed to early periods in the evolution of cave art (Gravettian and Solutrean) are distributed throughout the cave, with a clear preference for the main axis of the passage, the Magdalenian motifs are concentrated in the third of the cave nearest the entrance, with a large number of them in side passages and small chambers, some of which are literally full of paintings and engravings.

One aspect, in which the Lower Gallery at La Garma displays great potential for contributing to our knowledge of Palaeolithic societies, is that of the activities related to the graphic expression. First, there is the question of how the paintings and engravings were produced. The excellent state of preservation of the Magdalenian floors has allowed information to be recovered about the processing of the pigments. Thus, it has been possible to study some stains of paint on the floor, palettes with remains of crushed iron oxide, and even some sources of the raw materials used to make the paintings in the cave (Arias, Laval *et al.* 2011). No less interesting is the existence of some evidence which enables us to tackle the complex problem of ceremonial visits and activities related with the art. At La Garma, some evidence of this kind has been found (Arias *et al.* 2003). For instance, two metres up, above the Magdalenian camp located near the original entrance of the cave; there is a gallery which is difficult to reach, where nothing apart from a number of painted signs has

Figure 8.6: Middle Magdalenian floors and structures in Zone IV of the Lower Gallery.

been found. However, a path was worn along the floor of this gallery, consisting of very loose sandy sediment. It seems unlikely that this would have been a deliberate construction, rather the result of the frequentation of the gallery by the Palaeolithic people. It does, however, show that a group of paintings, located in a relatively difficult location, was visited with a certain frequency. This is a question about which many speculations have been made, but for which observable evidence appears to exist in this case.

Moreover, the distribution of Magdalenian floors and paintings at this site challenges the notion that areas with cave art and habitation areas were segregated spatially (Laming-Emperaire 1962; Leroi-Gourhan 1965). At La Garma, the large Magdalenian camp in the cave vestibule is situated next to numerous paintings and engravings on the walls, some of which can clearly be attributed to the same period (Fig. 8.7).

One particularly interesting point in the case of La Garma is the simultaneous use (at least with the precision that archaeology is able to determine the chronology) of two very close Palaeolithic settlements. This is quite clear in the middle Magdalenian, when the Lower Gallery and La Garma A were both occupied (possibly at the same time). The difference between both caves suggests a differential use of each site. Without doubt, the main settlement would have been the Lower Gallery, located near the valley floor, and with a considerably larger surface area and better habitation conditions. It is interesting to note, however, the analogy between the stone structures documented in the vestibule of the Lower Gallery and at the rear of the second chamber at La Garma A.

It is possible that the simultaneous use of both caves occurred in earlier periods of the Upper Palaeolithic. So far we have no information about the stratigraphy of the Lower Gallery. However, the cave art is proof of some kind of human activity in earlier periods, also represented at La Garma A (Lower Magdalenian, Solutrean, Gravettian). Yet the relationship between both caves in those periods is not necessarily identical. Bear in mind that the habitation conditions at La Garma A were better in the early and middle phases of the Upper Palaeolithic than in the Magdalenian. At the same time, the existence of cave art in the Lower Gallery does not necessarily imply that there were contemporary occupations. This is an issue that needs to be explored by future research.

The last hunter-gatherers at La Garma: the Mesolithic

We find a new example of apparently simultaneous human presence in two caves in La Garma hill during the late Mesolithic. In this case, the evidence comes from La Garma A, Cueva del Mar and El Truchiro. The dates obtained prove the existence of Mesolithic groups at El Truchiro in the mid eighth millennium cal BC, at La Garma A from the mid-seventh millennium cal BC and in Cueva del Mar from around 6100 cal BC; in the latter the occupation continues during at least the first quarter of the sixth millennium. The Mesolithic levels in the three caves include high densities of marine molluscs, so they could be described as 'shell-middens', a term often used for Mesolithic levels rich in shells, but whose definition is in general quite vague (e.g. Waselkov 1987; Muckle 2006, 70, 229). In the case of La Garma, it should be pointed out that although the site is not too far from the coast (about 5km to the present shoreline), the distance is considerable if we take into account the large amounts of seashells found at the sites. These invertebrate assemblages include mainly species typical of rocky shores (e.g. limpets of the species *Patella*

Figure 8.7: Palaeolithic rock art near the original entrance of the Lower Gallery (Zone I). A middle Magdalenian floor can be seen below the painted wall.

intermedia, *Patella vulgata* and *Patella ulyssiponensis* and the topshell *Osilinus lineatus*) (E. Álvarez pers. comm.). This suggests that the groups who occupied La Garma during the Mesolithic gathered the molluscs on the open shore situated to the north of the site, but did not use the resources of the Bay of Santander, which already existed at that time, located some 6.5km to the west.

The poor preservation of the deposits in Cueva del Mar (reduced to a few remains adhering to the wall of this large cave) and El Truchiro (mostly removed before 1903, in an excavation about which we possess very little information) makes it impossible to determine the function of those sites. At La Garma A, the deposit is rich in mammal fauna and stone implements, which suggests that it is associated with a temporary or permanent settlement, where the usual activities at a hunter-gatherers' camp were carried out. This is quite interesting, as most of the numerous shell middens on the coast in Cantabrian Spain appear to be areas where waste accumulated and it is unusual to find remains of everyday activities in them (González Morales 1982; Fano 2001). In any case, it is possible that also at La Garma A the main activity area was located outside the cave, as its sedimentary fill had now reached a considerable height, so that it was no longer possible to stand upright in much of the cave. Equally, it should be noted that the Mesolithic strata extend towards the outer part of the cave, but, in this case, there was obviously no appreciable difference between the position of the overhang of the entrance then and its current position. This suggests that, as at other Mesolithic sites in northern Spain, the main area of human activity was outside the cave and the presence of archaeological deposits in the cave derives from the accumulation of the waste produced by nearby activities, whether as a result of the deliberate cleaning up of rubbish, or the mere fortuitous consequence of the proximity of the cave to the settlement. Something similar can be said of El Truchiro, a very small cave at the bottom of La Garma hill that would have been practically full of sediment, and where it would have been necessary to stoop or crawl in order to enter.

In the sixth millennium cal BC, still in the Mesolithic, a new development took place, which was to mark the function of the hill for several millennia: the use of its caves as burial sites. As mentioned above, the information comes from El Truchiro, where a burial has been found, dug into the Mesolithic levels. The radiocarbon determination obtained for the human remains situate this burial in the first quater of the sixth millennium, a time when a large number of Mesolithic human remains have been found in the Iberian Peninsula (Arias and Álvarez-Fernández 2004). This is a single burial, of a juvenile individual, deposited on his side with his limbs bent (Fig. 8.8), a frequent posture in Cantabrian Spain in the Mesolithic (Arias *et al.* 2009). However, the burial at El Truchiro displays one highly interesting feature. The tomb was

Figure 8.8: Mesolithic burial in El Truchiro Cave.

affected by the action of fire, probably indirectly, which has resulted in the very poor condition of the skeletal remains but also in the preservation of the remains of a wooden container in which the body was deposited. This is a kind of coffin, or at least a base of oak bark, of which a large amount of evidence has been found beneath the skeleton.

The cave as a tomb: Chalcolithic and Bronze Age

In the second half of the fifth millennium cal BC, in Cantabrian Spain, as in many other parts of Europe, a profound change occurred in funerary practices: the generalisation of collective burials. In this region, this practise is seen in two types of burial spaces: megalithic monuments and natural caves (Arias 1999; Ontañón and Armendariz 2005–6). The latter are frequent in the eastern half of the region (Cantabria and the Basque Country), where around 160 sites have been catalogued. In general, for this funerary function, small caves were chosen (sometimes small fissures that are difficult to access) or narrow passages in larger caves. Although the poor preservation of these contexts leaves many important details in the dark, in general a small number of individuals were buried (usually less than ten). Associated with the bodies are objects of personal adornment and other artefacts that can often be interpreted as grave goods: pottery vessels (in some cases probably containers of some kind of drink or food), stone, bone or metal weapons, and everyday objects. It seems that the bodies were usually deposited directly on the cave floor, although some recent excavations have noted that shallow pits were dug to hold the burials.

La Garma hill provides one of the best examples of this kind of behaviour in the region. Six caves on the south side of the hill (most of the accessible openings in a relatively restricted area) were used as burial sites in the Chalcolithic and Bronze Age. The commencement of this type of funerary behaviour is dated in the early third millennium, when the burial use of La Garma A, La Garma B, El Truchiro, La Garma C and La Garma D has been documented.

In later moments of the third millennium, the funerary use of several of these caves continued (at least at La Garma A, La Garma B and La Garma C), and in the first half of the second millennium, in the Bronze Age, burials have been documented at La Garma A (in an area nearer the entrance), La Garma B and Peredo. In such a large number of sites, considerable variability has been seen in the laying-out of the bodies and the preparation of the burial area.

The vestibule of La Garma A, La Garma B and D, El Truchiro and Peredo are typical examples of underground spaces used for the successive burial of bodies without any apparent preparation of the space. However, in La Garma B a deliberate organization of the funerary remains has been found, including a particular location for the bodies and for large pottery vessels, probably intended for collective offerings (Gutiérrez Cuenca 2010). In turn, the inner chamber at La Garma A and the nearby cave of La Garma C are good examples of a larger investment in labour in the burial rituals in this phase of recent prehistory. At the former site, the stalagmitic crust on the floor was perforated in two places to open up pits in which at least five bodies (four adults and one child) were deposited. The bones were disarticulated, which suggests the possibility that these were secondary burials. The larger grave contained several arrowheads made with flat bifacial retouch and one outstanding object, a flint dagger (Fig. 8.9), apparently imported from the south of the Iberian Peninsula. This can be considered as one of the earliest signs of the development (or at least of the exhibition) of social inequalities in northern Spain (Arias *et al.* 1999).

In turn, the excavations at La Garma C documented the opening of a pit in the floor of the cave, in this case in a small chamber, which was reached from the outside by a narrow descending ramp. The human remains, which were poorly preserved, were found together with objects frequently used as grave goods at that time, such as pottery, a polished stone axe and a flint arrowhead. The main difference seen in this cave is that the entrance of the burial chamber was walled over, sealing off the chamber in a similar way to megalithic chambers.

Medieval cavers: explorations in La Garma Cave during the seventh and eighth centuries AD

Human visits to the cave system at La Garma did not end in prehistory. The interior of the cave, including its deepest parts, was visited several times during the Middle Ages, and some of the evidence of this exploration could hardly be more surprising.

The discovery of medieval pottery in the vestibule at La Garma A, with the characteristic forms of the local pottery manufacturers in the Middle Ages, such as jugs with a square mouth, witnessed the use of what was then a small cave as a shelter or even as a natural cold store, in connection with the everyday customs related to the farming activity of the local inhabitants.

In contrast, the archaeological record in the Intermediate and Lower Galleries, which is exceptionally well-preserved, displays traits of activities very different from subsistence tasks, and which are linked with ritual and funerary uses that are difficult to interpret. It must be borne in mind that the Lower Gallery is a cave passage of difficult access, reached by descending two shafts from the only entrance open to the cave system at the present time. Nonetheless, in both of the deep levels of La Garma, hundreds of pieces of charcoal have been found, the result of burning sticks or torches of hazel wood (*Corylus avellana*). These are spread on the cave floor and in some points form the remains of hearths. Pieces of charcoal are found throughout the Lower Gallery, in every part that can be reached by humans, even on the ledges overhanging the large shaft that marks the end in the deepest part of the cave. The amount of burnt matter and the characteristics of the wood used, which burns quickly, suggest that a large amount of wood was taken into the cave, which implies a certain preparation of the activity, which would not have been a mere underground 'adventure'. The radiocarbon determinations obtained on the charcoal indicate that the cave was visited at least twice, between the seventh and eighth centuries AD.

The presence of human remains is even more amazing than this evidence of 'medieval caving', and may, at least in part, explain why people visited such a sinister place. These comprise the skeletons of five young male individuals, lying on the floor in two sectors of the cave near each other, close to the shaft that descends from the Intermediate Gallery. Deposited in areas bounded by natural features of the cave, some of them are surrounded by pieces of stalagmites (Fig. 8.10). One of them was wearing, at the time of his death, a belt with a lyre-shaped belt buckle with damascene decoration, of typical Visigothic type. Only the blade of an iron knife or punch is associated with another of the bodies. The peculiarity of these human remains, as regards their position in such a hidden underground location, is complicated even further by another aspect, making their interpretation more difficult. All the skulls had been crushed, practically smashed to pieces when the bodies had already become skeletons (which require a period of over two years).

This behaviour, together with position of the bodies in an unusual place (normally at that time burials were in cemeteries next to churches), suggests that some other kind of explanation must be sought. One aspect that could orientate us is the bias in the age distribution in comparison with the probable source population. In this respect, the hypothesis recently put forward by J. A. Hierro (2008) for this and other coetaneous contexts in Cantabria and other parts of the Iberian Peninsula are particularly interesting. According to this author one cause which might explain these deposits is the consequence of an epidemic of a severe infectious disease on the community. The deceased, the bearers of the illness, would have been taken in the same clothes and accessories with which they died to a remote and hidden place to avoid the plague spreading further. In certain historical and cultural contexts, contagious diseases and their transmission are associated with the spirits of the dead. Perhaps the destruction of

Figure 8.9. Chalcolithic flint dagger from La Garma A.

the skulls could be connected with that belief, in an attempt to stop the deceased returning as ghosts, in the same way that the bodies of other condemned or accursed individuals are mutilated or have their heads and other anatomical parts cut off. This attempt at an interpretation, which needs to be verified, is a response to one of the fascinating challenges that the archaeological zone at La Garma poses for historical research.

Final remarks

The long sequence at La Garma provides an excellent panorama of the diachronic evolution of human activities in a cave environment. As we have seen, the different sites display evidence of the presence, even if it is occasional, of the societies who occupied the region throughout most of history. Of the periods into which we usually divide up the past, only the Iron Age, Antiquity and the Modern age are absent.

Figure 8.10: Early medieval funerary deposit in Zone V of the Lower Gallery.

However, the intensity and range of the uses made of the caves vary considerably. Most of the evidence is concentrated in the Upper Palaeolithic and the third and second millennia cal BC. The density and variety of the evidence is greatest in the former period. The sequence at La Garma A reveals an intermittent use of the cave vestibule as a habitat, with some periods of intense activity. Certainly, the middle Magdalenian is the period when the most intense use of the areas inside the cave entrances has been seen, with a large camp in Zone I in the Lower Gallery. It is interesting to highlight the possible simultaneous use of two very nearby caves in this period, while it remains to be seen whether the surface of the Lower Gallery hides earlier archaeological deposits. But human activity in the Upper Palaeolithic was not limited to dwelling in the entrances. The Lower Gallery also contains abundant evidence of a long history of human presence deeper in the cave. A first exploration of the available archaeological documentation suggests a wide variety of social activities, including art and other ritual kinds of phenomena.

In the Mesolithic, human groups were still present in the karst at La Garma, but now the centre of interest moves towards the exterior. No signs of human presence have been found in deep areas of the karst, and it may even be thought that the archaeological deposits in the cave entrances correspond simply to the use of these as places to accumulate the waste of camps located outside.

Apart from some evidence found in La Garma A, the neolithisation process involved abandoning the caves in La Garma hill as dwelling places. However, in two phases, the caves were used as burial sites: the long and intense stage in the third and second millennia cal BC, and the more limited and enigmatic episode in the seventh and eighth centuries AD.

It is not easy to give an exact explanation of the factors that influenced these changes in the use of caves by humans. Evidently, one factor that should be taken into account is the environment. It is probably not by chance that the periods of most intensity in the use of the caves as a habitat coincides with the Pleistocene. The cave vestibules provide natural shelter from the worst weather, particularly when the entrances face south, which is the case with the two galleries at La Garma that were occupied in the Palaeolithic.

The improvement in the climate in the Holocene freed, to a certain extent, the human groups from the limitations imposed by the harsh glacial conditions. However, at La Garma, the hunter-gatherers at the time of the post-glacial climatic optimum continued to maintain a link with the caves, which in this case is more difficult to interpret.

The domestic use of caves was abandoned with the development of peasant societies. In this case, it seems reasonable to resort to explanations connected with the new needs of the populations. These were probably settled in villages or farms near their fields, and thus the caves in La Garma hill became spaces specialised in symbolic and ritual functions, especially in relation with funerary practices. Another factor should be borne in mind: the intense accumulation of sediment during the Pleistocene and the early Holocene caused the entrances of most of the caves in the karst at La Garma to be practically filled to the roof, so they were no longer suitable as dwelling places. However, their new morphology, with low and narrow entrances, made them particularly suitable for a use as the last home of the dead.

In fact, symbolism appears to have been a factor of great significance throughout the long history of the relationship between humans and the cave environment documented at La Garma. It is clear that the large concentration of archaeological evidence in the Upper Palaeolithic is not only derived from utilitarian factors, such as shelter from inclement weather. Human activity inside the caves can be explained better by the profound symbolic significance this dark, mysterious and hostile environment must have held for those hunters. Times changed symbols too without doubt, but the choice of this environment for burials several millennia later is probably still connected with that attraction, inseparably linked with fear, that the world of caves has always held, even until the present time, for our species.

Acknowledgements

The research at La Garma has been funded through an agreement between the Government of Cantabria and the University of Cantabria. All the dates in this paper are given in calendar years. The calibration of the radiocarbon determinations was performed with the IntCal09 curve (Reimer *et al.* 2009), using the OxCal 4.1 program (Ramsey 2009).

References

Arias, P. (1999) Antes de los cántabros. Panorama del Neolítico y las Edades de los metales en Cantabria. In *I Encuentro de Historia de Cantabria*, 209–254. Santander, Servicio de Publicaciones de la Universidad de Cantabria-Consejería de Cultura y Deporte del Gobierno de Cantabria.

Arias, P. (2009) Rites in the dark? An evaluation of the current evidence for ritual areas at Magdalenian cave sites. *World Archaeology* 41(2), 262–294.

Arias, P. and Álvarez Fernández, E. (2004) Iberian foragers and funerary ritual: a review of Paleolithic and Mesolithic evidence on the Peninsula. In M. R. González Morales and G. A. Clark (eds.) *The Mesolithic of the Atlantic Façade: Proceedings of the Santander Symposium*, 225–248. Tempe, Arizona State University.

Arias, P., Armendariz, Á., Balbín, R., Fano, M. Á., Fernández-Tresguerres, J., González Morales, M. R., Iriarte, M. J., Ontañón, R., Alcolea, J. J., Álvarez Fernández, E., Etxeberria, F., Garralda, M. D., Jackes, M. and Arrizabalaga, Á. (2009) Burials in the cave: new evidence on mortuary practices during the Mesolithic of Cantabrian Spain. In S. B. McCartan, R. J. Schulting, G. Warren and P. Woodman (eds.) *Mesolithic Horizons: Papers Presented at the Seventh International Conference on the Mesolithic in Europe, Belfast 2005*, 650–656. Oxford, Oxbow Books.

Arias, P., González Sainz, C., Moure, A. and Ontañón, R. (2003) Unterirdischer Raum, Wandkunst und paläolithische Strukturen: einige Beispieleder Höhle La Garma (Spanien). In A. Pastoors and G. Weniger (eds.) *Höhlenkunst und Raum: Archäeologische und ArchitektonischePerspektiven*, 29–46. Mettmann, Neanderthal Museum.

Arias, P., Laval, E., Menu, M., González Sainz, C. and Ontañón, R. (2011) Les colorants dans l'art pariétal et mobilier paléolithique de La Garma (Cantabrie, Espagne). *L'Anthropologie*, 115 (3–4), 425–445.

Arias, P. and Ontañón, R. eds. (2005) *La Materia del Lenguaje Prehistórico: El Arte Mueble Paleolítico de Cantabria en su Contexto*. 2nd edition. Santander, Instituto Internacional de Investigaciones Prehistóricas de Cantabria.

Arias, P., Ontañón, R., Álvarez Fernández, E., Aparicio, T., Chauvin, A. M., Clemente, I., Cueto, M., González Urquijo, J. E., Ibáñez, J. J., Tapia, J. and Teira, L. C. (2005) La estructura magdaleniense de La Garma A: aproximación a la organización espacial de un hábitat paleolítico. In N. F. Bicho (ed.) *O Paleolítico: Actas do IV Congreso de Arqueologia Peninsular (Faro, 14 a 19 de Setembro de 2004)*, 123–141. Faro, Centro de Estudos de Património, Universidade do Algarve.

Arias, P., Ontañón, R., Álvarez Fernández, E., Cueto, M., Elorza, M., García-Moncó, C., Güth, A., Iriarte, M. J., Teira, L. C. and Zurro, D. (in press) Magdalenian floors in the Lower Gallery of LaGarma: a preliminary report. In S. Gaudzinski-Windheuser, O. Jöris, M. Sensburg, M. Street and E. Turner (eds.) *Come in ... and Find out: Opening a New Door into the Analysis of Hunter-Gatherer Social Organisation and Behaviour. Proceedings of Colloquium 58. 15th U.I.S.P.P. Congress, Lisbon*. Mainz, Verlag des Römisch-Germanischen Zentralmuseums.

Arias, P., Ontañón, R., González Urquijo, J. E. and Ibáñez, J. J. (1999) El puñal de sílex calcolítico de La Garma A (Omoño, Cantabria). In *Sautuola VI. Estudios en Homenaje al Profesor Dr. García Guinea*, 219–228. Santander, Consejería de Cultura y Deporte del Gobierno de Cantabria.

Cabrera, V. (1984) *El Yacimiento de la Cueva del Castillo (Puente Viesgo, Santander)*. Madrid, Ministerio de Cultura.

Fano, M. Á. (2001) Habitability of prehistoric settlements: proposal for the study of one of the elements involved, and first results for the Cantabrian Mesolithic (Northern Spain). *Journal of Iberian Archaeology* 3, 25–34.

Garralda, M. D. (2005–6) Los Neandertales en la Península Ibérica. In *Homenaje a Jesús Altuna. Tomo III: Arte, Antropología y Patrimonio Arqueológico*, 289–314. San Sebastián, Sociedad de Ciencias Aranzadi.

González Morales, M. R. (1982) *El Asturiense y otras Culturas Locales: La Explotación de las Áreas Litorales de la Región Cantábrica en los Tiempos Epipaleolíticos*. Santander, Ministerio de Cultura.

González Sainz, C. (2003) El conjunto parietal paleolítico de la Galería Inferior de La Garma (Cantabria): avance de su organización interna. In R. de Balbín and P. Bueno (eds.) *Arte Prehistórico Desde los Inicios del s. XXI: Primer Symposium Internacional de Arte Prehistórico de Ribadesella*, 201–222. Ribadesella, Asociación Cultural Amigos de Ribadesella.

Gutiérrez Cuenca, E. (2010) *Los Comportamientos Funerarios durante la Prehistoria Reciente en la Región Cantábrica: El DepósitoSepulcral de la Cueva de La Garma B (Omoño, Cantabria)*. Unpublished Masters dissertation, Santander, Universidad de Cantabria.

Hierro, J. Á. (2008) *La Utilización de las Cuevas en Cantabria en Época Visigoda: Los Casos de Las Peñas, La Garma y El Portillo del Arenal*. Unpublished Masters Dissertation. Santander, Universidad de Cantabria.

Julien, M., Audouze, F., Baffier, D., Bodu, P., Coudret, P., David, F., Gaucher, G., Karlin, C., Larriere, M., Masson, P., Olive, M., Orliac, M., Pigeot, N., Rieu, J. L., Schmider, B. and Taborin, Y. (1988) Organisation de l'espace et fonction des habitats magdaléniens du Bassin Parisien. In M. Otte (ed.) *De la Loire à l'Oder: Les Civilisations du Paléolithique Final dans le Nord-Ouest Européen. Actes*, 85–123. Oxford, British Archaeological Reports.

Laming-Emperaire, A. (1962) *La Signification de l'Art Rupestre Paléolithique: Méthodes et Applications*. Paris, Picard.

Leroi-Gourhan, A. (1965) *Préhistoire de l'Art Occidental*. Paris, Lucien Mazenod.

Muckle, R. J. (2006) *Introducing Archaeology*. Toronto, University of Toronto Press.

Olive, M., Audouze, F. and Julien, M. (2000) Nouvelles données concernant les campements magdaléniens du Bassin parisien. In B. Valentin Eriksen, P. Bodu and M. Christensen (eds.) *L'Europe Centrale et Septentrionale au Tardiglaciaire. Actes de la Table-Ronde Internationale de Nemours (14–16 mai 1997)*, 289–304. Nemours, Éditions de l'Association pour la Promotion de la Recherche Archéologique en Île-de-France.

Oliver, P. (1997) *Encyclopaedia of Vernacular Architecture of the World*. Cambridge, Cambridge University Press.

Ontañón, R. (2003) Sols et structures d'habitat du Paléolithique supérieur, nouvelles données depuis le Cantabres: la Galerie Inférieure de La Garma (Cantabrie, Espagne). *L´Anthropologie* 107, 333–363.

Ontañón, R. and Armendariz, Á. (2005–6) Cuevas y megalitos: los contextos sepulcrales colectivos en la Prehistoria reciente cantábrica. In *Homenaje a Jesús Altuna. Tomo II: Arqueología*, 275–286. San Sebastián, Sociedad de Ciencias Aranzadi.

Ramsey, C. B. (2009) Bayesian analysis of radiocarbon dates. *Radiocarbon* 51(1), 337–360.

Reimer, P. J., Baillie, M. G. L., Bard, E., Bayliss, A., Beck, J. W., Blackwell, P. G., Ramsey, C. B., Buck, C. E., Burr, G. S., Edwards, R. L., Friedrich, M., Grootes, P. M., Guilderson, T. P., Hajdas, I., Heaton, T. J., Hogg, A. G., Hughen, K. A., Kaiser, K. F., Kromer, B., McCormac, G., Manning, S. W., Reimer, R. W., Richards, D. A., Southon, J. R., Talamo, S., Turney, C. S.

M., van der Plicht, J. and Weyhenmeyer, C. E. (2009) IntCal09 and Marine09 Radiocarbon age calibration curves, 0–50,000 years cal BP. *Radiocarbon* 51(4), 1111–1150.

Rouzaud, F. (1997) La paléospéléologie. In *Proceedings of the 12th International Congress of Speleology. La Chaux de Fonds, 10–17 August 1997*, 49–52. Basel, Speleo Projects.

Straus, L. G. (1979) Caves: a palaeoanthropological resource. *World Archaeology* 10(3), 331–339.

Waselkov, G. A. (1987) Shellfish gathering and shell midden archaeology. *Advances in Archaeological Method and Theory* 10, 93–210.

Chapter 9

Shedding Light on Dark Places: Deposition of the Dead in Caves and Cave-Like Features in Neolithic and Copper Age Iberia

Estella Weiss-Krejci

The Iberian Peninsula is rich in caves, rockshelters, avens and rock fissures which were used for the deposition of the dead from the Palaeolithic until early historical times. This article discusses contexts with human remains which date to the Neolithic and Copper Age (mid sixth to late third millennia cal BC). Since cave deposits during this period often contain disarticulated multiple individuals ranging from a few to hundreds of people their analysis and interpretation is not easy. Ethnographic analogy with cave burials from all over the world shows that similar deposits can result from a range of diverse practices and processes and are imbued with different meanings.

Introduction

Caves and rockshelters are landscape features which have been used by people for a variety of purposes throughout time and the world (Barber and Hubbard 1997; Bonsall and Tolan-Smith 1997; Barker *et al.* 2005; White and Culver 2005; Angelucci *et al.* 2009). One common ritual use of caves and cave-like structures is for the deposition of the dead, a behaviour which apparently dates back as far as the Middle Palaeolithic (Mouret 2004; Bar-Yousef *et al.* 2009). However, what archaeologists generally call 'burials' in caves (e.g. Nelson and Barnett 1955; Hubbard and Barber 1997; Antón and Steadman 2003) can result from natural and cultural, ritualistic and non-ritualistic processes. Not only funerary rites but death by accident, murder, human sacrifice, witchcraft and post-funerary rituals can equally be held responsible (Pastron and Clewlow 1974, 310; Stone 1997, 204; Gargett 1999; Steadman *et al.* 2000; Simmons 2002, 264; Mouret 2004; Toussaint *et al.* 2004, 164; Lucero and Gibbs 2007).

Considering the intense natural and human disturbance agencies in caves and rockshelters, to reconstruct the formation processes of mortuary deposits and to understand their meaning is not easy. In this article I will discuss these problems using the example of Spain and Portugal. The Iberian Peninsula is rich in caves, rockshelters and rock fissures which were used for the deposition of the dead from the Palaeolithic until early historical times (e.g. González and Freeman 1978; Ontañón 2003; Benítez *et al.* 2007; Bischoff *et al.* 2007; Polo-Cerdá *et al.* 2007; Trinkaus *et al.* 2007; Ruiz *et al.* 2008; Zilhão 2009, 827). Below, I will focus on caves which were used during the Neolithic and Copper Age including a late Bell Beaker horizon. In Portugal these mainly occur in the Beira Litoral, Alto Ribatejo, Estremadura and Algarve; in Spain they are located in Cantabria, the Basque Country, Navarra, Catalonia, Madrid, Castile-La Mancha, the Valencian region, Murcia and the central and eastern parts of Andalusia (Fig. 9.1).

Mortuary contexts in Neolithic and Copper Age Iberia

The Iberian Neolithic and Copper Age lasts from the second half of the sixth to the late third millennium cal BC. Despite considerable regional chronological variation (e.g. Pellicer 1995; Zilhão 2000, 2003; Gibaja 2004, 679; Díaz del Río and García 2006; Chapman

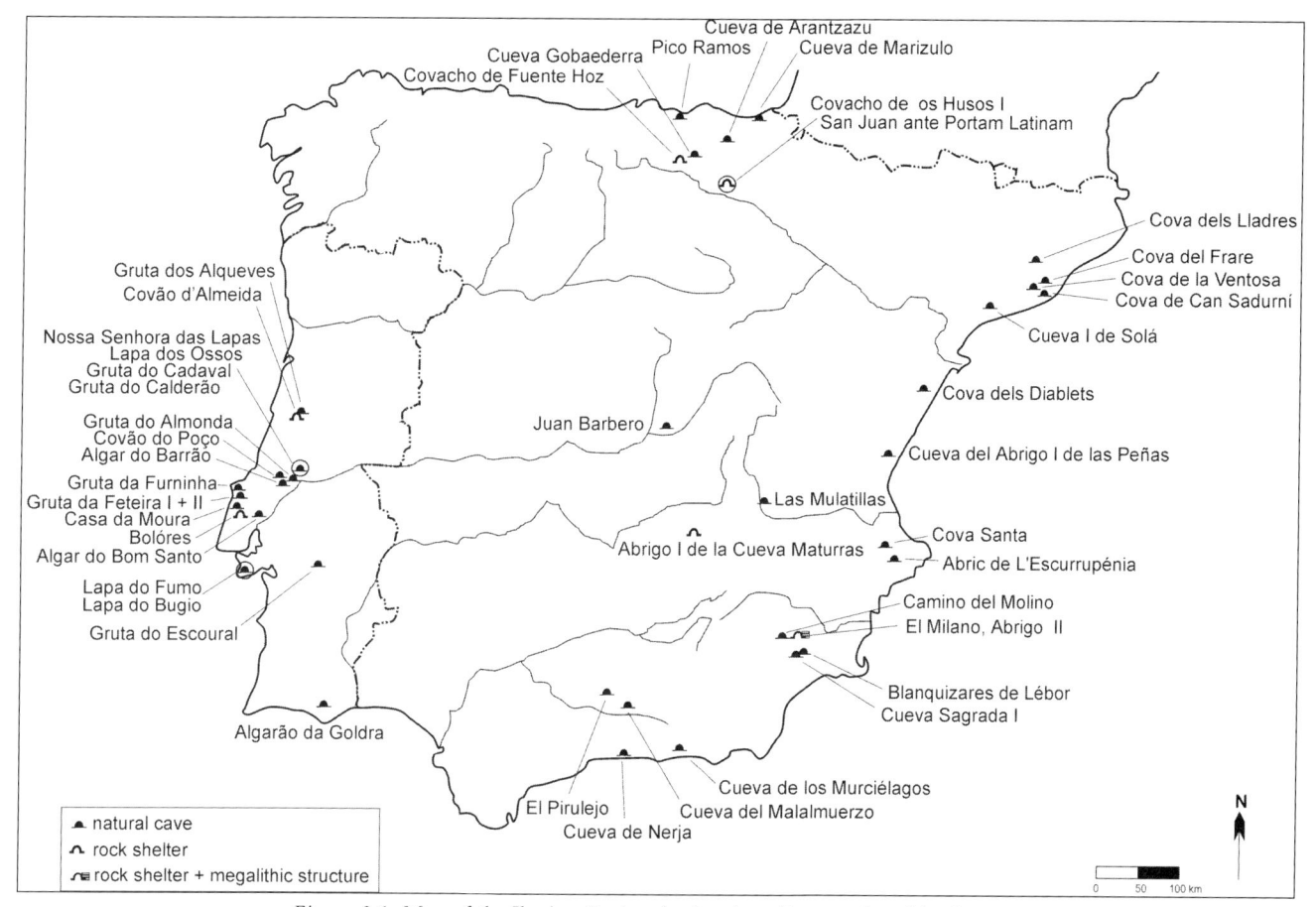

Figure 9.1: Map of the Iberian Peninsula showing sites mentioned in the text.

2008, 211) this period of over 3000 years can be divided into three major phases: from the mid sixth to the mid fifth millennia cal BC (6700/6500–5600/5400 BP), – from the mid fifth to the mid fourth millennia cal BC (5600/5400–4900/4700 BP), and from the mid fourth to the late third millennia cal BC (4900/4700–3700/3500 BP).

From the mid sixth to the mid fifth millennia cal BC

In the Early Neolithic a gradual adoption of a sedentary lifestyle accompanied the introduction of plant cultivation and animal husbandry (e.g. Arias 1999; Rojo and Kunst 1999, 48; Jorge 2000; Zilhão 2003; Zapata *et al.* 2004; Chapman 2008). In the sixth millennium cal BC Early Neolithic farming groups overlap with Late Mesolithic hunter-gatherers (e.g. Zilhão 2000; Cunha *et al.* 2002, 189; Alday 2005; Guilaine and Manen 2007, 25). For burial of their dead, Mesolithic groups used shell middens in Atlantic and Mediterranean river estuaries as well as cave entrances and rockshelters (Arias and Pérez 1990; Arias and Garralda 1996; González 1997; Jackes *et al.* 1997; Arias 1999, 415; Zilhão 2000, 2004a; Arias and Álvarez-Fernández 2004, 237; Garcia *et al.* 2006). Human remains pertaining to Early Neolithic groups have been encountered in simple pits at open air sites and in the interior of caves (Rojo and Kunst 1999, 24–35; Zilhão 2004b; Guilaine and Manen 2007, 27).

Three Portuguese and three Spanish caves hold more-or-less indubitable evidence for deposition of the dead during the Early Neolithic (see Table 9.1 and Fig. 9.1). In burial caves from the Valencian region (Asquerino 1976; Martí *et al.* 1980) and central and eastern Andalusia (Acosta 1995, 36–49; Pellicer 1995, 83–99) the relationship between the Early Neolithic materials and the human remains is not entirely clear (see also Soler 1997, 349; Zilhão 2001). Also not included in Table 9.1 is one problematic radiocarbon date from Casa da Moura, Portugal (5990±60 BP, Strauss *et al.* 1988, 70).

Cave/rockshelter	Region, province/Country	BP/material	Literature
Mid 6th to mid 5th millennium cal. BC			
Gruta do Almonda	Alto Ribatejo/Portugal	6445±45/bone beads	Zilhão 2009, 829
Can Sadurní	Barcelona/Spain	6405±50/cereal seed	Blasco *et al.* 2005, 628
Gruta do Caldeirão	Alto Ribatejo/Portugal	6130±90/human bone	Zilhão 1992, 78
Nossa S. das Lapas	Alto Ribatejo/Portugal	6100±70/human bone	Oosterbeek 1993, 55
Cueva de los Murciélagos	Granada/Spain	6086±45/sandal fiber	Cacho *et al.* 1996, 116
Cueva de los Murciélagos	Granada/Spain	5900±38/sandal fiber	Cacho *et al.* 1996, 116
Cueva de los Murciélagos	Granada/Spain	5861±48/fiber	Cacho *et al.* 1996, 116
Gruta do Caldeirão	Alto Ribatejo/Portugal	5810±70/human bone	Zilhão 1992, 78
Cueva de Nerja	Málaga/Spain	5785±80/human bone	Simón *et al.* 2005, 650
Mid 5th to mid 4th millennium cal. BC			
Gruta do Cadaval	Alto Ribatejo/Portugal	5350±50/human bone	Oosterbeek 1997, 72
Cova dels Lladres	Barcelona/Spain	5330±90/juniper seed	Ten 1989, 355
Cueva de Marizulo	Guipúzcoa/Spain	5285±65/human bone	Armendariz 1992, 16
El Milano, Abrigo II	Murcia/Spain	5220±280/human bone	Walker and San N. 1995, 111
Gruta do Cadaval	Alto Ribatejo/Portugal	5180±140/human bone	Cruz 1997, 219
Gruta do Cadaval	Alto Ribatejo/Portugal	5160±50/human bone	Oosterbeek 1997, 72
Covacho de Fuente Hoz	Álava/Spain	5160±110	Armendáriz 1992, 16
Nossa S. das Lapas	Alto Ribatejo/Portugal	5130±140/human bone	Cruz 1997, 213
Algarão da Goldra	Algarve/Portugal	4990±320/charcoal	Straus *et al.* 1992, 144
Gruta do Caldeirão	Alto Ribatejo/Portugal	4940±70/human bone	Zilhão 1992, 78

Table 9.1: Uncalibrated radiocarbon dates for 12 Iberian sites, which were used for the deposition of the dead during the Early and Middle Neolithic.

From the mid fifth to the mid fourth millennia cal BC

Apart from a few early exceptions, the emergence of megalithism characterises the second part of the fifth millennium cal BC (Jorge 2000, 59–61; González *et al.* 2004, 66; García 2006, 152). In most areas it post-dates the introduction of agriculture by hundreds of years (Zilhão 2000; Zapata *et al.* 2004, 294, 315). Only in the northern coastal fringe of the Bay of Biscay, where human groups introduced agro-pastoralism as the main basis of subsistence rather late (around 4600–4300 cal BC), does the construction of megalithic tombs form a more-or-less contemporaneous phenomenon (Peña-Chocarro *et al.* 2005, 585–586). In regions which are rich in caves these continue to play an important role as mortuary spaces (Rubio 1981; Armendariz 1992, 16; Cruz 1997; Lomba *et al.* 2009). In Catalonia, additionally cists in open settlements and the interior of mines are used to deposit the dead (Fernández and Pérez 1989; Gibaja 2004; Blasco *et al.* 2005).

Four Portuguese and four Spanish rock features are listed in Table 9.1. Escoural cave (Fig. 9.2) (5560±160 BP on a sample of human bone; Soares 1995, 115), Lapa do Fumo (5040±160 BP on charcoal, Soares and Cabral 1993, 230) and the rockshelter San Juan ante Portam Latinam (5070±140 and 5020±140 BP on human bone, Vegas 1992, 12) have been omitted, since their radiocarbon dates are considered as too early.

From the mid fourth to the late third millennia cal BC

The Late Neolithic and Copper Age is characterised by population increase and technological innovation. People started to settle in ditched and walled enclosures and develop gold and copper metallurgy (Díaz and

García 2006; Chapman 2008). At a few sites the dead are buried under Late Neolithic dwellings (Upper Guadalquivir valley: Lizcano *et al.* 1992) and within Chalcolithic walled enclosures (Portugal: Jorge *et al.* 1999). The majority of bones, however, appear in natural rock features, in megalithic monuments, artificial limestone caves (*hypogea*), *tholoi*, *tumuli*, silos and pits (e.g. Parreira and Serpa 1995; Silva 1999; Weiss-Krejci 2005, 2006a; Díaz-del-Río 2006, 72; García 2006, 154–160).

The number of caves with human remains dating to this time period is large (Fernández and Pérez 1989; Armendariz 1990; Ramos and Pita 1992; Etxeberria, 1994; Cardoso and Soares 1995; Cruz 1997; Rodanés 1997; Augustí 2002; Pascual 2002; Rubio 2002; Soler 2002) but in many instances bones and artefacts have been encountered in a very disturbed and messy condition. A sample of 22 caves, rockshelters and rock fissures with contextualised radiocarbon dates is provided in Table 9.2. A comparison of dates available for these mortuary contexts indicates that – especially in those areas where artificial caves and *tholoi* become increasingly important – burial in natural caves is less frequent in the second part of the third millennium cal BC (García 2006, 151–160, tables 11.1 to 11.4). Many caves are used (or reused) for the deposition of the dead by people of the Bell Beaker horizon (Llongueras *et al.* 1981; Oosterbeek 1993, 55; Pascual 2002).

The meaning and mortuary use of caves and rockshelters: cross-cultural insights

Walker (1983, 43) proposed that cave burials are part of a single tradition of vault-burial and the equivalent of burial in monumental tombs in non-karst areas. Oosterbeek (1997, 70–71) also considers megaliths as types of artificial caves with the cave or tomb chamber symbolizing the 'uterus of Mother Earth'. From an overall Iberian perspective it seems that megalithic

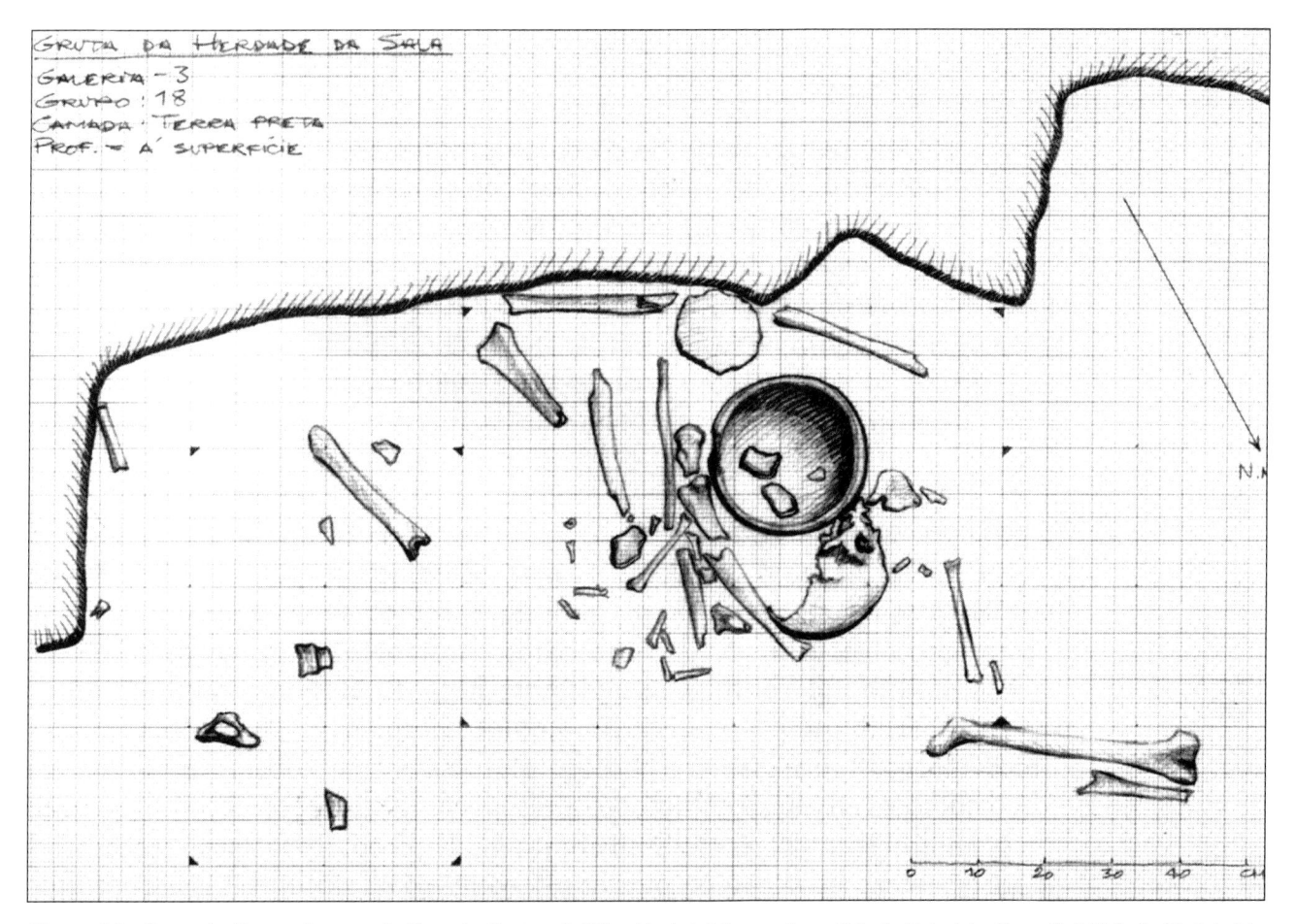

Figure 9.2: Gruta do Escoural, central Alentejo, Portugal. Disarticulated bones from Galería 7 (originally called Galería 3) dated to 5560±160 BP. The drawing was made in 1963 (after Araújo and Lejeune 1995, 76; original in the possession of the National Museum of Archaeology, Lisbon).

Cave/rockshelter	Region, province/Country	BP/material	Literature
Mid 4th to late 3rd millennium BC			
Algar do Bom Santo	Estremadura/ Portugal	4860±45/human bone	Duarte 1998, 113
Lapa do Bugio	Estremadura/ Portugal	4850±45	Monteiro *et al.* 1971, 117
Covacho de los Husos I	Álava/ Spain	4830±110	Álvarez *et al.* 1997, 297
Pico Ramos	Vizkaya/ Spain	4790±110/human bone	Zapata 1995, 59
Gruta da Feteira II	Estremadura/ Portugal	4760±80/human bones	Waterman 2006
Gruta do Escoural	Central Alentejo/ Portugal	4680±80/human bone	Soares 1995, 115
Algar do Barrão	Alto Ribatejo/ Portugal	4660±70/human bone	Carvalho *et al.* 2003, 117
Gruta dos Ossos	Alto Ribatejo/ Portugal	4630±80/human bone	Cruz 1997, 216
Algar do Bom Santo	Estremadura/ Portugal	4630±60/human bone	Duarte 1998, 113
Casa da Moura	Estremadura/ Portugal	4600±90/bone pin	Cardoso and Soares 1995, 11
S. Juan a. Portam Lat.	Álava/ Spain	4570±40/human bone	Armendáriz 2007
Gruta da Feteira I	Estremadura/ Portugal	4570±70/human bone	Zilhão 1995
Cova das Lapas	Estremadura/ Portugal	4550±60/human bone	Soares and Cabral 1993, 230
Gruta do Escoural	Central Alentejo/ Portugal	4500±60/human bone	Soares 1995, 115
Gruta dos Alqueves	Beira Litoral/ Portugal	4490±50/human bone	Cabral 1987, 50
Covão d'Almeida	Beira Litoral/ Portugal	4480±60/human bones	Gana and Cunha 2003, 132
Gruta dos Ossos	Alto Ribatejo/ Portugal	4460±110/human bone	Cruz 1997, 216
S. Juan a. Portam Lat.	Álava/ Spain	4460±70/human bone	Armendáriz 2007
Cova del Frare	Barcelona/ Spain	4450±100/charcoal	Martín *et al.* 1981, 108
Lapa do Bugio	Estremadura/ Portugal	4420±110/bone object	Cardoso and Soares 1995, 11
Lapa do Fumo	Estremadura/ Portugal	4420±45/bones	Soares and Cabral 1993, 230
Gruta do Escoural	Central Alentejo/ Portugal	4420±60/human bones	Soares 1995, 115
Cueva de Arantzazu	Guipúzcoa/ Spain	4390±55	Armendáriz *et al.* 1998, 119
Gruta da Feteira II	Estremadura/ Portugal	4370±45/human bones	Waterman 2006
Covão do Poço	Alto Ribatejo/ Portugal	4360±60/human bones	Carvalho *et al.* 2003, 117
Gruta da Furninha	Estremadura/ Portugal	4335±65/bone bead	Cardoso and Soares 1995, 11
Camino del Molino	Murcia/ Spain	4260±40/human bone	Lomba *et al.* 2009, 219
Pico Ramos	Vizkaya/ Spain	4210±110/human bone	Zapata 1995, 59
S. Juan a. Portam Lat.	Álava/ Spain	4200±95/human bone	Armendáriz 2007
Bolóres rock shelter	Estremadura/ Portugal	4150±40/human bone	Lillios *et al.* 2010
Gruta da Feteira I	Estremadura/ Portugal	4110±60/human bone	Zilhão 1995
Pico Ramos	Vizkaya/ Spain	4100±110/human bone	Zapata 1995, 59
Bolóres rock shelter	Estremadura/ Portugal	4050±40/human bone	Lillios *et al.* 2010
Algar do Bom Santo	Estremadura/ Portugal	4030±280/human bone	Duarte 1998, 113
Bolóres rock shelter	Estremadura/ Portugal	4000±40/human bone	Lillios *et al.* 2010
Camino del Molino	Murcia/ Spain	3990±40/human bone	Lomba *et al.* 2009, 219
Gruta dos Ossos	Alto Ribatejo/ Portugal	3970±140/human bone	Cruz 1997, 216
Camino del Molino	Murcia/ Spain	3950±40/human bone	Lomba *et al.* 2009, 219
Cueva Sagrada I	Murcia/ Spain	3870±100/fibre	Eiroa and Lomba 1998, 99
Bolóres rock shelter	Estremadura/ Portugal	3530±40/human bone	Lillios *et al.* 2010

Table 9.2: Uncalibrated radiocarbon dates for 22 Iberian cave sites, which were used for the deposition of the dead during the Late Neolithic and Copper Age. For San Juan Ante Portam Latinam five additional contemporary dates have been published (4520±75 BP, 4520±50 BP, 4510±40 BP, 4440±40 BP, 4325±70 BP) (Armendáriz 2007). For Gruta do Escoural another two dates have been published (4610±60 BP, 4460±70 BP) (Soares 1995, 115) and for Algar do Bom Santo three additional dates exist (4780±50 BP, 4705±65 BP, 4430±50 BP) (Duarte 1998, 113).

tombs as well as artificial and natural rock features were imbued with similar symbolic meanings. Agricultural communities first used natural cave chambers for the deposition of the dead, which gradually inspired the construction of megalithic tombs in non-cave areas. Once artificial caves and *tholoi* were constructed the importance of caves for burial declined. However, the relationship between these different types of mortuary structure is not that simple. If one considers the patterns of cave use on a smaller regional scale, variable and contrary developments become visible (e.g. Jimenez and Barroso 1995). Hence, it cannot be taken for granted that the symbolism attached to caves was necessarily the same everywhere.

Another important question concerns different types of natural landscape features. Not only large caves with multiple chambers (e.g. Almonda, Caldeirão and Escoural caves) but rockshelters (Bolóres, and San Juan ante Portam Latinam), avens (Algarão da Goldra) and rock fissures (e.g. Cueva I de Solá de la Vila de Pradell, and Pico Ramos) have been used for the deposition of the dead. The question is whether the ancient inhabitants of Spain and Portugal perceived these features as different types of natural places and buried different categories of people in them. Ethnographers and archaeologists working in areas where historic records and a certain cultural continuity with modern groups of people exist have addressed some of these problems. Therefore, starting out with a cross-cultural discussion may help to shed some light on the Iberian evidence.

Etic and emic perceptions of caves and other rock features

Caves and other types of rock features are classified by a variety of factors such as shape, size and geology (e.g. Klimchouk 2004; White and Culver 2005). According to one definition (Weaver 2008, 6) a cave generally has an opening that is deeper than it is wide and penetrates a hill. A rockshelter is a natural overhang on the side of a rocky slope or cliff and is wider than it is deep. Other types of borderline caves are crevices and rock fissures – long, narrow cracks or openings in the face of a rock – as well as dolines and sinkholes (Lewis 2000). Rockshelters and horizontal entrance areas of caves may appear as safe and bright places. Not only can they usually be seen from afar but they are also places with a view. On the other hand, the interior of a cave is dark. From the inside of deep and narrow caves the view towards the outer world is entirely obstructed. Caves are ideal to keep things in as well as to keep things out whereas rockshelters do not have the same qualities.

Despite these obvious differences, there may be a conceptual continuum between caves and rockshelters. Geography is a cultural construct and the elements of a landscape must be defined in emic terms (Brady 1997, 603). One example for an emic definition of cave derives from Mesoamerica where caves and cave-like features played and still play a crucial role in ceremonies associated with rituals of life, death, fertility and rebirth (Brady and Veni 1992; Hapka and Rouvinez 1997; Stone 1997; Colas *et al.* 2000; Moyes and Brady 2005). In the Maya area caves are considered as portals to the underworld. The Maya word for cave means hole or cavity and applies to anything that penetrates the earth or a mountain. Not only caves but also dolines, sinkholes, places where springs appear and disappear, crevices, artificial holes dug into the limestone and man-made caves fall under this category; even a rockshelter when it is seen as an entry point into the underworld is a cave (Brady 1997, 603).

Perry (1915, 149–150) suggests that among certain Indonesian groups cave burial is connected to myths of origin. Caves symbolize the underground where the ancestors came from and dead people are returning to. Though associations between caves, the dead and the underworld are known from other areas of the world (e.g. Bradley 2000, 27) they must not be taken as universal. In Tibetan sacred geography, which distinguishes between mountains (up), lowlands and water (down) and the middle world of living people, mountains are the abode of gods and deities. They lend power to landscape features such as rocks, spring, plants and caves. Those caves which are considered as power places often serve as pilgrimage destinations (Aldenderfer 2005, 10–11).

Who is buried in caves and rockshelters?

The example of the Maya shows that knowledge of cosmology does not necessarily explain mortuary deposits. A wide range of ritual cave-associated practices was followed by the ancient Maya (around 1000 BC to AD 1500). Caves and rockshelters hold between one and hundreds of buried individuals, including all age groups (Rue *et al.* 1989; Saul and Saul 2002; Weiss-Krejci 2006b, 76). Interpretation of the skeletal evidence ranges from inhumation and cremation burials, burials of shamans, to the deposition of sacrificial victims and murdered witches (e.g. McNatt 1996; Stone 1997, 204; Stone and Brady 2005; Lucero and Gibbs 2007; Prufer and Dunham 2009). That caves served as deposition places for so called 'deviant social personae' – i.e. people who died an unusual or bad death, outcasts, witches, as well as the very young and very old (e.g. Murphy 2008; Aspöck 2009) – has not only been suggested for Maya caves. Age patterns and skeletal evidence for bad health or non-natural death can indicate the presence of these specific types of people. Leach (2008), who examined bones from British Neolithic burial caves north of Manchester, reaches the conclusion that the people in the caves had been excluded from regular ceremonies. This assessment is based on the fact that all these individuals suffered either from arthritis or injuries.

Ethnographic examples
Ethnographic accounts from Southeast Asia and the Pacific, where caves have been used as burial places for thousands of years (Harrison 1957; Kusch 1982; Antón and Steadman 2003; Anderson 2005; Barker 2005; Chazine 2005; Treerayapiwat 2005), confirm that the meaning of natural places is culturally constructed and therefore highly variable. The same applies to the identity of the people buried in them. Fison (1881, 143) reports that on Kadavu, one of the Fiji islands, commoners were thrown into a deep rocky chasm while chiefs were treated with greater ceremony. Among the Sa'dan Toraja of highland South Sulawesi (Indonesia) most individuals are buried in cliffside graves. Low-status individuals are buried within two days after death whereas the final deposition of high-status individuals takes a much longer time and follows a period of body storage and exhumation (Hutchinson and Aragon 2002, 32). Among the Igorot of the Cordillera Region of Luzon, Philippines, cave burial was reserved only for rich people whose bodies were mummified. However, in times of cholera and small pox epidemics the disease victims were put altogether in one burial cave, regardless of their social status (Picpican 2003, 111). The use of caves to hide away corpses is known from the Pacific region. On Mangaia, Cook Islands, caves were one among a variety of possible burial places. To bury a body in a secret cave or throw it into a chasm – especially the body of a dead warrior – sometimes simply had the purpose of ensuring that enemies could not get to it and burn it (Antón and Steadman 2003, 134). In the Fiji islands a dead chief was first buried in his house. After a while the bones were taken up and carried by night to some far-away inaccessible cave in the mountains, only known by a few trustworthy people. Ladders were constructed in order to reach the cave, but removed after deposition. As on Mangaia, the purpose was to protect the corpse from destruction by enemies (Fison 1881, 141–142).

Ethnographers have observed that no description of what is done by any one group can be taken as applicable to all the others (Fison 1881, 137; Kroeber 1927, 313; Ucko 1969, 270–271). One village may be cave-buriers while their neighbours bury people in earth graves. One group buries their chiefs in caves while among others it is a burial place for commoners. Yet another group may use caves for both commoners and elites, with the latter receiving more elaborate treatment. Circumstances of death and practical reasons also come into play.

Formation processes of cave deposits with human remains

The use of ethnographic information does not solve the problem of the meaning of caves in prehistoric Iberia but it helps to come up with possible scenarios to explain the state of human bones.

Deposition of articulated corpses
Let us first discuss skeletons which appear fully or partially articulated with a large percentage of the smaller bones present. These are usually considered as evidence of 'inhumation' or 'primary burial'. Though one should note that accidental death, the placement of murder victims and 'deviant social personae' (Simmons 2002, 264; Toussaint *et al.* 2004, 164; Lucero and Gibbs 2007) involving burial of corpses in the flesh (i.e. fresh corpses without decomposition) in caves has been reported from a variety of places in the world. For example, on Mangaia it was a custom to wrap the corpse in layers of cloth and then place it in a cave or in a deep vertical crevice within a day of death (Antón and Steadman 2003). However, not every articulated corpse which is deposited in a cave during a funeral is necessarily a primary burial. The deposition of anthropogenically mummified corpses in caves is known from many parts of the world, e.g. the Canary Islands, South America, the Philippines and Australia (Aufderheide 2003, 9–100, 162, 276–278, 284). The Guanche of the Canary Islands stored their mummified dead in hidden caves. The corpses – mainly those of the elite members of society – were wrapped in a series of leather shrouds, strapped to a wooden bier and deposited in the cave in a horizontal or vertical position (Eddy 1997, 87; Aufderheide 2003, 162). The Guanche, as well as the Muisca of Colombia and the Igorot of the Philippines, fire-dried the corpses without evisceration before depositing them in the cave (Aufderheide 2003, 99; Picpican 2003, 75).

If mistaken for evidence of inhumation, the skeletal remains of mummies constitute a serious methodological problem. Mummies or mummy parts can be moved around and re-deposited many times without losing bodily articulation (Parker Pearson *et al.* 2007; Weiss-Krejci 2011a, 166; 2011b, 80). Additionally when corpses are mummified by smoking, the fire may leave some of the bones charred. Though the bones will show evidence of burning while still fresh (= 'green'), this should not be mistaken for evidence of cremation (Weiss-Krejci 2005).

Post-depositional processes can be responsible for the way in which corpses that are buried articulated and complete end up disarticulated and incomplete. When a cave or rock overhang is used for successive deposition of bodies the bones of the previously buried are often disturbed. These disturbances may be accidental or intentional. One also needs to consider non-ritual post-depositional processes. Whether an articulated corpse was in the flesh or mummified, wrapped or unwrapped, placed in a jar or in a coffin, exposed on the surface of the cave floor or buried in a pit, placed on a wooden shelf or even hung from a cliff wall, will have an impact on the final distribution of the body parts (Alt *et al.* 2003; Picpican 2003, 112; Willis and Tayles 2009). Naturally, the body in the pit has a much higher chance of survival in a more-or-less complete state than the corpse on the cave floor. Collapsing shelves and hanging coffins which tumble down once the wood has rotted away will also not result in articulated bodies.

Deposition of disarticulate skeletons and selected bones
Disarticulation of bones at cave sites may also be attributed to so-called 'secondary' or 'multi-stage' funerals. In a variety of societies the bodies of the dead are temporarily stored in or above the ground and later exhumed for the final rites (Hertz 1907; van Gennep 1909). During the final funerary ceremony – either held for one person or multiple people – the fleshless skeleton or parts of it are carried to the caves and rockshelters (Kusch 1982, 94–96; Flood 1997, 197; Aufderheide 2003, 100; Panell and O'Connor 2005, 195). Among the Sa'dan Toraja the corpse of a high status individual – in contrast to that of a low status individual which is buried in the cliff-side grave a few days after death – resides in the house for a period ranging from several months to several years. When the storage time is completed the desiccated corpse is cleaned and the disarticulated bones are brought to a collective rock cave consisting of one or more burial chambers hewn out of the face of a cliff (Hutchinson and Aragon 2002, 33). The human remains in the caves are also periodically cleaned and reorganized. New clothes are fabricated and offerings are left in the cliff-side grave. Some bodies may even be reburied from one cave to another (Waterson 1995, 208–210).

The Yuko of Venezuela place the fresh corpse on a platform or hut floor for one month. Then they collect the bones; sew them in a cloth bag which is hung from the roof of a hut for several years. Only then is the skeleton with all its parts disposed of in a cave (Aufderheide 2003, 100). The Asabano of New Guinea collect the entire skeleton of men from their exposure platforms into feather-decorated net bags and carry them either to rockshelters or to men's sacred houses to be stored and used for magic and prayer (Lohmann 2005, 189). Rockshelters and caves are also used for the disposal of skulls without other bones, e.g. in New Guinea or in the Pacific (Stodder 2005, 246). At Vella Lavella, Solomon Islands, the skulls of the deceased are removed from the bodies after some time of storage and placed in huts or rockshelters (Fig. 9.3). Among the Dowayo of North Cameroon, the process is the other way round. The corpse is first buried in a rock feature, but the head is removed some three weeks later, cleaned and placed in a pot in a tree (Barley 1981, 151).

Reburial of body parts into a cave from another place, or reorganization within the cave, not only can take place as part of the funerary cycle but also centuries later. Human bones, which are treated as relics, can be transported between various locations in the landscape (Chamberlain 1999). People may also decide to exhume and bring the deceased with them when moving from one territory to another.

Mortuary patterns in prehistoric Iberian caves from the mid sixth to the late third millennia cal BC

Single and multiple bodies in the Early Neolithic

In the Iberian Peninsula the mortuary record has played a special role in shaping ideas about ancient social structure and cultural change. A strong emphasis has been made to differentiate between 'individual' and 'collective' burials. Cave burials which hold a single body are a hallmark of the early parts of the Neolithic. Examples are the Early Neolithic cave Nossa Senhora das Lapas, Portugal (Oosterbeek 1993) and the Middle Neolithic Marizulo cave, Basque region (Armendariz 1992, 16). In Late Neolithic and Copper Age contexts, on the other hand, single depositions are rare. Single and double graves with articulated corpses only reappear during the Bell Beaker times and in the Early Bronze Age (Fig. 9.4; e.g. Oosterbeek 1993, 55; Chapman 2005, 29).

The absence of articulated, single burials from contexts of the fourth and early third millennia BC has led to the erroneous assumption that 'collective' replaced 'individual' burial after the Early Neolithic (e.g. Armendariz 1992, 16; Oosterbeek 1997, 71).

Figure 9.3: Skulls at a native cemetery at Vella Lavella, Solomon Islands. To the right there is a stone figure carved in the face of the rock (photo by R. A. Lever in the 1940s; courtesy R. Mittersakschmöller archive).

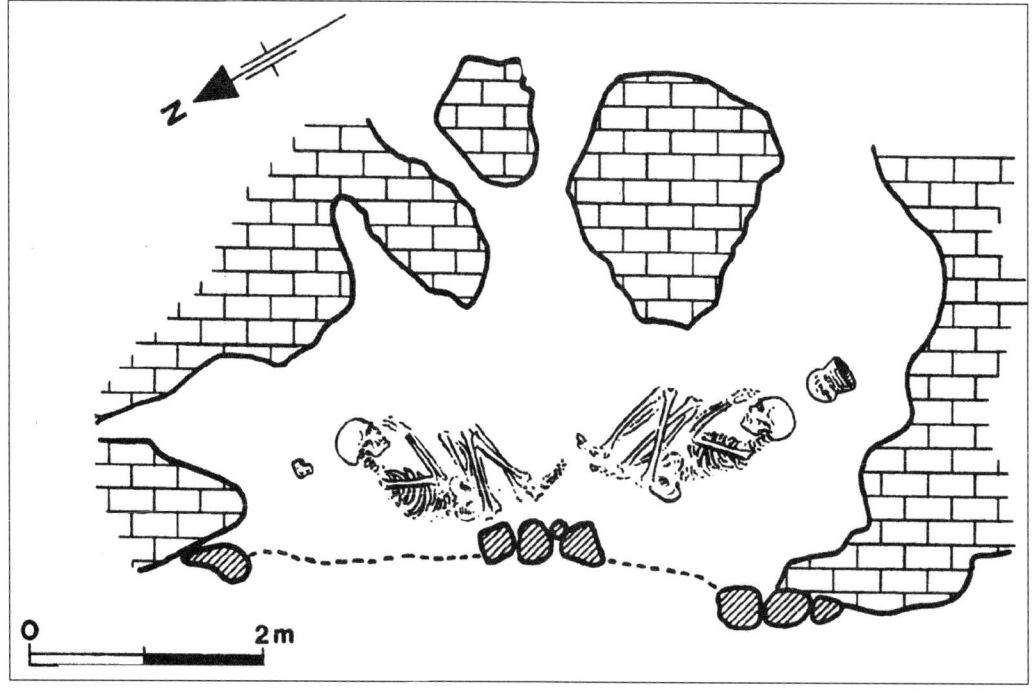

Figure 9.4: Two Bell Beaker burials at Cova de la Ventosa, Catalonia (after Llongueras et al. 1981, 99). The drawing was made by Antoni Bregante of the Catalonian Museum based on information provided by the site excavators. The excavation was directed by P. Ferrer; the materials turned over to M. Llongueras, the museum conservator. The study of the skeletons was conducted by D. Campillo and A. Martín (courtesy Museu d'Arqueologia de Catalunya and Araceli Martín; personal communication Araceli Martín).

Some have even gone as far as to suggest that these supposed transitions, first from individual to collective and then back to individual at the end of the third millennium BC, are evidence for increasing social differentiation (see Weiss-Krejci 2010a). Not only is this argument devoid of any logic but the re-evaluation of the evidence clearly shows that the practice of depositing multiple bodies is as old as the practice of single deposition. 'Collective' burials did not replace 'individual' ones. On the contrary, from the beginning of the Early Neolithic diverse burial practices existed (Jackes *et al.* 1997, 643).

One of the oldest 'collective' cave burial deposits has been found in the Cueva de los Murciélagos, Albuñol, Granada. Unfortunately the evidence from this cave – remarkable because of the excellent preservation of organic materials including baskets and sandals made from esparto grass – was already recovered in the nineteenth century. According to the reports many mummified bodies were found on the cave floor accompanied by grave goods (Alfaro 1980). The remains from two sandals and one burial shroud were dated to the Early Neolithic (see Table 9.1) (Cacho *et al.* 1996). Though it is now impossible to solve the problems of whether the dead had been put in the cave simultaneously or 'individually' one after another at different moments, of whether the mummification process was natural or artificial, and of how many dead people were included and for how long the cave was used, Murciélagos is an early example of the deposition of multiple bodies in one location.

Gruta do Calderão and Nossa Senhora das Lapas: two caves from the Alto Ribatejo, Portugal
A second contemporaneous and more reliable example derives from Portugal. Horizon NA2 of the Gruta do Caldeirão, which dates to 6130±90 BP (5296–4843 2σ cal BC), contained the scattered remains of one child and around five adults (at least two males and one female) distributed over a space of 9m². As in the Murciélagos cave, the bodies had been placed on the cave floor without protective features. One female was buried against the north wall together with a ceramic vessel fragment of the Cardial style. Two males were buried at the south wall, one associated with microliths and another with 120 beads made from freshwater snails and sea shells (Zilhão 1992, 75–78). The skeletal remains at the Caldeirão cave were located within a few meters from each other and, according to Jackes and Lubbell (1992, 260), were most likely deposited as entire bodies.

Not very far away from Gruta do Caldeirão – around 2km to the north – lies the small cave of Nossa Senhora das Lapas. In this cave the disturbed single burial of a child, which dates to 6100±70 BP (5320–4847 2σ cal BC), was encountered in layer B. In contrast to the individuals at Gruta do Caldeirão it had been buried in a simple stone-lined pit. The burial was associated with flint blades, one quartzite pebble, plain and linear incised pottery sherds, and beads of greenstone and *Glycymeris* shells (Oosterbeek 1993, 55; 1997, 71).

The existence of two contemporary but different types of mortuary deposits in two caves which are both located in the Nabão river valley in close proximity to each other could carry different implications. Apart from the possibility that we are looking at expressions of random mortuary practices of Early Neolithic people, these two mortuary deposits might belong to two different groups of people which followed different mortuary traditions – one occupying the area north of the Nabão river (Lapa dos Ossos), the other the area to the south (Caldeirão). The fact that they were neighbours and shared the same economic basis is in line with the observation that variability in mortuary behaviour between neighbouring groups is quite common in preindustrial societies throughout the world (e.g. Kroeber 1927). On the other hand, we also need to take into account the possibility that, as discussed above, specific types of caves may have been reserved for people who were not accorded proper funerary treatment. Whether this is the case at either the Caldeirão cave or at Nossa Senhora das Lapas or at both places cannot be determined at present. But one should not discount it as one among several possibilities. The problem is that, at the moment, nobody knows if there was an Early Neolithic burial norm and what it looked like, because we are probably missing out large parts of the population that received treatments and kinds of depositions which left no trace in the archaeological record (Guilaine and Manen 2007, 28).

Multiple bodies in caves and rockshelters of the Late Neolithic and Copper Age

Caves, rockshelters and rock fissures of the Iberian Late Neolithic and Copper Age are highly variable concerning the number of people, age patterns, the distribution and state of human skeletal elements, as well as associated animal bones and types and quantity of artefacts (see Weiss-Krejci 2005, 2006a, 2011a). The formation processes which created these deposits appear as complex as those described in ethnographic accounts from all over the globe. Deposition of bodies in the flesh, of mummies, of completely disarticulated but complete skeletons and of isolated body parts and bones could all be responsible. The alterations inflicted by ritual and non ritual post-depositional processes years or centuries after the funeral, and the fact that some of these formation processes have taken place in combination with each other, further complicate a satisfactory analysis and interpretation.

Number of individuals, age patterns and state of the bones

The numbers of individuals found in Late Neolithic/Copper Age caves and rock features range from one tooth at the Las Mulatillas cave, Valencia region (Paz de Miguel 2000) to over 1300 individuals at Camino del Molino, Murcia (Lomba *et al.* 2009). The rockshelter of San Juan ante Portam Latinam contained 338 individuals (Etxeberria and Herrasti 2007), Algar do Bom Santo a minimum of 121 (Duarte 1998) and the rock fissure of Pico Ramos 104 individuals (Baraybar and de la Rua 1995). At the Gruta da Feteira II a minimum of 68 people were distributed in two separate layers (Waterman 2006). Lapa do Bugio contained 28 individuals (Monteiro *et al.* 1971; Cardoso 1992), Cueva del Abrigo I de las Peñas at least 27 (Palomar 1983) and Algar do Barrão 20 (Carvalho *et al.* 2003). Fourteen were found at the Bolóres rockshelter and at Abric de l'Escurrupénia (Pascual 2002; Lillios *et al.* 2010), 13 in the Copper Age level of the Lapa do Fumo (Serrão and Marques 1971), 11 at Juan Barbero (Bermúdez and Pérez 1984, 113–117; Martínez 1984; Aliaga 2008, 25), between five and 10 at the Cueva Sagrada I (Doménech *et al.* 1987; Eiroa 1987), nine at Cova Santa (Martí 1981), around eight at Cueva I de Solá de la Vila de Pradell (Vilaseca 1972), seven at Cova del Frare (Martín *et al.* 1981), four individuals each at El Pirulejo (Asquerino 1999), Cova dels Diablets (Aguilella, Gusi and Olària 1999), Abrigo I de la Cueva Maturras (Gutiérrez, Gómez and Ocaña 2002) and Cueva del Malalmuerzo (Carrión and Contreras 1979), and two at Cueva de Arantzazu (Armendariz and Etxeberria 1996, 57).

Caves and rockshelters almost always hold a combination of sub-adults (individuals younger than 20 or 21 years of age) and adults though, as the following examples show, the ratios are variable. At San Juan ante Portam Latinam 201 individuals (60 per cent) are younger than 20 years of age and 137 (40 per cent) are older (Etxeberria and Herrasti 2007). At Pico Ramos 72 individuals are younger than 21 years (69 per cent) and 32 (31 per cent) older (Baraybar and de la Rua 1995). The Bolóres rock shelter held eight sub-adults and six adults (Lillios *et al.* 2010) and Cova del Frare five infants and two adults (Martín *et al.* 1981). At Gruta da Feteira II there are fewer sub-adults (26) than adults (42) (Waterman 2007).

Within the category of sub-adults in several instances specific age groups are under- or over-represented. This unbalanced representation suggests that at some sites specific age groups were selected or omitted from deposition in, or removal from, a cave or rockshelter. Lillios *et al.* (2010) think that the Bolóres rockshelter (Fig. 9.5) may have been a location where the bodies of at least some of the dead were placed for a period of time to allow for biological decomposition. While some of the adults were removed to another location for final burial, sub-adults (at Bolóres six individuals [43 per cent] are younger than seven years of age) were generally not provided with a secondary burial.

Not only age patterns but also the state and completeness of the bones and the distribution of artefacts and animal bones indicate if human bones accumulated through successive deposition in one place or reburial from elsewhere. Let us look at a few examples.

Successive deposition of corpses and post-depositional processes

The Neolithic rockshelter of San Juan ante Portam Latinam (around 338 individuals, Ebro Valley, 3300–3000 cal BC) was probably used by the local population as a so called 'primary' burial place. Though certain individuals were probably killed and placed there at the same time (*en masse*) – arrows were embedded in human bones – the overall evidence suggests that the tomb was used over a longer period, cleaned periodically and the bones reorganized and skulls stacked up. Among the grave goods there were more than 100 flint tools, including arrowheads, pottery fragments, two polished axes, and ornaments such as 200 *Dentalium* shells, necklace beads, wild-boar canines and drilled marine shells. There is no evidence that bones were burnt and only a few animal bones have been encountered (Guilaine and Zammit 2005:152–157; Vegas 2007). All age groups are present though only 95 individuals (28 per cent) are younger than seven years but 32 per cent died between seven and 20 (Etxeberria and Herrasti 2007).

Abrigo I de la Cueva Maturras (four individuals, Upper Guadiana, fourth to third millennia BC) forms part of a system of three small connected cavities, one of which was used for the deposition of the dead (Gutiérrez *et al.* 2002). Three individuals were found in an articulated position, buried on their right sides with the legs flexed and the hands in front of the faces. They were probably deposited in the flesh directly on the cave floor. The death of the fourth individual most likely preceded the death of the three others. It consisted of a jumble of bones which had been put over the body of one of the articulated corpses. Gutiérrez *et al.* (2002) think that the skeleton had been disturbed and rearranged when the other individuals were buried. Post-depositional manipulation of the bones is also indicated by one of the ceramic containers, which

Figure 9.5: Bolóres rock shelter, Estremadura, Portugal. Adult 1 is the most skeletally complete individual and dates to 4050 ± 40 BP (photo by E. Weiss-Krejci 2007, reproduced by courtesy of K. Lillios).

were found at the feet of the three articulated skeletons. All three pots contained charcoal but one also held two burnt human phalanges (Gutiérrez *et al.* 2002, 118). Apart from the ceramics various other objects such as bone tools, arrow points and flint blades accompanied the dead. The arrow points were not embedded in the dead people's flesh but probably formed part of a hunter/warrior tool kit (Gutiérrez *et al.* 2008). There was evidence for at least one burning episode which caused considerable damage to the human bones and artefacts. After the fire had burnt, the entire mortuary deposit was sealed and intentionally covered by a layer of calcarenite and stone blocks.

Deposition of disarticulated and incomplete skeletons
Cut marks on the bones and evidence of gnawing in combination with the almost complete absence of foot and hand bones, femurs, and so on suggests that the bones from the small cave of El Pirulejo (Andalusia, four individuals, Late Neolithic) had been exhumed and re-deposited together with the bones of hares (Asquerino 1999, 34–37). In some cases, e.g. Blanquizares de Lébor, lithic tools and bones showed evidence of burning whereas other artefacts were entirely unaffected by fire (Arribas 1953). Bones which belong to 13 individuals in the Copper Age layer of the Portuguese Lapa do Fumo (Estremadura, late fourth, early third millennia BC) were burnt, cut, gnawed and covered with red ochre. According to Serrão and Marques (1971, 136–138) a fire had been set in a specially prepared part of the cave with the disarticulated bones deposited on top. After the fire had died the bones were covered with grave goods, among them schist plaques and bone objects. These artefacts were not burnt. Afterwards the whole compound was sprinkled with red ochre. Ochre was found on top of burnt bones, inside bone cracks, on charcoal and on the objects.

The analysis of osteological material from the cave surface of Algar do Bom Santo, Portugal (at least 121 individuals, second half of the fourth and third millennia BC) reveals a clear dominance of human cranial and long bone fragments, mainly femurs and tibias. In this case, no evidence for cutting or burning was detected (Duarte 1998). At Algar do Barrão (Alto Ribatejo, 20 individuals, end of the fourth millennium BC) bodies were brought into the cave from the outside together with sediments after a phase of passive excarnation. The cave held animal bones pertaining to goats, deer, rabbits and dogs (Carvalho *et al.* 2003).

Ambiguous evidence
The mortuary deposit of the Cueva de Juan Barbero, Madrid (Bermúdez and Pérez 1984, 113–117) is thought to result from accumulation through successive deposition (Aliaga 2008, 25). However, there is an alternative explanation. Fragmentary bones, which are out of position, piled up to form packages, burnt, covered with ochre and deposited with animal bones can also result from exhumation and reburial.

At the rock fissure of Pico Ramos (104 individuals) the human bones were fragmentary and commingled, but a few body parts showed correct anatomical positioning. One individual was burnt in a state of advanced decomposition (Baraybar and de la Rua 1995, 164–167). The cave contained sherds from eight fragmentary ceramic vessels as well as flint, polished stone, bone, copper artefacts and a variety of animal bones (Zapata 1995, 44–51). Elsewhere I have argued that the Pico Ramos deposit is probably not the result of sequential deposition of bodies in the flesh, as suggested by Zapata (1995, 44), but the product of simultaneous deposition of exhumed corpses in different stages of decomposition (Weiss-Krejci 2005; 2011a). Sherds pertaining to one vessel and animal bones belonging to one animal were found distributed through three stratigraphic levels. Many long bones were missing and the cave has a very unbalanced age distribution. Only six individuals (5 per cent) are younger than seven whereas 66 individuals (64 per cent) are between seven and 20 years of age (Baraybar and de la Rua 1995, 155).

Though burnt bones are often found in 'secondary' burial contexts (Weiss-Krejci 2005, 2006a) they do not always represent evidence of reburial. Post-depositional processes (e.g. Cueva Maturras) are also responsible. In one instance, Abric de l'Escurrupénia, fire affected bodies still in the flesh (Pascual 2002). In this specific case the burning may have had the purpose to carbonize the entrails in order to better preserve the corpses. The fire was clearly centred on the trunk (Weiss-Krejci 2005, 51).

Post-funerary ritual use of human bones and problems of chronology

Especially in caves and rockshelters in central Portugal, fragmentary, disarticulated and partially burnt human bones from Copper Age/Bell Beaker contexts have produced dates which are too early for their contexts. At Casa da Moura a human ulna from the Copper Age level 1a was radiocarbon dated to 5990±60 BP and thus clearly too old for its context (Straus *et al.* 1988, 69–70). In Room 1 of the Gruta do Cadaval human remains consisting in cranial fragments and phalanges from the Bell Beaker level C provided a date of 5180±140 BP (Cruz and Oosterbeek 1985, 72; Cruz 1997). At the Gruta dos Ossos the disarticulated bones from upper levels I–III turned out to be much older than the articulated bones from the lower level IV. The bone jumble from the upper level dates to the end of the fourth millennium cal BC (4630±80 BP and 4460±110 BP) whereas the burial from the lower level IV dates to the middle of the third millennium (3970±140 BP) (Cruz 1997, 216). A layer of disarticulated and burnt bones from the small rockshelter of Covão d'Almeida contained a Bell Beaker sherd while the other ceramics, bone artefacts and human bones date to the end of the fourth millennium (4480±60 BP on human bone), representing a time gap of 1000 years (Corrêa and Teixeira 1949; Vilaça 1990; Gana and Cunha 2003, 132).

In the absence of better stratification and a larger sample of radiocarbon dates, at present the conclusions remain highly hypothetical. The pattern suggests that during the second part of the third millennium cal BC and associated with the Bell Beaker horizon some caves and rockshelters in central Portugal were used for new types of rituals which involved the exhumation, manipulation and re-deposition of much older bones (Weiss-Krejci 2011a). Possibly at those times caves had been transformed into kinds of sanctuaries or shrines which were used for the veneration of the remains of people of the past. Instead of disregarding dates which are too old, we should look at these cave deposits from a new perspective. What appears as the result of so called 'secondary' burial rites actually could be evidence for the deposition of relics or, alternatively, evidence for the use of bones for witchcraft and magic (e.g. Pastron and Clewlow 1974, 310). Considering the substantial economic, technological and social changes, as well as the innovations in and proliferation of mortuary structures, which took place between the sixth and third millennium BC it would not be surprising if the perception and mortuary use of caves by the prehistoric inhabitants of the Iberian Peninsula also changed over time.

Conclusions

Archaeologists working in the Iberian Peninsula have oversimplified the interpretation of mortuary deposits by reducing the evidence to 'primary burial' or 'inhumation' (Aguilella *et al.* 1999, 25), 'secondary burial' or 'ossuary' (Martí, 1981, 181; Palomar 1983, 132; Strauss 1997, 5) as well as 'individual' and 'collective' (Oosterbeek 1997; Molina-Burguera and Pedraz 2000, 12). These terms mask a range of diverse practices and processes with very different meanings. Caves and rockshelters in the Iberian Neolithic and Copper Age were probably used in many variable ways: single immediate deposition; *en masse* simultaneous deposition of people who died at the same time; deposition of people who died at different moments and whose corpses were buried in the same cave successively one after another; multi-stage rites where the dead were temporarily stored in or above the ground and later exhumed and entire skeletons or parts re-deposited in the cave individually or collectively; reburial and manipulation of bone relics during post-funerary rituals years or centuries later. The variability in behaviours is large and hard to detect because these processes can occur in combination with each other. These problems have also concerned archaeologists working outside the Iberian Peninsula (e.g. Vanderveken 1997; Chamberlain 1999; Polet and Cauwe 2002; Santoro *et al.* 2005).

Ethnographic analogy confirms the assumption that the presence of a single articulated individual in a cave is not an indication of a higher attention given to 'the individual'. On the contrary, in many parts of the world mortuary rites are extremely lengthy for individuals of higher ranked families, involving storage, exhumation and collective reburial of disarticulated bones, whereas poorer or low status individuals are deposited quickly and in the flesh (Weiss-Krejci 2011b, 75). Collective burial of disarticulated bones is not necessarily an expression of collective identity, communalism, collectivism and solidarism as suggested by several researchers (e.g. Criado 1989, 91; Aguado 2008, 14), but covers a variety of unrelated funerary and post-funerary processes which lead to the accumulation of multiple bodies in one place.

Revisiting some of the evidence deriving from older excavations, a stronger future focus on fine-grained stratigraphy and making an effort to determine in the field which state the bodies were in when first disposed and whether the bodies were introduced individually or collectively could yield new and unexpected insights. Such an effort is a painstaking procedure (e.g. excavations at Bolóres) but in the end it is the only method to gain those types of information which hold the clue to truly understanding the behaviours of people in the distant past.

Acknowledgements

Part of this research was funded by the Portuguese Science Fund FCT (SFRH/BPD/8608/2002). I would like to thank the editors of this volume for organising the EAA session in Malta 2008 and the Austrian Science Fund FWF (P18949-G02) for funding my trip to the EAA meeting. My special thanks go to Katina Lillios for inviting me to join the excavations at the Bolóres rockshelter in 2007 and Anna Waterman for her help. I also would like to thank Reinhold Mittersakschmöller for providing a photo from his archive.

References

Acosta Martínez, P. (1995) Las culturas del neolítico y calcolítico en Andalucía Occidental. *Espacio, Tiempo y Forma. Serie I, Nueva Época. Prehistoria y Arqueología* 8, 33–80.

Aguado Molina, M. (2008) Del orden social y del orden del universo: la llamada religión megalítica y su uso ideológico por las comunidades de los milenios IV–III a.c. a través del análisis del significado de sus monumentos funerarios. *Cuadernos de Prehistoria y Arqueología de la Universidad Autónoma de Madrid* 34, 7–21.

Aguilella i Arzo, G., Gusi i Jener, F. and Olària i Puyoles, C. (1999) El jaciment prehistòric de La Cova dels Diablets (Alcalà de Xívert, Castelló). *Quaderns de Prehistòria i Arqueologia de Castelló* 20, 7–35.

Alday Ruiz, A. (2005) The transition between the last hunter-gatherers and the first farmers in southwestern Europe: the Basque perspective. *Journal of Anthropological Research* 61, 469–494.

Aldenderfer, M. (2005) Caves as sacred places on the Tibetan plateau. *Expedition* 47(3), 8–13.

Aliaga Almela, R. (2008) El mundo funerario calcolítico de la Región de Madrid. *Cuadernos de Prehistoria y Arqueología* 34, 23–39.

Alfaro Giner, C. (1980) Estudio de los materiales de cestería procedentes de la Cueva de los Murciélagos (Albuñol, Granada). *Trabajos de Prehistoria* 37, 109–162.

Alt, K. W., Burger, J., Simons, A., Schön, W., Gruppe G., Hummel, S., Grosskopf, B., Vach, W., Buitrago Téllez, C., Fischer, C.-H., Möller-Wiering, S., Shrestha, S. S., Pichler, S. L. and von den Driesch, A. (2003) Climbing into the past: first Himalayan mummies discovered in Nepal. *Journal of Archaeological Science* 30, 1529–1535.

Álvarez Clavijo, P., Ceniceros Herreros, J. and Ilarraza Tejada, J. A. (1997) Nuevos datos para la definición del Calcolítico en el Valle Alto-Medio del Ebro. In R. de Bálbin Behrmann and P. Bueno Ramírez (eds.) *II Congreso de Arqueología Peninsular: Neolítico, Calcolítico y Bronce. Tomo II. Zamora, del 24 al 27 de Septiembre de 1996.* Volume 2, 291–300. Zamora, Fundación Rei Afonso Henriques.

Anderson, D. (2005) The use of caves in peninsular Thailand in the Late Pleistocene and Early and Middle Holocene. *Asian Perspectives* 44 (1), 137–153.

Angelucci, D. E., Boschian, G., Fontanals, M., Pedrotti, A. and Vergès, J. M. (2009) Shepherds and karst: the use of caves and rock-shelters in the Mediterranean region during the Neolithic. *World Archaeology* 41, 191–214.

Antón, S. C. and Steadman, D. W. (2003) Mortuary patterns in burial caves on Mangaia, Cook Islands. *International Journal of Osteoarchaeology* 13, 132–146.

Apellániz, J. M. (1974) *El Grupo de Los Husos durante la Prehistoria con Cerámica en el País Vasco.* Estudios de Arqueología Alavesa 7. Vitoria, Diputación Foral de Alava.

Apellániz, J. M., Llanos A. and Fariña, J. (1967) Cuevas sepulcrales de Lechon, Arralday, Calaveras y Gobaederra (Alava). *Estudios de Arqueología Alavesa* 2, 21–47.

Araújo, A. C. and Lejeune, M. (1995) *Gruta do Escoural. Necrópole Neolítica e Arte Rupestre Paleolítica*. Trabalhos de Arqueologia 8. Lisbon, Instituto Português do Património Arquitectónico e Arqueológico.

Arias Cabal, P. (1999) The origins of the Neolithic along the Atlantic coast of Continental Europe: a survey. *Journal of World Prehistory* 13, 403–464.

Arias Cabal, P. and Álvarez-Fernández, E. (2004) Iberian foragers and funerary ritual: a review of Paleolithic and Mesolithic evidence on the Iberian Peninsula. In M. González Morales and G. A. Clark (eds.) *The Mesolithic of the Atlantic Façade: Proceedings of the Santander Symposium*. Arizona State University Anthropological Research Paper 55, 225–248. Tempe, Arizona State University.

Arias Cabal, P. and Garralda, M. D. (1996) Mesolithic burials in Los Canes cave (Asturias, Spain) *Human Evolution* 11, 129–138.

Arias Cabal, P. and Pérez Suárez, C. (1990) Las sepulturas de la cueva de Los Canes (Asturias) y la neolitización de la región cantábrica. *Trabajos de Prehistoria* 47, 39–62.

Armendariz, A. (1990) Las cuevas sepulcrales en el País Vasco. *Munibe (Antropologia-Arkeologia)* 42, 153–160.

Armendariz, A. (1992) La idea de la muerte y los ritos funerarios durante la Prehistoria del País Vasco. *Actas del I Congreso Nacional de Paleopatología. IV Reunión de la Asociación Española de Paleopatología. Enfermedad y Muerte en el Pasado. Donostia-San Sebastián 21–23 Junio 1991*. Munibe, Suplemento 8, 13–32. San Sebastián.

Armendariz, A. (2007) Las dataciones radiocarbónicas. In J. I. Vegas Aramburu (ed.) *San Juan ante Portam Latinam: Una Inhumación Colectiva Prehistórica en el Valle Medio del Ebro. Memoria de las Excavaciones Arqueológicas 1985,1990 y 1991*, 101–106. Vitoria, Fundación José Miguel de Barandiarán, Diputación Foral de Álava.

Armendariz, A., and Etxeberria Gabilondo, F. (1996) Excavación de la cueva sepulcral de Arantzazu (Oñati, Gipuzkoa). *Munibe (Antropologia-Arkeologia)* 48, 53–58.

Armendariz, A., Etxeberria Gabilondo, F. and Herrasti Erlogorri, L. (1998) Excavación de la cueva sepulcral Nardakoste IV (Oñati, Gipuzkoa). *Munibe (Antropologia-Arkeologia)* 50, 111–120.

Arribas Palau, A. (1953) El ajuar de las cuevas sepulcrales de Los Blanquizares de Lébor (Murcia). *Memorias de los Museos Arqueológicos Provinciales* 13–14, 78–125.

Aspöck, E. (2009) *The Relativity of Normality: An Archaeological and Anthropological Study of Deviant Burials and Different Treatment at Death*. Unpublished doctoral thesis, Reading, University of Reading.

Asquerino Fernández, M. D. (1976) Vasos cardiales ineditos de la Cueva de la Sarsa (Bocairente, Valencia). *Trabajos de Prehistoria* 33, 339–350.

Asquerino Fernández, M. D. (1999) Sepulturas de la prehistoria reciente en Priego de Córdoba. *Anales de Prehistoria y Arqueología* 15, 29–39.

Aufderheide, A. C. (2003) *The Scientific Study of Mummies*. Cambridge, Cambridge University Press.

Augustí i Farjas, B. (2002) Depósitos funerarios con cremación durante el Calcolítico y el Bronce en el nordeste de Catalunya. In M. A. Rojo Guerra and M. Kunst (eds.) *Sobre el Significado del Fuego en los Rituales Funerarios del Neolítico*. Studia Archaeologica 91, 65–82. Valladolid, Secretariado de Publicaciones e Intercambio Editorial, Universidad de Valladolid.

Bar-Yosef Mayer, D. E., Vandermeersch, B. and Bar-Yosef, O. (2009) Shells and ochre in Middle Paleolithic Qafzeh Cave, Israel: indications for modern behavior. *Journal of Human Evolution* 56, 307–314.

Baraybar, J. P. and de la Rua, C. (1995) Estudio antropológico de la población de Pico Ramos (Muskiz, Bizkaia): consideraciones sobre la demografía, salud y subsistencia. *Munibe (Antropologia-Arkeologia)* 47, 151–175.

Barber, M. B. and Hubbard, D. A. Jr. (1997) Overview of the use of caves in Virginia: a 10,500 year history. *Journal of Cave and Karst Studies* 59, 132–136.

Barker, G. (2005) Burial rituals of prehistoric forager-farmers in Borneo: the Neolithic cemeteries of Niah cave, Sarawak. *Expedition* 47, 14–19.

Barker, G., Reynolds, T. and Gilbertson, D. (2005) The human use of caves in peninsular and Island Southeast Asia: research themes. *Asian Perspectives* 44 (1), 1–15.

Barley, N. (1981) The Dowayo dance of death. In S. C. Humphreys and H. King (eds.) *Mortality and Immortality: The Anthropology and Archaeology of Death*, 149–159. London, Academic Press.

Benítez de Lugo Enrich, L., Álvarez García, H. J., Molina Cañadas, M. and Moraleda Sierra, J. (2007) Consideraciones acerca del Bronce de La Mancha a partir de la investigación en la cueva prehistórica fortificada de Castillejo de Bonete (Terrinches, Ciudad Real): campañas 2003–2005. In J. M. Millán Martínez and C. Rodríguez Ruza (eds.) *Archqueología de Castilla-La Mancha. Actas de las Primeras Jornadas (Cuenca 13–17 Diciembre 2005)*, 231–262. Cuenca, Ediciones de la Universidad de Castilla-La Mancha.

Bermúdez de Castro, J. M. and Pérez, P. J. (1984) Apendice 4. Restos humanos de la cueva del cerro de Juan Barbero (Tielmes de Tajuña, Madrid); estudio antropológico. *Trabajos de Prehistoria* 41, 113–119.

Bischoff, J. L., Williams, R. W., Rosenbauer, R. J. Aramburu, A., Arsuaga, J. L., García, N. and Cuenca-Bescós, G. (2007) High-resolution U-series dates from the Sima de los Huesos hominids yields 600 kyrs: implications for the evolution of the early Neanderthal lineage. *Journal of Archaeological Science* 34, 763–770.

Blasco, A., Edo, M., Villalba, M. J. and Saña, M. (2005) Primeros datos sobre la utilización sepulchral de la Cueva de Can Sadurní (Begues, Baix Llobregat) en el Neolítico Cardial. In P. Arias Cabal, R. Ontañón Peredo and C. García-Moncó Piñeiro (eds.) *III Congresso de Neolítico en la Península Ibérica*, 625–633. Santander, Servicio de Publicaciones de la Universidad de Cantabria.

Bonsall, C. and Tolan-Smith, C. eds. (1997) *The Human Use of Caves*. British Archaeological Reports International Series 667. Oxford, Archaeopress.

Bosch Lloret, A. (1994) Las primeras sociedades neolíticas del extremo nordeste de la Península Ibérica. *Archivo de Prehistoria Levantina* 21, 9–31.

Bradley, R. (2000) *An Archaeology of Natural Places*. London, Routledge.

Brady, J. E. (1997) Settlement configuration and cosmology: the role of caves at Dos Pilas. *American Anthropologist* 99, 602–618.

Brady, J. E. and Veni, G. (1992) Man-made and pseudo-karst caves: the implications of subsurface features within Maya centers. *Geoarchaeology: An International Journal* 7, 149–167.

Cabral, J. M. P. (1987) Certificado de datação pelo radiocarbon. *Trabalhos de Antropologia e Etnologia* 27, 50.

Cacho Quesada, C., Papi Rodes, C., Sánchez-Barriga Fernández, A. and Alonso Mathias, F. (1996) La cestería decorada de la Cueva de los Murciélagos (Albuñol, Granada) *Complutum Extra* 6, 105–122.

Cardoso, J. L. (1992) A Lapa do Bugio. *Setúbal Arqueológica* 9–10, 89–225.

Cardoso, J. L., and Soares, A. M. (1995) Cronologia absoluta das grutas artificiais da Estremadura Portuguesa. *Al-Madan* 4, Series 2, 10–13.

Carrión, F. and Contreras Cortés, F. (1979) Yacimientos neolíticos en la zona de Moclín, Granada. *Cuadernos de Prehistoria de la Universidad de Granada* 4, 21–56.

Carvalho, A. F., Antunes-Ferreira, N. and Valente, M. J. (2003) A gruta-necrópole neolítica do Algar do Barrão (Monsanto, Alcanena). *Revista Portuguesa de Arqueologia* 6, 101–119.

Chamberlain, A.T. (1999) Carsington Pasture Cave, Brassington, Derbyshire: a prehistoric burial site. Capra 1. <http://capra.group.shef.ac.uk/1/carsing.html>, accessed 25 February 2010.

Chapman, R. (2005) Mortuary analysis: a matter of time? In G. F. M. Rakita, J. E. Buikstra, L. A. Beck and S. R. Williams (eds.) *Interacting with the Dead: Perspectives on Mortuary Archaeology for the New Millennium*, 25–40. Gainesville, University Press of Florida.

Chapman, R. (2008) Producing inequalities: regional sequences in later prehistoric southern Spain. *Journal of World Prehistory* 21, 195–260.

Chazine, J.-M. (2005) Rock art, burials, and habitations: caves in East Kalimantan. *Asian Perspectives* 44 (1), 219–230.

Colas, P. R., Reeder, P. and Webster, J. (2000) The ritual use of a cave on the Northern Vaca Plateau, Belize, Central America. *Journal of Cave and Karst Studies* 62, 3–10.

Corrêa, A. M. and Teixeira, C. (1949) *A Jazida Pré-Histórica de Eira Pedrinha*. Lisbon, Publicações do Serviço Geológico de Portugal.

Criado Boado, F. (1989) Megalitos, espacio, pensamiento. *Trabajos de Prehistoria* 46, 75–98.

Cruz, A. R. (1997) *Vale do Nabão. Do Neolítico à Idade do Bronze*. Arkeos 3. Tomar, Centro Europeu de Investigação da Pré-História do Alto Ribatejo.

Cruz, A. R. and Oosterbeek, L. (1985) A Gruta do Cadaval: elementos para a pré-história do Vale do Nabão. *Arqueologia na Região de Tomar (da Pré-História à Actualidade). Suplemento ao Boletím Cultural e Informativo da Câmara Municipal de Tomar* 1, 61–77.

Cunha, E., Umbelino, C. and Cardoso, F. (2002) New anthropological data on the Mesolithic communities from Portugal: the shell middens from Sado. *Human Evolution* 17, 187–198.

Díaz del Río, P. (1996) El enterramiento colectivo de El Rebollosillo (Torrelaguna). *Reunión de Arqueología Madrileña*, 98–200.

Díaz del Río, P. (2006) An appraisal of social inequalities in central Iberia (c. 5300–1600 cal. BC). In P. Díaz del Río and L. García Sanjuán (eds.) *Social Inequality in Iberian Late Prehistory*. British Archaeological Reports International Series 1525, 67–79. Oxford, Archaeopress.

Díaz del Río, P. and García Sanjuán, L. eds. (2006) *Social Inequality in Iberian Late Prehistory*. British Archaeological Reports International Series 1525. Oxford, Archaeopress.

Doménech Ratto, G., Moreno Cascales, M., Fernández-Villacañas M. and Ruiz Ibáñez, T. (1987) Apéndice I. Estudio preliminar de los restos óseos procedentes del enterramiento colectivo localizado en la 'Cueva Sagrada'. *Anales de Prehistoria y Arqueología* 3, 25–30.

Duarte, C. (1998) Necrópole neolítica do Algar do Bom Santo: contexto cronológico e espaço funerário. *Revista Portuguesa de Arqueologia* 1, 107–118.

Eddy, M. R. (1997) Symbolism of space in Guanche culture cave sites of the Canary Islands. In C. Bonsall and C. Tolan-Smith (eds.) *The Human Use of Caves*. British Archaeological Reports International Series 667, 87–89. Oxford, Archaeopress.

Eiroa García, J. J. (1987) Noticia preliminar de la primera campaña de excavaciones arqueológicas en el poblado de La Salud y en Cueva Sagrada I (Lorca), Murcia. *Anales de Prehistoria y Arqueología* 3, 53–76.

Eiroa García, J. J. and Lomba Maurandi, J. (1998) Dataciones absolutas para la prehistória de la región de Murcia: estado de la cuestión. *Anales de Prehistoria y Arqueología* 13–14, 81–118.

Etxeberria Gabilondo, F. (1994) Aspectos macroscópicos del hueso sometido al fuego. Revisión de las cremaciones descritas en el País Vasco desde la arqueología. *Munibe (Antropologia-Arkeologia)* 46, 111–116.

Etxeberria Gabilondo, F. and Herrasti Maciá, L. (2007) Los restos humanos del enterramiento de SJAPL: caracterización de la muestra, tafonomía, paleodemografía y paleopatología. In J. I. Vegas Aramburu (ed.) *San Juan ante Portam Latinam: Una Inhumación Colectiva Prehistórica en el Valle Medio del Ebro. Memoria de las Excavaciones Arqueológicas 1985, 1990 y 1991*, 159–282. Vitoria, Fundación José Miguel de Barandiarán, Diputación Foral de Álava.

Fernández Vega, A. and Pérez Cañamares, E. (1989) Enterramientos en cueva, sepulcros megalíticos y sepulcros en fosa en Cataluña: estudio comparativo. *Espacio, Tiempo y Forma. Serie I, Nueva Época. Prehistoria y Arqueología* 2, 131–152.

Fison, L. (1881) Notes on Fijian burial customs. *The Journal of the Anthropological Institute of Great Britain and Ireland* 10, 137–149.

Flood, J. (1997) Australian Aboriginal use of caves. In C. Bonsall and C. Tolan-Smith (eds.) *The Human Use of Caves*. British Archaeological Reports International Series 667, 193–200. Oxford, Archaeopress.

Gana, R. P. and Cunha, E. (2003) A Neolithic case of cranial trepanation (Eira Pedrinha, Portugal). In R. Arnott, S. Finger and C. U. M. Smith (eds.) *Trepanation: History – Discovery –Theory*, 131–136. Amsterdam, Swets and Zeitlinger.

Garcia Guixé, E., Richards, M. P. and Subirà, M. E. (2006) Palaeodiets of humans and fauna at the Spanish Mesolithic site of El Collado. *Current Anthropology* 47, 549–556.

García Sanjuán, L. (2006) Funerary ideology and social inequality in the Late Prehistory of the Iberian South-West. In P. Díaz del Río and L. García Sanjuán (eds.) *Social Inequality in Iberian Late Prehistory*. British Archaeological Reports International Series 1525, 149–169. Oxford, Archaeopress.

Gargett, R. H. (1999) Middle Paleolithic burial is not a dead issue; the view from Qafzeh, Saint-Ce´saire, Kebara, Amud and Dederiyeh. *Journal of Human Evolution* 37, 27–90.

Gibaja Bao, J. F. (2004) Neolithic communities of the northeastern Iberian Peninsula: burials, grave goods, and lithic tools. *Current Anthropology* 45, 679–685.

González Echegaray, J. and Freeman, L. G. (1978) *Vida y Muerte en Cueva Morín*. Santander, Institución Cultural de Cantabria.

González Morales, M. (1997) Changes in the use of caves in Cantabrian Spain during the Stone Age. In C. Bonsall and C. Tolan-Smith (eds.) *The Human Use of Caves*. British Archaeological Reports International Series 667, 63–69. Oxford, Archaeopress.

González Morales, M., Strauss L. G., Diez Castillo, A. and Ruiz Cobo, J. (2004) Postglacial coast and inland: the Epipaleolithic-Mesolithic-Neolithic transitions in the Vasco-Cantabrian region. *Munibe (Antropologia-Arkeologia)* 56, 61–78.

Guilaine, J. and Manen, C. (2007) From Mesolithic to Early Neolithic in the western Mediterranean. *Proceedings of the British Academy* 144, 21–51.

Guilaine, J. and Zammit, J. (2005) *The Origins of War: Violence in Prehistory*. Oxford, Blackwell Publishing.

Gutiérrez Sáez, C., Gómez Laguna, A. and Ocaña Carretón, A. (2002) Fuego y ritual en el enterramiento colectivo de Cueva Maturras (Argamasilla de Alba, Ciudad Real). In M. A. Rojo Guerra and M. Kunst (eds.) *Sobre el Significado del Fuego en los Rituales Funerarios del Neolítico*. Studia Archaeologica 91, 99–126. Valladolid, Secretariado de Publicaciones e Intercambio Editorial, Universidad de Valladolid.

Gutiérrez Sáez, C., Martín Lerma, I. Marín de Espinosa Sánchez, J. A. and Márquez Mora, B. (2008) Industria lítica tallada del ajuar funerario del Abrigo I de Cueva Maturras (Argamasilla de Alba, Ciudad Real): análises tecnológico y functional. *Espacio, Tiempo y Forma. Serie I, Nueva Época. Prehistoria y Arqueología* 1, 257–274.

Hapka, R. and Rouvinez, F. (1997) Las Ruinas Cave, Cerro Rabon, Oaxaca, Mexico: a Mazatec Postclassic funerary and ritual site. *Journal of Cave and Karst Studies* 59, 22–25.

Harrison, T. (1957) The Great Cave of Niah: a preliminary report on Bornean prehistory. *Man* 57, 161–166.

Hertz, R. (1907) Contribution à une étude sur la représentation collective de la mort. *Année Sociologique* 10 (1905–1906), 48–137.

Hubbard, D. A. Jr. and Barber, M. B. (1997) Virginia burial caves: an inventory of a desecrated resource. *Journal of Cave and Karst Studies* 59, 154–159.

Hutchinson, D. L. and L. V. Aragon (2002) Collective burials and community memories: interpreting the placement of the dead in the southeastern and mid-Atlantic United States with reference to ethnographic cases from Indonesia. In H. Silverman and D. B. Small (eds.) *The Space and Place of Death*. Archaeological Papers of the American Anthropological Association 11, 27–54. Arlington, American Anthropological Association.

Jackes, M., and Lubell, D. (1992) The early Neolithic human remains from Gruta do Caldeirão. In J. Zilhao *Gruta do Caldeirão. O Neolítico Antigo*. Trabalhos de Arqueologia 6, 259–295. Lisbon, Instituto Português do Património Arquitectónico e Arqueológico.

Jackes, M., Lubell, D. and Meiklejohn C. (1997) Healthy but mortal: human biology and the first farmers of western Europe. *Antiquity* 71, 639–658.

Jiménez Sanz, P. J. and Barroso Bermejo, R. M. (1995) El fenómeno funerário durante la prehistória reciente en el centro de La Meseta: la provincia de Guadalajara. In V. O. Jorge (ed.) *1º Congresso de Arqueologia Peninsular (Porto, 12–18 de Outubro de 1993). Actas VI*. Trabalhos de Antropologia e Etnologia, 35(2), 211–223. Oporto, Sociedade Portuguesa de Antropologia e Etnologia.

Jorge, S. O. (2000) Domesticating the land: the first agricultural communities in Portugal. *Journal of Iberian Archaeology* 2, 43–98.

Jorge, S. O., Oliveira, M. L., Nunes S. A. and Gomes, S. R. (1999) Uma estructura ritual com ossos humanos no sítio pré-histórico de Castelo Velho de Freixo de Numão (Vila Nova de Foz Côa). *Portugália* 19–20, new series, 29–70.

Klimchouk, A. (2004) Caves. In J. Gunn (ed.) *Encyclopedia of Caves and Karst Science*, 203–205. London, Routledge.

Kroeber, A. L. (1927), Disposal of the dead. *American Anthropologist* 29, 308–315.

Kusch, H. (1982) Die Bestattungshöhlen der Südtorajas im zentralen Hochland der Insel Sulawesi (Indonesien). *Die Höhle* 3, 91–100.

Leach, S. (2008) Odd one out? Earlier Neolithic deposition of human remains in caves and rockshelters in the Yorkshire Dales. In E. M. Murphy (ed.) *Deviant Burial in the Archaeological Record*, 17–34. Oxford, Oxbow Books.

Leitão, M., North, C. T., Norton, J., Ferreira, O. V. and Zbyszewski, G. (1987) A gruta pré-histórica do Lugar do Canto, Valverde (Alcanede). *O Arqueólogo Português* 5, series 4, 37–65.

Lewis, J. (2000) Upwards at 45 degrees: the use of vertical caves during the Neolithic and Early Bronze Age on Mendip, Somerset. *Capra* 2. <http://capra.group.shef.ac.uk/2/upwards.html>, accessed 5 March 2010.

Lillios, K. T., Waterman, A. J., Artz, J. A. and Josephs, R. L. (2010) The Neolithic-Early Bronze Age mortuary rockshelter of Bolóres, Torres Vedras, Portugal: results from the 2007 and 2008 excavations. *Journal of Field Archaeology* 35(1), 16–36.

Lizcano, R., Camara, J. A., Riquelme, J. A., Cañabate, M. L., Sanchez, A. and Afonso, J. A. (1992) El polideportivo de Marto: producción económica y símbolos de cohesión en un asentamiento del neolítico final en las Campiñas del Alto Guadalquivir. *Cuadernos de Prehistoria de la Universidad de Granada* 16–17, 5–101.

Llongueras, M., Ferrer, P., Campillo, D. and Martín, A. (1981) Enterrament campaniforme a la Cova de la Ventosa (Piera, Anoia). *Ampurias* 43, 97–111.

Lohmann, R. I. (2005) The afterlife of Asabano corpses: relationships with the deceased in Papua New Guinea. *Ethnology* 44, 189–206.

Lomba Maurandi, J., López Martínez, M. and Ramos Martínez, F. (2009) Un excepcional sepulcro del calcolítco: Camino del Molino (Caravaca de la Cruz). In J. A. Melgares Guerrero, P. E. Collado Espejo and J. A. Bascuñana Coll (session coordinators) *XX Jornadas de Patrimonio Cultural de la Región de Murcia, 6 de Octubre al 3 de Noviembre 2009*, 205–219. Murcia, Ediciones Tres Fronteras.

Lucero, L. J. and Gibbes, S. A. (2007) The creation and sacrifice of witches in Classic Maya society. In V. Tiesler and A. Cucina (eds.) *New Perspectives on Human Sacrifice and Ritual Body Treatments in Ancient Maya Society*, 45–73. New York, Springer.

Martí, B. (1981) La Cova Santa (Vallada, Valencia). *Archivo de Prehistoria Levantina* 16, 159–196.

Martí Oliver, B., Pascual Pérez, V., Gallart Martí, M. D., López García, P. Pérez Ripoll, M., Acuña Hernández, J. D. and Robles Cuenca, F. (1980) *Cova de l'Or, Beniarrés, Alicante*. Servicio de Investigación Prehistórica, Serie de Trabajos Varios 65. Valencia, Servicio de Investigación Prehistórica.

Martín, A., Guilaine, J., Thommeret, J. and Thommeret, Y. (1981) Estratigrafía y dataciones C14 del yacimiento de la 'Cova del Frare' de St. Llorenç del Munt (Matadepera, Barcelona). *Zephyrus* 32–33, 101–111.

Martínez Navarrete, M. I. (1984) El comienzo de la metalurgia en la provincia de Madrid: la cueva y cerro de Juan Barbero (Tielmes, Madrid). *Trabajos de Prehistoria* 41, 17–91.

McNatt, L. (1996) Cave archaeology of Belize. *Journal of Cave and Karst Studies* 58, 81–99.

Molina-Burguera, G. and Pedraz, T. (2000) Nuevo aporte al eneolítico valenciano: la Cueva de las Mulatillas (Villargordo del Cabrile, Valencia). *Anales de Prehistoria y Arqueología* 16, 7–14.

Monteiro, R., Zbyszewski, G. and Ferreira, O. (1971) Nota preliminar sobre a lapa pré-historica do Bugio (Azóia-Sesimbra). In *Actas do II Congresso de Arqueologia (Coimbra 1970)*, Volume 1, 107–120. Coimbra, Ministerio da Educação Nacional, Junta Nacional da Educação.

Mouret, C. (2004) Burials in caves. In J. Gunn (ed.) *Encyclopedia of Caves and Karst Science*, 167–169. London, Routledge.

Moyes, H. and Brady, J. E. (2005) The heart of creation, the heart of darkness: sacred caves in Mesoamerica. *Expedition* 47(3), 31–36.

Murphy, E. M. ed. (2008) *Deviant Burial in the Archaeological Record*. Oxford, Oxbow Books.

Nelson, W. H. and Barnett, F. (1955) A burial cave on Kanaga Island, Aleutian Islands. *American Antiquity* 20, 387–392.

Ontañón Peredo, R. (2003) *Caminos Hacia la Complejidad: El Calcolítico en la Región Cantábrica*. Santander, Servicio de Publicaciones de la Universidad de Cantabria.

Oosterbeek, L. (1993) Nossa Senhora das Lapas: excavation of prehistoric cave burials in Central Portugal. *Papers of the Institute of Archaeology* 4, 49–62.

Oosterbeek, L. (1997) Back home! Neolithic life and the rituals of death in the Portuguese Ribatejo. In C. Bonsall and C. Tolan-Smith (eds.) *The Human Use of Caves*. British Archaeological Reports International Series 667, 70–78. Oxford, Archaeopress.

Palomar, V. (1983) La Cueva del Abrigo I de las Peñas (Navajas, Castellón). *Cuadernos de Prehistoria y Arqueología Castellonenses* 9, 123–134.

Pannell, S. and O'Connor, S. (2005) Toward a cultural topography of cave use in East Timor: a preliminary study. *Asian Perspectives* 44 (1), 193–206.

Parker Pearson, M., Chamberlain, A. Collins, M. Cox, C., Craig, G., Craig, O., Hiller, J., Marshall, P., Mulville, J. and Smith, H. (2007) Further evidence for mummification in Bronze ge Britain. *Antiquity* 81, 313. Project Gallery <www.antiquity.ac.uk>, accessed 30 May 2010.

Parreira, R. and Serpa, F. (1995) Novos dados sobre o povoamento da região de Alcalar (Portimão) no IV e III milénios a.c. In V. O. Jorge (ed.) *1º Congresso de Arqueologia Peninsular (Porto, 12–18 de Outubro de 1993). Actas VII*. Trabalhos de Antropologia e Etnologia, 35(3), 233–256. Oporto, Sociedade Portuguesa de Antropologia e Etnologia.

Pascual Benito, J. L. (2002) Incineración y cremación parcial en contextos funerarios neolíticos y calcolíticos del este peninsular al sur del Xúquer. In M. A. Rojo Guerra and M. Kunst (eds.) *Sobre el Significado del Fuego en los Rituales Funerarios del Neolítico*. Studia Archaeologica 91, 155–189. Valladolid, Secretariado de Publicaciones e Intercambio Editorial, Universidad de Valladolid.

Pastron, A. G. and Clewlow, C. W. Jr. (1974) The ethno-archaeology of an unusual Tarahumara burial cave. *Man*, new series 9, 308–311.

Paz de Miguel, M. (2000) Anexo I. Restos humanos de la Cueva de las Mulatillas. *Anales de Prehistoria y Arqueología* 16, 15.

Pellicer Catalán, M. (1995) Las culturas del neolítico-calcolítico en Andalucía Oriental. *Espacio, Tiempo y Forma. Serie I, Nueva Época. Prehistoria y Arqueología* 8, 81–134.

Peña-Chocarro, L., Zapata, L., Iriarte, M. J. González Morales, M. and Straus, L. G. (2005) The oldest agriculture in northern Atlantic Spain: new evidence from El Mirón Cave (Ramales de la Victoria, Cantabria). *Journal of Archaeological Science* 32, 579–587.

Perry, W. J. (1915) Myths of origin and the home of the dead in Indonesia. *Folklore* 26, 138–152.

Picpican, I. (2003) *The Igorot Mummies: A Socio-Cultural and Historical Treatise*. Quezon City, Rex Book Store.

Polet, C. and Cauwe, N. (2002) Les squelettes mésolithiques et néolithiques de l'abri des Autours (province de Namur, Belgique). *Comptes Rendus Palevol* 1, 43–50.

Polo-Cerdá, M., Romero, A., Casabó, J. and De Juan, J. (2007) The Bronze Age burials from Cova Dels Blaus (Vall d'Uixó, Castelló, Spain): an approach to palaeodietary reconstruction through dental pathology, occlusal wear and buccal microwear patterns. *Homo: Journal of Comparative Human Biology* 58, 297–307.

Prufer, K. M. and Dunham, P. S. (2009) A shaman's burial from an Early Classic cave in the Maya Mountains of Belize, Central America. *World Archaeology* 41, 295–320.

Ramos, C. R. and Pita, L. T. (1992) Os horizontes de ocupação humana das grutas de Alcobaça Algar. *Boletim da Sociedade Portuguesa de Espeleologia* 3, 27–40.

Rodanés Vicente, J. M. (1997) La cuevas sepulcrales en la Rioja. Estudio histórico-arqueológico. *Munibe (Antropologia-Arkeologia)* 49, 77–93.

Rojo, M. A. and Kunst, M. (1999) Zur Neolithisierung des Inneren der Iberischen Halbinsel: Erste Ergebnisse des interdisziplinären spanisch-deutschen Forschungsprojekts zur Entwicklung einer prähistorischen Siedlungskammer in der Umgebung von Ambrona (Prov. Soria). *Madrider Mitteilungen* 40, 1–52.

Rubio de Miguel, I. (1981) Enterramientos neolíticos de la Península Ibérica. *Cuadernos de Prehistoria y Arqueología de la Universidad Autónoma de Madrid* 7–8, 39–73.

Rubio de Miguel, I. (2002) El mundo funerario neolítico peninsular: algunas reflexiones sobre su trasfondo social. *Anales de Prehistoria y Arqueología* 16–17, 53–66.

Rue, D. J., Freter, A. C. and Ballinger, D. A. (1989) The caverns of Copan revisited: Preclassic sites in the Sesesmil river valley, Copan, Honduras. *Journal of Field Archaeology* 16, 395–404.

Ruiz Cobo, J. Eagan, P. T., Bandres, A., Etxeberria, F. and Herrasti, L. (2008) Los restos humanos de la Cueva del Torno (Fresnedo, Solórzano) en el contexto de las cuevas sepulcrales del valle de Asón (Cantabria). *Munibe (Antropologia-Arkeologia)* 59, 157–170.

Santoro, C. M., Standen, V. G., Arriaza, B. T. and Dillehay, T. D. (2005) Archaic funerary pattern or postdepositional alteration? The Patapatane burial in the highlands of South Central Andes. *Latin American Antiquity* 16, 329–346.

Saul, J. M. and Saul, F. P. (2002) Forensics, archaeology, and taphonomy: the symbiotic relationship. In W. D. Haglund and M. H. Sorg (eds.) *Advances in Forensic Taphonomy. Method, Theory, and Archaeological Perspectives*, 71–97. Boca Raton, CRC Press.

Serrão, E. C. and Marques, G. (1971) Estrato pré-campaniforme da Lapa do Fumo (Sesimbra). In *Actas do II Congresso de Arqueologia (Coimbra 1970)*, Volume 1, 121–142. Coimbra, Ministerio da Educação Nacional, Junta Nacional da Educação.

Silva, A. M. (1999) Human remains from the artificial cave of São Pedro do Estoril II (Cascais, Portugal). *Human Evolution* 14, 199–206.

Simmons, T. (2002) Taphonomy of a karstic cave execution site at Hrgar, Bosnia-Herzegovina. In W. D. Haglund and M. H. Sorg (eds.) *Advances in Forensic Taphonomy: Method, Theory, and Archaeological Perspectives*, 263–275. Boca Raton, CRC Press.

Simón Vallejo, M. D., Fernández Domínguez, E., Turbón Borrega, B., Cortés Sánchez, M., Lozano Francisco, M. C., Vera Peláez, J. L., Riquelme Cantal, J. A. and Sanchidrián Torti, J. L. (2005) Aportaciones al conocimiento de la utilización de la Cueva de Nerja como necropolis durante el Neolítico. In P. Arias Cabal, R. Ontañón Peredo and C. García-Moncó Piñeiro (eds.) *III Congreso de Neolítico en la Península Ibérica*, 643–652. Santander, Servicio de Publicaciones de la Universidad de Cantabria.

Soares, A. M. M. (1995) Datação absoluta da Necrópole 'Neolítica' da Gruta do Escoural. In A. C. Araújo and M. Lejeune (eds.) *Gruta do Escoural: Necrópole Neolítica e Arte Rupestre Paleolítica*. Trabalhos de Arqueologia 8, 111–122. Lisbon, Instituto Português do Património Arquitectónico e Arqueológico.

Soares, A. M. M. and Cabral, J. M. P. (1993) Cronologia absoluta para o Calcolítico da Estremadura e do Sul de Portugal. *1º Congresso de Arqueologia Peninsular (Porto, 12–18 de Outubro de 1993), Actas II*. Trabalhos de Antropologia e Etnologia 33, 217–235.

Soler Diaz, J. A. (1997) Cuevas de inhumación múltiple en el País Valenciano: una aproximación al rito desde la significación de los distintos elementos del registro. In R. de Bálbin Behrmann and P. Bueno Ramírez (eds.) *II Congreso de Arqueología Peninsular. Neolítico, Calcolítico y Bronce. Tomo II. Zamora, del 24 al 27 de Septiembre de 1996*, Volume 2, 347–358. Zamora, Fundación Rei Afonso Henriques.

Soler Diaz, J. A. (2002) *Cuevas de Inhumación Múltiple en el País Valenciano*. Alicante, Real Academía de História, Diputación Provincial de Alicante.

Steadman D. W., Antón S. C. and Kirch, P. V. (2000) Ana Manuku: a prehistoric ritualistic site on Mangaia, Cook Islands. *Antiquity* 74, 873–883.

Stodder, A. L. W. (2005) The bioarchaeology and taphonomy of mortuary ritual on the Sepik Coast, Papua New Guinea. In G. F. M. Rakita, J. E. Buikstra, L. A. Beck and S. R. Williams (eds.) *Interacting with the Dead. Perspectives on Mortuary Archaeology for the New Millennium*, 228–250. Gainesville, University Press of Florida.

Stone, A. (1997) Precolumbian cave utilization in the Maya area. In C. Bonsall and C. Tolan-Smith (eds.) *The Human Use of Caves*. British Archaeological Reports International Series 667, 201–206. Oxford, Archaeopress.

Stone, A. and J. E. Brady (2005) Maya caves. In W. B. White and D. C. Culver (eds.) *Encyclopedia of Caves*, 366–369. Amsterdam, Elsevier.

Straus, L. G. (1997) Convenient cavities: some human uses of caves and rockshelters. In C. Bonsall and C. Tolan-Smith (eds.) *The Human Use of Caves*. British Archaeological Reports International Series 667, 1–8. Oxford, Archaeopress.

Straus, L. G., Altuna, J., Ford, D., Marambat, L., Rhine, J. S., Schwarcz, J-H. P. and Vernet, J-L. (1992) Early farming in the Algarve (Southern Portugal): a preliminary view from two cave excavations near Faro. In *Homenagem a Ernesto Veiga de Oliveira*. Trabalhos de Antropologia e Etnologia 32, 141–172. Oporto, Sociedade Portuguesa de Antropologia e Etnologia.

Straus, L. G., Altuna, J., Jackes, M. and Kunst, M. (1988) New excavations in Casa da Moura (Serra d'el Rei, Peniche) and at the Abrigos de Bocas (Rio Maior), Portugal. *Arqueologia* 18, 65–95.

Ten i Carné, R. (1989) La cova sepulcral neolítica epicardial dels Lladres (Vacarisses, Vallès Occidental). *Empúries* 48–50, 352–355.

Toussaint, M., Lacroix, P., Lambermont, S., Lemaire, J.-F. and Beaujean, J.-F. (2004) La sépulture d'enfant néolithique des nouveaux réseaux du Trou du Moulin, à Goyet (Gesves, prov. de Namur): rapport préliminaire. *Notae Praehistoricae* 24, 159–166.

Treerayapiwat, C. (2005) Patterns of habitation and burial activity in the Ban Rai Rock Shelter, northwestern Thailand. *Asian Perspectives* 44 (1), 231–245.

Trinkaus, E., Maki, J. and Zilhão, J. (2007) Middle Paleolithic human remains from the Gruta da Oliveira (Torres Novas), Portugal. *American Journal of Physical Anthropology* 134, 263–273.

Ucko, P. (1969) Ethnography and archaeological interpretation of funerary remains. *World Archaeology* 1, 262–280.

Vanderveken, S.(1997) Les ossements humains néolithiques de Maurenne et Hastière (Province de Namur). *Notae Praehistoricae* 17, 177–184.

Van Gennep, A. (1909) *Les Rites de Passage*. Paris, Nourry.

Vegas Aramburu, J. I. (1992) El enterramiento de San Juan ante Portam Latinam: las más numerosas señales de violencia de la Prehistoria peninsular. *Kultura* 5, 9–20.

Vegas Aramburu, J. I. (ed.) (2007) *San Juan ante Portam Latinam. Una Inhumación Colectiva Prehistórica en el Valle Medio del Ebro. Memoria de las Excavaciones Arqueológicas 1985, 1990 y 1991*. Vitoria, Fundación José Miguel de Barandiarán, Diputación Foral de Álava.

Vilaça, R. (1990) Sondagem arqueológica no Covão d'Almeida (Eira Pedrinha, Condeixa-a-Nova). *Antropologia Portuguesa* 8, 101–131.

Vilaseca, S. (1972) Las cuevas sepulcrales I y II de Solá de la Vila de Pradell (Bajo Priorato). *Trabajos de Prehistoria* 29, 31–54.

Walker, M. J. (1983) Laying a mega-myth: dolmens and drovers in prehistoric Spain. *World Archaeology* 15, 37–50.

Walker, M. J. and San Nicolás del Toro, M. (1995) Disposal of the dead and dispersal of the living in preargaric S.E. Spain. Abrigo 2 de El Milano and the revision of the dynamics of cultural change. 'Little Big Men' and no growth in population? In W. H. Waldren (ed.) *Ritual, Rites and Religion in Prehistory*. British Archaeological Reports International Series 611(2), 110–169. Oxford, Tempus Reparatum.

Waterman, A. J. (2006) *Health Status in Prehistoric Portugal: Dental Pathology and Childhood Mortality Patterns from the Late Neolithic Burials of Feteira (Lourinhã)*. Unpublished Masters Paper, Iowa City, The University of Iowa.

Waterman, A. J. (2007) Health status in prehistoric Portugal: dental pathology and childhood mortality patterns from the Late Neolithic burials of Feteira (Lourinhã). *American Journal of Physical Anthropology* 132 (S44), 245.

Waterson, R. (1995) Houses, graves and the limits of kinship groupings among the Sa'dan Toraja. *Bijdragen tot de Taal-, Land- en Volkenkunde* 151, 194–217.

Weaver, H. D. (2008) *Missouri Caves in History and Legend*. Columbia, University of Missouri Press.

Weiss-Krejci, E. (2005) Formation processes of deposits with burned human remains in Neolithic and Chalcolithic Portugal. *Journal of Iberian Archaeology* 7, 37–73.

Weiss-Krejci, E. (2006a) Animals in mortuary contexts of Neolithic and Chalcolithic Iberia. In E. Weiss-Krejci, C. Duarte and J. Haws (session coordinators.), N. Bicho (ed.) *Animais na Pré-história e Arqueologia da Península Ibérica. Actas do IV Congresso de Arqueologia Peninsular, Faro 2004*, 35–45. Faro, Universidade do Algarve, Faculdade de Ciências Humanas e Sociais.

Weiss-Krejci, E. (2006b) The Maya corpse: body processing from Preclassic to Postclassic times in the Maya highlands and lowlands. In P. R. Colas, G. LeFort and B. Liljefors Persson (eds.) *Jaws of the Underworld: Life, Death, and Rebirth Among the Ancient Maya. 7th European Maya Conference, The British Museum, London, November 2002*. Acta Mesoamericana 16, 71–86. Markt Schwaben, Anton Saurwein.

Weiss-Krejci, E. (2011a) Changing perspectives on mortuary practices in Late Neolithic/Copper Age and Early Bronze Age Iberia. In K. Lillios (ed.) *Comparative Archaeologies. The American Southwest (AD 900–1600) and the Iberian Peninsula (3000–1500 BC)*, 153–174. Oxford, Oxbow Books.

Weiss-Krejci, E. (2011b) The formation of mortuary deposits: implications for understanding mortuary behavior of past populations. In S. C. Agarwal and B. Glenncross (eds.) *Social Bioarchaeology*, 68–106. New York, Wiley-Blackwell.

White, W. B. and Culver, D. C. (2005) Cave, definition of. In W. B. White and D. C. Culver (eds.) *Encyclopedia of Caves*, 81–85. Amsterdam, Elsevier.

Willis, A. and Tayles, N. (2009) Field anthropology: application to burial contexts in prehistoric Southeast Asia. *Journal of Archaeological Science* 36, 547–554.

Zapata, L. (1995) La excavación del depósito sepulcral calcolítico de la Cueva Pico Ramos (Muskiz, Bizkaia): las industria ósea y los elementos de adorno. *Munibe (Antropologia-Arkeologia)* 47, 35–90.

Zapata, L., Peña-Chocarro, L., Pérez-Jordá, G. and Stika, H.-P. (2004) Early Neolithic agriculture in the Iberian Peninsula. *Journal of World Prehistory* 18, 283–325.

Zilhão, J. (1992) *Gruta do Caldeirão: O Neolítico Antigo*. Trabalhos de Arqueologia 6. Lisbon, Instituto Português do Património Arquitectónico e Arqueológico.

Zilhão, J. (1995) Primeiras datações absolutas para os níveis neolíticos das Grutas do Caldeirão e Feteira: origens, estruturas e relações das culturas Calcolíticas da península Ibérica. In *Actas das I Jornadas Arqueológicas de Torres Vedras (3–5 de Abril 1987)*, 113–122. Lisbon, Instituto Português do Património Arquitectónico e Arqueológico.

Zilhao, J. (2000) From the Mesolithic to the Neolithic in the Iberian Peninsula. In T. D. Price (ed.) *Europe's First Farmers*, 144–182. Cambridge, Cambridge University Press.

Zilhão, J. (2001) *Radiocarbon evidence for maritime pioneer colonization at the origins of farming in west Mediterranean Europe*. Proceedings of the National Academy of Sciences USA 98, 14180–14185.

Zilhão, J. (2003) The Neolithic transition in Portugal and the role of demic diffusion in the spread of agriculture across West Mediterranean Europe. In A. J. Ammerman and P. Biagi (eds.) *The Widening Harvest. The Neolithic Transition in Europe: Looking Back, Looking Forward*, 207–223. Boston, Archaeological Institute of America.

Zilhão, J. (2004a) The Mesolithic of Iberia. In P. Bogucki and P. J. Crabtree (eds.) *Ancient Europe 8000 B.C.–A.D. 1000. Encyclopedia of the Barbarian World*, Volume 1, 157–164. New York, Charles Scribner's Sons.

Zilhão, J. (2004b) Gruta do Caldeirão. In P. Bogucki and P. J. Crabtree (eds.) *Ancient Europe 8000 B.C.–A.D. 1000. Encyclopedia of the Barbarian World*, Volume 1, 255–258. New York, Charles Scribner's Sons.

Zilhão, J. (2009) The Early Neolithic artifact assemblage from the Galeria da Cisterna (Almonda karstic system, Torres Novas, Portugal). In *De Méditerranée et d'Ailleurs. Mélanges Offerts à Jean Guilaine*, 821–835. Toulouse, Archives d'Écologie Préhistorique.

Chapter 10

The Bronze Age Use of Caves in France: Reinterpreting their Functions and the Spatial Logic of their Deposits through the *Chaîne Opératoire* Concept

Sébastien Manem

Caves constitute an important – sometimes major – archaeological support for understanding French Bronze Age cultures. Over several decades, many studies have demonstrated a diversity of cave functions, independently of cultural context. Caves were used for ritual and funerary practices but also in everyday life, or occasionally for economic functions, as dwelling place or for protection during periods of insecurity. Different arguments are used to support these interpretations but paradoxically the same elements are generally used to argue for a ritual or dwelling function. This complex question of cave function hides a fundamental debate about these European Bronze Age societies: were they really troglodyte? Why did people deposit and abandon so many goods in the caves? This article demonstrate the real difficulties in establishing hypotheses of function, and propose another reference point based on the study of technical behaviour, using the concept of chaîne opératoire. The study is focused on the Duffaits Culture and the analysis of pottery which constitutes the main material deposited by Bronze Age people in these caves.

Introduction

Caves have been a major topic of discussion, central to European Bronze Age studies, for several decades. But they are still a complex archaeological entity. Generally dark, damp and difficult to access, they have a strange ambiance, which we feel with all our senses when we enter them. It was undoubtedly not trivial for Bronze Age people to use these underground worlds (Pétrequin *et al.* 1985, 216; Harding 2005). Caves are, however, important archaeological resources given the often abundant and varied materials deposited in them by Bronze Age people (Thauvin-Boulestin 1998; Harding 2000, 54–55, 317–320; 2005). These deposits can be linked to a diverse range of functions: domestic, economic, or ritual and funerary (The economic function – as sheepfolds, for example – will not be discussed here.). Given that this diversity of functions is observed in archaeology (Bonsall and Tolan-Smith 1997), the relatively large proportion of caves used for ritual or funerary processes in the European Bronze Age becomes all the more notable (Harding 2000, 54).

Paradoxically, this ritual function is rarely given prominence by French prehistorians, who tend to see caves as 'refuge dwelling places', 'annex dwelling places' or temporary/permanent dwelling places. According to the model proposed by P. Pétrequin (Pétrequin *et al.* 1985), refuge caves would have been used temporarily, in periods of 'insecurity', while 'annex dwelling places' refer to caves that would have been dependent upon a hypothetical, nearby, above-ground dwelling place. The arguments put forward to support these hypotheses focus upon four interlinked aspects: access/environment, topography, interior fittings and the archaeological material. How significant are the arguments proposed in favour of a domestic function? Are they distinct from those used to identify a ritual function? Will we get the same results and interpretations of we use another reference point as we would for technical behaviour?

Cave functions: current interpretations

Cave dwellings

The arguments used by archaeologists to discern a temporary or permanent domestic function are primarily based on the ease of access to the cave. The natural protection offered by a canopy, coupled with a sufficiently large entrance to light the cave, provided a comfortable domestic situation. This was the case for the Roucadour cave in the Lot region during the Middle Bronze Age, whose use can be – hypothetically – linked to the exploitation of the surrounding landscape (Gascó 2004). J. M. Treffort notes that the largely open caves of the southern Jura were occupied in the Late Bronze Age I–IIa phase (Treffort 2005). Cavities with smaller openings were more commonly used as temporary or seasonal stopping points (Treffort 2005, 405). But cave that hard of access caves were also used as dwelling place (*ibid.*). And, paradoxically, during the Late Bronze Age IIb phase, only those caves which were hard of access continued to be used in the southern Jura (*ibid.*). This was also the case in the Middle Bronze Age at the Noyer cave in the Lot region, whose entrance is located in a vertical cliff face, 35m from the ground (Clottes and Lorblanchet 1972; Giraud 1989). The orientation of the opening is sometimes put forward as a factor determining the use of caves as dwelling places, as in the case of the Perrats cave, in Charente, which is oriented to the West (Gomez de Soto 1996a). But some openings facing directly North, such as those in La Roche Noire (Treffort 2005, 410), were also used.

The topography of the caves is also an important factor relating to their use as dwelling places. Large spaces appear favourable to the installation of a household, such as in the Roucadour cave where the main chamber is 15–20m wide (Gascó 2004, 523). The preferential usage of the areas nearest to the cave's entrance has also been noted (Gomez de Soto 1996a; Gascó 2004).

Interior features are also seen to be a significant indicator of a domestic function in the literature. They include structures which have been hollowed out (silo pits) or constructed (posts, walls, hearths), as in Roucadour, under the porch, dating to the Late Bronze Age IIIa–b phases (Gascó 2004, 529), and also at La Roche Noire (Treffort 2005, 410). A palisade was found in the Gigny cave, in the Jura (Pétrequin *et al.* 1988). The presence of large post holes with a diameter of 20cm and fragments of a wall have been interpreted as the remains of a wooden house constructed in the interior of the cave (*ibid.*). Other features have also been noted, such as an oven to roast cereals in the Noyer cave, which also explains the presence of a large ceramic storage unit, used to store cereals (Giraud 1989, 436).

Refuge caves

Refuge caves are characterised by difficulty of site access (Pétrequin *et al.* 1985). The entrance to the Cloches cave, at Saint-Martin-d'Ardèche, is situated 35m above the Ardèche. Its first chamber is on a sharp incline, which renders it doubly hard to access (Vital 1986). This position is similar to that of the Noyer cave whose second potential use is as a refuge (Clottes and Lorblanchet 1972). A unobtrusive entrance is felt to be an important feature of refuge caves by P. Pétrequin *et al.* (1985, 220).

In general, the interior of such caves is not particularly comfortable for habitation. These caves can be damp and their floor was often not suited to the installation of a household (Pétrequin *et al.* 1985). The literature therefore tends to indicate that these locations were primarily used for their defensive aspect, with little regard given to their interior comfort, as seen in the Cloches cave.

P. Pétrequin describes defensive features in two cases: the Source du Dard rock-shelter, in the Jura, and la Baume de Gonvillars, in la Haute-Saône (Pétrequin *et al.* 1985, 219). The objects found inside are still abundant and varied. J. Gomez de Soto offers a pragmatic explanation for this: human needs remain the same in everyday life or during period of insecurity (Gomez de Soto and Kerouanton 1991, 387). The Cloches cave is equipped with hearths, pieces of wattle and material including pottery, faunal remains, grinding stones and a range of bronzes. The interior were used for the manufacture and firing of pottery, as shown by the discovery of tools and piles of un-fired coils (Vital 1986).

Annex dwelling caves

Caves known as 'annex dwelling places' are those thought to be dependent on above-ground sites. The Quéroy cave, in Charente, is accessible via a narrow vertical swallow-hole. Late Bronze Age IIIb occupation here have been interpreted as the annex of a dwelling place. The difficulties of access to the site and the amount of fallen rocks covering the floor eliminated the possibility of this site being interpreted as either a dwelling or refuge cave, despite these features being used by P. Pétrequin to justify a refuge function. In this particular case, other arguments have been put forward to justify the function of the cave as an annex dwelling place: the presence of sherds on the surface and the large number of structures found in the cave (hearths, and light structures represented by postholes). Several hundred ceramic vessels, various bronzes (tools and ornaments) and evidence of bone industry (tools, ornaments and horse harness pieces) were found in the cave. Terracotta fragments were also found, such as spindle whorls or loom weights. Foundry waste was also found (Gomez de Soto and Kerouanton 1991).

Most of the sites interpreted as refuges, annex dwelling places, or dwelling places contained a large range of artefacts including both fine and storage pottery, bronze and evidence of crop plants, such as cereals, which were often carbonised (Pétrequin *et al.* 1988; Gascó 2004) or harvest products (Marinval 1983). The importance of the storage pottery and the cereals which accompanied them has been interpreted in terms of the use of caves for agricultural storage, as with the Noyer cave (Giraud 1989), la Roche Noire cave (Treffort 2005, 412), as well as the cave in Gigny (Pétrequin *et al.* 1988). The presence of burnt grain has been interpreted in terms of accidental fires at La Roche Noire (Treffort 2005, 412) or storage problems, as in the Noyer cave (Giraud 1989). The possibility of grain roasting areas has also been suggested (Clottes and Lorblanchet 1972). The cereals remains have been interpreted as evidence of the caves' important agricultural position in society (Giraud 1989) or linked to occupation by farmers (Pétrequin *et al.* 1988). Faunal remains are, however, common, and, in addition to their significance as food, were used in various ways as raw materials for artefacts. The caves in the Noyer group mainly contained the bones of ovicaprines and pigs, often slaughtered at a young age. Horses were also found, and significant number of bones from wild fauna.

Questions over these interpretations

In many cave publications, a domestic function is taken for granted, as if it were obvious. There is rarely any discussion of the issue of 'domestic versus ritual'. Since above-ground ceremonial sites are generally identified through architectural criteria, caves engender immediate problems of identification. Furthermore, the parallel between ritual and domestic items is such that we can not sidestep this discussion. The Sindou cave contained – as do dwelling places – a wide range of artefacts, cereals and fauna (Briois *et al.* 2000). Storage pottery is generally present at ritual sites. Almost 80 per cent of the Middle Bronze Age pottery deposited in the ritual chamber of the Duffaits necropolis cave, in Charente, is storage pottery (Gomez de Soto 1973, 1995; Manem 2008).

Fauna is equally present in domestic situations as in ritual cases, and has always played a major role in cult practices (Harding 2000). Agriculture is also well integrated in cult practices and regulates certain ritual cycles (Bradley 2000, 2005). This is equally true of metallurgy, given the symbolically loaded nature of the necessary skills and the specific nature of the transformation of the source material (Bradley 2005, 23). Ritual and funerary sites are also characterised by particular modification of the cave. Hearths are present, such as in the Duffaits cave (Gomez de Soto 1973). Structures are built and circulation zones are laid out, such as in the Khépri funerary cave near Ganties in Haute-Garonne. These spaces were maintained and a large quantity and variety of materials was deposited (Le Guillou *et al.* 2000).

The access to these ritual and/or funerary caves, or even their visibility from the surface, are also similar to sites interpreted as dwelling places. Sites hard of access and those largely open to the exterior are used. The Pendule cave, in the Ain region, is situated 30m above the ground level (Treffort 2005, 413) and is reminiscent of the Noyer cave's situation. Other caves, accessible via a fracture, as seen in the Duffaits necropolis caves, could very well have fulfilled the criteria of refuge caves as suggested by P. Pétrequin. Rockshelters were also used for ritual and funerary practices, as in the Early Bronze Age at the Pins site, in Charente (Boulestin and Gomez de Soto 2003). The Perrats cave, in Charente, is largely open to the hillside, facing West. It was used as a collective burial site in the Early Bronze Age (Boulestin 1996; Gomez de Soto 1996a) and the burials were accompanied by large quantities of storage pottery (Funay 2005). A mainly open cave provides enough light for everyday living, but also for carrying out cult activities.

The same arguments about access/environment, topography, interior fittings and the archaeological material can be used to discern a domestic function on the one hand and to establish a ritual and/or funerary function on the other. It can also be noted that no significant features – apart from those suggested for sheepfolds (Brochier *et al.* 1999; Carozza *et al.* 2005; Galop *et al.* 2007; Carozza and Galop 2008) – distinguish a ritual site from a living space, either for caves, or for other sites (Bradley 2005, 35).

In addition, the contradiction between the functions of the artefacts found together is never brought to the fore. In other words, one space cannot be both a site for the storage of foodstuffs, or even precious objects, and a space used for rubbish disposal, particularly in a confined space. The presence of wild and domestic fauna and pottery containing cereals is problematic. If these sites are living spaces, one would perhaps think that the people would have removed the food waste from the site, similarly the broken ceramics, which would have spoilt the inhabitants' living space. Rotting food remains, the process accelerated by the humid environment, could have attracted pests, in addition to the smell and lack of hygiene. Various Bronze Age farms and villages show evidence of effective waste management (Marcigny and Ghesquière 2003). The storage of cereals in these more or less damp caves was furthermore less than ideal (Boudy *et al.* 2005). If part of these groups' subsistence was dependent upon these products, this storage technique was very risky.

The concept of an annex dwelling place poses another problem linked to an early preconception: why would a cave be more dependent upon a hypothetical exterior site than the other way around? The deposits recovered in caves are often both quantitatively rich (several hundred ceramic vessels for example) and qualitatively rich (with bronzes, amber, etc.). They demonstrate the principal craft activities known to have been practiced during this period – a paradoxical concentration – while those traces which exist on the surface are much less significant. It would, therefore, seem much more logical to interpret the cave as a major site, upon which eventual exterior sites were dependant. A ritual place can be divided into usage both outside and inside, as El Pedroso in Castilla (Bradley *et al.* 2005).

These examples demonstrate that the question of the function of caves remains particularly complex for the Bronze Age. Beyond even of the interpretation of these places, it is important to know if Bronze Age people really were troglodytes (cave-dwellers) – was this a characteristic for this period? – or if this lifestyle can be refuted.

Cave functions: a question of technical behaviour

The chaîne opératoire *concept: a different perspective on the question of dwelling versus ritual places*

The objects or environments studied here do not necessarily provide the most pertinent information to use when seeking to understand these Bronze Age societies. When faced with the ambiguity of the criteria used to distinguish a living space from a ritual site, it seems essential to change the frame of reference. Studies of caves and the things they contain focus essentially on the objects actually found there. They rarely include that which is invisible, namely the physical and cognitive processes implicit in the manufacturing, that is, as the suite of operations, from raw material to finished object (Cresswell 1976; Pelegrin *et al.* 1988). A *chaîne opératoire* is described 'in terms of (1) techniques (physical modalities by which raw material is acquired and transformed), (2) methods (the particular sequence followed), and (3) tools' (Roux 2003, 9–10; Roux 1994). In other words, the studies focus on the last step of a long process: the result. If the result can be identical in either a ritual or a domestic context, the manufacturing processes will certainly not present the same diversity in both contexts. For pottery, a classic domestic context infers a limited number of potters (one or two for example), derived from the family circle; while a meeting place – as ritual site – suggests the participation of more people from different households, different social environment, etc. In terms of *chaîne opératoire*, this distinction between household context and ritual meeting place can to be translated and studied in term of technical variability.

Ethnology now recognises a profound link between the individual, his family and social context, cast, gender, culture, group ethnolinguistic and the technical behaviours he adopts (e.g. Herbich 1987; Lemonnier 1993; Kramer 1997; Stark 1998; Wallaert-Pêtre 1999; Gosselain 2002; Gelbert 2003; Gallay 2007; Degoy 2008; Stark *et al.* 2008; Roux 2010). It is particulary the case for shaping techniques, which are considered as the most stable stages in the *chaîne opératoire* (Gosselain 2002), even if one should not consider them 'closed technical units' (Gosselain 2008, 170). Three fundamental elements – affecting the individual in his socio-cultural environment – allow us to make these observations: learning, motor habits, and transmission.

Learning processes often begin at a very young age, according to the physical, psychological and social maturity of the child (Gosselain 2002; Wallaert 2008). The novice never learns through invention (Bril 2002; Roux 2007). For a number of years, a child will learn manufacturing processes from an adult. It is during this time that the individual acquires motor habits: operational automatisms which will be hard to modify over time because of the routine nature of the undertaking as fixed since childhood (Arnold 1985; Bril 2002; Gosselain 2002; Roux 2007). The process of learning manufacturing techniques profoundly affects the individual, implying a correlation of body and mind, as observed in psychology (Bril 2002) and ethnology (Arnold 1985; Gosselain 2002, 2008; Wallaert 2008).

The household is the basic social unit of production and reproduction,. This refers to the *habitus* (Bourdieu 1977) and perfectly translates the relationship between social boundaries and technical behaviour (Stark 1998).

In other words, the materials made by and for this domestic circle (Arnold 1985, 25) will be the result of identical *chaîne opératoire* of shaping. The production should be particularly homogenous if there is only one producer, because of the stability of the motor habits. Production would still be homogenous with several producers at the centre of the household, if the knowledge transmission from one generation to another is 'vertical' in nature – that is to say from parent to child, which is the type of transmission considered most conservative (Cavalli-Sforza 2000). Finally, the production of pottery found in a domestic context will be all the more homogenous or mildly diversified if no objects or few products are exchanged between households, or if knowledge is homogenous at the heart of a community.

From an archaeological perspective, the Bronze Age cultures settled around the English Channel (Marcigny *et al.* 2007) offer a particularly pertinent frame of reference for the study of technical behaviours at the centre of households. Work undertaken on the ceramic production of several sites – farms and hamlets – in Cornwall (UK) and in Lower Normandy (France) tends to confirm several of these elements (Manem 2008, 2010, forthcoming). The domestic production is generally weakly diversified: the number of *chaînes opératoires* (operational chains) of shaping and finish identified range between one and five, but qualitatively speaking, just one *chaîne opératoire* is always dominant and the variability concerns the operations of finish of the pottery. The number of *chaîne opératoires* generally increases with the number of houses or the economic importance of the site, but remains low. As such, the variability of technical behaviours is greater at Trevisker, where three successive/contemporary households are grouped (ApSimon and Greenfield 1972), than at the Nonant farm, where between one and two families were settled (Marcigny and Ghesquière 2008). But a more important economic position, as in the Tatihou farm (Marcigny and Ghesquière 2003) marks a more notable diversity of 'manières de faire', while still weak, from which we can infer possible exchanges (Manem 2008). The same results are found in the Southern Levant (Roux and Courty 2007).

Contrary to this, meeting places with a ritual function can be understood in terms of the participation of many individuals, a diversity which is perceptible in the level of know-how implied in the deposited material, where no one practice is dominant. This diversity of technical know-how would be all the greater when the participating individuals are trained at the heart of different learning networks. As such, we find this diversity in the Chalcolithic sanctuary sites of the Southern Levant (Roux and Courty 2007).

All in all, the application of the *chaîne opératoire* concept to the study of caves comes down to reversing the priority of the analysis: that is to say, studying the objects last (Roux 2010) and to first exploring the technical behaviours.

The caves of the Duffaits culture

Situated in the Centre-West of France, the territory of the Duffaits culture occupied a larg area between the Atlantic and continental worlds. This particularity makes this cultural group a centrepiece of the French Middle Bronze Age (Gomez de Soto 1973, 1978, 1980, 1995, 1996a, 1996b). The Duffaits culture extended from the western part of the Massif Central to the Charente region. This region is the best known and the richest in artefacts in Middle Bronze Age France (Fig. 10.1) (Gomez de Soto 1973, 1978, 1996a). This zone is also the most atypical of the territory, for the principal known sites are the La Rochefoucauld karst caves (Tournepiche 1998). Some have been interpreted as dwelling place/annex dwelling place along with the Perrats cave (Gomez de Soto 1996a) and Quéroy caves (Fig. 10.1.1 and 10.1.4) (Gomez de Soto 1978) and the lower rock-shelters of Bois-du-Roc (Gomez de Soto 1995). The caves also have a ritual and funerary function, as is the case with the Duffaits cave (Fig. 10.1.2 and 10.1.3) where 33 bodies and a significant and varied range of items were found (Gomez de Soto 1973; Boulestin 1988). The Quéroy and Perrats caves seem to have been used in several ways. In the first case, domestic occupations preceded an eventual ritual and funerary use indicated by the deposition of a female head covered by a number of domestic items (Gomez de Soto 1978, 1995; Boulestin 1994). In the second case, domestic occupations succeeded an earlier funerary function (Gomez de Soto 1996a).

Although (following the examples outlined in the first section of this article) the dwelling and annex dwelling function is the hypothesis put forward for the Quéroy and Perrats caves, we can see profound parallels with the items found in the cult chamber of the Duffaits necropolis cave. There we find large amounts of storage pottery, and the ones associated with the decapitated head at Quéroy show that this type of item was fully integrated in ritual practice. There is, however, a more significant proportion of storage pottery at the Duffaits necropolis cave (78 per cent) than in the Perrats cave (63 per cent) (Manem 2008). In these two caves, the deposits are varied, rich in bronzes, amber, faunal remains and cereals.

The identification of the shaping and finishing *chaînes opératoires* of the pottery found in the cult chamber at Duffaits, according to the techniques,

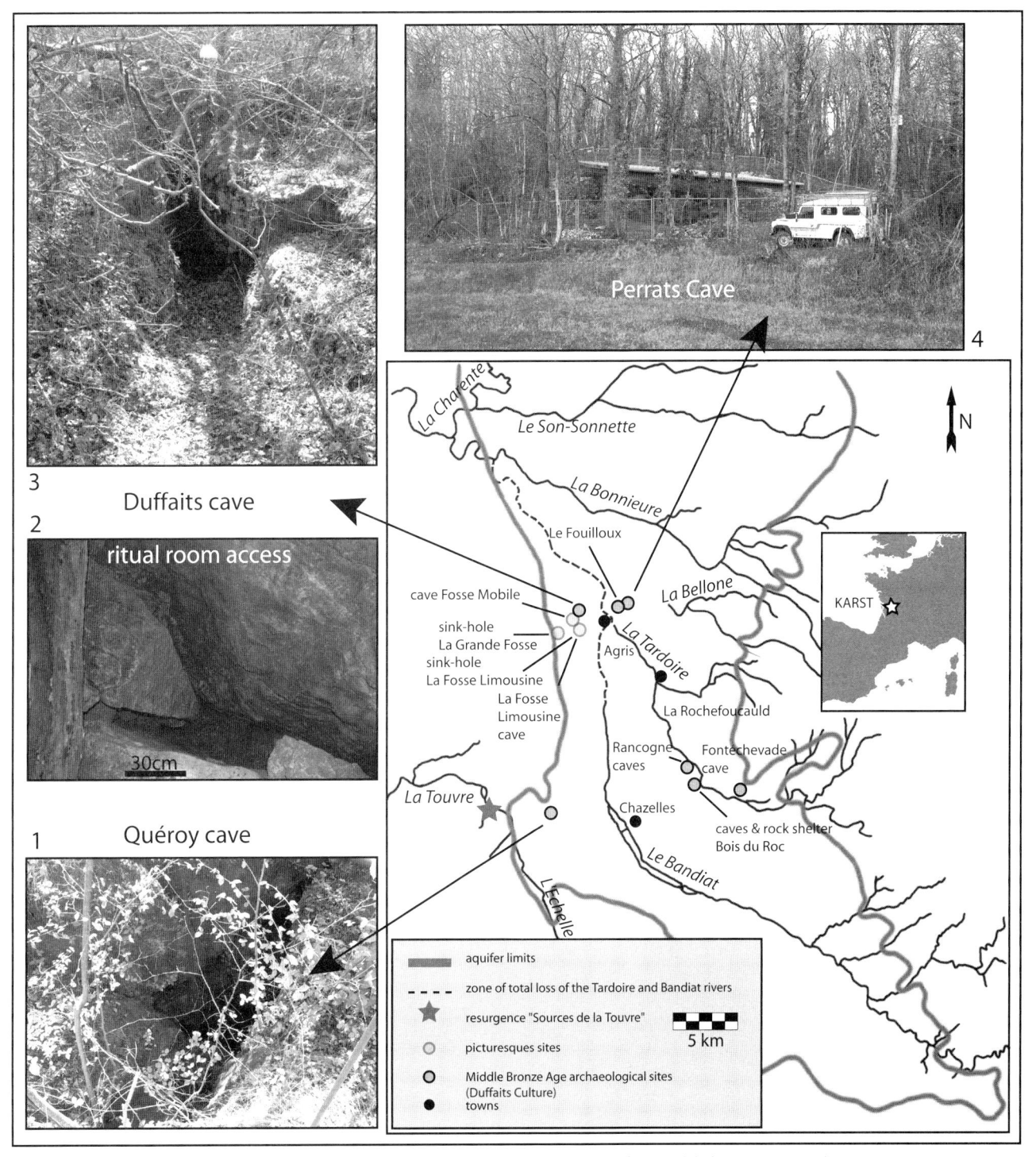

Figure 10.1: Middle Bronze Age caves, karst of La Rochefoucauld (Charente, France).

methods and study of macrotrace-features (Roux 1994), reveals a strong heterogeneity of technical behaviour (Manem 2008). The 16 *chaînes opératoires* identified distinguish different know-how, both on the level of certain techniques employed for shaping, and in methods and finishes. Certain potters used the mould technique to form the base and body while others employed more complex know-how and combined two techniques to shape the body: the primary forming was by coiling while the secondary forming was by beating.

The methods vary greatly. Certain potters shaped the base by modelling a mass of clay, while others used the technique of spiralling coils. The pressures exerted on the clay were essentially discontinuous but certain potters employed discontinuous pressure (primary forming) before using continuous pressure (secondary forming) for the shaping of the neck. The diversity of *chaînes opératoires* was also the result of diverse finishing techniques, such as smoothing, scraping, planing and burnishing. These techniques were not necessarily used on the entire surface of the pot and could be combined.

Finally, we are able to distinguish behaviours which are particular to each potter, even when they have mastered the same know-how. Contrary to the pottery studied from the farms around the Channel – where the site and the totality of the items found there – 904 pots – allowed the identification of five Middle Bronze Age occupational phases, and the identification of the minimum number of pots per phase (Manem 2008). This first analysis allows us to isolate the diversity of technical behaviours for each phase.

The technological analysis revealed a strong heterogeneity of technical behaviour (Manem 2008). The 904 pots were formed from 48 *chaînes opératoires*, including those identified in the Duffaits necropolis material (Fig. 10.2). Four techniques were recognised in the formation of the base: spiral coiling, moulding, modelling, and pinching. This strong diversity of 'manières de faire' is perceptible not only in the totality of the material but also at the heart of each occupation, irrespective of the number of pots. These

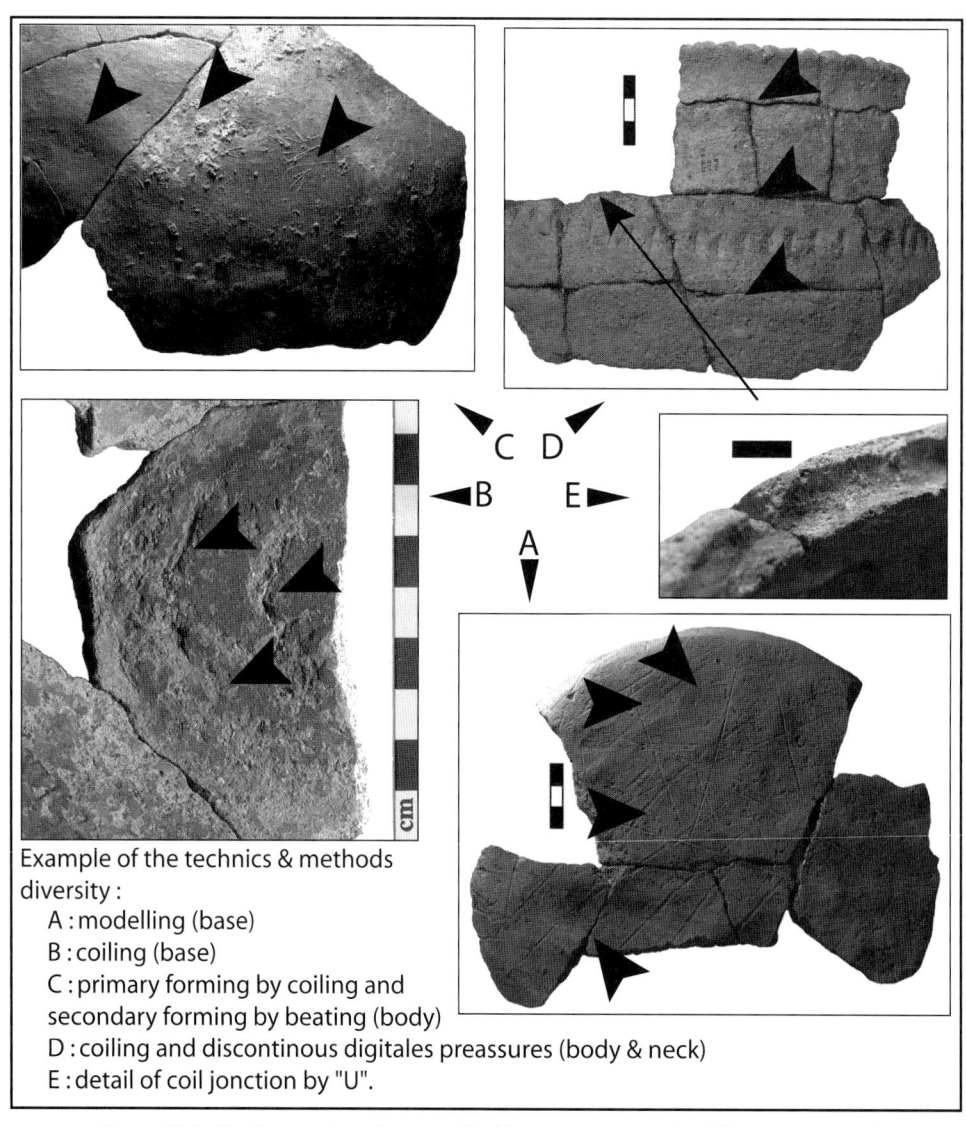

Example of the technics & methods diversity:
A : modelling (base)
B : coiling (base)
C : primary forming by coiling and secondary forming by beating (body)
D : coiling and discontinous digitales preassures (body & neck)
E : detail of coil jonction by "U".

Figure 10.2: Chaînes opératoires *details, Perrats cave ceramics (Charente, France).*

results resemble what we observe in a ritual and funeral setting, such as the Duffaits cave, much more closely than what we see in a true domestic situation. The diversity observed in the Perrats cave indicates a more significant number of participants, originating from a higher number of learning networks than was the case in the Duffaits cave. Additionally, the same observations can be made of the Quéroy cave, which confirms this tendency.

The ritual space: what I made, where I left it ...

The comparison between the implied technical behaviours at the heart of true households and in ritual contexts allow us, therefore, to conclude that the Perrats and Quéroy caves were not used as dwelling places or parts of dwelling places, but correspond rather only with ritual and/or funerary functions. Interpreting these caves as exclusively ritual sites means that no material is now functionally contradictory. The deposited and/or consumed meat at the sites was not left outside the cave, far from the cereal stores, because it is part of a cult practice. We can now understand the presence of rare bronze, amber and bone items (horse bits), and particularly why these items were not removed from the cave. We can also understand why all the craft and economic activities that one cultural group could practice are represented in these caves – weaving, metallurgy, amber trading, bone industry, pottery, hunting, livestock farming, harvesting and agriculture – a concentration of activities which seems incompatible with a household context. In other words, these caves are symbolic mirrors of real or maximised activities: agriculture cannot be a widespread economic practice in this group – no structure of plots or fields is yet known – but sublimated in a ritual context.

In view of the significant number of individuals involved in the production of material destined for these caves, either as offerings or as tools involved in cult practices, how was this practice organised? What is the relationship between the individual and these places? Are ritual practices wholly community-based or does representations of the individual emerge at the heart of this common space?

The classic spatial study of the ceramic deposits found in the Perrats cave initially offers a depositional logic which is very hard to understand (Manem 2008, Figs. 192–195). There is a significant mix of large, storage pottery and small, richly decorated vessels. This phenomenon can be seen in all of the Bronze Age occupations phases. This observation is not new within this cultural context. We see the same 'chaotic' mixture in the ritual chamber in the Duffaits necropolis cave (Gomez de Soto 1973, Fig. 2).

However, the nature of the deposits' organisation can be fully understood when analysed through the spatial distribution of not the objects but the individuals, through the technical know-how implied in the production of each ceramic (Manem 2008, Figs. 197–202). The pottery made using the same *chaîne opératoire* was deposited together, in one limited area, in more or less significant batches (Fig. 10.3). This phenomenon is observable in all of the occupations phases. When a number of pots were made using the same *chaîne opératoire*, for example the coiling technique, an analysis of the coil size used in each pottery allows for further specification of the individuals involved. In certain cases, pots have been found deposited in pairs in one limited space.

Ritual deposits are, as such, part of an individual logic. As the *chaîne opératoire* is not a visible element, it is possible that each potter deposited their own ceramic(s) in the cavity. If the men and women of the Duffaits culture shared the same ritual spaces (only a small proportion of the karst caves were used), they nonetheless did not combine the goods deposited in the caves. These spaces were both communal and individualised. This leads us to understand why nearly 57 per cent of the total corpus is decorated: the individualisation of the space and of the goods is visually reinforced by the richness and originality of the decorations. The ability to infer a relationship between a technical know-how and a space implies a respect for these deposits by the people who frequented these places, in addition of course for those objects used as cult tools. These deposits immediately take on a definitive character. This would explain the abundance of these deposits and why they were never cleared away to make room, as would have been the case had this been a dwelling place. This is certainly not the only indication of well-established community rules. The presence of certain foods and high value goods implies the centralised organisation of ritual practices which took place inside, as the internal features and the presence of hearths would seem to indicate.

Following the flow of water: a karstic aquifer transformed into a ritual landscape

The idea that the men and women of the Duffaits culture were not troglodytes calls for a complete revision of our understanding of their ways of life (where are the dwelling places?) and of our perception of the landscape of this period. When we draw these ritual and funerary cavities into the general karst context, we notice that they are centred near fossil, semi-active or strikingly active karstic phenomena, which give an atypical character to the Charente basin landscape (Fig. 10.4) (Tournepiche 1998). In particular,

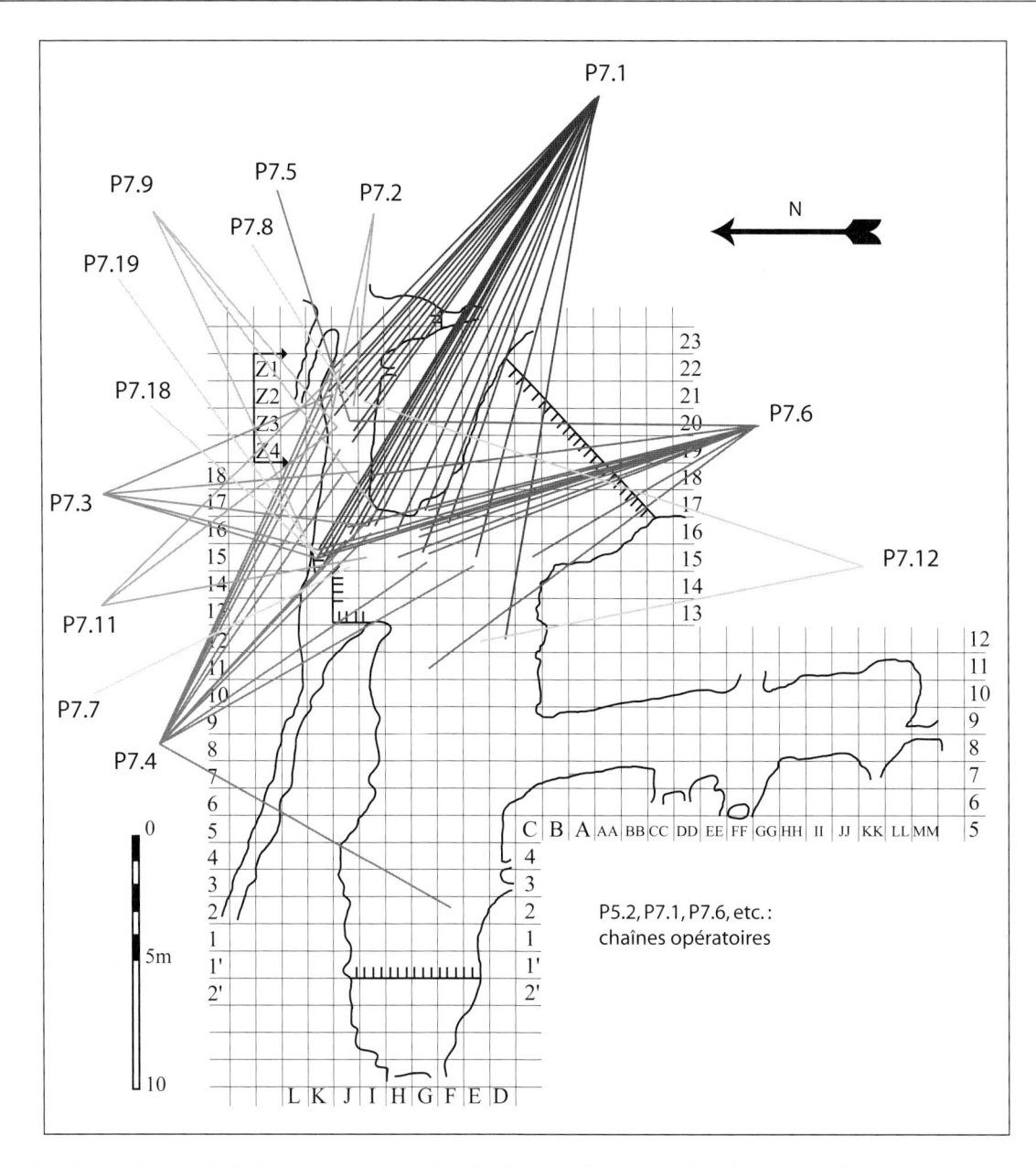

Figure 10.3: Perrats Cave, Middle Bronze Age, occupation 2: the ceramics made using the same chaîne opératoire *were deposited together here.*

the Duffaits necropolis is situated at the heart of the most visible phenomena. Two notable sinkholes, of which the largest is karst based, are 2km from the necropolis. The link between these sinkholes and the ritual/funerary world is demonstrated by the burial sites found in the Fosse Limousine cave, situated inside the sinkhole of the same name (Gomez de Soto 1995). Only 800m from the necropolis, the Fosse Mobile is the entrance to a vast network of cavities, stretching over 8km, in which a significant amount of water is present. This geographical zone is, in effect, a karstic aquifer where powerful infiltrations of water are essentially linked to partial or total losses of the flows of four rivers (Tournepiche 1998; Larocque *et al.* 1999, 2000; Kurtulus and Razack 2007). And yet, the Perrats and Duffaits' caves, at 4km from each other, are approximately equidistant from the junction of the Bandia and Tardoire rivers, which are subject to total cyclical losses. The water filters into the sub-soil and the numerous chasms which run alongside rivers (Fig. 10.5), sometimes making a deafening noise (as in the case of 'le gouffre de Chez Robi' – Enjalbert 1947).

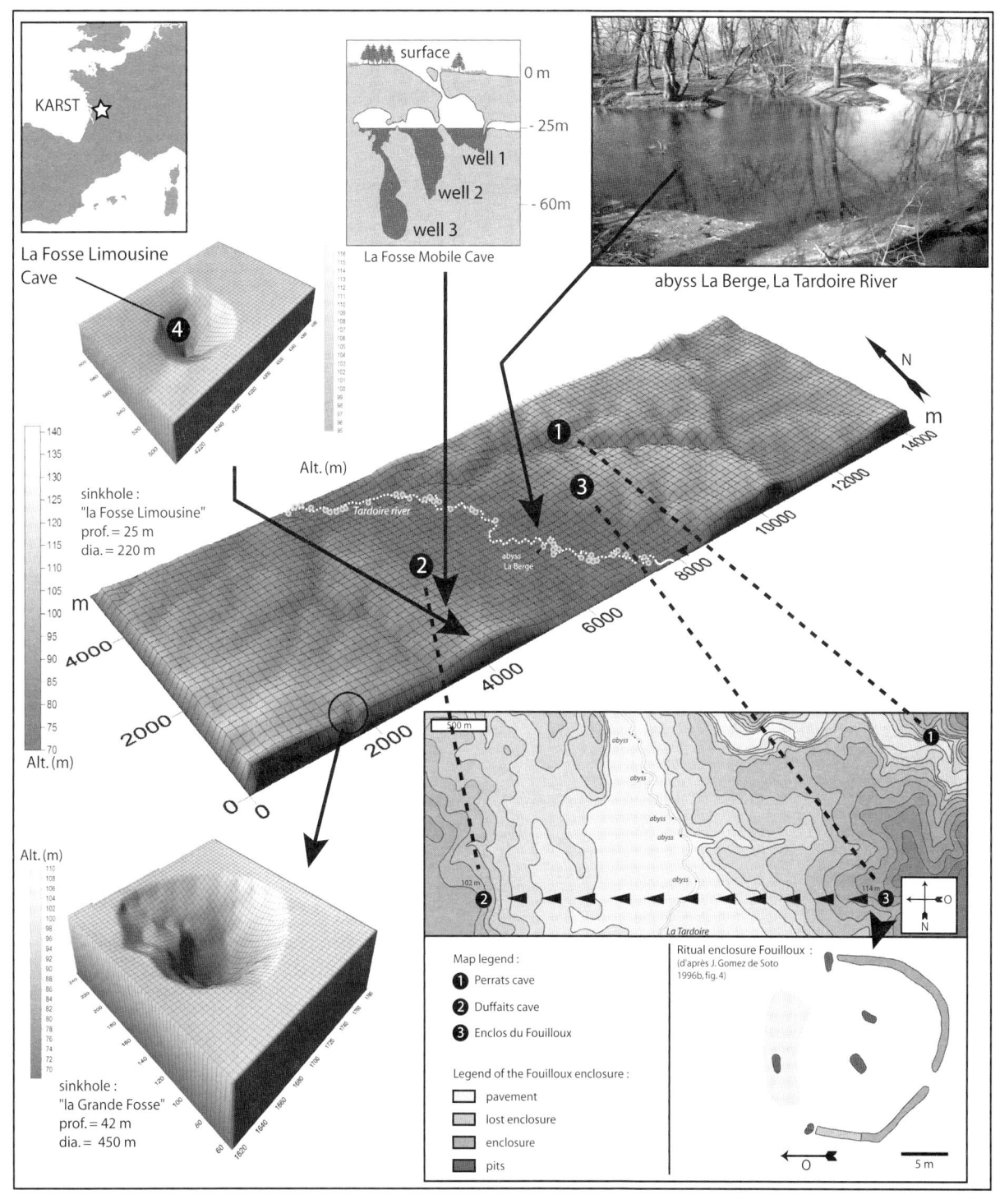

Figure 10.4: Ritual landscape: interaction between karst landscape, ritual caves and ritual enclosure (Charente, France).

The ritual and funerary practices therefore appear connected to two important European Bronze Age themes: the underground world, and the river (Bradley 1990, 2000; Harding 2000). Numerous bronze objects and sometimes human bones have been found in the European river-beds, for example the skulls from the Thames (Bradley and Gordon 1988), interpreted as 'river burial' by R. Bradley (2007, 202). However, in the case of Duffaits Culture and this specific karstic aquifer context, these themes are associated in the

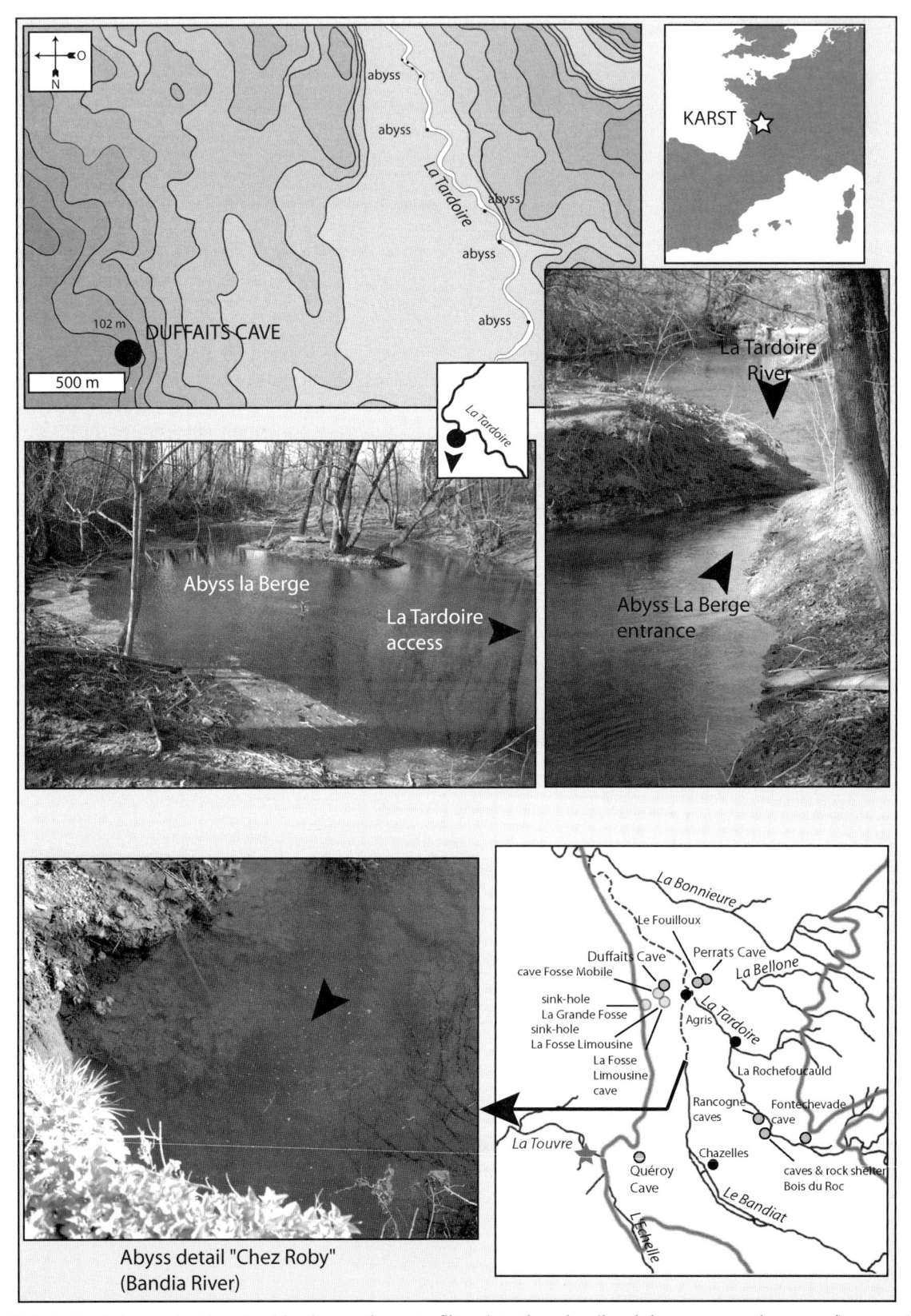

Figure 10.5: A river's impact inside a ritual landscape: the water filters into the sub-soil and the numerous abysses or fissures which run alongside them, like the 'gouffre de la Berge' for La Tardoire River, sometimes with long-recognised deafening sounds, like the 'gouffre de Chez Robi' for the Bandia River (Charente, France).

same place: the water of the river which rushes into the underground world. We can now understand the concentration of funerary and ritual sites around these geological phenomena and the exclusivity of the practices in these spaces. The unique structure of the Duffaits culture as known and constructed on the surface of this karst can be largely related back to this observation. The circular ditched enclosure Fouilloux is situated 800m from the Perrats cave. It contained some sherds and has been interpreted as a ritual structure because of its open shape, oriented towards the West (Gomez de Soto 1996b). It dominates the valley where the 'lost' (underground) flow of the Tardoire River is located. The presence of this enclosure is not a chance matter. Its entrance faces exactly towards the Duffaits necropolis (Fig. 10.4), situated at approximately the same height, on the opposite plateau, 4km away (Manem forthcoming). The entrance to this necropolis is almost invisible from a distance of 15m. This ditch may thus constitute the link between the symbolic importance of this river, whose water completely disappears underground, and the dead bodies that were placed in these cavities, in this same sub-soil. It is also indicative of the interaction between these different sites. Lastly, the absence of recognised dwelling places leads us to see at least this part of the karst as a ritual landscape (Manem forthcoming). This ritual landscape was passed from generation to generation, as demonstrated in the successive occupations of the Perrats cave throughout the Middle Bronze Age. The maintenance of certain technical traditions through the Middle Bronze Age indicated that the people who frequented these places passed on not only the memory of the landscape, but also of technical traditions. In all, ritual landscape and technical behaviours constituted an invisible and intimate part of this culture: a part which people from other cultures could not see directly. Passing this ritual landscape from one generation to another infers the ideas of heritage and territorial anchorage. If the notion of original space is marked by seasonal rites among certain cultures (Hirsch 2006), the question of seasonal rites can be raised for the Duffaits Culture, particularly as these caves were used since the origins of this culture. The agricultural and animal-related goods deposited in these caves are a further testament to the seasonal aspect of the ritual practices. The total losses of the flows of the river in the ritual landscape are also cyclical (Larocque *et al.* 1999, 2000). It is possible that the Duffaits people used caves situated upstream of the zone of total water loss of the Tardoire River, where the karstic chasms gradually remove the river water, but no site has yet been recognised beyond this zone (Fig. 10.1), where only the river-bed remains. We can, then, imagine that the ritual and funerary practices of the Duffaits culture followed the path of the water.

Conclusion

This approach demonstrates, first and foremost, that it is essential to study technical behaviours used to product objects as we find them and before a typological analysis (Roux 2010), particularly in the identification of the function of caves and of the spatial ordering of their deposits. The *chaîne opératoire* concept allows us to reconsider the function of caves and of the troglodytic nature of Bronze Age communities. The caves seem only to have been used for ritual and funerary functions, implying a large number of participants. If this seems to be the case for the Duffaits culture, the same question can then be applied to all French caves hitherto interpreted as dwelling places or as refuges, except those used as sheepfolds. In other words, it becomes clear that an exhaustive re-examination of these caves, using the *chaîne opératoire* concept, will allow us to better understand the function of these complex sites and the perception of the landscape of this period.

Acknowledgements

I wish to express my thanks to: Knut Andreas Bergsvik (University of Bergen), Robin Skeates (Durham University), Fondation Fyssen (Paris), Zoé Adams (University of Exeter), Richard Bradley (University of Reading), José Gomez de Soto (Centre National de la Recherche Scientifique), Anthony Harding (University of Exeter; postdoctoral supervisor), Cyril Marcigny (Institut National de Recherches Archéologiques Préventives), Claude Mordant (Université de Bourgogne), Catherine Perlès (Université de Paris 10; PhD director), Valentine Roux (Centre National de la Recherche Scientifique; PhD tutor), and the staff of the Musée d'Angoulême, Musée de Bayeux, Musée Maritime de Tatihou, and Royal Cornwall Museum.

References

ApSimon, A. M. and Greenfield, E. (1972) The excavations of Bronze Age and Iron Age settlements at Trevisker, St. Eval, Cornwall. *Proceedings of the Prehistoric Society* 38, 302–381.

Arnold, D. E. (1985) *Ceramic Theory and Cultural Process*. Cambridge, Cambridge University Press.

Bonsall, C. and Tolan-Smith, C. eds. (1997) *The Human Uses of Caves*. British Archaeological Reports International Series 667. Oxford, British Archaeological Reports.

Boudy, L., Fages, G. and Treffort, J.-M. (2005) Food storage in two Late Bronze Age caves of southern France: palaeoethnobotanical and social implications. *Vegetation History and Archaeobotany* 14, 313–328.

Bourdieu, P. (1977) *Outline of a Theory of Practice*. Cambridge, Cambridge University Press.

Boulestin, B. (1988) *Étude Anthropologique des Restes Humains de la Grotte des Duffaits (La Rochette, Charente)*. Unpublished doctoral thesis, Université de Bordeaux II.

Boulestin, B. (1994) La tête isolée de la grotte du Quéroy: nouvelles observations, nouvelles considérations. *Bulletin de la Société Préhistorique Française* 91/6, 440–446.

Boulestin, B. (1996) Le niveau funéraire Bronze ancien de la grotte des Perrats (Agris, Charente), premières observations. In C. Mordant and O. Gaiffe (eds.) *Cultures et Sociétés du Bronze Ancien en Europe*, 503–508. Paris, Éditions du Comité des Travaux Historiques et Scientifiques.

Boulestin, B. and Gomez de Soto, J. (2003) Le complexe funéraire des Renardières (Les Pins, Charente): regards sur la mort et la société au Bronze ancien. *Bulletin de la Société Préhistorique Française* 100/4, 757–790.

Bradley, R. (1990) *The Passage of Arms: an Archaeological Analysis of Prehistoric Hoard and Votive Deposits*. Cambridge, Cambridge University Press.

Bradley, R. (2000) *An Archaeology of Natural Places*. London and New York, Routledge.

Bradley, R. (2005) *Ritual and Domestic Life in Prehistoric Europe*. London and New York, Routledge.

Bradley, R. (2007) *The Prehistory of Britain and Ireland*. Cambridge, Cambridge University Press.

Bradley, R. and Gordon J. (1988) Human skulls from the River Thames, their dating and significance. *Antiquity* 62/236, 503–509.

Bradley, R., Fabregas, R., Alves, L., Vilaseco, X. (2005) El Pedroso – a prehistoric cave sanctuary in Castille. *Journal of Iberian Archaeology* 7, 125–156.

Bril, B. (2002) L'apprentissage de gestes techniques : ordre de contraintes et variations culturelles. In B. Bril and V. Roux (eds.) *Le Geste Technique: Réflexions Méthodologiques et Anthropologiques*, 113–150. Ramonville Saint-Agne, Éditions Erès.

Briois, F., Crubezy, E. and Carozza, L. (2000) La grotte Sindou (Lot): une sépulture familiale du Bronze final. *Bulletin de la Société Préhistorique Française* 97/4, 553–559.

Brochier, J.-L., Beeching, A., Maamar, H. S. and Vital, J. (1999) Les grottes bergeries des Préalpes et le pastoralisme alpin, durant la fin de la préhistoire. In A. Beeching (ed.) *Circulations et Identités Culturelles Alpines à la Fin de la Préhistoire: Matériaux pour une Étude*, 77–114. Valence, Travaux du Centre d'Archéologie Préhistorique de Valence, 2.

Carozza, L., Galop. D., Marembert, F. and Monna, F. (2005) Quel statut pour les espaces de montagne durant l'âge du Bronze? Regards croisés sur les approches société-environnement dans les Pyrénées occidentales. *Documents d'Archéologie Méridionale* 28, 7–23.

Carozza, L. and Galop, D. (2008) Le dynamisme des marges: peuplement et exploitation des espaces de montagne durant l'âge du Bronze. In J. Guilaine (ed.) *Villes, Villages, Campagnes de l'Âge du Bronze. Séminaire du Collège de France*, 226–253. Paris, Éditions Errance.

Cavalli-Sforza, L. L. (2000) *Genes, Peoples, and Languages*. New York, North Point Press.

Clottes, J. and Lorblanchet, M. (1972) La grotte du Noyer (Esclauzels, Lot): note préliminaire. In *Congrès Préhistorique de France*, 145–164. Paris, Société Préhistorique Française.

Cresswell, R. (1976) Techniques et culture, les bases d'un programmede travail. *Techniques et Culture* 1, 7–59.

Degoy, L. (2008) Technical traditions and cultural identity: an ethnoarchaeological study of Andhra Pradesh Potters. In M. T. Stark, B. J. Bowser and L. Horne (eds.) *Cultural Transmission and Material Culture: Breaking down Boundaries* 199–222. Tucson, The University of Arizona Press.

Enjalbert, H. (1947) Le karst de La Rochefoucauld (Charente). *Annales de Géographie* 56/302, 104–124.

Funay, L. (2005) *Étude de la Céramique du Bronze Ancien de la Grotte des Perrats à Agris (Charente): Apport à l'Interprétation de l'Occupation Funéraire du Site*. Unpublished thesis, Université de Poitiers.

Gallay, A. (2007) The decorated marriage jars of the inner delta of the Niger (Mali): essay of archeological demarcation of an ethnic territory. *The Arkeotek Journal* 1/1: <http://www.thearkeotekjornal.org>, accessed June 1st 2010.

Galop, D., Carozza, L., Marembert, F. and Bal, M.-C. (2007) Activité agropastorales et climat durant l'âge du Bronze dans les Pyrénées : l'état de la question à la lumière des données environnementales et archéologiques. In H. Richard, M. Magny and C. Mordant (eds.) *Environnements et Cultures à l'Âge du Bronze Occidental*, 107–142. Paris, Éditions du Comité des Travaux Historiques et Scientifiques.

Gasco, J. (2004) La stratigraphie de l'âge du Bronze et de l'âge du Fer à Roucadour (Thémines, Lot): analyse culturelle et incidence paléographiques. *Bulletin de la Société Préhistorique Française* 101/3, 521–545.

Gelbert, A. (2003) *Traditions Céramiques et Emprunts Techniques dans la Vallée du Fleuve Sénégal*. Paris, Éditions de la Maison des Sciences de l'Homme and Épistèmes.

Giraud, J.-P. (1989) L'âge du Bronze moyen en Quercy. In C. Mordant (ed.) *Dynamique du Bronze Moyen en Europe*, 429–442. Paris, Éditions du Comité des Travaux Historiques et Scientifiques.

Gomez de Soto, J. (1973) La grotte sépulcrale des Duffaits (La Rochette, Charente). *Bulletin de la Société Préhistorique Française* 70, 401–444.

Gomez de Soto, J. (1978) La stratigraphie chalcolithique et protohistorique de la grotte du Quéroy à Chazelle, Charente. *Bulletin de la Société Préhistorique Française* 75/10, 394–421.

Gomez de Soto, J. (1980) *Les Cultures de l'Âge du Bronze dans le Bassin de la Charente*. Périgueux, Éditions Fanlac.

Gomez de Soto, J. (1995) *Le Bronze Moyen en Occident: La Culture des Duffaits et la Civilisation des Tumulus*. Paris, Éditions Picard.

Gomez de Soto, J. (1996a) *Grotte des Perrats à Agris (Charente), 1981–1994. Étude Préliminaire*. Chauvigny, Éditions A.P.C..

Gomez de Soto, J. (1996b) Réflexions sur un possible nécromantion du Bronze moyen. *Bulletin de la Société Préhistorique Française* 93/4, 566–578.

Gomez de Soto, J. and Kerouanton, I. (1991) La grotte du Quéroy à Chazelles (Charente): le Bronze final IIIb. *Bulletin de la Société Préhistorique Française* 88/10–12, 341–392.

Gosselain, O. (2002) *Poteries du Cameroun Méridional. Styles Techniques et Rapports à l'Identité*. Paris, Éditions du Centre National de la Recherche Scientifique.

Gosselain, O. (2008) Mother Bella was not a Bella: inherited and transformed traditions in Southwestern Niger. In M. T. Stark, B. J. Bowser and L. Horne (eds.) *Cultural Transmission and Material Culture: Breaking down Boundaries* 150–177. Tucson, The University of Arizona Press.

Harding, A. F. (2000) *European Societies in the Bronze Age*. Cambridge, Cambridge University Press.

Harding, A. F. (2005) The Bronze Age use of caves. In R. Laffineur, J. Driessen and E. Warmenbol (eds.) *L'Âge du Bronze en Europe et en Méditerranée. Actes du XIVème Congrès de l'Union Internationale des Sciences Préhistoriques et Protohistoriques*. British Archaeological Reports International Series 1337, 1–4. Oxford, British Archaeological Reports.

Herbich, I. (1987) Leaning patterns, pottery interaction and ceramic style among the Luo of Kenya. *The African Archaeological Review* 5, 193–204.

Hirsch, E. (2006) Landscape, myth and time. *Journal of Material Culture* 11/1–2, 151–165.

Kramer, C. (1997) *Pottery in Rajasthan: Ethnoarchaeology of Two Indian Cities*. Washington D.C. and London, Smithsonian Institution Press.

Kurtulus, B. and Razack, M. (2007) Evaluation of the ability of an artificial neural network model to somulation the input-output responses of a large karstic aquifer: the La Rochefoucauld aquifer (Charente, France). *Hydrogeology Journal* 15, 241–254.

Larocque, M., Banton, O., Ackerer, P. and Razack, M (1999) Determining karst transmissivities with inverse modeling and an equivalent porous media. *Ground Water* 37/6, 897–903.

Larocque, M., Banton, O. and Razack, M. (2000) Transient-State history matching of a karst aquifer ground water flow model. *Ground Water* 38/6, 939–946.

Lemonnier, P. (ed.) (1993) *Technological Choices. Transformation in Material Cultures since the Neolithic*. London and New-York, Routledge.

Le Guillou, Y., Boës, E., Lecomte, N. and Paulin, J. (2000) Grotte de Khépri à Ganties, Haute-Garonne. *Bulletin de la Société Préhistorique Française* 97/4, 539–541.

Manem, S. (2008) *Etude des Fondements Technologiques de la Culture des Duffaits (Âge du Bronze Moyen)*. Unpublished doctoral thesis, Université de Paris 10, Laboratoire de Préhistoire et Technologie, UMR 7055 du Centre National de la Recherche Scientifique.

Manem, S. (2010) Des habitats aux sites de rassemblement à vocation rituelle: l'âge du Bronze selon le concept de chaîne opératoire. *Les Nouvelles de l'Archéologie* 119, 30–36.

Manem, S. (forthcoming) Les lieux naturels atypiques, source du paysage rituel: le karst de La Rochefoucauld et la Culture des Duffaits (Charente, France). In S. Wirth and S. Beranger (eds.) *Paysages Funéraires de l'Âge du Bronze*. Münster, Landschaftsverband Westfalen-Lippe - Archäologie für Westfalen.

Marcigny, C. and Ghesquière, E. (2003) *L'Île de Tatihou (Manche) à l'Âge du Bronze: Habitats et Occupation du Sol*. Paris, Éditions de la Maison des Sciences de l'Homme.

Marcigny, C. and Ghesquière, E. (2008) Espace rural et systèmes agraires dans l'Ouest de la France à l'âge du Bronze: quelques exemples normands. In J. Guilaine (ed.) *Villes, Villages, Campagnes de l'Âge du Bronze*, 256–278. Paris, Éditions Errance.

Marcigny, C., Ghesquière, E. and Kinnes, I. (2007) Bronze Age cross-Channel relations. The Lower-Normandy (France) example: ceramic chronology and first reflections. In C. Burgess, P. Topping and F. Lynch (eds.) *Beyond Stonehenge: Essays on the Bronze Age in Honour of Colin Burgess*, 255–267. Oxford, Oxbow Books.

Marinval, P. (1983) Étude de quelques semences archéologiques provenant de niveaux de l'âge du Bronze de la grotte du Quéroy, Chazelles (Charente). *Bulletin de la Société Archéologique et Historique de la Charente* 4, 203–214.

Pelegrin, J., Karlin, C. and Bodu, P. (1988) Chaînes opératoires: un outil pour le préhistorien. In J. Tixier (ed.) *Technologie Préhistorique*, 55–62. Paris, Éditions du Centre National de la Recherche Scientifique.

Pétrequin, P., Chaix, A. M., Pétrequin, A. M. and Piningre, J.-F. (1985) *La Grotte des Planches-près-Arbois (Jura): Proto-Cortaillod et Âge du Bronze Final*. Paris, Éditions de la Maison des Sciences de l'Homme.

Pétrequin, A. M., Pétrequin, P. and Vuillemey, M. (1988) Les occupations néolithiques et protohistoriques de la Baume de Gigny (Jura). *Revue Archéologique de l'Est* 39/1–2, 3–39.

Roux, V. (1994) La technique du tournage: définition et reconnaissance par les macrotraces. In D. Binder and J. Courtin (eds.) *Terre Cuite et Société. La Céramique: Document Technique, Économique, Culturel*, 45–58. Juan-les-Pins, Éditions APDCA.

Roux, V. (2003) A dynamic systems framework for studying technological change: application to the emergence of the potter's wheel in the Southern Levant. *Journal of Archaeological Method and Theory* 10/1, 1–30.

Roux, V. (2007) Ethnoarchaeology: a non historical science of reference necessary for interpreting the past. *Journal of Archaeological Method and Theory* 14/2, 153–178.

Roux, V. (2010) Classification des assemblages céramiques selon le concept de 'chaîne opératoire': une approche anthropologique de la variabilité synchronique et diachronique. *Les Nouvelles de l'Archéologie* 119, 4–9.

Roux, V. and Courty, M.-A. (2007) Analyse techno-pétrographique céramique et interprétation fonctionnelle des sites: un exemple d'application dans le Levant Sud chalcolithique. In A. Bain, J. Chabot and M. Moussette (eds.) *La Mesure du Passé: Contributions à la Recherche en Archéométrie (2000–2006)*. British Archaeological Reports International Series 1700, 153–167. Oxford, British Archaeological Reports.

Stark, M. T. ed. (1998) *The Archaeology of Social Boundaries*. Washington, D.C., Smithsonian Institution Press.

Stark, M. T., Bowser, B. J. and Horne, L. eds. (2008) *Cultural Transmission and Material Culture: Breaking down Boundaries*. Tucson, The University of Arizona Press.

Thauvin-Boulestin, E. (1998) *Le Bronze Ancien et Moyen des Grands Causses du Quercy*. Document Préhistorique 11. Souillac, Éditions du Comité des Travaux Historiques et Scientifiques.

Tournepiche, J.-F. (1998) *Géologie de la Charente*. Angoulême, Germa.

Treffort, J.-M. (2005) La fréquentation des cavités naturelles du Jura méridional au Bronze final: état de la question, nouvelles données et perspectives. *Bulletin de la Société Préhistorique Française* 102/2, 401–416.

Vital, J. (1986) La grotte des Cloches à Saint-Martin-d'Ardèche. *Bulletin de la Société Préhistorique Française* 83/11, 503–545.

Wallaert-Pêtre, H. (1999) Potières et apprenties Vere du Cameroun: style techniques et processus d'apprentissage. *Techniques et Culture* 33, 89–116.

Wallaert, H. (2008) The way of the potter's mother: apprenticeship strategies among Dii potters from Cameroon, West Africa. In M. T. Stark, B. J. Bowser and L. Horne (eds.) *Cultural Transmission and Material Culture: Breaking down Boundaries*, 178–198. Tucson, The University of Arizona Press.

Chapter 11

Caves in Context: The Late Medieval Maltese Scenario

Keith Buhagiar

Areas of the Maltese countryside containing a brittle limestone deposit known locally as Mtarfa Member contain numerous cave-settlements of unknown antiquity. Even though remaining in use until the early modern period, and some until the first few decades of the twentieth century, the available archaeological and landscape evidence hints at a twelfth century date, and might indeed be related to a period of agricultural expansion experienced by Maltese rural areas during this period. The various late medieval Maltese cave typologies, the rock-cut churches – several of which were situated within the precincts of palaeochristian hypogea, and the hydraulic strategies employed by the cave occupants in order to retrieve a perennial water source from the occupied landscape are all discussed. This study forms part of a broader research project, aimed at investigating Maltese late medieval settlement location and any related water management systems. Landscape and settlement analysis is based on available archaeological, archival and historical evidence, together with extensive personal non-invasive field research conducted in various parts of the archipelago.

Introduction

The roots of Maltese late medieval troglodytism probably lie in the twelfth and thirteenth centuries and possibly reflect coordinated attempts at increasing the agricultural output of specifically designated countryside areas of Malta. It was formerly observed that the north-west sector of Malta is 'strangely bare' of late medieval aboveground village-type settlements known locally as *raħal* (Wettinger 1975, 190). Blame was tentatively placed on a defect in the surviving documentation, or the fact that the countryside in this part of Malta had been depopulated for such a long period of time, that the surviving place names dropped their *raħal* prefix (Wettinger 1975, 190).

Following the identification of numerous man-excavated cave-settlement sites and water galleries tapping the perched aquifer in the north and west of Malta, this study proposes the hypothesis that it was the *għar* (cave) settlements which prevailed in this region and not the *raħal* ones. The geographical parameters for the field investigation were determined by the natural distribution of Upper Coralline Limestone deposits.

The geological and climatic context

The Maltese archipelago lies in the central Mediterranean Sea south of Sicily. Its largest islands (listed in descending order according to size) are Malta, Gozo and Comino. Central to this paper's discussion is the island of Malta, which occupies a total land surface area of 153km^2, and has a maximum length and width of 27.4km and 14.5km respectively (Fig. 11.1). Geological deposits are almost exclusively sedimentary in formation, and started to form in a marine environment between 30 and 6 million years before present. The archipelago owes its origin to prolonged stress between the European and African continents, where plate tectonic activity completely reshaped the central Mediterranean basin into a series of horst and graben formations. Tectonic activity also uplifted several portions of the Sicilian-Tunisian Platform on which Malta lies, a few hundred metres above sea level (Zammit-Maempel 1977, 18; Pedley *et al.* 2002, 1, 18–29).

Four distinct rock layers constitute the basic geology (Fig. 11.2), and when undisturbed by land faulting, the

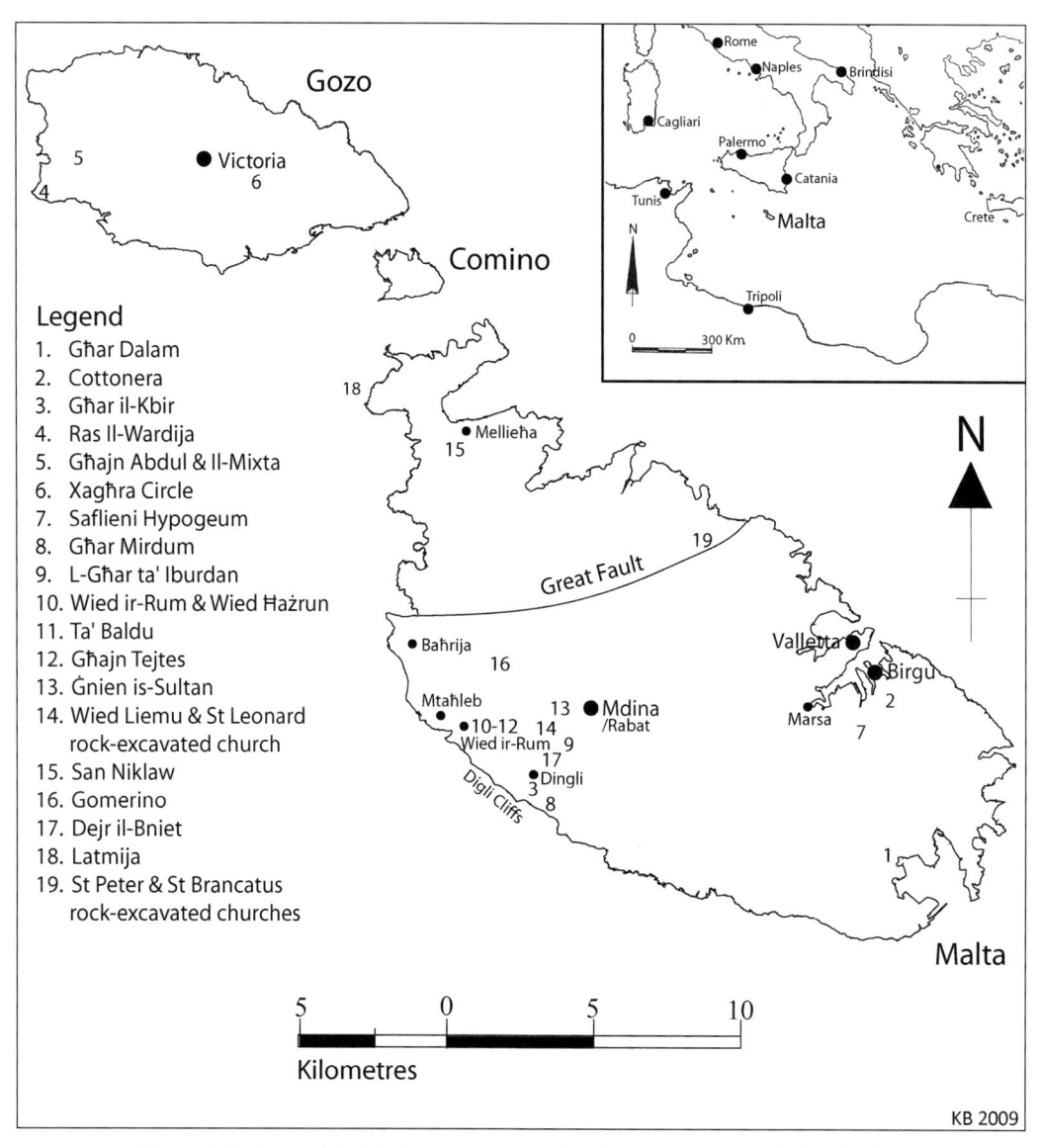

Figure 11.1: Map of the Maltese archipelago showing toponym spatial distribution.

horizontal stratification from bottom to top reads as follows: (1) Lower Coralline Limestone; (2) Globigerina Limestone; (3) Blue Clay; and (4) Upper Coralline Limestone (Pedley *et al.* 2002, 35). Exposed Lower Coralline and Globigerina Limestone deposits mainly cover areas of central and southern Malta. Late medieval cave-settlement location was nonetheless mainly determined by Upper Coralline Limestone and Blue Clay distribution.

From a technical perspective, Blue Clay is the most important rock horizon, and it is due to its presence that an easily accessible water table, locally referred to as the *perched aquifer*, exists. Water stored above this impermeable rock deposit has, since antiquity, been recognized as a vital and easily accessible resource. In the absence of the required technical expertise to extract water from the *mean-sea-level aquifer*, the perched water table was the only reliable water source available in the archipelago until the mid nineteenth century.

Upper Coralline Limestone is the youngest rock formation, four subdivisions of which have been identified. The most important of these is Mtarfa Member, which is composed of massive to thickly bedded carbonate mudstones and wackstones. The thickness of this stratum varies from 12 to 16m and, in contrast to other Upper Coralline Strata with their characteristically hard deposits, Mtarfa Member deposits can be cut and quarried with relative ease. It is within this rock deposit that almost all cave settlements in north and north-west Malta are located (Buhagiar, K. 2007, 110–112).

Malta's climate is typically Mediterranean and is characterised by hot dry summers and warm wet winters (Bowen-Jones *et al.* 1961, 48–49). The annual temperature range is of approximately 15°C and an average precipitation of 560mm makes rainfall insufficient and erratic and creates regular drought conditions (Skinner *et al.* 1997, 188). The Maltese landscape is, furthermore, characterised by the almost complete absence of woodland vegetation and scarce soil deposits leaving the bare rock exposed. The prevailing environmental conditions and the local geology and topography provided an ideal springboard for the widespread diffusion of troglodytism.

The historical context

The adaptation of caves into houses and cultic shrines represents an ancient Mediterranean practice. Places such as Granada in Spain, Matera in Basilicata (Italy), Matmata in Tunisia and Cava d'Ispica in Sicily amongst others, show that, 'Mediterranean people have always chosen caves and grottoes, natural and excavated, as providing convenient, cool and often defensible dwellings, stores, stalls, cisterns, churches, burial places and catacombs' (Luttrell 1979, 461).

Strabo (XVI, 7, 25–260), amongst other ancient authors, noted the habit of some African people of using caves as houses. When the Mediterranean coast of Africa was colonized by the Romans, they too adapted themselves to the scarcity of timber and

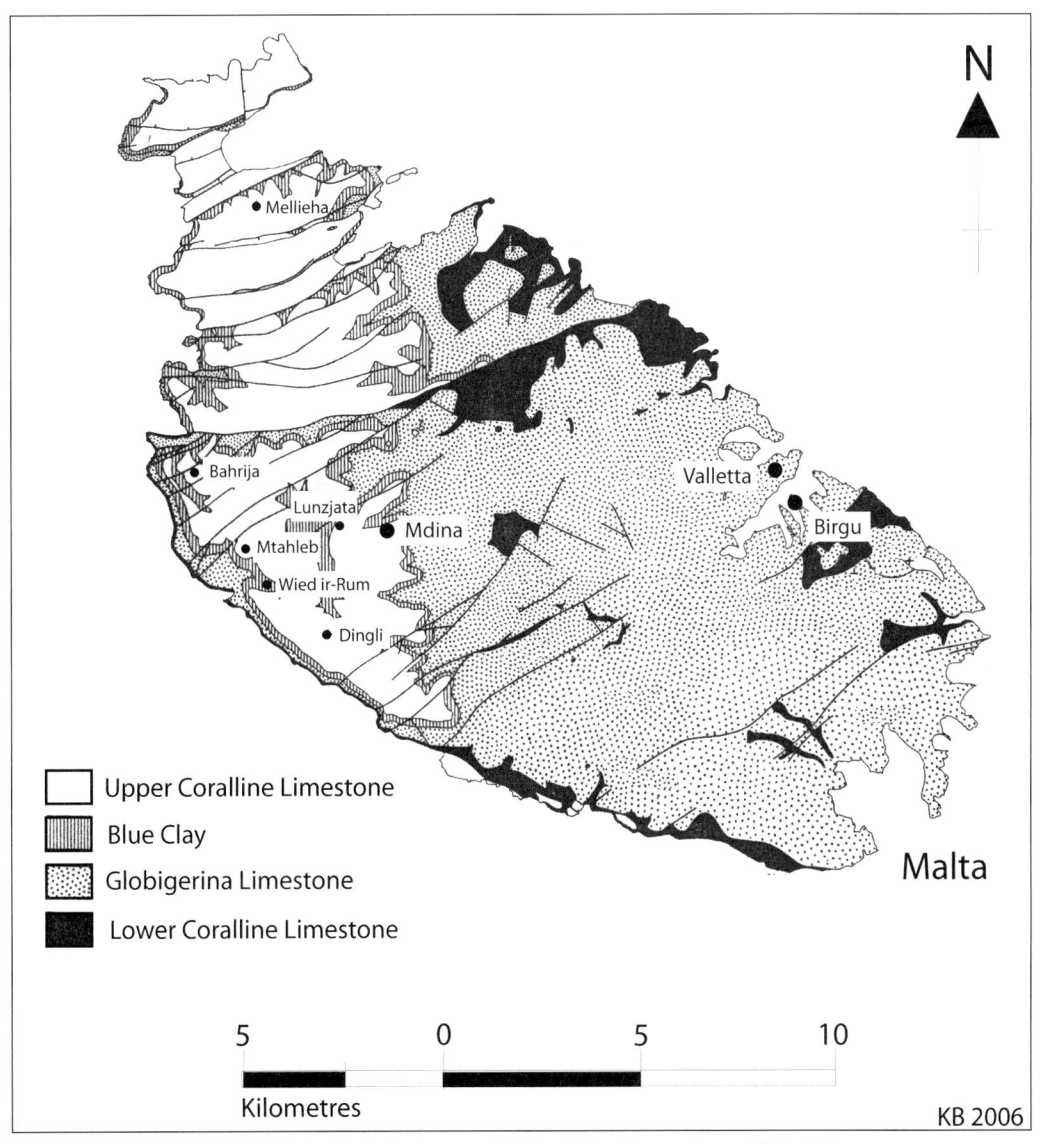

Figure 11.2: Geological map of Malta (after Bowen-Jones 1961, 24).

the availability of easily quarried rock, and there are examples at Cyrenaica, Bulla Regis and elsewhere of villas and other structures that are wholly or partially rock-cut (Buhagiar, M. 1984, 17). Malta was no exception, and the local archaeological record gives testimony to the widespread use of caves for dwelling, burial and cultic purposes since prehistoric times (Fig. 11.1). Worth mention are the human remains unearthed at Għar Dalam dating to around 5000 BC. Archaeological material dating to the Għar Dalam phase has also been discovered in the caves at il-Mixta, close to Għajn Abdul in Gozo. The Saflieni and the Xagħra Circle prehistoric burial complexes, besides being multi-period burial places, also had a cultic significance (Trump 2000, 90–2, 169, 184, 67–74, 177–8). The discovery of substantial amounts of Borġ in-Nadur type pottery in Għar Mirdum, in the territory of of Dingli, points to the occupation of the cave by a Borġ in-Nadur type, Bronze Age community (Mallia 1965, 9; see also http://www.shurdington.org/gharmirdum/index.html).

Excavations by the Italian, *Missione Archeologica Italiana a Malta* (the Italian Archaeological Mission at Malta), at Ras il-Wardija in Gozo in the late 1960s revealed the extensive use of a rock-hewn cave during the Punico-Hellenistic era (Buhagiar M. 1988, 69–87). Fragments of late-Roman coarse pottery and medieval glazed ware at l-Għar ta' Iburdan in the territory of Rabat (Malta), denote the cave's probable use for habitation purposes in late Roman and Byzantine times (Hägglund 1976–7, 397; Buhagiar, M. 1988, 17).

In the late Middle Ages many Maltese were cave dwellers. Jean Quintin is the first known author to mention troglodytism in Malta (Quintin d'Autun 1536, 31). His *Insulae Melitae Descriptio* (A Description of the Maltese Islands), published in 1536, shows Quintin's surprise at the great number of cave-dwellers inhabiting the rural section of the island – a trend which probably reflects a long-established medieval life-pattern in the Maltese countryside. Even the maritime settlement of Birgu appears to have been, partially at least, troglodytic in nature. Toponyms starting in *Għar* (cave) as is the case with Għar il-Kbir in the territory of Dingli and l-Għar ta' Iburdan, territory of Rabat (Malta), also hint at the diffused nature of troglodytism in Malta (Wettinger 1975, 181–216). Late medieval documentation also shows how caves were used during this period for the purpose of animal pens (Wettinger 1982, 34).

The troglodytic phenomenon was widespread in the Mediterranean region throughout the middle ages whenever environmental conditions proved favourable. Arid and semi-arid areas which suffered from a lack of timber, but which on the other hand provided plentiful natural rock shelters and an abundance of easily quarried stone, were instrumental in conditioning a type of architecture which was entirely stone oriented besides encouraging cave-dwelling.

The cave-dwelling phenomenon during the late medieval period

The roots of Maltese late medieval troglodytism probably lie in the twelfth and thirteenth centuries, and are the result of new attitudes adopted following the Norman reconquest of 1127 (Buhagiar, M. 2005, 40). A strong troglodytic tradition during this period might possibly reflect coordinated attempts at increasing the agricultural output of specifically designated countryside areas of Malta.

A reconstruction of the Maltese landscape in late antiquity and the medieval period at large is still a work in progress, but the available archaeological and historical documentation hints at a clear-cut break between the Byzantine period, which ended in 870 AD, and the Norman occupation of Malta, which commenced in 1091 AD. The definite Muslim conquest of 870 was marked by bloodshed and destruction, probable retaliations against the Christian inhabitants and an orchestrated demographic shifting programme which included the death or exile of the local bishop with the island being reduced to an uninhabited ruin (Brown 1975, 81–84; Luttrell 1975, 21–28; 1992, 100; Wettinger 1986, 90–91). The post-870 phase is unfortunately sparsely documented and the majority of the tenth century Muslim sources concerning Malta are silent on the period between 870 and 1048. One of the handful of known sources dating to this period hints at Malta being an uninhabited island containing large flocks of sheep, wild donkeys and an abundance of honey (Dalli 2006, 27). Malta was, moreover, visited by ship builders, because its wood was of the strongest kind, by fishermen, due to the tastiness of its fish, and there is also the mention of trees of pine, juniper and olive (Brincat 1995, 11–12).

It is unlikely that Malta was totally depopulated during this period (Luttrell 2002, 100; Molinari and Cutajar 1999, 9–16; Cutajar 2004, 58), but a drastic decline in population numbers is not improbable. By way of hypothesis, it is likely that Muslim retaliation against the local population would be primarily directed towards the urban centres of the archipelago, namely Mdina and Birgu in Malta, and the Citadel in Gozo. Even if an ethnic cleansing policy was employed, it is doubtful that this would efficiently target more remote countryside locations which could still harbour small communities accustomed to living at subsistence or near-subsistence level. Cliff-face settlements excavated in remote areas of the Maltese countryside provide an excellent case in point. Anyone unfamiliar with the topography and terrain is almost certain to overlook the presence of inconspicuous troglodytic settlements which blend extremely well with their natural surroundings. Nonetheless, even if this was the case, the remaining inhabitants appear to have been too few in number to influence the subsequent course of events – even to leave any trace of their existence in the spoken language (Wettinger 1986, 95). A linguistic analysis of the Maltese language excludes any signs of language stratification, with the linguistic basis appearing to be solely of an Arabic origin (Brincat 1995, 1–7). There are furthermore, close parallels between Sicilian and Maltese toponyms which suggest intimate Sicilian – Maltese linguistic connections (Brincat 1995, 27).

Historic documentary sources claim that Malta was once more repopulated by the Muslims in 1048–9, in time to ward off a Byzantine invasion – a fact which has been tentatively interpreted not as being a sure sign of the island's depopulation in earlier centuries, but from the perspective of a large-scale Muslim colonisation which possibly took place as late as the early eleventh century (Dalli 2006, 58–62). The fact that women and daughters formed part of this early eleventh century wave of migration makes it improbable that the sole purpose behind the colonisation of Malta during this period was the establishment of a garrison. Demographic expansion connected with a period of economic prosperity experienced by Sicily during the first half of the eleventh century, the Sicilian civil war between the different Muslim caliphates which commenced in 1038, or even fear of the Normans, whose territorial ambitions in the Southern Italian peninsula certainly made them a force to be reckoned with, are all possibilities which must be given their due weight. The uncertainty caused by the impending Norman conquest of Sicily in the eleventh century led a section of the wealthier Muslim families to emigrate from Sicily (Brincat 1995, 20–21; Von Falkenhausen 2002, 262; Metcalfe 2003, 28), and there is the remote possibility that several opted to settle in Malta.

Malta was far from a deserted place around a century later, when the Muslim geographer Al Idrisi speaks in terms of the archipelago as being, '...away from the island of Pantelleria at a distance of 100 miles towards the east one finds the island of Gozzo with a secure port. From Gozzo one goes to a small isle named Kamuna. From there going eastwards one finds the island of Malta. It is large and has a sheltered place on the east side. Malta has a town and abounds in pastures, sheep, fruit and honey' (Wettinger 1986, 97; for a slightly different translation of Al Idrisi see Amari 1880, vol. i, 53, 75). Investment in agricultural intensification, which seemingly took place in the post-1127 period, appears to have been generally successful, and succeeded in placing to the forefront cotton cultivation, an item of luxury trade, mention of which is made in a property inventory drawn up in Genoa in June 1164 (Buhagiar, M. 2007, 18, 40). Furthermore, it is significant that twelfth century pollen samples retrieved from the Marsa plain adjoining the east coast of the island indicate an increase in wood, cereal and flax vegetation (Fenech 2007, 112), and appear to confirm the agricultural intensification process the island is proposed to have been experiencing during this period.

There is no direct archaeological evidence for any major settlement outside Mdina and its suburb of Rabat throughout most of the Norman and early post-Norman period, but it is my suspicion that several rock-cut settlements in the north-west sector of Malta associated with *giardini*-type cultivations (orchards), might have already made their appearance during this period. Similarly, landscape evolution in South Italy during the Norman period appears to have initially centred on the development of *giardini*-type cultivation and was only subsequently followed by the development of dry-farming (Martin 2002, 19). Furthermore, it is probably within the context of population recovery experienced in the western Mediterranean basin, which commenced during the eight and ninth centuries and gained momentum by the eleventh century (Martin 2002, 17), that the local troglodytic phenomenon has to be assessed. Population recovery was indeed the main driving force for fuelling an agricultural intensification process during this period.

It is possible that fertile valleys in the north-west sector of Malta, such as Wied Ħażrun, Wied ir-Rum, Ġnien is-Sultan and Wied Liemu, which contain extensive field terraces, water galleries, and rock-cut settlements, all formed part of this post-1127 intensification process. Field terracing construction coupled with the hydrological intensification of an area entails a labour intensive input, and it may take decades to completely transform a previously uncultivated landscape into an agriculturally productive one. The *Gumerin* (Gomerino) and *Deir Ilbniet* (Dejr il-Bniet) estates, both in the territory of Rabat (Malta), were already listed as *giardini* in 1317 and 1351 respectively (Bresc 1975, 152). Due to the availability of perennial water springs, *giardini* are capable of producing a summer crop in an otherwise arid season, thus increasing the economic value of such land.

Water galleries (Fig. 11.3), the life-source of *giardini*-type cultivations, are hewn into Mtarfa Member deposits at a right angle to the rock-face and are commonly located in the upper terraced sections of valleys, a short distance below troglodytic settlement sites and above agricultural land. Galleries are generally easily identifiable from their rectangular-shaped rock-cut entrance that is on average 0.8m wide and a little more than 1.5m high and are located in areas of northern and north-western Malta and areas of Gozo which possess the necessary geological stratification (Buhagiar, K. 2007, 119–121). There are instances; however, where the gallery entrance lies in a cave's interior.

The depth of such galleries is unknown, but several of the recorded water tunnels may well be over half a kilometre deep. A gallery partially investigated at Lunzjata in the territory of Rabat (Malta) is over 90m long. The investigation of difficult to access water galleries has in recent years been facilitated by the use of a remotely operated experimental submersible camera (Fig. 11.4), equipped with video and sonar sensors, digital compass, robotic arm and a Global Positioning System (GPS) locator (see <http://users.csc.calpoly.edu/~cmclark/MaltaMapping/index.html>). Remote gallery exploration was indeed a breakthrough when it comes to the mapping and investigation of submerged gallery sections.

All galleries provide the surrounding area with a perennial water source, though the volume of retrieved

Figure 11.3: Rock-hewn water gallery at Santi, territory of Rabat (Malta). The canal at the base of the gallery channels the extracted water towards the gallery entrance.

Figure 11.4: The Remotely Operated Vehicle (ROV) in action whilst investigating and mapping the Lunzjata water gallery in the territory of Rabat (Malta).

water varies from one gallery to the other. In order to ease the flow of water retrieved from the perched aquifer, a canal is often carved into the gallery floor. Galleries are generally level with the highest terraced field on the valley side, with water being transported from the gallery's entrance to any adjoining and underlying fields by means of stone canals. The dating of the Maltese galleries is a task which requires caution, but they closely resemble Qanat-type water extraction systems which probably filtered into Malta through neighbouring Sicily (Buhagiar, K. 2007, 118–122). Field trips to Sicily have so far succeeded in the identification of three such galleries at Enna, S. Lucia di Mendola in the territory of Palazzolo Acreide, and Ferla in the territory of Syracuse. The first recorded instance for water galleries in Sicily is the early 1300s, but they are likely to have been present in the landscape at a much earlier stage.

It has been observed by Wettinger that the north-west sector of Malta is 'strangely bare' of known *rahal* (village) settlement sites, either due to a defect in the surviving documentation, or because the Maltese countryside had been depopulated for so long that the surviving placenames dropped their *rahal* prefix (Wettinger 1975, 190; Fiorini 1993, 118–119). Personal fieldwork and non-invasive field surveys carried out in north and north-west Malta were in many ways an eye-opener and make it probable that it was the *ghar*-type settlements which prevailed in these areas. Sicilian cave-sites were also distinguished by the *ghar* prefix and it is frequent for Maltese caves to derive their names from Muslim personal nomenclature. A translation of *Ghar Dalam* is not the 'Cave of Darkness', but the 'Cave of Dalam' – Dalam being a surname which survived right into the fifteenth century (Fiorini 1988, 14).

The majority of such settlements, even though presently lying in an abandoned state, still survive to a fair degree. There are two distinct types of medieval cave-settlement in Malta: first, the adaptation of natural karst depressions; and, second, cliff-face settlements. Cave usage varies from cultic worship to human habitation, animal pens or storage spaces often connected to agricultural usage, animal-driven mills (*centimolo*), and apiaries. Examples of karst feature settlements are Għar il-Kbir and Latmija. These involve the occupation of one of more caves hewn into the sides of an open-air, natural rock-hollow, and in the case of Għar il-Kbir, it appears that the settlement was geared towards a pastoral economy (Buhagiar, K. 1997, 64–70). Because karst feature settlements occur in Upper Coralline deposits, it is probable that the hard geological formations frequently restricted cave enlargement.

Cliff-face settlements are located within the sides of ridges and valleys and involve the occupation of a series of natural or artificially enlarged caves. The majority of the caves surveyed fall under this category, and are often hewn into a surprisingly brittle Upper Coralline deposit locally referred to as the Mtarfa Member. This formation is very easy to excavate, and does not make the process of cave excavation and enlargement as labour intensive and time consuming as commonly argued. Indeed, the location of most cave-settlements suggests that their occupants possessed a sound knowledge of the local geology. Mtarfa Member deposits are commonly located only a few metres above the perched aquifer, often successfully tapped by means of an underground gallery, ensuring the settlement and the underlying fields had a perennial water source.

The majority of the local troglodytic settlements was probably subject to an organic type of development and appears to be closely associated with the development of *giardini*. Cave re-occupation and enlargement often involved the destruction of previous occupation phases, but, in the absence of stratified deposits, dating is a difficult task. The presence of animal troughs easily identifies caves utilised in their last phase of occupation as animal pens. Caves containing no water or feeding troughs were used either for human habitation or for storage purposes. Tethering holes are common features associated with both the human and animal occupation of caves.

Caves frequently cluster together into units, but isolated caves containing evidence of human or animal habitation are fairly frequent. A terrace, often present on the outside of cliff-face settlements, is a common addition aimed at linking together two or more adjoining caves (Buhagiar, K. 1997, 72–86). Dry- and wet-stone walls commonly screen a large section of the caves' entrance, leaving an arched or square-headed doorway as the only means of access to the dimly lit cave interior (Fig. 11.5). Narrow slits in the upper section of the cave screening wall sometimes act as windows, and dry-stone constructions frequently partition cave interiors into a series of individual spaces. There are instances where cave screening walls and interior partitioning walls were occasionally plastered and whitewashed, but the widespread nature of this practice in the late medieval period has still to be appropriately verified.

Maltese cave-settlements are accessed by means of one or more well-defined footpaths, some of which have a cobble-type paving. Numerous other similarly surfaced paths probably lie buried beneath modern concrete paving. In areas of difficult terrain, dry-stone ramps facilitated access to troglodytic settlements. The ramps, built parallel to the cliff-face, are similar in method of construction to dry-stone walls, and a soil and rubble infill, sometimes capped by means of a cobbled surface, bridges the gap between the cliff-face and the rubble wall.

Two distinct types of roofing strategies were recorded. When the dry-stone screening walls were built around 0.6m apart from the overhanging rock-ledge roofing the cave, the intermediate gap was bridged and sealed by means of roughly sawn, thin ashlar slabs. These rested against the rock-face at an angle that generally varies between 20° and 40°, secured in place by means of mortar, and made watertight with a *deffun* or cocciopesto covering (in both instances this consists in a mixture of ground pottery and lime mortar) (Fig. 11.6). This technique was probably resorted to in an attempt to gain more internal cave space.

A light roof structure often covered those caves with screening walls, and was built at a distance of over 0.7m from the overhanging cave roof. It is unlikely that the roof structures observed by the author predate the first decades of the twentieth century, but the materials utilised and the construction methods employed remained essentially unchanged since the late medieval period, and are probably similar to those recorded by Quintin in the 1530s (Quintin de Autun 1536, 31). Only dead vegetal material, easily obtainable from the surrounding countryside, was utilised. Unrefined carob and fig tree branches were often used as load-bearing members instead of timber beams, and were normally spaced between 0.6 and 0.8m apart. Large quantities of dried bamboo reeds bridged the gap between each beam. A thick, compact layer of hay finally capped the roof, presumably tied to the beams in order to prevent its dispersal during rough weather (Fig. 11.5).

Several caves to the north of the island were also used as bee hives for honey production. Known locally as *mġiebaħ*, most of these are probably of an early modern date, and belong to the phase following the abandonment of a number of cave-units in the interim late sixteenth to early eighteenth century, when a number of these were converted into apiaries. The entrance of such caves is likewise screened off by means of a dry-stone wall, the only access to the interior often being via two square-headed doorways. Bee-apertures, usually in the shape of rectangular perforations are usually neatly arranged in one or more registers in the cave screening wall. The arrangement of the interior is rather simple, often consisting of two registers of shelves abutting the interior screening wall on which are mounted the earthenware hive-jars.

For centuries, the rural inhabitants of the island lived at an almost subsistence level. Caves portray a marked absence of unnecessary ornamentation and were primarily conceived as being practical rather than fashionable. Cave-dwelling remained a common feature of the Maltese rural landscape until the first decades of the twentieth century, but it is likely that troglodytism drastically decreased in popularity following the appearance of the *razzett* (farmhouse) structure, which even though of an unknown antiquity, probably dates to the early modern period, and was a direct response to the economic well-being generated by the Hospitaller Order of the Knights of St John of Jerusalem in seventeenth century Malta (Buhagiar, M. 2005, 49).

The emerging scenario based on personal research shows how, in the early modern period and possibly earlier, a number of peasant families residing in the countryside area surrounding the *Civitas* (Mdina), owned a small property, often consisting of one or two rooms in the suburb of Rabat (Malta), which were meant to function as an overnight temporary shelter whilst visiting or conducting agriculture-related business there (Mary Grace Vella pers. comm. 22/4/2008). Should this have been the case, it might also help explain why the militia registers compiled during the course of the fifteenth century speak in terms of a complete demographic vacuum in the north-west section of Malta, whereas it is more than evident that, due to the well-managed water resources, field terracing and the overall fertile nature of the land, it is highly improbable that such territory could be allowed, at any stage, to remain vacant. The complete absence of *rahal*-type toponyms in the north and north-west sectors of Malta can probably be explained by the fact that the exposed Mtarfa Member deposits and the easily accessible perched aquifer water sources, tapped by means of galleries, encouraged the presence of troglodytic settlements.

The sighting of Maltese troglodytic settlements also parallels the Sicilian model, where the location of most cave-settlements is likewise conditioned by the available geological profile. Sicilian cave-sites are often sited in naturally defendable, difficult-to-reach places (Messina 1989, 109–11), and troglodytic settlements such as Scicli, Cava d'Ispica and Pantalica were observed by the author to be excavated within a friable sedimentary rock deposit which is visually identical to the Maltese Mtarfa Member rock stratum. The caves' setting in relation to their surrounding landscape also closely parallels the Maltese model, the majority of which command unobstructed views of the surrounding area and underlying fields, which in the past were probably tilled by the cave occupants themselves.

The rock-cut churches and oratories

Maltese rock-cut churches can be divided into two different categories: urban; and rural. Whilst the setting in which they are located is different, both share a number of common characteristics and are the product of the same religious pressures and social-cultural conditions. The urban churches lie within the precincts of palaeochristian hypogea and tend to show a greater preoccupation with architectural enhancement and elaboration than their rural counterparts (Buhagiar, M. 2005, 58). Rural rock-cut churches are more simplistic than their urban counterparts and form an integral part of the troglodytic landscape in which they are located. Cave churches adjoin cliff-face settlement sites and are almost exclusively excavated into Mtarfa Member rock-deposits. As is the case with the cave church dedicated to St Nicholas at Mellieħa, which gave its name to the underlying valley, troglodytic churches must have been a landmark within the context of the late medieval Maltese landscape. The same applies to the cave churches of St Peter in the territory of Naxxar and St Brancatus at Għargħur, the dedication of which was assimilated in the toponomy of the surrounding area (Buhagiar, K. 1997, 72–77, 84–86).

Rural cave churches were of a rather intimate size, had a dimly lit interior, and were frequently accessed from the rock-terrace which connected two or more cave settlement units. The cave church exterior was commonly enclosed by a dry-stone wall, the only means of interior access being a narrow square-headed doorway. The diffused nature of the troglodytic phenomenon and rock-excavated churches in late medieval north and north-west Malta is perhaps illustrated by the fact that the principal cave church of Mellieħa occupied the status of a parish in the

Figure 11.5: Cave cluster, fronted by a wet-rubble wall and partially roofed over by a light-roof wooden structure at Il-Baħrija, territory of Rabat (Malta).

fourteenth century, and assisted the spiritual needs of nearby cliff-face settlements (Wettinger 2002, 41–47).

Both rural and urban churches were probably fitted with either masonary or wooden altars and in the instance of the cave church of St Leonard in the territory of Rabat (Malta), and that dedicated to St Peter at Naxxar, there is evidence of a flagstone floor in the interior, or a cobbled passageway facilitating access to the often difficult-to-reach entrance. Cave church interior furnishing was probably sparse, with *dukkien* type benches (rock-excavated benches aimed at providing seating accommodation) sometimes excavated into the rock-face along the side walls. Several of the surviving churches were decorated by murals which survive in a precarious state of preservation. The surviving murals speak a common iconographic language and are Siculo-Byzantinesque in tradition and inspiration (Buhagiar M. 2007, 98).

The dating of both urban and rural churches is difficult and constant cave reutilisation makes it improbable that any archaeologically relevant deposits survive within. In many instances the area fronting caves, including cave churches, has been too disturbed to make its archaeological assessment a viable exercise. The most reliable dating source remains the art historical analysis of the preserved murals, with those surviving at Abbatija tad-Dejr and the rock-cut oratory of St Agatha's catacombs, both at Rabat (Malta), being the most important. The analysis of the sinopia of a possible Deësis (the representation of a blessing Christ, the Virgin Mary and St John the Baptist) located in the apsed niche of the east wall in Oratory I at Abbatija tad-Dejr, coupled with data furnished by a nineteenth century report on the painting, hint at a work of art which stylistically comes from the milieu of Siculo-Norman Sicily, where similar representations

Figure 11.6: Detail of a wet-rubble wall enclosing access to a cave-compound at Bahrija, territory of Rabat (Malta).

of Christ the Pantokrator are frequently encountered in the rock-cut and built churches (Messina 1979, 49; Buhagiar M. 2005, 59–61). Within the same oratory at Abbatija tad-Dejr are two other murals, probably showing the archangel Michael and St John the Evangelist. The icons are contained within a deep red frame and probably carried the legend with the saints' names in Latin characters. The combination of the Byzantine style with Latin text is another reliable dating element, even though such an artistic style remained a standard practice locally till at least the fifteenth century (Buhagiar M. 2005, 61).

Another mural painting surviving within the adjoining Oratory II at Abbatija tad-Dejr, today on display at the National Museum of Fine Arts at Valletta, shows the fusion of the Crucifixion and the Annunciation scenes – the amalgamation of both themes suggesting a fourteenth century date (Buhagiar M. 2005, 63), and parallels closely a painting at the *Grotta dei Santi* at *Monterosso Almo* at Ragusa, Sicily, which also seems to be of a coeval date (Messina 1989, 117). Similarly, another two Siculo-Byzantinesque icons survive in an oratory at St Agatha in Rabat. It is probable that the icons belonged to a more extensive fresco-cycle destroyed in the late-fifteenth century or early sixteenth century to make way for a series of Late Gothic devotional images (Buhagiar M. 2005, 65–66).

Even though there is no direct historical documentary evidence, it has been proposed that in post-Islamic times, several of the Maltese cave-churches were administered by Greek-rite Basilian monks (Buhagiar M. 2007, 317–338). This hypothesis is based on Sicilian and Pantallerian models where Greek-rite monks operating from rural monasteries and anchoritic stations saw to the spiritual needs of rural, often isolated communities (Luttrell 1975, 37–38). Much of the argument centres round the meaning of the word *Dejr*, which can mean either a cow shed, or animal pen or a monastic building (Wettinger 2000, 107–108). Used all over the Islamic East, Muslim Sicily and Spain, *dejr* is often associated with a Christian monastic establishment. Whilst making no distinctions between Greek and Latin-rite clergy, historical Sicilian documentation favours the presence of Greek monks who carried out their evangelisation programme amongst the Muslim communities of the island.

Local archaeological evidence does hint in an indirect manner towards the presence of Greek-rite monasticism, but the issue is still the subject of debate and necessitates further scholarly investigation. So far, none of the known cave churches and troglodytic settlements can be associated with monastic establishments, but of particular interest is the site of Abbatija tad-Dejr, where an impressive palaeochristian burial complex was re-utilised as a cult centre during the post-Muslim period and is the foremost contender for a Greek-rite site during this period (Buhagiar M. 2005, 58–60). *Raheb* toponyms may also offer valuable clues on this issue, with the word either meaning monk or hermit. The modern Maltese word for monk is *patri*, derived from the Italian *padre*. There is the possibility that *raheb* was used to denote an Augustinian friar during the late medieval period, but, in all probability, the word's linguistic origins recall a pre-late fourteenth century date. It is furthermore significant that the name of a field located close to the Abbatija tad-Dejr site is *Bir Rhiebu* (the monk's well) (Buhagiar M. 2005, 58–61; 2007, 95).

Conclusions

The emergent scenario, based on the available archaeological, historical and toponomastic evidence eludes to the ever increasing probability that present day settlement and field patterns owe their origin to twelfth century agricultural intensification efforts. Landscape transformation must have entailed significant capital investment, and was, in its initial stages, a labour intensive process which consisted in the construction of terraced land and the excavation of troglodytic dwellings and water galleries in areas of exposed Mtarfa Member deposits. Both caves and water galleries are difficult to date, but it appears that in the latter instance these are twelfth or thirteenth century efforts in improving the hydrological potential of selected valley-sites. The resultant agricultural setup became eventually known as *giardino*.

The Maltese troglodytic phenomenon and landscape evolution appears to have close parallels with neighbouring Sicily and particular areas of South Italy. The Puglia region for instance has a semi-arid type climate, water-carved valleys, and garigue areas of karst formation and *terra rossa* (red-coloured soil deposits) which closely parallel the Maltese scenario. Landscape evolution in South Italy during the Norman period appears to have first centred on the development of *giardini*-type cultivations, then dry-farming, and, only at a later stage, the utilisation of waste-land for rough grazing and wood gathering (Martin 2002, 19). Within a Maltese context, dry-farming involved field construction, mainly on formerly Upper Coralline and Globigerina plains and appears to be closely associated with the widespread appearance during the late medieval period of *raħal*-type settlements.

For centuries, the rural inhabitants of the island lived almost at a subsistence level. They did not try to conquer or crush nature, but attuned to the challenge posed by topography. Cave-settlements portray a marked absence of unnecessary ornamentation and, together with rural structures when available, were conceived to be practical rather than fashionable. Rectangular recesses hewn into the rock walls of caves were frequently utilised for storage purposes. Habitable caves were presumably sparsely furnished and perhaps contained a table and door, apart from a couple of other wooden furnishings. Settlements were sometimes spread out on different levels, making the best possible use of the limited space available.

Each settlement is unique, and size, asymmetry and usage create endless combinations. Troglodytic settlements and any adjoining above ground vernacular architectural elements are often the result of successful human interaction with the landscape. These frequently complement and form an integral part of their natural surroundings. Cave-dwelling was probably not limited to rural areas. Carlo Castone Della Torre di Rezzonico wrote in a 1793 travel account that a number of families preferred to dig caves in the sides of the ditch surrounding Valletta and in the Cottonera area (located in the south Grand Harbour area), rather than having to pay a housing rent (Eynaud 1989, 61).

Future research will explore possibilities by which the agricultural landscape, cave-settlements and the water galleries discussed above can be scientifically dated. Of hindrance to this challenging exercise is the possibility that many of the caves included in the field survey may have been abandoned as late as in the first half of the twentieth century. This would have resulted in major disturbance of older layers. Moreover, it is likely that most cave-units only preserve shallow internal deposits and probably lack stratification. The investigation of terraced land fronting cave-settlements might on the other hand prove to be a more fruitful exercise and surface counts of potsherds can perhaps lead to the identification of a settlement's dumping ground. The employment of this method of investigation is likely to yield encouraging results in two troglodytic sites in particular: the St Nicholas cave-settlement in Mellieħa and a cave settlement site located below the cliff-face at Rdum Dikkiena in the locality of Siġġiewi (Buhagiar K. 2003, 242). In both instances, concentrations of ceramic scatters were located in the terraced land underlying these settlements. Such an exercise would however require the availability of a more reliable Maltese medieval pottery typology (Luttrell 2002, 1–17; Molinari and Cutajjar 1999, 9–15).

References

Amari, M. (1880) *Biblioteca Arabo-Sicula* 1. (Reprinted in 1982 with appendix of 1880).

Bowen-Jones, H., Dewdney, J. C. and Fisher W. B. (1961) *Malta: Background for Development*. Durham, University of Durham.

Bresc, H. (1975) The secrezia and the royal patrimony in Malta: 1240–1450. In A. Luttrell (ed.) *Medieval Malta: Study on Malta before the Knights*, 243–265. London, The British School at Rome.

Brincat, J. M. (1995) *Malta 870–1054: Al-Himyarī's Account and its Linguistic Implications*. Malta, Said International.

Brown, T. S. (1975) Byzantine Malta: a discussion of the sources. In A. Luttrell (ed.) *Medieval Malta: Study on Malta before the Knights*, 71–87. London, The British School at Rome.

Buhagiar, K. (1997) The Għar Il-Kbir settlement and the cave-dwelling phenomenon in Malta. Unpublished thesis, University of Malta.

Buhagiar, K. (2007) Water management strategies and the cave-dwelling phenomenon in late-medieval Malta. *Medieval Archaeology* 51, 103–131.

Buhagiar, M. (1984) Mediterranean architecture: medieval cave-dwellings and rock-cut churches in Malta. *Atrium: Mediterranean and Middle East Architectural and Construction Review* 3, 17–22.

Buhagiar, M. (1988) Two archaeological sites: Ras Ir-Raħeb, Malta, and Ras Il-Wardija, Gozo. *Melita Historica* 10(1), 69–87.

Buhagiar, M. (2005) *The Late Medieval Art and Architecture of the Maltese Islands*. Malta, Fondazzjoni Patrimonju Malti.

Buhagiar, M. (2007) *The Christianisation of Malta: Catacombs, Cult Centres and Churches in Malta to 1530*. British Archaeological Reports International Series 1674. Oxford, Archeopress.

Cistern Exploration Project (2009) available from:<http://users.csc.calpoly.edu/~cmclark/MaltaMapping/index.html>

Cutajar, N. (2004) The archaeology of Malta's Middle Ages. In K. Gambin (ed.) *Malta Roots of a Nation: The Development of Malta from an Island People to an Island Nation*. Malta, Midsea Books.

Dalli, C. (2006) *Malta: The Medieval Millennium*. Malta, Midsea Books.

Eynaud, J. (1989) *Carlo Castone Della Torre Di Rezzonico – Viaggio Di Malta Anno 1793*. Malta, Midsea Books.

Fenech, K. (2007) *Human-Induced Changes in the Environment and Landscape of the Maltese Islands from the Neolithic to the 15th Century AD*. British Archaeological Reports International Series 1682. Oxford, Archaeopress.

Fiorini, S. (1988) Sicilian connections of some medieval Maltese surnames. In G. Brincat (ed.) *Incontri Siculo-Maltesi*, 104–138. Malta, Malta University Press.

Fiorini, S. (1993) Malta in 1530. In V. Mallia-Milanes (ed.) *Hospitaller Malta 1530–1798*, 111– 198. Malta, Mireva.

Hägglund, R. V. (1976–1977) Għar Ta' Iburdan: un insediamento in eta tardo-Romana. *Kokalos*, 22–23 (Atti del iv Congresso Internazzionale di Studi sulla Sicilia Antica, Rome), 396–399.

Luttrell, A. T. (1975) Approaches to medieval Malta. A. T. Luttrell (ed.) *Medieval Malta: Study on Malta before the Knights*, 1–70. London, The British School at Rome.

Luttrell, A. T. (1979) Malta troglodytica: Għar il-Kbir. *Heritage* 2, 461–465.

Luttrell, A. T. (2002) *The Making of Christian Malta*. Aldershot, Ashgate.

Mallia, F. S. (1965) A dagger from the second millennium before Christ at Dingli. *The Sunday Times* [Malta], 7th March 1965, 9.

Martin, J. M. (2002) Settlement and the agrarian economy. In G. A. Loud and A. Metcalfe (eds.) *The Society of Norman Italy*, 17–45. Boston, Brill.

Messina, A. (1979) *Le Chiese Rupestri del Siracusano*. Palermo, Istituto Siciliano di Studi Bizantini e Neoellenici,

Messina, A. (1989) Trogloditismo medievale a Malta. *Melita Historica* 10(2), 109–120.

Metcalfe, A. (2003) *Muslims and Christians in Norman Sicily: Arabic Speakers and the End of Islam*. London, Routledge.

Molinari, A. and Cutajar, N. (1999) Of Greeks and Arabs and of feudal knights. *Malta Archaeological Review* 3, 9–16.

Pedley, M., Clarke, M. H. and Galea, P. (2002) *Limestone Isles in a Crystal Sea*. Malta, P.E.G.

Quintin d'Autun, J. (1536) Insulae Melita descriptio. In H. C. R. Vella (1980) *The Earliest Description of Malta (Lyons 1536)*. Malta, DeBono Enterprises.

Skinner, M., Redfern, D. and Farmer, G. (1997) *The Complete A–Z Geography Handbook*. London, Hodder & Stoughton.

Trump, D. (2000) *Malta: An Archaeological Guide*. Malta, Progress Press.

Von Falkenhausen, V. (2002) The Greek presence in Norman Sicily: the contribution of the archival material in Greek. In G. A. Loud and A. Metcalfe (eds.) *The Society of Norman Italy*, 253–287. Boston, Brill Leiden.

Wettinger, G. (1975) Lost villages and hamlets of Malta. In A. Luttrell (ed.) *Medieval Malta: Study on Malta before the Knights*, 181–216. London, The British School at Rome.

Wettinger, G. (1982) Agriculture in Malta in the late Middle Ages. In M. Buhagiar (ed.) *Proceedings of History Week 1981*, 1–48. Malta, The Historical Society of Malta.

Wettinger, G. (1986) The Arabs in Malta. *Malta: Studies of its Heritage and History*, 87–104, Malta, Mid-Med Bank Ltd.

Wettinger, G. (2000) *Place-Names of the Maltese Islands ca. 1300–1800*. Malta, P.E.G.

Wettinger, G. (2002) Mellieħa in the Middle Ages. In J. Catania (ed.) *Mellieħa though the Tides of Time*, 41–47. Malta, Mellieħa Local Council.

Zammit-Maempel, G. (1977) *An Outline of Maltese Geology*. Malta, G. Zammit-Maempel.

Chapter 12

Caves in Need of Context: Prehistoric Sardinia

Robin Skeates

This chapter attempts to synthesize the available data relating to the human use of over 100 natural caves in Sardinia during prehistory, and to contextualize these caves and their occupations in relation to wider landscapes, lifeways and beliefs, over space and time. In the Upper Palaeolithic and Mesolithic, a few caves were used by mobile groups as base-camps and shelters within wide socio-economic territories. In the Early Neolithic, large inland caves continued to serve as residential bases for communities practicing a mixed economy, while other caves sheltered living – and now also deceased – members of mobile task groups. Between the Middle Neolithic and the Early Iron Age, the ritual use of the interiors of selected caves was elaborated, whereas large caves were increasingly abandoned as long-term dwelling places as the Sardinian landscape became more extensively settled, although a few caves continued to be used as convenient shelters.

Introduction

Cave archaeology in Sardinia began in 1873. In that year, Giovanni Spano, a Sardinian priest, linguist and archaeologist, excavated some Neolithic deposits in Grotta Sa Rocca 'e Ulari, near Borutta (Spano 1873; see also De Waele 2005). He was soon followed by other Italian scholars interested in geology, palaeontology and archaeology (e.g. Francesco Orsoni, Arturo Issel and Filippo Vivanet), who undertook excavations in caves such as Grotta di Sant'Elia and Grotta di San Bartolomeo, Grotta s'Oreri, and Grotta di Genna Luas. After something of a lull in the first half of the twentieth century, with the notable exception of excavations in the Grotta di San Michele ai Cappuccini, Sardinian cave studies revived following the 1955 National Speleological Congress, which was held in Sardinia. New discoveries and excavations have since been made by a variety of groups, including members of Sardinian speleological societies, the staff of the Sardinian Soprintendenze per i Beni Archeologici, local archaeology enthusiasts, and scholars from universities in Sardinia, northern Italy, the UK and the Netherlands. Their fieldwork has resulted, to date, in the discovery of prehistoric remains in some 114 natural caves in Sardinia (see Appendix), including numerous karst caves (ranging from extensive underground systems to small caves and rockshelters), caves in granites (especially in north-east and central Sardinia), and a few caves resulting from the erosion or alteration of volcanic rocks (such as Sa Grutta de is Caombus, formed by a tectonic fissure in basaltic rock).

With the exception of a few monographs and articles detailing the results of excavations and specialist analyses at particular cave sites (e.g. Agosti *et al.* 1980; Carta 1966–7; Ferrarese Ceruti – Pitzalis 1987; Klein Hofmeijer *et al.* 1987–8; Lo Schiavo and Usai 1995; Pitzalis 1988–9; Taramelli 1915; Trump 1983), the majority of this work has only been published superficially, as preliminary reports and notes in conference proceedings and journals, or as inclusions in grand narratives of Sardinian prehistory (e.g. Lilliu 2003) – most dominated by a concern with the refinement of relative chronologies based upon artefact typology. This patchwork of incomplete information has consequently undermined previous attempts to synthesize the available data relating to prehistoric caves in Sardinia (e.g. Fadda 1991; Usai 2008). It has also led to a decontextualization of the archaeology of the caves, particularly geographically.

The aim of this chapter, then, is not only to produce a synthesis of the human uses of caves (including rockshelters) in prehistoric Sardinia, but also to attempt to contextualize those caves as meaningful 'lived in' places, appropriated and abandoned by real people who participated in dynamic cultural practices and social relations. This approach complements my on-going field project in and around a group of prehistoric caves in the territory of Seulo in central Sardinia, which is seeking to evaluate ideas about the ritual transformation of persons, objects and caves using a range of modern scientific techniques (Skeates 2009–10). My chapter focuses on natural caves (as opposed to artificial caves – such as rock-cut tombs), and extends over a 19,000 year period, from the Upper Palaeolithic to the Middle Bronze Age, ordered, as much as possible, by an absolute chronology based upon calibrated radiocarbon dates (Stuiver *et al.* 2005; Tykot 1994). I do, however, exceed these limits for comparative purposes.

Upper Palaeolithic and Mesolithic caves: durable places in a dynamic islandscape

The human use of caves in Sardinia during the Upper Palaeolithic (*c.* 20,000–11,000 cal BC) is currently represented by just one archaeological site: Grotta Corbeddu (e.g. Klein Hofmeijer 1997; Klein Hofmeijer *et al.* 1987–8; 1989; Sondaar *et al.* 1984; 1995; Spoor 1999; Spoor and Sondaar 1986) (Fig. 12.1).

This cave can be interpreted as a significant base-camp and landmark, situated on the edge of the Supramonte uplands in the Valle di Lanaittu, in east-central Sardinia. The valley contains numerous large karstic caves, and may have comprised a key route (for humans and migrating deer) through a more extensive hunting and gathering territory that incorporated a variety of ecological zones, camp-sites and activity areas. Grotta Corbeddu is a large cave system, extending more-or-less horizontally over some 130m, with a small but accessible main entrance leading to a succession of four relatively commodious 'halls', connected by more restricted spaces. Speleotherms are present, particularly in the interior, and a water channel runs between two of the halls. Knowledge and use of this durable structure in the landscape evidently passed between successive human generations, leading to the gradual accumulation of deep, stratified, cultural deposits, which have been radiocarbon dated to a timespan of

around 22,000–11,000 years cal BC. The human use of the cave system at this time extended beyond the naturally lit entrance area: since relevant deposits were found not only in Hall 1, adjacent to the entrance, but also in Hall 2, in the interior. Processing activities are indicated by the large proportion of scrapers in the Epigravettian style lithic industry. Such tools were produced mainly on silicified limestone, although the presence of other rock types, such as flint, quartz and goethite, indicates that raw materials were obtained from a variety of sources. The bones of hunted animals, butchered and consumed in the cave, were dominated by two now extinct members of Sardinia's native Late Pleistocene fauna, the deer (*Megaceros cazioti*) and the Sardinian pika (*Prolagus sardus*). Some of the deer

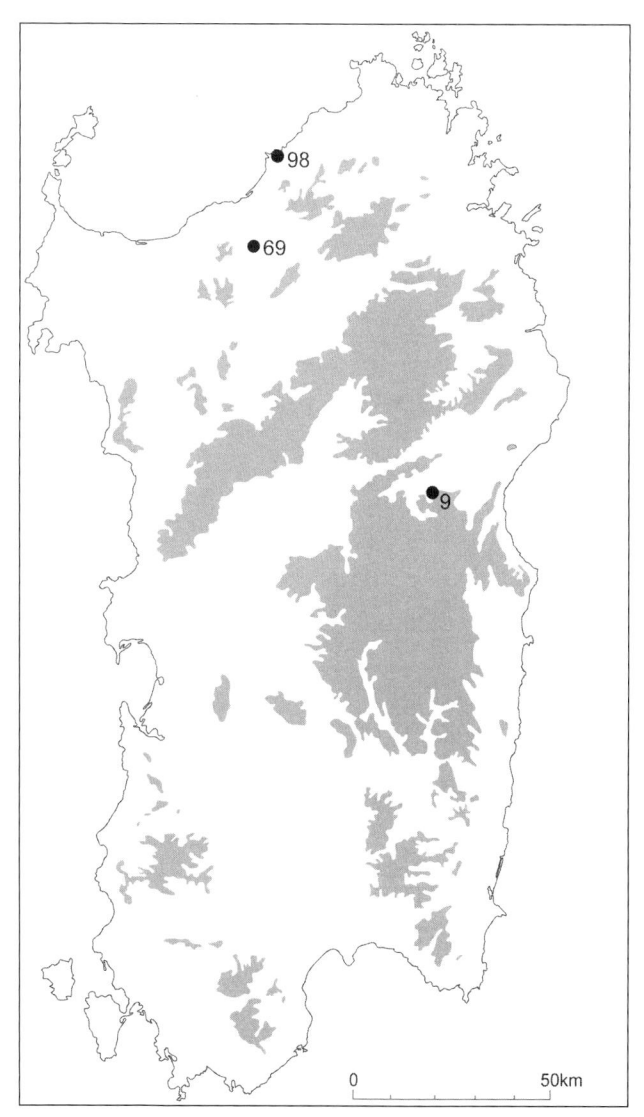

Figure 12.1: Map of Upper Palaeolithic and Mesolithic cave sites in Sardinia.

bones, especially the ulnae and mandibles, appear to have been used opportunistically as artefacts. A few disarticulated human bones were also found, although it is unclear precisely how they were deposited in the cave.

A slightly expanded number and variety of caves were used by human groups in Sardinia during the Final Upper Palaeolithic and Mesolithic (c. 11,000–6000 cal BC), including: Grotta Corbeddu; Grotta di Su Coloru (Fenu *et al*. 1999–2000; 2002; Martini *et al*. 2007; Pitzalis *et al*. 2003); and Riparo di Porto Leccio (Aimar *et al*. 1997) (Fig. 12.1).

In Grotta Corbeddu, the relevant stratified cultural deposits (Stratum 2, Hall 2) have been radiocarbon dated to 11,100–6600 cal BC. Charcoal derived from fires lit in the cave was now a characteristic component of these deposits, in addition to the continued presence of bones of the Sardinian pika (*Prolagus sardus*), and a few disarticulated human bones. Grotta di Su Coloru is comparable in a number of ways. It is a large karstic cave complex, extending for a total of around 640m. One large and one much smaller entrance lead to a wide and sinuous main gallery, from which depart two long corridors and various other spaces. Speleotherms are present, as well as an active spring. Stratified deposits, the base of which has not yet been reached by on-going excavations, contained a Late Mesolithic stratum (L). This has been radiocarbon dated to 7000–6200 cal BC. It contained charcoal, a lithic industry dominated by scrapers, and the bones of microfauna. The cave is located on the interfluvial Tanna Manca plateau, at an altitude of 340m, about 12km inland from the north coast of Sardinia.

By contrast, Riparo di Porto Leccio is a small rock-shelter, situated in a bay on the north coast of Sardinia. Here the Mesolithic style lithic industry was produced on local pebbles. The choice of this site presumably reflects a more maritime orientation for the groups of 'trapper-fisher-coastal-nomads' who sheltered here, whose annual territory may have extended overseas as much as inland (Costa *et al*. 2003).

Early Neolithic caves: multi-purpose structures

A significantly larger number and variety of caves were occupied during the Early Neolithic throughout Sardinia, spanning the earlier 'Cardial Impressed Ware' phase (5700–5300 cal BC) and the later 'Filiestru' ceramic phase (5300–4700 cal BC) (Fig. 12.2). Some 21 Sardinian caves have been reported to contain pottery of this period, ranging from large caves previously occupied during the Palaeolithic and/or Mesolithic, to a small rockshelter used as a shelter and burial place for hunter-herders, to large inland caves inhabited by early farming communities, to coastal caves with more specialized economic and ritual uses (see Appendix). In addition, a few Early Neolithic open-air sites are known in Sardinia, at places such as Sella del Diavolo and Su Stangioni. Their relationship to the cave-sites is unclear, but may have been complementary, particularly in the case of Sella del Diavolo, which is located on the Capo S. Elia promontory on the south coast of Sardinia, close to the Early Neolithic cave-sites of Grotta di Sant'Elia and Grotta di San Bartolomeo (Atzeni 1962; Orsoni 1879; Patroni 1901).

The large Upper Palaeolithic and Mesolithic cave complexes of Grotta Corbeddu and Grotta di Su Coloru

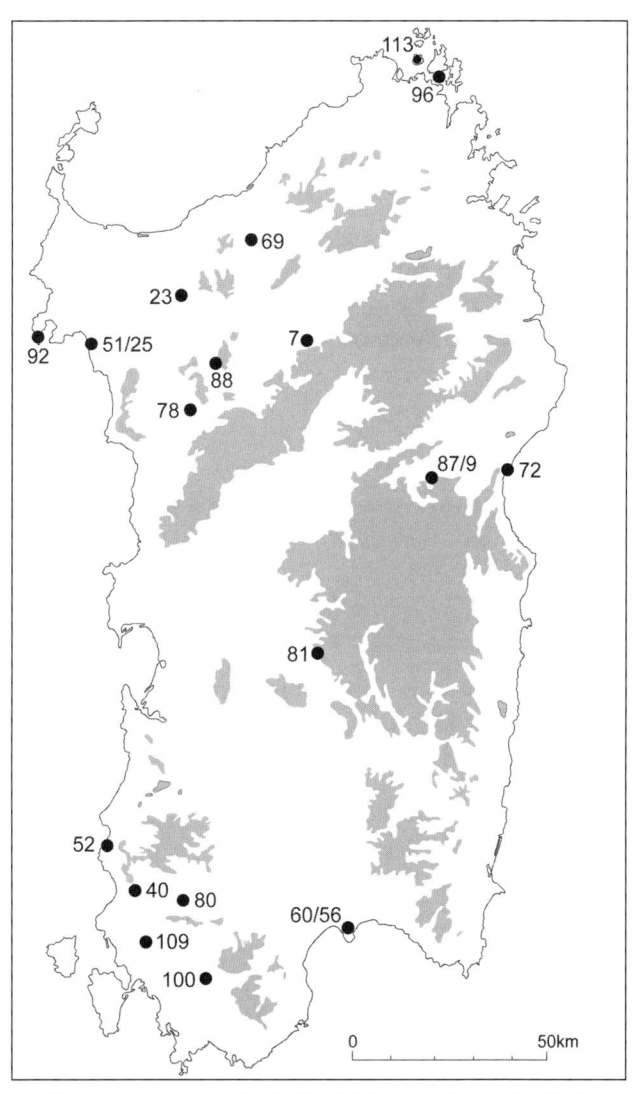

Figure 12.2: Map of Early Neolithic cave sites in Sardinia.

continued to be occupied during the Cardial Impressed Ware phase. However, the degree of demographic and cultural continuity indicated by these sites remains ambiguous. Certainly, at Grotta Corbeddu, the Early Neolithic deposits (in the lower part of Stratum 1a, Hall 2), which contained the remains of hearths and wild animals, are indicative of the maintenance of a mainly hunting and gathering economy focussed on locally available species (Sanges 1987). But they also contained Cardial Impressed Ware, obsidian artefacts, and seashells reflecting new practices and long distance connections.

The Riparo sotto roccia Su Carroppu represents another category of Early Neolithic cave-site (Atzeni 1972; 1977; Lugliè *et al.* 2007). It has been interpreted both as a refuge (possibly for groups of hunter-herders) and as a burial place. The site is located some 10km inland from the south-west coast of Sardinia, at an altitude of 350m, in a zone of rocky hills. It overlooks a small valley. It is a small but deep limestone rockshelter. Its lowest deposits contained: fragments of Cardial Impressed Ware; remains of domestic and wild animal species, the latter including deer, boar, Sardinian pika (*Prolagus sardus*), and fish; and chipped stone artefacts, including geometric microliths, of local chert and quartzite and imported jasper and obsidian, the latter divided into blades on Sardinian C type material (from sources on the northeast side of Monte Arci) and flakes on Sardinian A and Sardinian B2 type material (from sources in the southwestern part of Monte Arci and along the lower western slopes of Monte Arci). The contracted skeletons of two buried individuals were also found in the interior of the cave, accompanied by a perforated disk of schist and by perforated seashells of the rustica dove shell (*Columbella rustica*) and of the scallop (*Pectunculus glycymeris*).

Large inland caves, such as Grotta Filiestru and Grotta Sa Korona di Monte Majore, were also used as dwelling places for early farming communities. Grotta Filiestru, for example, is located at 410m above sea level, on the north side of a small valley, on the edge of the Bonu Ighinu basin, some 20km inland from the north-west coast of Sardinia (Trump 1983). The site has justifiably been interpreted as the dwelling place of a small group of early farmers, who presumably cultivated their crops and herded their animals in the catchment area of the well-placed cave, which lay just 30m from a permanent spring. The cave is composed of a large and well-lit main chamber (21m long, and 6–2m high), from which depart three restricted passages. Excavations in the large chamber uncovered Cardial Impressed Ware deposits overlying a culturally sterile basal deposit. Radiocarbon determinations provide a date-range of around 5700–5350 cal BC for this first phase of occupation. The deposits contained: ashes; remains of wheat (*Triticum monococcum* and *Triticum dicoccum*) and peas (*Pisum* sp.); domesticated animal bones (sheep/goat, pigand cattle); wild animal bones, including those of the red fox (*Vulpes vulpes*) and Sardinian pika (*Prolagus sardus*); pottery fragments; chipped stone tools and debitage of Sardinian obsidian, jasper and flint; grindstone fragments; and bone awls. The obsidian was used especially for cutting meat and hides (Hurcombe and Phillips 1998). Overlying these deposits was a stratum containing predominantly plain pottery, assigned to the later 'Filiestru' ceramic phase of the Early Neolithic, and radiocarbon dated to around 5200–4700 cal BC.

Some small coastal caves, such as Tafone di Cala Corsara and Riparo di Santo Stefano, appear to have been used as shelters during the course of primarily economic-related activities focussed on coastal resources. Tafone di Cala Corsara, for example, lies on the Cala Corsara bay on the south coast of the Isola di Spargi, an island in the Maddalena archipelago, situated off the north coast of Sardinia (Ferrarese Ceruti and Pitzalis 1987). Its excavators have interpreted it as a key stepping stone and resting place along an important maritime communication route between Corsica and Sardinia, although its local significance as a base for the exploitation of coastal food resources should also be considered. It is a small cave, roughly 4m in diameter, and 2–3m high, and was produced by weathering of a granitic rock face. It contained an intact basal deposit with three sherds of Cardial Impressed Ware, some chipped stone artefacts of flint and obsidian, and mollusc shells.

By contrast, the deep cave of Grotta Verde was probably used during the Early Neolithic primarily for the performance of mortuary rituals (e.g. Antonioli *et al.* 1994; Lamberti *et al.* 1986; Lilliu 1978; Lo Schiavo 1987; Tanda 1987). Its entrance is situated 75m above present-day sea level, on the steep east slope of Capo Caccia, a coastal promontory (containing several karstic caves) flanking the bay of Porto Conte. The entrance leads down to a large chamber containing huge speleotherms, from which depart various branches, including one that descends down a slope of 45 degrees to a terminal chamber. Today, this chamber is filled by a small freshwater lake, but in the Early Neolithic, when the sea level was at least 10m lower, it would have led down to a small chamber with niches, and then down again to a larger submerged chamber and other spaces. Underwater excavations in the small chamber revealed a series of human burials, placed along a wall of the chamber, in the natural niches and in rock-cut hollows. The human remains were badly preserved and have not been directly dated, but included skull fragments and vertebrae, three of the latter in anatomical connection. Three whole Filiestru style vessels accompanied the burials, the smallest of which had been placed inside the largest one. Other vessels were found in the larger submerged chamber, perhaps having been dropped during the course of collecting drinking water from (what would then have been) an underlying freshwater lens overlying seawater.

Middle and Late Neolithic caves: continuity and ritualization

Selected caves continued to be exploited by human groups in Sardinia for a variety of purposes during the Middle and Late phases of the Neolithic, some now with even more overtly ritual dimensions (Fig. 12.3). The Middle Neolithic (or Bonu Ighinu ceramic phase) dates to around 4700–4000 cal BC, while the poorly defined Late Neolithic (or San Ciriaco phase) can only be placed somewhere around 4000 cal BC. Some 40 caves in Sardinia have been reported to contain pottery of this period (see Appendix), which is almost double the number of caves occupied during the Early Neolithic (21). The caves range from large inland cave dwellings, to small coastal shelters, to more unique ritual caves. In addition, a growing number of open sites in Sardinia can be assigned to the Middle Neolithic, one third located in the Campidano lowlands of the south-west Sardinia. The best-known of these is Cuccuru S'Arriu, a village with a cemetery of rock-cut tombs containing inhumations and secondary burials accompanied by grave goods, including a series of stone figurines.

Large inland caves continued to be occupied, inhabited and abandoned during the Middle Neolithic, many having previously been occupied during the Early Neolithic. For example, at Grotta di Su Coloru, the presence of three strata (E–C) containing Bonu Ighinu pottery is indicative of a long-term series of occupations (one with cobbling) and abandonments. By contrast, at Grotta Filiestru cultural deposits containing Bonu Ighinu pottery continued to accumulate in a single stratum, radiocarbon dated to around 4500–4350 cal BC. It was probably at this time that a pit was dug down into the underlying Early Neolithic deposits and filled with three pottery jars (which remained intact), one containing around 2kg of red ochre, the other two a black pigment. These pigments, if not also the special deposit itself, might have been associated with the performance of some kind of colourful rituals in (and even around) the cave, although they might also have been used in manufacturing processes. A fragment of a rare 'lithic ring' (perhaps used as a bracelet), made of a polished, dark grey stone, measuring about 10cm in diameter, was also found in the Middle Neolithic deposits. Animal bones from this phase of the cave's history indicate increasing numbers of cattle at the expense of pig, while plant remains indicate the (presumably local) cultivation of a wider range of crop plants, including emmer wheat (*Triticum dicoccum*), faeroese barley (*Hordeum hexasticum*), lentils (*Lens esculenta*) and broad beans (*Vicia faba*). The nearby Grotta di Sa 'Ucca de Su Tintirriòlu (Mara) was also occupied during the Middle Neolithic, probably for the first time (Alessio *et al.* 1978; Contu 1970; Loria and Trump 1978; Trump 1990, 18–19). The site lies on the opposite side of the valley to Grotta Filiestru. It is a 301m long karst cave. A relatively narrow entrance leads to a 10m long chamber, which is separated by a

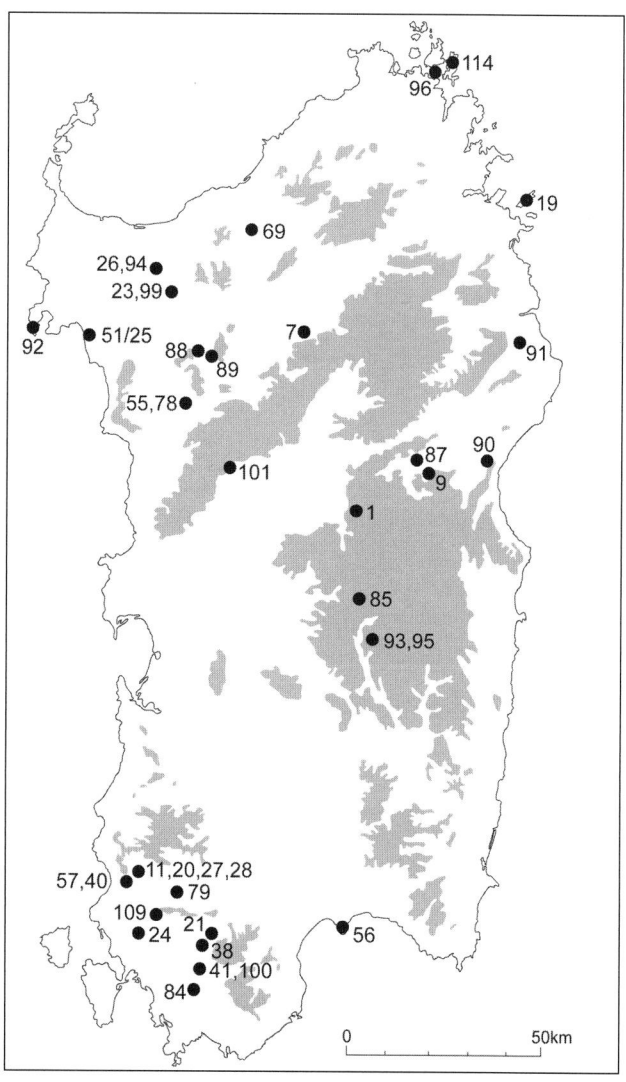

Figure 12.3: Map of Middle and Late Neolithic cave sites in Sardinia.

low area from a spacious hall (around 60m long, 6m wide and 3.5m high) and a narrow horizontal passage. The deposits in the main chamber contained a large quantity of ash and other cultural material. A sample of charcoal has been radiocarbon dated to around 4700–4350 cal BC. The precise nature of this Middle Neolithic occupation of the cave is unclear. So too is the precise relationship between Grotta di Sa 'Ucca de Su Tintirriòlu and Grotta Filiestru, which lie just 400m apart, and which – according to the radiocarbon dates – were occupied at the same time. But they presumably formed part of a more intensively inhabited landscape incorporating the Bonu Ighinu basin, within which caves comprised integral elements of dwelling, ritual and cosmology.

Small caves also continued to be occupied during the Middle Neolithic, probably as shelters used in relation to a narrower range of economic activities linked to the exploitation of local resources. For example, the small caves formed by weathering (known to geologists as 'tafoni') eroded into the granite of the Maddalena archipelago continued to be used: in this case, one lying on Cala Serena, a small bay on the north-west coast of the Isola di Caprera, was found to contain two decorated sherds of Bonu Ighinu pottery and some chipped stone artefacts (Difraia 2007).

Some other Sardinian caves appear to have been used exclusively for the performance of rituals during the Middle-Late Neolithic, including burial rites, the production of anthropomorphic cave paintings, and the deposition of figurines, although the dating of the art in particular is problematic.

Grotta Rifugio, for example, was used as a burial cave during the Middle Neolithic (Agosti *et al.* 1980; Biagi and Cremaschi 1978; Carta 1966-7; Germanà 1978). It may also have been perceived by its users as a durable natural monument, strategically situated in a cultural landscape whose use remained characterised by a degree of human mobility and connectivity, despite the adoption of agriculture. It is located in a significant crossing-place in the landscape: on the northern edge of the Sopramonte uplands, at the mouth of the Gonogósula gorge, about 600m from the point at which the tributaries of the Oliena and Frattale streams join to form the River Cedrino. The limestone cave is of modest dimensions (16m long), and is composed of three spaces, which lie on three different levels. The entrance chamber, which contains speleotherms, is followed, after a 2m drop, by a chamber with an adjoining passage, from which one descends to a passage of around 11m in length, access to which was widened by the breakage of some stalactites (presumably in prehistory). Disarticulated human remains, belonging to an estimated nine adults and three children, were found in this deepest part of the cave system, possibly having been carried and deposited there during the course of secondary burial rites. They were accompanied by a deposit of dark soil, containing ashes and charcoal, the latter derived from plant species indicative of a local vegetation of both Mediterranean maquis and sub-mountainous brushwood (reflecting the ecotonal nature of the cave's location). This burnt deposit contained a large quantity and wide range of ritual offerings, many made of raw materials obtained (either directly or though exchange) from more distant sources, including the coast of Sardinia, situated about 20km to the east. The most numerous category of artefact was body ornaments. These included: 1284 small cylindrical necklace beads – 860 of the green silicate mineral chlorite, 424 of the carbonate mineral aragonite; 410 tubular tusk shells (*Dentalium* sp.); 127 perforated rustica dove shells (*Columbella rustica*); 12 bracelets of European thorny oyster (*Spondylus gaederopus*); four perforated dog cockle shells (*Glycymeris glycymeris*); one miniature ring made from a small top shell (*Gibbula divaricata*); seven perforated boar's tusks; and one bone pendant. Other artefacts comprised: fragments from at least 17 pottery vessels, including a deep cup with zoomorphic handles; 24 chipped stone artefacts – of obsidian, flint and basalt; a greenstone axe-head; three bone points and one bone awl; and a pebble with coloured traces of ochre. Faunal remains included: numerous wild species; the bones of at least four sheep or goat and one cow; and mollusc shells.

Grutta I de Longu Fresu is an example of a Neolithic painted cave, recently discovered by M. G. Gradoli, containing a set of material features of symbolic and ritual significance (Gradoli and Meaden in prep; Skeates 2009–10). It lies on the edge of a small stream, the Riu Longu Fresu, at an altitude of around 750 metres, in the uplands of the Barbagia di Seulo. It is a relatively small, 15m long, karst cave. Its restricted entrance leads to a low tunnel with eight lateral niches, some containing traces of former springs. At the back of this cave, four probably related Neolithic features have been identified, close together in an area just two metres long. The first is the skull of a human adult, cemented to the cave wall by flowstone, and radiocarbon dated to around 4250–4050 cal BC. This date places the skull either towards the end of the Sardinian Middle Neolithic or in the poorly defined Late Neolithic (San Ciriaco phase). The second feature is represented by a few disarticulated human bones, dated to the same period, found scattered in the adjacent floor deposits and in some of the niches. These probably represent the original deposition of at least one whole human body on the floor of this cave, and the secondary placement of some of its bones along the side walls and niches, but also the possible the removal of other bones from the cave (J. Beckett pers. comm. 2009). The third feature is a small, semi-circular, stone structure, formed by a modified group of stalagmites, within whose delimited area was found a small trapezoidal axe-head of smoothed greenstone. The fourth feature is a small group of paintings in an adjacent niche, just to the side of a now extinct spring (Fig. 12.4). The paintings were produced with a dark grey pigment. They are difficult to decipher, being covered by flowstone, but the general consensus is that at least two schematic linear representations of anthropomorphic (or combined human-animal) figures can be seen, with legs, arms and either an elongated head or horns. The style of these paintings ties in well with the corpus of Neolithic cave paintings in the central Mediterranean (e.g. Graziosi 1973). However, samples of the flowstone overlying the paintings are currently being dated using the Uranium-series method, in order to obtained a more precise estimate of the age of the paintings, particularly compared to the radiocarbon dated skull.

Grotta del Papa is another example of a painted prehistoric cave with schematic anthropomorphic 'stick' figures, which *might* also date to the Middle Neolithic (D'Arragon 1997). Even more problematic are some corpulent, anthropomorphic figurines, made of stone and bone, imprecisely provenanced to a few Sardinian caves, but comparable in style to the series of stone figurines found in the Middle Neolithic cemetery of Su Cùccuru s'Arriu (e.g. Lilliu 1999, 180–1, 194, 209–12; 2003, 45, 51). Examples come from Riparo sotto roccia di Su Monte, Grotta I di Su Concali de Coróngiu Acca, and Grotta di Monte Meana. Although these objects are clearly of symbolic significance, it is difficult to ascribe ritual significance to the caves in which they were found: particularly given the loss of context caused by their poorly recorded discoveries.

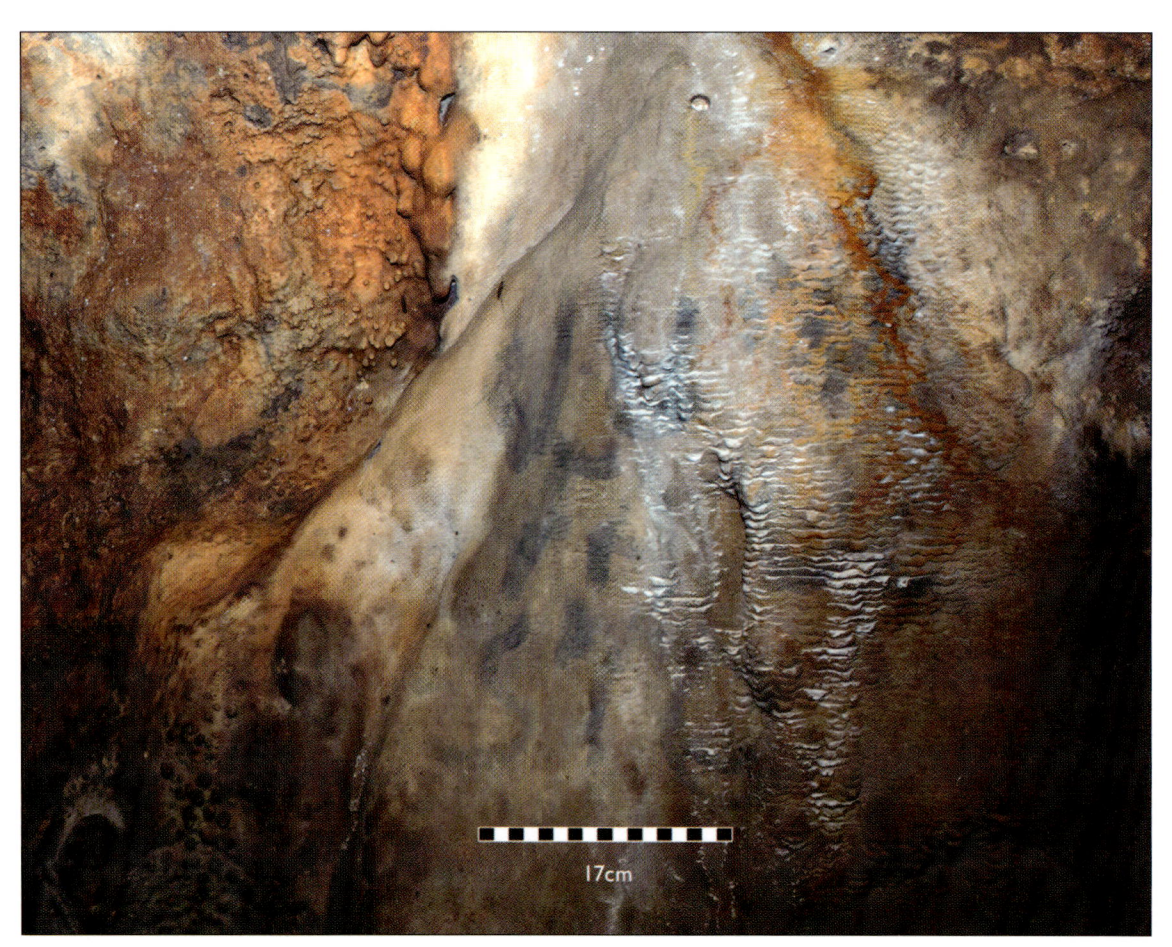

Figure 12.4: Neolithic paintings in Grutta I de Longu Fresu (Seulo). Photo: G. Farci.

Final Neolithic and Earlier Copper Age caves: continuity

In general, the pattern of cave use established in the Middle-Late Neolithic remained very much the same in the Final Neolithic and Early Copper Age, with the human uses of caves ranging from dwelling to burial and other ritual practices (Fig. 12.5). The Final Neolithic (primarily associated with the Ozieri culture and ceramic style, but also, in north-east Sardinia, with the Arzachena culture) dates to around 4000–3200 cal BC, while the Earlier Copper Age (or Sub-Ozieri/Abealzu-Filigosa ceramic phase) can only be tentatively dated to around 3200–2700 cal BC. Some 46 caves in Sardinia have been reported to contain pottery of this period (see Appendix). Taken at face-value, this again represents an increase in cave use compared to the previous period (from 40 to 46). However, the Final Neolithic and Earlier Copper Age was almost twice as long as the Middle and Late Neolithic. Certainly, then, any increase in cave use was not on the same scale as the great expansion of open villages, rock-cut tombs (or *domus de janas*) and other monuments (including menhirs, dolmens, cists, megalithic circles, and the unique ceremonial mound and shrine at Monte d'Accoddi) in Sardinia during the Final Neolithic, which outnumbered cave sites by a ratio of about 8 to 1 (Lilliu 2003). Natural caves certainly comprised an integral part of the rich cultural and cosmological landscape of Final Neolithic Sardinia: as indicated, for example, by the various material dimensions of their human uses, including certain symbolic details, which echo features of the artificially constructed structures for the living, the dead and the supernatural. But, given the numerical dominance of other types of site, the significance of caves may now have become primarily local (particularly compared to their apparent regional significance in the Upper Palaeolithic and Early Neolithic).

Large inland dwelling caves, first occupied during the Early or Middle Neolithic, continued to be inhabited during the Final Neolithic. Grotta di Sa 'Ucca de Su Tintirriòlu, for example, appears to have been used more intensively during the Final Neolithic, since the bulk of the prehistoric material excavated at this site can be assigned to the Ozieri culture, radiocarbon dated here to around 3950–3550 cal BC (Trump 1989). The deposits contained outstanding quantities of cultural material, including: much ash, a few human bones, animal bones and mollusc shells, 2,400 classifiable Ozieri style pottery sherds from 11 types of vessel (some engraved with schematic female figures), three ceramic figurines, polished stone axe heads, projectile points of flint and obsidian, flint blades, and a bone handle. The excavators have suggested that the cave may have served as some kind of ritual site and/or as a habitation site (Contu 1970; Trump 1990), although these two dimensions do not need to be regarded as mutually exclusive, particularly in the context of the generally more ritualized Ozieri culture. The nearby cave-site of Grotta Filiestru also continued to be inhabited during the Final Neolithic (Trump 1989), radiocarbon dated here to around 4250–2950 cal BC. Faunal remains were dominated by the bones of sheep/goat, while the artefacts were generally less elaborate than the material found in the Ozieri culture deposits of Grotta di Sa 'Ucca de Su Tintirriòlu. This has led Trump (1983) to suggest that the site may now have served as a pastoralist out-station of a nearby village.

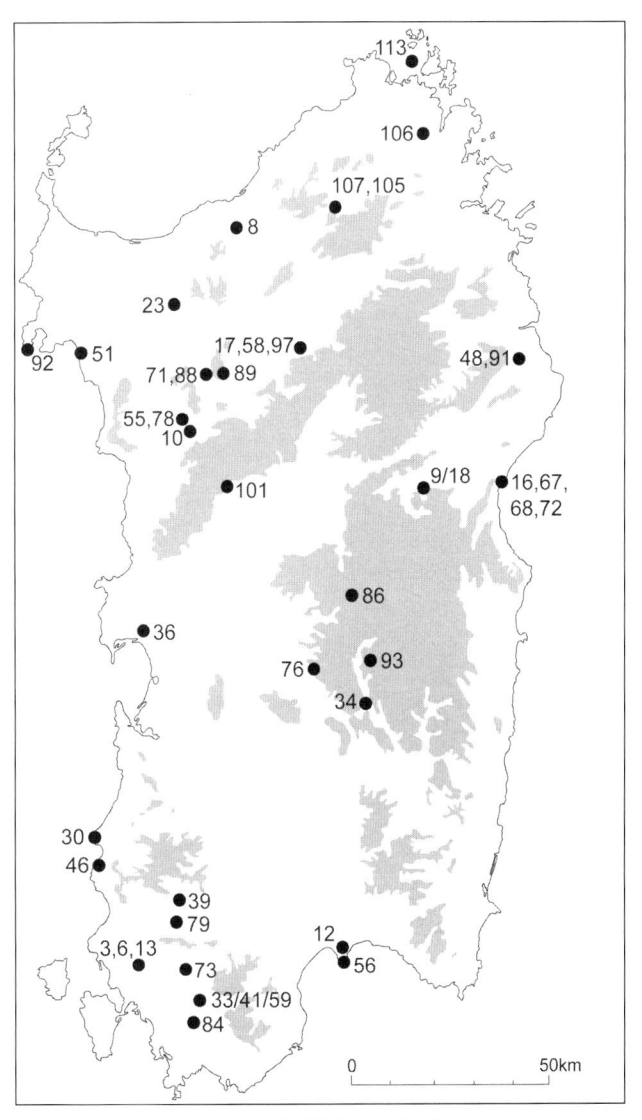

Figure 12.5: Map of Final Neolithic and Earlier Copper Age cave sites in Sardinia.

Grotta del Guano was also used as a dwelling place during the Final Neolithic, perhaps having first been occupied during the Middle Neolithic, or exclusively in the Ozieri phase, radiocarbon dated here to around 3750–3400 cal BC (Alessio *et al.* 1971; Castaldi 1972; 1980; 1987; Lo Schiavo 1978b). It is situated close to the Middle Neolithic burial cave of Grotta Rifugio, at an altitude of 138m, on the River Cedrino. It is a large karst cave, with five entrances leading to a series of chambers and tunnels. In addition to some burnt deposits containing charcoal and carbonized remains of barley, wheat and legumes, and a range of artefacts (including a broken ceramic figurine), fragments of 'plaster' were found which might represent the remains of structures built in the cave. A concreted skull was also found here. Lo Schiavo (1978b) has argued that this was washed into the cave by the river: an interpretation which ties in with her argument that the cave was used exclusively as a dwelling place. However, it could be that the exterior chambers of this cave complex were used primarily for dwelling purposes, given that the majority of the grindstones were found in them, while other spaces in the complex could have served other purposes.

An exception to this pattern is provided some large granitic rock-shelters in north-east Sardinia, which were adopted and adapted by members of the Arzachena culture as a new form of naturally defended settlement (Lilliu 2003, 77–8). At Punta Candela, for example, the basal deposit in one of the natural cavities comprised a 'domestic' deposit of fragments of pottery vessels and artefacts of obsidian and flint.

Other caves were used, perhaps more exclusively, for the performance of rituals; material dimensions of which included the deposition of special assemblages of human remains and artefacts, and the decorating of cave walls with anthropomorphic and abstract symbols.

Ritual deposits were found, for example, in the type-site of the Ozieri culture, Grotta di San Michele ai Cappuccini (Porro 1915; Taramelli 1915; Lilliu 1950). This large cave system, which contains numerous speleotherms, comprises two chambers and an irregular series of more restricted spaces, which extend for about 80m. Access to the cave would originally have been difficult, being entered by a 6–7m deep shaft. This, and the nature of the cave deposits, suggests that the cave was not used as a dwelling place, but rather as a place of mortuary and votive deposition. A 40–50cm deep deposit was excavated in the first chamber, which lies below the entrance shaft. It contained: some human remains, including a skull and long bones (perhaps from just one individual); a large quantity of finely decorated Ozieri style pottery sherds; some red ochre; a spindle whorl; a polished miniature greenstone axe head; a stone polisher; flint arrowheads and blades; an obsidian core; imported pebbles (one incised with anthropomorphic designs); bone points; a bone pin or spatula; and two stone figurines. Some of this material, and in particular the smashed pottery vessels, might have been dropped into the cave from outside.

Possibly votive deposits of pottery vessels were also found in the Grotta Conca Niedda (Melis 1993). This is a large cave system, composed of chambers and corridors which extend over a distance of about 500m, located at an altitude of 280m in the tributary valley of Conca Niedda. It contained fragments of large storage vessels and bowls, which have been assigned to the Early Copper Age.

Grutta de is Janas is a large cave complex with two interconnected branches, located on a middle-upper hill slope at 797 metres (Skeates 2009–10). The north branch is about 100m long, and the west branch about 75m. The cave contains numerous speletherms. Later prehistoric ritual deposits have been found throughout the complex. Excavations in a low but wide chamber (between 1 and 1.5m high, and 11m long by 5m wide), situated at the inner end of the entrance corridor leading to the West branch of the cave complex, revealed rich ritual deposits of the Final Neolithic, radiocarbon dated here to around 3800–3650 cal BC (Fig. 12.6). These comprised a homogeneous, 12–24cm deep, burnt layer, composed of stones and fine dark grey ashey soil. This contained relatively large pottery sherds, animal bones, obsidian artefacts (including arrowheads and flakes, all originally deposited with very sharp – and potentially unused – edges), a perforated seashell pendant, and a polished red bead. Many of the artefacts bore signs of having been intensively burnt, and even the bedrock was burnt in places.

Deposits of human remains have been found in other Ozieri culture caves. For example, the inland cave of Grotta Sa Rocca Ulari also contained human burials, associated with Ozieri style pottery, stone axe heads, flint and obsidian artefacts (Lilliu 1957, 76; Spano 1873). This large cave system, whose main chamber measures 190m long, is located on the hill slope of Colle di Sorres, just 50m along from the Ozieri culture *domus de janas* cemetery at San Pietro di Sorres (Soro 2009). Unfortunately, any more precise relationship between these two sites is unclear, since the cave deposits had been disturbed by antiquarian excavations and have only been published superficially.

Further examples of poorly contextualized figurines have been provenanced to a number of caves known to have been occupied during the Final Neolithic. For example, an unusual basalt figurine (14cm long), known as 'the Venus of Macomer', found in

Riparo s'Adde could either date to the Ozieri phase or to the Upper Palaeolithic (Lilliu 1999, 175–7; Mussi 2010; Pesce 1949). This small rockshelter was discovered in 1949, but its cultural deposits had been almost completely emptied before the archaeological authorities intervened.

Some examples of parietal cave art – both engraved and painted – have also been assigned to the Final Neolithic. The best (but still questionably) dated are the engravings in the Grotta del Bue Marino (Dorgàli) (Lo Schiavo 1978a; 1980). This is a large limestone karst cave located on the Gulf of Orosei. A group of at least 14 schematic anthropomorphic figures, and two circles with a dot in the middle, were engraved on one of the curved rock surfaces at the entrance to the cave. The figures are around 30cm long, with arms pointing up and legs down in the form of a double U. They have been compared stylistically to the figures carved in the 'Domus Branca' *domus de janas* in the Final Neolithic cemetery of Moseddu. Furthermore, Ozieri culture material was found in this cave, in the remains of a hearth, adjacent to the point at which the art was placed, including fragments of pottery, animal bones, and a greenstone polisher. Riparo di Luzzanas, by contrast, is a small painted rockshelter (Basoli 1992; Dettori Campus 1989). It has narrow tunnel with a sloping roof, on which a group of red ochre painted anthropomorphic stick figures and concentric circles occupy an area of 2 by 1m. The 10 anthropomorphic figures, which have downward pointing arms, legs and penises, have been compared to others found in a range of archaeological contexts in Sardinia, Corsica, Sicily and south-east Italy, and have consequently been assigned stylistically (and therefore not securely) to the end of the Ozieri phase.

Full Copper Age, Early Bronze Age and Middle Bronze Age caves: expansion and diversification

The Full Copper Age, Early Bronze Age and Middle Bronze Age (or 'Proto-Nuraghic' period) are associated in Sardinia with the Monte Claro, Bell Beaker, Bonnànnaro A (or Corona Moltana) and Bonnànnaro B (or Sa Turricula) styles of pottery, and can be loosely dated to between around 2700 and 1300 cal BC, although the absolute and relative dating of this period remains problematic. The number of caves occupied in Sardinia now increased significantly (to some 74 caves, as opposed to some 46 in the Final Neolithic and Earlier Copper Age) (see Appendix) (Fig. 12.7). In the Monte Claro culture, for example, 28 per cent of the 90 known archaeological sites are natural caves (Lilliu 2003, 143–4). And in the process, some caves that had been used and abandoned in previous periods were now reoccupied, to be used both as dwelling places and – increasingly – as places for human burial and

Figure 12.6: Excavation in Grutta de is Janas (Seulo). Photo: R. Skeates.

for the performance of other underground rituals. Spatially, there was a significant concentration and growth in the human uses of caves in the Carbonia-Iglesias province in south-west Sardinia (a rise from around 20 per cent of all known occupied caves in the Final Neolithic and Earlier Copper Age to over 40 per cent in this period). Given the agricultural marginality of this area, particularly compared to the adjacent Campidano, which comprised the settlement heartland of the Monte Claro culture (but then a sparsely settled area in the Early Bronze Age), it could be that this increase was related to a territorial expansion into this area by Monte Claro groups who practiced mainly pastoralism and hunting, although their ability to cultivate crops and their connectivity to other socio-economic groups should not be underestimated (cf. Webster 1996, 53, 69–72).

The indications that large caves continued to be used as dwelling places, but often less intensively and more intermittently, might also offer some support for this economic hypothesis in relation to other parts of Sardinia. For example, Grotta Filiestru continued to be occupied, although less intensively than in earlier periods, as indicated by the presence of Monte Claro, Bell Beaker, Bonnànnaro and Sa Turricula style pottery in its upper three stratigraphic levels, radiocarbon dated to around 2300–1700 cal BC (Trump 1983). An increase of sheep/goat and pig, at the expense of cattle, was also noted in the faunal assemblage of this period at this site.

The human use of caves as burial places increased significantly in this period, as did the quantity and range of associated grave goods, at a time in which social display and differentiation, in life and in death, became even more prominent features of the Sardinian cultural landscape. According to Lilliu (2003, 321) 25 per cent of Early Bronze Age Bonnànnaro culture burials were in natural caves, the rest being in artificial rock-cut hypogea, megalithic tombs, cists, and other types of structure. Primary and secondary burials have been identified. For example, Grotta Sisaia, a small cave (about 18m deep), located on the rocky side of the Valle di Lanaittu, contained a hearth and a mortuary deposit placed within a niche formed by two stalagmitic columns and a cavity in the cave wall (Fadda 1991; Ferrarese Ceruti and Germanà 1978). Charcoal from the hearth was radiocarbon dated to around 2450–2050 cal BC. The mortuary deposit included an articulated human skeleton, accompanied by two Bonnànnaro style pottery vessels and a granite grindstone. The skeleton was of a woman, aged 25–30 years, and had a healed trepanned skull as well as a healed fracture on the left arm. Riparo sotto roccia Su Cannisoni is a wide rockshelter, situated at an altitude of 808m in a highly visible area of cliffs on the edge of the limestone Pissu is Ilippas hill (Gradoli and Meaden in prep; Skeates 2009–10). Toward the eastern end of the rockshelter, directly below a now-extinct spring, a secondary burial deposit was covered by a pile of stones, which was later cemented by flowstone (Figs 12.8–12.9). The intact secondary mortuary deposit comprised: a pair of adult human skulls (one radiocarbon dated to around 1550–1450 cal BC); and an adjacent semi-circle of stones containing a large group of disarticulated human bones (especially long-bones, but also fragments of a child skull), some animal bones (including sheep/goat), five relatively large fragments of later prehistoric pottery, and some fragments of charcoal. More extensive secondary burial deposits have been identified in Grotta di Tanì (Lilliu 2003, 144–8). This is a 34m deep, limestone cave of tectonic origin. Groups of human remains, some burnt, others delimited by lines of stones, were

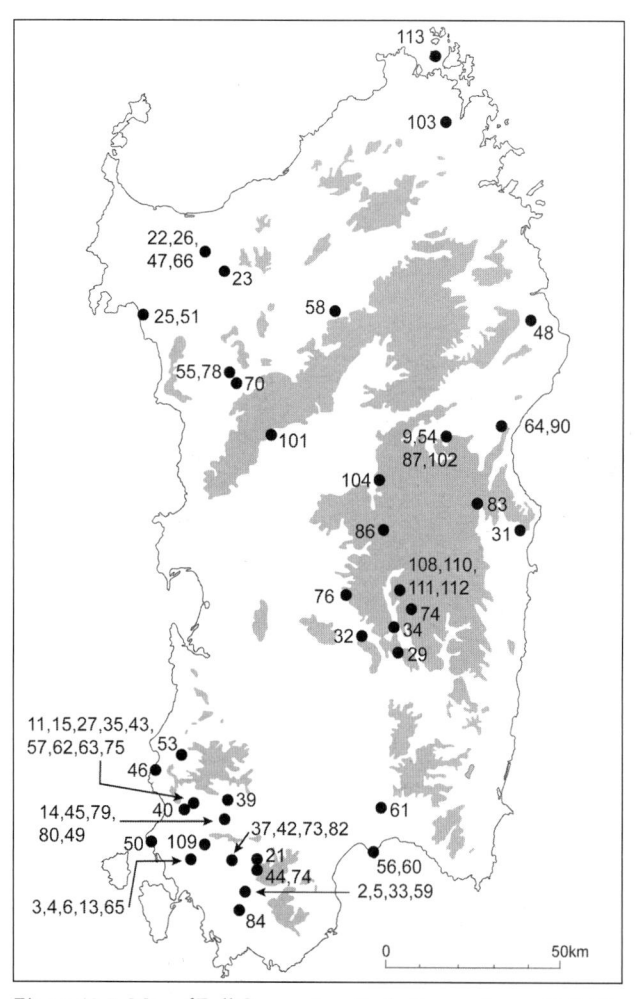

Figure 12.7: Map of Full Copper Age, Early Bronze Age and Middle Bronze Age cave sites in Sardinia.

Figure 12.8: Excavation of a secondary burial deposit at Riparo sotto roccia Su Cannisoni (Seulo). Photo: R. Skeates.

Figure 12.9: Secondary burial deposit of skulls and long bones at Riparo sotto Roccia Su Cannisoni (Seulo). Photo: P. van Carsteren.

found throughout the cave system, together with a range of grave goods. The latter included: mainly Monte Claro style pottery, ranging from large storage vessels to miniature vessels; 10 copper awls; a copper ring; a rectangular bone button; a necklace composed of tubular bones, a stalactite pendant, and animal canines; cockle shells (*Cardium* sp.); olive stones; and pieces of cork.

Some particularly small caves were also used as burial places or ossuaries at this time. They might be loosely compared, in form and contents, to the earlier and contemporary, artifical, *domus de janas* rock-cut tombs of Sardinia. Indeed, when located in suitable positions in the cultural landscape, they might have provided a symbolically valid and technologically convenient – if finite – alternative to either the re-use of existing rock-cut tombs or the construction of new tombs, at least by certain sections of society. One example is the natural cave of Grotta di Baraci, which was used as a burial place during the Full Copper Age and Early Bronze Age (Pitzalis 1988–9). It is located at an altitude of 685m on the north-east slope of Monte Guzzini, about 5km from the River Flumendosa. The cave comprises two chambers. The small lower chamber, measuring 5.1m wide, was repeatedly used as a collective tomb. This was found to contain: numerous human bones; fragments of pottery vessels, including some miniature bowls; and some tools of obsidian and flint, including a flint arrowhead, whose fresh edges, together with the presence of debitage, suggested to the excavator that they had been manufactured then deposited *in situ* as grave goods. Another example is Su Grutta 'e is Bittuleris (Gradoli and Meaden in prep; Maxia and Cossu 1952; Skeates 2009–10). This small, single-chambered, natural cave is just 8m wide and 6.5m deep (Fig. 12.10). It lies at an altitude of 826m, just below the summit of a rocky spur on the edge of the limestone hill of Pissu is Ilippas. The disturbed cultural deposits contained Bronze Age material, radiocarbon dated to around 1750–1600 cal BC), including: the skull of an old adult male that had been trepanned three times; substantial quantities of other human bones; a few animal bones; some small fragments of pottery; obsidian artefacts; a bone pendant; and a bead. Specialist study of the human remains points to successive primary inhumations in this cave, of male and female adults and children, followed by significant disturbance and fragmentation of the bones. In fact, one of a pair of vertebrae from an arthritic individual appears to have been removed from this cave and incorporated in the secondary burial deposit of Su Cannisoni rockshelter (see above), which lies immediately below Su Grutta 'e is Bittuleris.

Votive deposits of artefacts, especially pottery, were also placed in the inner spaces of selected cave systems, perhaps as offerings to chtonic spirits. For example, a few Full Copper Age artefacts were deposited in Grotta Murroccu (Fadda 1991; Sanges 1985). The cave is located in the river canyon known as Codula di Luna. It comprises a small and poorly accessible curving corridor, a little over 7m long. In the deepest part of the corridor, a Monte Claro style four-handled jar was found *in situ*, partly cemented to the cave wall by flowstone.

Cave art may also have continued to be produced during this period, although dating remains a serious problem. The eroded granite rockshelter known as Riparo sotto roccia di Frattale (Oliena) contains an anthropomorphic engraving which has been assigned to the Full Copper Age/Early Bronze Age (Moravetti 1980). The schematic figure, of double U form, was engraved on the back wall of the small cavity situated at the base of the rockshelter. It is comparable in form

Figure 12.10: Excavation of Su Grutta 'e is Bittuleris (Seulo). Photo: R. Skeates.

to the figures from the Grotta del Bue Marino (assigned, above, to the Final Neolithic). However, soundings in the entrance to the chamber identified fragments of Bell Beaker style vessels, and so the excavator has also assigned the figure to this period.

Another occasional ritual practice associated with caves might have comprised sealing their entrances. For example, Grotta di San Michele ai Cappuccini, best-known for its Final Neolithic special deposits, might have been ritually sealed during the Bronze Age. Certainly, when the cave was discovered in the early twentieth century, the entrance shaft had been carefully blocked by two tall cylindrical slabs (each around 2m long and 0.8m in diameter), made of granite. Given that the Bonnànnaro style pottery fragments found in the cave represents the last use of the cave, it is possible that the cave was sealed during the Bronze Age (rather than the Final Neolithic, as has sometimes been assumed). This practice might be compared to the sealing of at least some of the contemporary rock-cut tombs, such as Tomb VII at Serra Is Aràus, which was sealed with a sandstone statue-stela, and to the restriction of access to the interior of contemporary megalithic 'Giant's Tombs' via portals.

Later Bronze Age and Early Iron Age caves: underground cult places

The late-final phases of Sardinia's Nuragic Bronze Age and the Early Iron Age, which date to around 1300–730 cal BC, lie outside the focus of this paper, and will therefore only be considered briefly here, in order to close the discussion of the human uses of caves in prehistory. Although as many as 90 per cent of caves previously occupied during the Full Copper Age, Early Bronze Age and Middle Bronze Age appear to have been abandoned at this time, caves certainly continued to be used, but primarily for the ritual purposes of human burial and votive deposition.

Continuing a trend begun in the previous period, numerous small caves were adapted into tombs, often in the vicinity of Nuragic villages (Lilliu 2003, 452–4). In north-east Sardinia, they took the form of small natural fissures in the granite, measuring 2–4m long, which were now often subdivided and closed by stone walls. They contained primary and secondary burials, and some grave goods (especially pottery vessels), although these were much less rich than those found in contemporary megalithic tombs. Lilliu (2003, 453) has consequently suggested these funerary caves were established by relatively weak and impoverished groups.

Better known but less numerous are the Nuraghic 'cave-shrines', whose roots can be traced back (at least in part) to the growing use of caves as places of votive deposition in the Full Copper Age. In addition to nuraghi, santuaries, Giants' Tombs, and sacred wells and springs, these caves comprised significant places of ritual performance and votive deposition in the Sardinian cultural landscape. For example, Grotta Pirosu is best known as an Early Iron Age Nuraghic cave sanctuary or votive place, although it was previously occupied between the Final Neolithic and Middle Bronze Age (Alessio et al. 1970; Lo Schiavo and Usai 1995). It is a karst cave, whose main branch extends for about 180m. 120m in from, and 95m below, the entrance, a wide chamber, containing speleotherms and water pools, leads to a round recess, delimited by speleotherm columns. Within this was found: a stalagmite 'altar'; a small water pool; a hearth containing charcoal radiocarbon dated to around 900–800 cal BC; and an extraordinary accumulation of votive objects, including around 1500 pottery vessels and 109 metal objects of copper, bronze and gold. By contrast, Sa Grutta de is Caombus was modified architecturally, probably in the Final Bronze Age/Early Iron Age, into what has been described as a Nuraghic hypogean temple linked to a chthonic cult (Lilliu 2003, 626). The cave, now partly collapsed, was originally a fissure in basaltic rock. It would have been entered via a 1m wide stairway of 46 steps, comparable to those constructed at sacred well sites.

Conclusion

This chapter has attempted to present – in context – the archaeological evidence for the human uses of natural caves across prehistoric Sardinia. On reflection, I have to admit that the task of contextualizing remains incomplete, primarily due to the limitations of the available published data. Nevertheless, I hope that this chapter represents a step in the right direction. When seen in context, these caves and the archaeological remains found within them can be understood from two major perspectives. When seen from the outside, especially on a regional scale, they can be understood as diverse places variably connected to wider lifeways, landscapes and beliefs. And when considered from the inside, on a much more local scale, they can be understood as selected, malleable, and multi-sensory architectural spaces, whose occupation involved the conscious installation of assemblages of (inter-related) practical and symbolic resources (cf. Skeates 1997; in press).

In the Upper Palaeolithic and Mesolithic, at least two large, inland caves and one small, coastal rockshelter were repeatedly occupied and abandoned by mobile groups of hunter-gatherers as base-camps and shelters within wider socio-economic territories. This pattern of cave use continued into the Early Neolithic, albeit with some significant cultural modifications: with large inland caves continuing to serve as residential bases for the exploitation of local resources, both wild and domesticated; while an expanded variety of other kinds of caves (ranging from small rockshelters to a deep coastal cave) continued to shelter living – and now also deceased – members of mobile task groups who practiced a combination of hunting, herding, foraging and farming, in conjunction with occasional burial rites. Continuity is again evident in the Middle and Final phases of the Neolithic, although the secret-sacred use of the deep, dark and wet interiors of selected caves was elaborated, particularly through the accumulation of richer mortuary and votive deposits and of colourful cave paintings within them, whose symbolism was both distanced from and connected to the rich cultural and cosmological landscape of Neolithic Sardinia and the wider Central Mediterranean region (cf. Whitehouse 1992). This trend towards the sacralization and control of the underworld, at the same time as the elaboration of above-ground monuments, developed further between the Full Copper Age and Middle Bronze Age, and indeed on into the Early Iron Age. Now, ritual performances in a larger number and wider range of caves, including some particularly small burial caves, led to the deposition of even richer mortuary and votive deposits at selected natural places, and, in some cases, to their architectural modification. This trend, and a wider settling of formerly marginal parts of the Sardinian landscape (including the widespread establishment of nuraghic villages and towers), led to a decrease in the use of large caves as long-term dwelling places by residential communities, although a few of them continued to be intermittently occupied and abandoned as convenient shelters by mobile herders and hunters.

What this synthesis offers, at least, is a challenge to future research on the human use of caves in prehistoric Sardinia: to either fill in, or re-stitch, the contextual web of relations outlined here.

Acknowledgements

I am very grateful to Dott.ssa Giusi Gradoli and Dott. ssa Simona Losi for commenting on an earlier draft of this chapter, and for supplying me with copies of some publications. I would also like to thank Dott. ssa Gradoli, who (re)discovered the archaeology of the Seulo caves, mentioned above, and Terry Meaden, who undertook a preliminary contextualization of them: they kindly invited me to direct research on the Seulo caves – the results of which have significantly enhanced my understanding of prehistoric Sardinia and its underworld. And finally, a big thank you to Yvonne Beadnell for producing the maps.

Appendix: prehistoric caves in Sardinia

Note: the data presented here should be treated with some caution, since they are drived from a wide variety of sources, most of which cannot be critically evaluated. They do, however, provide a general indication of the human use of caves in different periods of Sardinian prehistory.

Site #	Site name	Locality	Major period(s) of occupation *	Selected bibliographic references
1	Grande Riparo sotto roccia di Sa Conca Fravihà	Ollolai	MN-LN	Fadda 1991, 1993
2	Grotta A. di S. Pantaleo	Santadi	FCu-MBA	Melis n.d.
3	Grotta A.C.A.I.	Carbonia	FN-ECu, FCu-MBA	Lilliu 2003
4	Grotta Antico	Carbonia	FCu-MBA	Bettini 2000
5	Grotta B di S. Pantaleo	Santadi	FCu-MBA	Melis n.d.
6	Grotta Barbusi	Carbonia	FN-ECu, FCu-MBA	Lilliu 2003
7	Grotta Bariles	Ozieri	EN, MN-LN	Tanda 1982
8	Grotta Conca Niedda	Sedini	FN-ECu	Melis 1993

Site #	Site name	Locality	Major period(s) of occupation *	Selected bibliographic references
9	Grotta Corbeddu	Oliena	Pal, Meso, EN, MN-LN, FN-ECu, FCu-MBA	Eisenhauer and Kalis 1999; Klein Hofmeijer 1997; Klein Hofmeijer *et al.* 1987, 1987–8, 1989; Sanges 1985–6; 1987; Sondaar and Sanges 1993; Sondaar *et al.* 1984, 1986, 1995; Spoor 1999; Spoor and Sondaar 1986
10	Grotta de Su Guanu	Pozzomaggiore	FN-ECu	Lilliu 2003
11	Grotta degli Scheletri	Iglesias	MN-LN, FCu-MBA	Floris 2007
12	Grotta dei Colombi	Cagliari	FN-ECu	Lilliu 2003
13	Grotta dei Fiori	Carbonia	FN-ECu, FCu-MBA	Lilliu 2003
14	Grotta dei Pipistrelli	Villamassàrgia	FCu-MBA	Lilliu 2003
15	Grotta del Bandito	Iglesias	FCu-MBA	Lilliu 2003
16	Grotta del Bue Marino	Dorgàli	FN-ECu	Fadda 1991; Lilliu 1957; Lo Schiavo 1978a, 1980
17	Grotta del Carmelo	Ozieri	FN-ECu	Lilliu 2003
18	Grotta del Guano	Oliena	FN-ECu	Alessio *et al.* 1971; Castaldi 1972, 1980, 1987; Lo Schiavo 1978b
19	Grotta del Papa	Isola di Tavolara	MN-LN	D'Arragon 1997
20	Grotta del Sorcio	Iglesias	MN-LN	Floris 2007
21	Grotta dell'Acqua Calda	Acquacadda	MN-LN, FCu-MBA	Alessio *et al.* 1970
22	Grotta dell'Anfora	Sassari	FCu-MBA	Usai 2006
23	Grotta dell'Inferno	Muros	EN, MN-LN, FN-ECu, FCu-MBA	Contu 1970a
24	Grotta della Campana	Carbonia	MN-LN	Deidda 2008
25	Grotta della Medusa	Alghero	EN, MN-LN, FCu-MBA	Usai 2006
26	Grotta della Scala di Giocca	Sassari	FCu-MBA	Lilliu 2003
27	Grotta della Volpe	Iglesias	MN-LN, FCu-MBA	Atzeni 2005
28	Grotta delle Scalette	Iglesias	MN-LN	Floris 2007
29	Grotta di Baraci	Nurri	FCu-MBA	Pitzalis 1988–9
30	Grotta di Capo Pecora	Arbus	FN-ECu	Lilliu 2003
31	Grotta di Coa 'e Serra	Baunei	FCu-MBA	Sanges 1985
32	Grotta di Crabilis	Isili	FCu-MBA	Lilliu 2003
33	Grotta di Forresu	Santadi	FN-ECu, FCu-MBA	Melis n.d.
34	Grotta di Frommosa I	Villanova Tulo	FN-ECu, FCu-MBA	Lilliu 2003
35	Grotta di Genna Luas	Iglesias	FCu-MBA	Lilliu 2003
36	Grotta di Is Aruttas	Cabras	FN-ECu	Lilliu 2003
37	Grotta di Is Ollargius	Narcào	FCu-MBA	Bettini 2000
38	Grotta di Is Piras	Nuxis	MN-LN	Deidda 2008
39	Grotta di Monte Acqua	Domusnovas	FN-ECu, FCu-MBA	Lilliu 2003
40	Grotta di Monte Casula	Monteponi	EN, MN-LN, FCu-MBA	Tanda 1982
41	Grotta di Monte Meana	Santadi	MN-LN, FN-ECu	Lilliu 2003
42	Grotta di Narcào	Narcào	FCu-MBA	Lilliu 2003
43	Grotta di Nicolai 'e Nébida	Iglesias	FCu-MBA	Lilliu 2003
44	Grotta di Nuxis	Nuxis	FCu-MBA	Lilliu 2003
45	Grotta di Orbai	Villamassàrgia	FCu-MBA	Bettini 2000

Site #	Site name	Locality	Major period(s) of occupation *	Selected bibliographic references
46	Grotta di Padre Nocco	Buggerru	FN-ECu, FCu-MBA	Lilliu 2003
47	Grotta di Palmaera	Sassari	FCu-MBA	Lilliu 2003
48	Grotta di Perapala	Siniscola	FN-ECu, FCu-MBA	Lilliu 2003
49	Grotta di Pitzu Asimus	Villamassàrgia	FCu-MBA	Lilliu 2003
50	Grotta di Punta Niedda	Portoscuso	FCu-MBA	Lilliu 2003
51	Grotta di Rureu	Alghero	EN, MN-LN, FN-ECu, FCu-MBA	Lilliu 1978
52	Grotta di S'Acqua Gelara	Buggerru	EN	Tanda 1982
53	Grotta di S'Orreri	Fluminimaggiore	FCu-MBA	Floris 2007
54	Grotta di Sa Oche	Oliena	FCu-MBA	Museo Nazionale Archeologico di Nuoro 2006
55	Grotta di Sa Ucca 'e Su Tintirriòlu	Mara	MN-LN, FN-ECu, FCu-MBA	Alessio et al. 1978; Contu 1970b; Loria and Trump 1978; Trump 1989, 1990
56	Grotta di San Bartolomeo	Cagliari	EN, MN-LN, FN-ECu, FCu-MBA	Atzeni 1962; Orsoni 1879 Patroni 1901
57	Grotta di San Lorenzo	Iglesias	MN-LN, FCu-MBA	Atzeni 2005
58	Grotta di San Michele ai Cappuccini	Ozieri	FN-ECu, FCu-MBA	Lilliu 1950; Porro 1915; Taramelli 1915
59	Grotta di San Paolo	Santadi	FN-ECu, FCu-MBA	Lilliu 2003
60	Grotta di Sant'Elia	Cagliari	EN, FCu-MBA	Orsoni 1879
61	Grotta di Sant'Isidoro	Sinnai	FCu-MBA	Lilliu 2003
62	Grotta di Santa Lucia	Iglesias	FCu-MBA	Lilliu 2003
63	Grotta di Santa Vita	Iglesias	FCu-MBA	Bettini 2000
64	Grotta di Sas Formicas	Dorgàli	FCu-MBA	Lilliu 2003
65	Grotta di Serbariu	Carbonia	FCu-MBA	Lilliu 2003
66	Grotta di Serra di Lioni	Sassari	FCu-MBA	Gruppo Speleo Ambientale Sassari n.d.
67	Grotta di Sos Dorroles	Cala Gonone	FN-ECu	Lilliu 2003
68	Grotta di Sos Sirios	Dorgàli	FN-ECu	Moravetti 1998
69	Grotta di Su Coloru	Laerru	Meso, EN, MN-LN	Fenu et al. 1999–2000, 2002, 2003, 2007; Martini et al. 2007; Pitzalis et al. 2001, 2003
70	Grotta di Su Guanu	Pozzomaggiore	FCu-MBA	Lilliu 2003
71	Grotta di Su Idighìnzu	Thiesi	FN-ECu	Lilliu 1957
72	Grotta di Su Marináiu	Cala Gonone	EN, FN-ECu	Lilliu 1957
73	Grotta di Su Moiu	Narcào	FN-ECu, FCu-MBA	Lilliu 2003
74	Grotta di Tamara	Nuxis	FCu-MBA	Lilliu 2003
75	Grotta di Tanì	Iglesias	FCu-MBA	Lilliu 2003
76	Grotta di Villahermosa	Laconi	FN-ECu, FCu-MBA	Lilliu 2003
77	Grotta Domus de is Janas	Sadali	FCu-MBA	Fadda 1991
78	Grotta Filiestru	Mara	EN, MN-LN, FN-ECu, FCu-MBA	Hurcombe and Phillips 1998; Trump 1982, 1983, 1989, 1990
79	Grotta I di Su Concali de Còrongiu Acca	Villamassàrgia	MN-LN, FN-ECu, FCu-MBA	Lilliu 2003
80	Grotta II di Su Concali de Còrongiu Acca	Villamassàrgia	EN, FCu-MBA	Lilliu 2003
81	Grotta Maimòne	Laconi	EN	Lilliu 2003
82	Grotta Montega	Narcào	FCu-MBA	Lilliu 2003
83	Grotta Murroccu	Urzuléi	FCu-MBA	Fadda 1991; Sanges 1985

Site #	Site name	Locality	Major period(s) of occupation *	Selected bibliographic references
84	Grotta Pirosu	Su Benatzu	MN-LN, FN-ECu, FCu-MBA	Alessio *et al.* 1970; Lo Schiavo and Usai 1995
85	Grotta Pitzu 'e Pranu	Belvì	MN-LN	Lilliu 2003
86	Grotta Pitzu 'e Toni	Tonara	FN-ECu, FCu-MBA	Lilliu 2003
87	Grotta Rifugio	Oliena	EN, MN-LN, FCu-MBA	Agosti *et al.* 1980; Biagi and Cremaschi 1980; Carta 1966–7; Germanà 1978
88	Grotta Sa Korona di Monte Majore	Thièsi	EN, MN-LN, FN-ECu	Foschi 1982; Foschi Nieddu 1987; 1989; Lilliu 1957
89	Grotta Sa Rocca 'e Ulari	San Pietro di Sorres	MN-LN, FN-ECu	Lilliu 1957; Soro 2009
90	Grotta Sisaia	Dorgàli	FCu-MBA	Fadda 1991; Ferrarese Ceruti and Germanà 1978
91	Grotta Su Anzu	Siniscola	MN-LN, FN-ECu	Lilliu 2003
92	Grotta Verde	Alghero	EN, MN-LN, FN-ECu	Antonioli *et al.* 1994; Lamberti *et al.* 1986; Lilliu 1978; Lo Schiavo 1987; Tanda 1987; Usai 2006
93	Grutta de is Janas	Seulo	FN-ECu	Skeates 2009–10
94	Grutta di lu Sorigu Antigu	Sassari	MN-LN	Sanna and Sanna 2002
95	Grutta I de Longu Fresu	Seulo	MN-LN	Skeates 2009–10
96	Riparo di Cala di Vela Marina	Isola di Santo Stefano	EN, MN-LN	Difraia 2007; Lilliu 1957
97	Riparo di Luzzanas	Ozieri	FN-ECu	Basoli 1992; Dettori Campus 1989
98	Riparo di Porto Leccio	Trinità d'Agultu	Meso, FN-ECu	Aimar *et al.* 1997
99	Riparo di Su Monte	Muros	MN-LN	Lilliu 2003
100	Riparo di Tatinu	Santadi	EN, MN-LN	Tanda 1982
101	Riparo s'Adde	Macomer	MN-LN, FN-ECu, FCu-MBA	Mussi 2010; Pesce 1949
102	Riparo sotto roccia di Frattale	Oliena	FCu-MBA	Moravetti 1980
103	Riparo sotto roccia di Monte Incappiddatu	Arzachena	FCu-MBA	Lo Schiavo 1991
104	Riparo sotto roccia di Sa Conca Fravihà	Ollolai	FCu-MBA	Fadda 1991, 1993
105	Riparo sotto roccia Monte di Deu	Tempio	FN-ECu	Lilliu 2003
106	Riparo sotto roccia Punta Candela	Arzachena	FN-ECu	Lilliu 2003
107	Riparo sotto roccia Santa Chiara	Tempio	FN-ECu	Lilliu 2003
108	Riparo sotto roccia Su Cannisoni	Seulo	FCu-MBA	Skeates 2009–10
109	Riparo sotto roccia Su Carròppu	Sirri	EN, MN-LN, FCu-MBA	Atzeni 1972, 1977; Lugliè *et al.* 2007
110	Sa Forada de Gastea	Seulo	FCu-MBA	Lilliu 2003
111	Su Grutta' e is Bittuleris	Seulo	FCu-MBA	Maxia and Cossu 1952; Skeates 2009–10
112	Su Stampu Erdi	Seulo	FCu-MBA	Unpublished data from Seulo caves project
113	Tafone di Cala Corsara	Isola di Spargi	EN, FN-ECu, FCu-MBA	Ferrarese Ceruti and Pitzalis 1987
114	Tafone di Cala Serena	Isola di Caprera	MN-LN	Difraia 2007

* Key to major periods of occupation: Pal – Palaeolithic; Meso – Mesolithic; EN – Early Neolithic; MN-LN – Middle to Late Neolithic; FN-ECu – Final Neolithic and Earlier Copper Age; FCU-MBA – Full Copper Age, Early Bronze Age and Middle Bronze Age.

References

Agosti, F., Biagi, P., Castelletti, L., Cremaschi, M. and Germanà, F. (1980) La Grotta Rifugio di Oliena (Nuoro): caverna ossario neolitica. *Rivista di Scienze Preistoriche* 35, 75–124.

Aimar A., Giacobini G. and Tozzi C. (1997) Trinità d'Agultu (Sassari): Località Porto Leccio. *Bollettino di Archeologia* 43–45, 82–87.

Alessio, M., Bella, F., Improta, S., Belluomini, G., Cortesi, C. and Turi, B. (1970) University of Rome carbon-14 dates VIII. *Radiocarbon* 12/2, 599–616.

Alessio, M., Bella, F., Improta, S., Belluomini, G., Cortesi, C. and Turi, B. (1971) University of Rome carbon-14 dates IX. *Radiocarbon* 13/2, 395–411.

Alessio, M., Allegri, L., Bella, F., Improta, S., Belluomini, G., Calderoni, G., Cortesi, C., Manfra, I. and Turi, B. (1978) University of Rome carbon-14 dates XVI. *Radiocarbon* 20/1, 79–104.

Antonioli, F., Ferranti, L. and Lo Schiavo, F. (1994) The submerged Neolithic burials of the Grotta Verde at Capo Caccia (Sardinia, Italy): implication for the Holocene sea-level rise. *Memorie Descrittive del Servizio Geologico Nazionale* 52, 290–312.

Atzeni, E. (1962) The Cave of San Bartolomeo, Sardinia. *Antiquity* 36, 184–189.

Atzeni, E. (1972) Su Carroppu di Sirri (Carbonia). *Rivista di Scienze Preistoriche* 27, 478–479.

Atzeni, E. (1977) Riparo sotto roccia di 'Su Carròppu' (Sirri–Carbonia). *Rivista di Scienze Preistoriche* 32, 357–358.

Atzeni, E. (2005) *Ricerche Preistoriche in Sardegna*. Cagliari, Edizioni AV.

Basoli, P. (1992) Dipinti preistorici nel Riparo di Luzzanas (Ozieri, Sassari): techniche di rilevamento, esame iconografico ed inquadramento culturale'. In *Atti della XXVIII Riunione Scientifica dell'Istituto Italiano di Preistoria e Protostoria. L'Arte in Italia dal Paleolitico all'Età del Bronzo. Firenze, 20–22 Novembre 1989. In Memoria di Paolo Graziosi*, 495–506. Firenze, Istituto Italiano di Preistoria e Protostoria.

Bettini, C. (2000) Le zone archeologiche più interessanti del sud ovest Sardo. <http://www.sardegnadelsudovest.it/archeo_zone.htm>, accessed 14 Jan 2010.

Biagi, P. and Cremaschi, M. (1980) La Grotta Rifugio di Oliena. In *Sardegna Centro-Orientale dal Neolitico alla fine del Mondo Antico. Nuoro – Museo Civico Speleo-Archeologico. Mostra in Occasione della XXII Riunione Scientifica dell'Istituto Italiano di Preistoria e Protostoria*, 11–15. Sassari, Soprintendenza ai Beni Archeologici per le Provincie di Sassari e Nuoro.

Carta, E. (1966–7) Documenti del Neolitico Antico nella Grotta 'Rifugio' di Oliena (Nuoro). *Studi Sardi* 20, 48–67.

Castaldi, E. (1972) La datazione con il C-14 della Grotta del Guano o Gonagosula (Oliena–Nuoro): considerazioni sulla cultura di Ozieri. *Archivio per l'Antropologia e la Etnologia* 102, 233–275.

Castaldi, E. (1980) Relazione preliminare sullo scavo della Grotta del Guano o Gonagosula (Oliena–Nuoro). In *Atti della XXII Riunione Scientifica dell'Istituto Italiano di Preistoria e Protostoria nella Sardegna Centro-Settentrionale*, 149–160. Firenze, Istituto Italiano di Preistoria e Protostoria.

Castaldi, E. (1987) Grotta del Guano di Oliena: relazione preliminare dello scavo 1978. In *Atti della XXVI Riunione Scientifica dell'Istituto Italiano di Preistoria e Protostoria. Il Neolitico in Italia. Firenze, 7–10 Novembre 1985. Volume II*, 831–844. Firenze, Istituto Italiano di Preistoria e Protostoria.

Contu, E. (1970a) Grotta dell'Inferno (Muros). *Rivista di Scienze Preistoriche* 25, 435.

Contu, E. (1970b) Sa Ucca de Su Tintirriòlu (Mara). *Rivista di Scienze Preistoriche* 25, 434–435.

Costa, L., J.-D. Vigne, Bocherens, H., Desse-Berset, N., Heinz, C., de Lanfranchi, F., Magdeleine, J., Ruas, M.-P., Thiebault, S. and Tozzi, C. (2003) Early settlement on Tyrrhenian islands (8th millennium cal. BC): Mesolithic adaptation to local resources in Corsica and northern Sardinia. In L. Larsson (ed.) *Mesolithic on the Move: Papers Presented at the Sixth International Conference on the Mesolithic in Europe, Stockholm 2000*, 3–10. Oxford, Oxbow Books.

D'Arragon, B. (1997) Nuovo figure schematiche antropomorfe dalla Sardegna prenuraghica: le pitture rupestri della Grotta del Papa, Isola di Tavolara (SS – I). *Tracce* 9. <http://www.rupestre.net/tracce/TAVOLAR.htm>, accessed 26 Sept. 2008.

Deidda, L. (2008) The middle Neolithic San Ciriaco phase archaeology. <http://www.ab-origine.it/articoli_dettaglio_ING.asp?idcollegamento=30&collegamento=CATEGORIA&pageelenco=1&id=73>, accessed 17 Sept 2010.

Dettori Campus, L. (1989) Dipinti rupestri schematici in loc. Luzzanas, Ozieri. In L. Dettori Campus (ed.) *La Cultura di Ozieri: Problematiche e Nuove Acquisizioni*, 103–111. Ozieri, Edizioni Il Torchietto.

De Waele, J. (2005) The speleological bibliography of Sardinia (Italy). Paper presented at the 14th International Congress of Speleology. <http://www.ese.edu.gr/media/lipes_dimosiefsis/14isc_proceedings/o/164.pdf>, accessed 26 Jan 2011.

Difraia, T. (2007) Arcipelago di La Maddalena (La Maddalena, Prov. Di Sassari). *Rivista di Scienze Preistoriche* 57, 467–468.

Eisenhauer, U. and Kalis, A. J. (1999) Die Holozänen schichten der Grotta Su Corbeddu (Sardinien): ein archäologischer und paläoökologischer bericht. *Archäologisches Korrespondenzblatt* 29, 489–504.

Fadda, M. A. (1991) *Il Museo Speleo–Archeologico di Nuoro*. Sassari, Carlo Delfino editore.

Fadda, M. A. (1993) Ollolai (Nuoro): località Monte San Basilio. *Bollettino di Archeologia* 19–21, 162–163.

Fenu, P., Martini, F. and Pitzalis, G. (1999–2000) Gli scavi nella grotta Su Coloru (Sassari): primi risultati e prospettive di ricerca. *Rivista di Scienze Preistoriche* 50, 165–187.

Fenu, P., Martini, F., Pitzalis, G. and Sarti, L. (2002) Le datazioni radiometriche della Grotta Su Coloru (Sassari) nella transizione Mesolitico – Neolitico. *Rivista di Scienze Preistoriche* 52, 327–335.

Fenu, P. Martini, F. and Pitzalis, G. (2003) Grotta Su Coloru (Laerru, Prov. Di Sassari). *Rivista di Scienze Preistoriche* 53, 636.

Fenu, P., Martini, F. and Pitzalis, G. (2007) Grotta di Su Coloru (Laerru, Prov. Di Sassari). *Rivista di Scienze Preistoriche* 57, 443–444.

Ferrarese Ceruti, M. L. and Germanà, F. eds. (1978) *Sisaia: Una Deposizione in Grotta della Cultura di Bonnanaro*. Quaderni 6. Sassari, Soprintendenza ai Beni Archeologici per le Provincie di Sassari e Nuoro.

Ferrarese Ceruti, M. L. and Pitzalis, G. (1987) Il Tafone di Cala Corsara nell'Isola di Spargi (La Maddalena – Sassari). In *Atti della XXVI Riunione Scientifica dell'Istituto Italiano di Preistoria e Protostoria. Il Neolitico in Italia. Firenze, 7–10 Novembre 1985. Volume II*: 871–886. Firenze, Istituto Italiano di Preistoria e Protostoria.

Floris, F. ed. (2007) *La Grande Enciclopedia della Sardegna, Volume 5*. Sassari: La Nuova Sardegna.

Foschi, A. (1982) Il Neolitico antico della Grotta Sa Korona di Monte Majore (Thiesi, Sassari): nota preliminare. In J. Bousquet (ed.) *Le Neolithique Ancien Mediterraneen: Actes du Colloque International de Prehistoire, Montpellier 1981. Archeologie en Languedoc No. Special 1982*, 339–346. Lattes, Fédération Archéologiques de l'Hérault.

Foschi Nieddu, A. (1987) La Grotta Sa Korona di Monte Majore (Thiesi, Sassari): primi risultati dello scavo 1980. In *Atti della XXVI Riunione Scientifica dell'Istituto Italiano di Preistoria e Protostoria. Il Neolitico in Italia. Firenze, 7–10 Novembre 1985. Volume II*: 859–870. Firenze, Istituto Italiano di Preistoria e Protostoria.

Foschi Nieddu, A. (1989) Documenti di cultura Ozieri provenienti dalla grotta di Sa Korona di Monte Majore – Thiesi e dalla necropolis di Janna Ventosa – Nuoro. In L. Dettori Campus (ed.), *La Cultura di Ozieri: Problematiche e Nuove Acquisizioni*, 145–152. Ozieri: Edizioni Il Torchietto.

Germanà, F. (1978) Dettagli di paleopatologia traumatica in un osso carpale proveniente dalla grotta 'rifugio' di Oliena–Nuoro (Neolitico medio). *Archivio per l'Antropologia e la Etnologia* 108, 323–331.

Gradoli, M. G. and Meaden, G. T. (in prep) Sacred sites and symbolism in the Neolithic landscape of Barbagia di Seulo, Central Sardinia. In G. T. Meaden (ed.) *Archaeology of Mother Earth Sites and Sanctuaries through the Ages: Rethinking Symbols and Images, Art and Artefacts from History and Prehistory*. Oxford, British Archaeological Reports.

Graziosi, P. (1973) *L'Arte Preistorica in Italia*. Firenze, Sansoni.

Gruppo Speleo Ambientale Sassari (n.d.) Grotta di Serra di Lioni, Sassari. <http://www.gsas.it/lioni.html>, accessed 5 Feb 2010.

Hurcombe, L. and Phillips, P. (1998) Obsidian usage at the Filiestru cave, Sardinia: choices and functions in the Early and Middle Neolithic periods. In M. S. Balmuth and R. H. Tykot (eds.) *Sardinian and Aegean Chronology: Towards a Resolution of Relative and Absolute Dating in the Mediterranean. Proceedings of the International Colloquium 'Sardinian Stratigraphy and Mediterranean Chronology', Tufts University, Medford, Massachusetts, March 17–19, 1995*, 93–102. Oxford, Oxbow Books.

Klein Hofmeijer, G. (1997) *Late Pleistocene Deer Fossils from Corbeddu Cave: Implications for Human Colonization of the Island of Sardinia*. British Archaeological Reports International Series 663. Oxford, Tempus Reparatum.

Klein Hofmeijer, G., Sondaar, P. Y., Alderliesten, C., Van Der Borg, K. and De Jong, A. F. M. (1987) Indications of Pleistocene man on Sardinia. *Nuclear Instruments and Methods in Physics Research B* 29, 166–168.

Klein Hofmeijer, G., Martini, F., Sanges, M., Sondaar, P. Y. and Ulzega, A. (1987–8). La fine del Pleistocene nella Grotta Corbeddu in Sardegna: fossili umani, aspetti paleontologi e cultura materiale. *Rivista di Scienze Preistoriche* 41, 29–64.

Klein Hofmeijer, G., Alderliesten, C., van der Borg, K., Houston, C. M., de Jong, A. F. M., Martini, F., Sanges, M and Sondaar, P. Y. (1989) Dating of the Upper Pleistocene lithic industry of Sardinia. *Radiocarbon* 31/3, 986–991.

Lamberti, L., Lo Schiavo, F., Pallarés, F. and Riccardi, E. (1986) Lo scavo del laghetto interno della Grotta Verde di Alghero (campagna 1979). *Rivista di Studi Liguri*, 51, 545–552.

Lilliu, G. (1950) Ozieri. *Studi Sardi* 9/1–2, 440–444.

Lilliu, G. (1957) Religione della Sardegna Prenuragica. *Bullettino di Paletnologia Italiana* 2, 7–96.

Lilliu, G. (1978) Le grotte di Rureu e Verde nella Nurra d'Alghero (Sassari). *Archivio per l'Antropologia e l'Etnologia* 108, 629–690.

Lilliu, G. (1999) *Arte e Religione della Sardegna Prenuragica: Idoletti, Ceramiche, Oggetti d'Ornamento*. Sassari, Carlo Defino editore.

Lilliu, G. (2003) *La Civiltà dei Sardi dal Paleolitico all'Età dei Nuraghi*. Nuoro, Edizioni Il Maestrale.

Loria, R. and Trump, D. H. (1978) *Le Scoperte a 'Sa 'Ucca de su Tintirriòlu' e il Neolitico Sardo*. Monumenti Antichi serie generale 49, serie miscellanea 2/2. Roma, Accademia Nazionale dei Lincei.

Lo Schiavo, F. (1978a) Figurazioni antropomorfe nella grotta del Bue Marino: Cala Gonone (Dorgàli). In *Sardegna Centro–Orientale dal Neolitico alla fine del Mondo Antico. Nuoro – Museo Civico Speleo–Archeologico. Mostra in Occasione della XXII Riunione Scientifica dell'Istituto Italiano di Preistoria e Protostoria*, 53–55. Sassari, Soprintendenza ai Beni Archeologici per le Provincie di Sassari e Nuoro.

Lo Schiavo, F. (1978b) La Grotta di Gonagósula o del Guano, Oliena. In *Sardegna Centro–Orientale dal Neolitico alla fine del Mondo Antico. Nuoro – Museo Civico Speleo–Archeologico. Mostra in Occasione della XXII Riunione Scientifica dell'Istituto Italiano di Preistoria e Protostoria*, 17–40. Sassari, Soprintendenza ai Beni Archeologici per le Provincie di Sassari e Nuoro.

Lo Schiavo, F. (1980) La Grotta del Bue Marino a Cala Gonone. In *Dorgali: Documenti Archeologici*, 39–45. Sassari, Chiarella.

Lo Schiavo, F. (1987) Grotta Verde 1979: un contributo sul Neolitico antico della Sardegna. In *Atti della XXVI Riunione Scientifica dell'Istituto Italiano di Preistoria e Protostoria. Il Neolitico in Italia. Firenze, 7–10 Novembre 1985. Volume II*, 845–858. Firenze, Istituto Italiano di Preistoria e Protostoria.

Lo Schiavo, F. (1991) *Il Museo Archeologico di Sassari 'G. A. Sanna'*. Sassari: Carlo Defino.

Lo Schiavo, F. and Usai, L. (1995) Testimonianze cultuali di età nuragica: la grotta Pirosu in località su Benatzu di Santadi. In V. Santoni (ed.) *Carbonia e il Sulcis: Archeologia e Territorio*, 145–186. Oristano, Editrice S'Alvure.

Lugliè, C., Le Bourdonnec, F.-X., Poupeau, G., Atzeni, E., Dubernet, S., Moretto, P. Serani, L. (2007) Early Neolithic obsidians in Sardinia (Western Mediterranean): the Su Carroppu case. *Journal of Archaeological Science* 34, 428–439.

Martini, F., Sarti, L., Pitzalis, G. and Fenu, P. (2007) La grotte de Su Coloru en Sardaigne dans le cadre culturel de la haute mer Tyrrhénienne au Mésolithique et au Néolithique ancien. A. D'Anna, J. Cesari, L. Ogel and J. Vaquer (eds.) *Corse et Sardaigne Préhistoriques: Relations et Échanges dans le Contexte Méditerranéen. Actes des 128e Congrès Nationaux des Sociétés Historiques et Scientifiques, Bastia, 2003*, 49–58. Paris, Le Comité des Travaux Historiques et Scientifiques.

Maxia, C. and Cossu, D. (1952). Cranio dell'epoca nuragica con segni di trapanazione sincipitale in vita: studio anatomo-radiografico. *Rivista di Antropologia*, 39: 232–235.

Melis, M. G. (n.d.) L'area del Parco del Sulcis dalla preistoria al medioevo. <http://www.isolasarda.com/storiaparco.htm>, accessed 31 Mar 2009.

Melis, P. (1993) Sedini (Prov. di Sassari). *Rivista di Scienze Preistoriche* 47, 469–471.

Moravetti, A. (1980) Riparo sotto roccia con petroglifi in Località Frattale (Oliena – Nuoro). In *Atti XXII Riunione Scientifica dell'Istituto Italiano di Preistoria e Protostoria in Sardegna Centro–Settentrionale*, 199–226. Firenze, Istituto Italiano di Preistoria e Protostoria.

Moravetti, A. (1998) *Serra Orrios e i Monumenti Archeologici di Dorgali*. Sassari, Carlo Delfino.

Museo Nazionale Archeologico di Nuoro (2006) Il Supramonte: villaggio nuragico di Tiscali. <http://www.museoarcheologiconuoro.it/index.php?it/31/itinerario/4/11>, accessed 17 Sept 2010.

Mussi, M. (2010) The Venus of Macomer: a little-known prehistoric figurine from Sardinia. In P. G. Bahn (ed.) *An Enquiring Mind: Studies in Honor of Alexander Marshack*, 193–209. Oxford, Oxbow Books.

Orsoni, F. (1879). Sur les fouilles pratiquées dans les grottes de Cagliari (Sardaigne). *Bulletins et Mémoires de la Société d'Anthropologie de Paris* 2, 44–45.

Patroni, G. (1901) S. Bartolomeo presso Cagliari: grotta preistorica rinettata nell'aprile 1901. *Notizie degli Scavi di Antichità*, 381–389.

Pesce, G. (1949) La 'Venere' di Macomer. *Rivista di Scienze Preistoriche* 4.3–4, 123–133.

Pitzalis, G. (1988–9) La Grotta Preistorica di Baraci (Nurri–Nuoro). *Studi Sardi* 28, 161–201.

Pitzalis, G., Fenu and P. Martini (2001) Grotta Su Coloru (Laerru, Prov. Di Sassari). *Rivista di Scienze Preistoriche* 51, 503.

Pitzalis, G., Fenu, P., Martini, F. and Sarti, L. (2003) Grotta Su Coloru: primi dati sui contesti culturali mesolitici e neolitici (scavi 1999–2003). *Sardinia, Corsica et Baleares Antiquae: An International Journal of Archaeology* 1, 31–39.

Porro, G. G. (1915) La Grotta di S. Michele in Ozieri in provincial di Sassari. *Bullettino di Paletnologia Italiana* 41, 97–123.

Sanges, M. (1985) Le culture di Monte Claro e di Bonnanaro in alcune grotte delle Codule di Ilune e di Sisine, nella costa orientale della Sardegna. In W. H. Waldren, R. Chapman, J. Lewthwaite and R.-C. Kennard (eds.) *The Deya Conference of Prehistory. Early Settlement in the Western Mediterranean Islands and their Peripheral Areas*. British Archaeological Reports, International Series 229ii, 611–622. Oxford, British Archaeological Reports.

Sanges, M. (1985–6). Grotta Corbeddu (Oliena, Prov. di Nuoro). *Rivista di Scienze Preistoriche* 40, 387–388.

Sanges, M. (1987) Gli strati del Neolitico antico e medio nella Grotta Corbeddu di Oliena (Nuoro): nota preliminare. In *Atti della XXVI Riunione Scientifica dell'Istituto Italiano di Preistoria e Protostoria. Il Neolitico in Italia. Firenze, 7–10 Novembre 1985*. Volume II, 825–830. Firenze, Istituto Italiano di Preistoria e Protostoria.

Sanna, L. and Sanna, I. (2002) La Grutta di lu Sorigu Antigu: sulle orme dell'uomo primitivo. *Sardegna Speleologica* 19, 48–54.

Skeates, R. (1997) The human uses of caves in east-central Italy during the Mesolithic, Neolithic and Copper Age. In C. Bonsall and C. Tolan-Smith (eds.) *The Human Use of Caves*, 79–86. British Archaeological Reports Interbational Series 667. Oxford, Archaeopress.

Skeates, R (2009–10) Archaeological discoveries in the caves of Seulo, Central Sardinia. *The European Archaeologist* 32, 4–5.

Skeates, R. (in press) Constructed caves: transformations of the underworld in prehistoric Southeast Italy. In H. Moyes (ed.) *Journeys into the Dark Zone: A Cross Cultural Perspective on Caves as Sacred Space. Volume I*. Bounler, CO, University of Colorado Press.

Sondaar, P. Y. and Sanges, M. (1993) Oliena (Nuoro): Grotta Corbeddu. Campagna di scavo 1992–1993. *Bullettino di Archeologia* 1993, 19–21.

Sondaar, P. Y., De Boer, P. L., Sanges, M., Kotsakis, T. and Esu, D. (1984) First report on a Palaeolithic culture in Sardinia. In W. H. Waldren, R. Chapman, J. Lewthwaite and R.-C. Kennard (eds.) *The Deya Conference of Prehistory: Early Settlement in the Western Mediterranean Islands and the Peripheral Areas*. British Archaeological Reports International Series 229i, 29–59. Oxford, British Archaeological Reports.

Sondaar, P. Y., Sanges, M., Kotsakis, T. and De Boer, P. L. (1986) The Pleistocene deer hunter of Sardinia. *Geobios* 19/1, 17–25.

Sondaar, P. Y., Elburg, R., Klein Hofmeijer, G., Martini, F., Sanges, M., Spaan, A. and de Visser, H. (1995) The human colonization of Sardinia: a Late-Pleistocene human fossil from Corbeddu Cave. *Comptes Rendus de l'Academie des Sciences*, série 2a, 320, 145–150.

Soro, P.P. (2009) La necropoli neolitica a *domus de janas* di S. Pietro di Sorres in Comune di Borutta – Sassari. *Lanx* 2, 150–168.

Spano, G. (1873). *Scoperte Archeologiche Fattesi in Sardegna in tutto l'Anno 1873*. Cagliari, Tipografia Alagna.

Spoor, F. (1999) The human fossils from Corbeddu Cave, Sardinia: a reappraisal. In J. W. F. Reumer and J. De Vos (eds.) *Elephants have a Snorkel! Papers in Honour of Paul Y. Sondaar*. *Deinsea*, 7: 297–302.

Spoor, C. F. and Sondaar, P. Y. (1986) Human fossils from the endemic island fauna of Sardinia. *Journal of Human Evolution* 15, 399–408.

Stuiver, M., Reimer, P. J. and Reimer, R. W. (2005) CALIB 6.0. <http://intcal.qub.ac.uk/calib/calib.html>, accessed 10 Feb. 2009

Tanda, G. (1982) Il Neolitco antico della Sardegna. In J. Bousquet (ed.) *Le Neolithique Ancien Mediterraneen: Actes du Colloque International de Prehistoire, Montpellier 1981. Archeologie en Languedoc No. Special 1982*, 333–337. Lattes, Fédération Archéologiques de l'Hérault.

Tanda, G. (1987) Nouveaux éléments pour une définition culturelle des matériaux de la Grotta Verde (Alghero, Sassari, Sardaigna). In J. Guilaine, J. Courtin, J.-L. Roudil and J.-L. Vernet (eds.) *Premières Communautés Paysannes en Méditerranée Occidentale. Actes du Colloque International du C.N.R.S. Montpellier, 26–29 Avril 1983*, 425–431. Paris: Centre National de la Recherche Scientifique.

Taramelli, A. (1915) *Ozieri: Grotte Sepolcrali e Votive di S. Michele ai Cappuccini*. Cagliari.

Trump, D. (1982) The Grotta Filiestru, Bonu Ighinu, Mara (Sassari). In J. Bousquet (ed.), *Le Neolithique Ancien Mediterraneen: Actes du Colloque International de Prehistoire, Montpellier 1981. Archeologie en Languedoc No. Special 1982*, 327–346. Lattes, Fédération Archéologiques de l'Hérault.

Trump, D. (1983) *La Grotta di Filiestru a Bonu Ighinu, Mara (SS)*. Sassari, Ministero per i Beni Culturali e Ambientali.

Trump, D. (1989) La cultura di Ozieri a Sa Ucca de Su Tintirriòlu e Filiestru in loc. Bonu Ighinu–Mara. In L. Dettori Campus (ed.) *La Cultura di Ozieri: Problematiche e Nuove Acquisizioni*, 153–157. Ozieri, Edizioni Il Torchietto.

Trump, D. (1990) *Nuraghe Noeddos and the Bonu Ighinu Valley: Excavation and Survey in Sardinia*. Oxford, Oxbow Books.

Tykot, R. H. (1994) Radiocarbon dating and absolute chronology in Sardinia and Corsica. In R. Skeates and R. Whitehouse (eds.) *Radiocarbon Dating and Italian Prehistory*, 115–145. London, Accordia Research Centre and The British School at Rome.

Usai, L. (2006) Materiali prenuragici da alcune grotte del territorio di Alghero (Sassari). *Sardinia, Corsica et Baleares: An International Journal* 4, 29–41.

Usai, L. (2008) Testimonianze di età neolitica nelle grotte della Sardegna. Unpublished paper presented at the XXI Convegno del Gruppo Archeologico Neapolis, 'Sardegna Preistorica: Aspetti Culturali del Neolitico', Marzo–Aprile 2008, Guspini.

Webster, G. S. (1996) *A Prehistory of Sardinia: 2300–500 BC*. Sheffield, Sheffiled Academic Press.

Whitehouse, R. D. (1992) *Underground Religion: Cult and Culture in Prehistoric Italy*. London, Accordia Research Centre, University of London.

Chapter 13

Discovery and Exploratory Research of Prehistoric Sites in Caves and Rockshelters in the Barbagia di Seulo, South-Central Sardinia

Giusi Gradoli and Terence Meaden

Previously unrecognized prehistoric sites, including caves and rockshelters, have been identified in the archaeological palimpsest of the Barbagia di Seulo region of mountainous South-Central Sardinia. The search for and discovery of fresh sites was helped using place-name studies, folklore and traditions while also considering visual landscape imagery typical of Neolithic and Bronze Age symbolism and concepts from elsewhere in continental and island Europe. Such concepts include figurative and topographic symbols and images deemed to have fertility and Mother Earth connotations appropriate to the Neolithic and Bronze Age periods. The investigations led to the finding of several sites. Three are introduced in this paper. One is a rockshelter called Su Cannisoni, near to which is a small bone-filled cave or ossuary in a geographical area known as 'Sa Omu 'e is ossus', which means 'The House of the Bones'. A second is an oval, cavernous, low-ceilinged section along a branch of a major limestone cave, known as Is Janas. It was found to contain numerous hearths, bone fragments and pottery with dotted decoration and a triangular zone suggestive of the fifth millennium BC Bonu Ighinu culture. Thirdly, pieces of human skulls and important parietal art were found in another limestone cave, Grutta I de Longu Fresu, in a mountainside forest known as the 'Foresta di Addolì', which translates as 'The Trees of the Dead'. This cave appears to be a Neolithic shrine. One skull covered by flowstone was radiocarbon dated to 4259–4042 cal BC. This is Late Middle Neolithic and among the oldest Neolithic human remains known for central and southern Sardinia. A panel of rock art is preserved beneath transparent, calcite flowstone. It is arguably the earliest painted parietal art known in a natural cave in Sardinia.

Introduction

The Barbagia di Seulo is a mountainous limestone part of southern Central Sardinia which is cut by wooded valleys including the River Flumendosa and its tributaries. A preliminary study of place-names using linguistics and a knowledge of local traditions, raised the possibility that unknown prehistoric sites might be located. This led to the discovery of sites of varying character, including caves and rockshelters (Gradoli *et al.* 2006). At some of them concepts appropriate to pan-European Neolithic and Bronze Age symbolism were present in the form of landscape imagery arguably having fertility and Mother Earth connotations. Four of the sites are introduced in this paper: three are caves and one a rockshelter. In one cave, Grutta I De Longu Fresu, the earliest rock-art *painting* in a natural cave in Sardinia was discovered. Other sites are discussed elsewhere (Gradoli and Meaden in prep.).

Sacred landscapes

By the start of the fifth millennium BC the hunter-gatherer Mesolithic period (11,000 to around 6000 cal BC was giving way to the Neolithic in Sardinia. For proposed chronologies of prehistoric Sardinia see, for example, Tykot (1994, 129, table 10) and Ugas (1998, 253, table 25.1). Note that Lilliu (2003, 14, 79) and Tanda

(2004, 32–33) remark that there is little evidence for a visible Mesolithic in Sardinia.

There are several reasons why the full Neolithic did not develop rapidly. Only partly was it an economic phenomenon. During the initial phase when Neolithic practices were developing there were major regional variations in timing across Sardinia – because the coastline is very long, mountainous areas very rocky, and only part of the interior adequately fertile. The climate would, as now, have varied considerably across the diverse landscapes in terms of rainfall, snowfall, temperature, and wind strength and direction.

The sequence of Neolithic cultures, as expressed by the nature of ceramic manufacture and design, and in artefact development and use including obsidian tooling, begins with the Su Carroppu and Filiestru cultures of the Early Neolithic, and continue with the Ighinu Bonu (Middle Neolithic) and the Late Neolithic Ozieri cultures (cf. Tykot 1994). In *toto* the Neolithic lasted for about three millennia before the transition to the Chalcolithic or Eneolithic Age at about the end of the fourth millennium cal BC, to be followed by the Early Bronze Age which lasted to about 1900 cal BC.

Generally in Central and Northern Sardinia, many caves and rockshelters are known to have been used, and some occupied, in the Neolithic, Early Bronze Age and later periods to judge by archaeological studies of surviving material culture. This includes engravings or paintings on walls, megaliths, pebbles, pottery and other artefacts (Skeates, this volume). Rock-cut tombs – which are essentially small artificial caves – proliferated, as did freestanding, galleried or cave-like, megalithic constructions called dolmens and 'tombe di giganti'. At various times natural caves served as regular dwellings, bad-weather temporary shelters, durable shrines, and for funeral rites and interment purposes. Whatever the occasion, caves are protective environments with ambient temperatures that change little through the year, and are secure from extremes of climate and weather.

In the long time interval from the earliest Neolithic to the Early Bronze Age – around 6000 to 1900 cal BC – food production and settlement patterns would for many people have changed and improved, but supporting detail from archaeological evidence is scant. It is from burial practices – although atypical of life's cardinal routines – that we can learn much, because material indications of a funerary character can better survive the millennia compared with a homestead's impermanent perishable artefacts, foods and goods. However, caution is needed to ensure that this does not skew the interpretation of cultural processes and habits. Fortunately, and more especially where caves were occupied for domestic reasons, some details about living conditions and eating habits continue to emerge.

Non-native domesticated cereals (early wheat and barley varieties) and animals (bovine, porcine and caprine) may have been imported by indigenous hunter-gatherers or introduced by immigrant groups together with husbandry knowledge and practical experience, although Albarella *et al.* (2006, 193–227) have alternatively proposed that Sardinia may have been a region of independent animal domestication. Probably some combination of both processes happened on Sardinia during the centuries when agriculture gradually became more important. Nonetheless, the hunting and gathering of wild provisions would have continued to supplement significantly the foods resulting from the labours of pastoral and agrarian methods and customs.

At the same time other factors to do with religion and culture were developing, because landscapes were more than just geographic and inhabitable zones with land for hunting and planting. This is because prehistoric landscapes progressively allocated sectors and sites that were held to be sacred, in which specific locations served to store a community's wisdom, myths and histories for the purpose of expressing moral tales for human conduct, tribal-origin stories and cosmologies. We know this through anthropological studies of native communities worldwide (e.g. Basso 1996a, 1996b, 53–90 for the native-Indian Western Apache of Nevada), and as inferable for Palaeolithic/Neolithic/Bronze Age Europe and later periods using folklore and studies like those initiated by Frazer (1911), Eliade (1957, 153) and Campbell (1959, 396–399). Caves partake in such myths or legends, as do landscape areas and sites like mountains, outcropping rocks, springs, cascades, water courses, swamps and even trees whose lives can span many human generations. All can serve, sometimes through supposed occupation by spirits, as story-telling links with ancestors. Such geographical features were held to be 'liminal places', or meeting-points, in social and community terms with humankind. It has been suggested (Bradley 1991, 77–101; 2000, 35) that some locations, held to be significant, were explicitly 'labelled' by actual rock carvings or, as still perpetuated in the Americas and in Hindu India today, treated as places where offerings are made (McEwan and Guchte 1992, 359–371), shrines constructed (Aldenferfen 1990, 479–493) or art painted (Gheorghiu 2002).

Carmichael *et al.* (1994, 3) explained it thus: 'What is known as a sacred site carries with it a whole range of rules and regulation regarding people's behaviour in relation to it and implies a set of beliefs to do with the non-empirical world, often in relation to the spirits of the ancestors, as well as more remote or powerful gods or spirits'. Natural places in the landscape become vested with supernatural beliefs and spiritual symbolism which renders them sacred (Ucko 1994; Bradley 1997, 215; Taçon 1999, 33–57) – perhaps appreciated and venerated all the more because they have been named by the ancestors. Generally speaking, many rituals and ceremonies have been related to: (a) fertility cults, including those taking place at pilgrimage centres or shrines and caves; (b) initiation-type rites and ceremonies; and (c) ancestor cults, with rituals providing access to the deceased in the 'afterworld' for descendants wanting to legitimize social, political and economic relationships among the living and to validate and buttress lawful land and cave ownership. In this way memory, myth and history have been literally fed into named features of the landscape including caves, and story-telling aided by mnemonic recall (Gosden and Lock 1998, 2–12; Van Dyke 2008).

Caves – together with artificial caves like rock-cut hypogea, freestanding megalithic chamber-tombs and the later slab-built 'Giants Tombs' or 'tombe di giganti' – eminently served as places for rites and repositories associated with ancestor cults. Similarly, many natural places were deemed suitable for fertility rites via Earth Mother or Goddess supplication and adoration (Gimbutas 1989, 141–159). Prime among natural sites are those to do with water (springs, streams, waterfalls, lakes, swamps) and rocky regions or cliffs presenting holes, fissures or cracks of vulva-like character. The latter are particularly prominent at some of the karst caves of Barbagia di Seulo (like Grotta de Longu Fresu) and certain cliff overhangs (Su Cannisoni and Su Stampu de Su Turrunu).

Prehistoric Sardinian cave art of possible Neolithic and Bronze Age date

Several examples of prehistoric cave art dating from the Neolithic and Bronze Age are known for Sardinia.

There is a Late Neolithic cave at Tisiennari, Bortigiadas, in the north of Sardinia that displays at its entrance a flattish area carved with a panorama of about 40 vulvas aligned in three long rows. Among them are a couple of simplified anthropomorphic forms (Gimbutas 1989, 240, fig. 375). The engraved panel has a height of 0.84m.

Engraved vulvas at cave entrances are typical of Upper Palaeolithic southern France (Abri Blanchard, La Ferrassie, Pergouset, Laussel and Lalinde, some of them reddened with ochre: cf. Meaden 1997, 22). The five half-metre long vulvas at the Spanish cave El Castillo cave were all painted red. One of the authors has previously described this site as follows: 'Besides colouring engraved vulvas in this fashion, entire niches and fissures in caves used in prehistoric times have been found painted in red ochre. This suggests that natural clefts and niches were seen as likenesses of the human vulva, which stressed their intended use as "Goddess" vulvas' (Meaden 1997, 22–23). Furthermore, actual cave entrances were looked upon as vulvar entrances to the womblike cavern (Cyriax 1921, 205). These comments are appropriate to the astounding natural pink-coloured fissures and niches at Su Cannisoni and Grutta I di Longu Fresu, and at Is Janas (discussed below).

At the Grotta del Papa (north-east Sardinia) three anthropomorphic stick-like figures, painted black, survive – found by D'Oriano in 1992. According to D'Arragon (1997) they could be Middle Neolithic.

Additional examples of engraved and/or painted parietal cave art have been assigned to the Sardinian Final Neolithic. At the entrance to the Grotta del Bue Marino on the eastern coast of Central Sardinia (Dorgàli, Gulf of Orosei) (Lo Schiavo1978; 1980) are painted engravings of 0.3m long anthropomorphic figures. At the *domus de janas* (rock-cut tomb) of Branca at Moseddu there are complex engraved art murals of likely Late Neolithic (Ozieri culture) date.

At the rockshelter of Riparo di Luzzanas are paintings of 10 anthropomorphic stick figures and concentric circles rendered in red ochre. Again, they might be from the Ozieri Neolithic period (Basoli 1992), if one can judge on stylistic grounds through similarity with art elsewhere on the rocks of natural caves, rockshelters, and man-made caves of *domus de janas* type in Sardinia, Corsica, Sicily and South-East Italy. Tanda (1998) has ordered and classified the art known for 148 constructions of *domus de janas* type in Sardinia. These carvings are more or less stylized anthropomorphic, animal and geometrical petroglyphs. The only *painted* art is on the engraved figures at the Luzzanas rockshelter, the Grotta del Papa (D'Arragon 1997) and, as reported below, at Grotta I di Longu Fresu.

D'Arragon (1995) wrote that '...the phenomenon of Sardinian petroglyphs, and subsequent strong stylization of the human figure, is to be included as part of the art of the Central Mediterranean (Southern Italy and islands, [including] Corsica) for the early age of metals ... at the time of 'important developments in the Eneolithic's final moments and the early Bronze Age'. Another rockshelter – Riparo sotto roccia di Frattale (Oliena–Nuoro) – exhibits at the back of a cavity at the shelter's base geometric marks and an anthropomorphic engraving that may date to the Chalcolithic/Early Bronze Age (Moravetti 1980).

Further carvings of likely Late Neolithic/Bronze Age date are known from necropoli at Bonorva, as with the *domus de janas* of Sa Pala Larga (Stefano 2010) and spiral paintings at a rock-cut hypogeum (Meozzi 2010).

Statuettes of stone and bone representing female human figures in a rigid standing position are typically Sardinian. Lilliu (1999) describes many, and the figurines from rock-cut tombs of the sixth and fifth millennia cal BC are noteworthy. The statuettes are often called 'dea madre' (e.g. Atzeni 1978, 1–69; Antona 1998, 111–119). Gimbutas (1989, 200–202; 1991, 228–236, 240–243) provides illustrated discussions saying that they 'occur for thousands of years from the Upper Palaeolithic through the Bronze Age in every culture in Europe' (Gimbutas 1990, 240). Dating is unclear for many Sardinian statuettes but some do appear to be Middle Neolithic, like those from Su Cùccurus'Arriu (Lilliu 1999, 180–1, 194, 209–12; 2003, 45, 51). Sites include Riparo sotto rocciadi Su Monte, Grotta I di Su Concali de CoróngiuAcca, and Grotta di Monte Meana.

Sacred symbols in Neolithic and Bronze Age cultures

Pastoral and agrarian peoples are noted for a devotional interest in symbolism and images pertinent to the divine feminine. Caves, rock-cut hollows, and *domus de janas* or tomb constructions exemplify this interest. In addition, natural cracks and fissures have attracted similar attention, as have natural hollows, holes and tunnels in cliffs and outcrops. The phenomenon of cupmark creation may be a symbolic figuration related to this circumstance: '…their meaning suggested by the fact that such cupmarks to this day retained some of their symbolic significance in the European peasant subculture which attributes healing powers to the rainwater which collects in them…' (Gimbutas 1989, 61). Starting in the Upper Palaeolithic and continuing in the Neolithic and Bronze Age, some cracks, fissures and elongated niches in caves were painted with red ochre if not already naturally red or pink (Meaden 1997, 22–23). They would appear to relate to the mythology of the times. For instance, consider the case of the peak sanctuary of Jouktas on Crete where there is a fine natural rock fissure that was so well respected before and during the Aegean Bronze Age that an enclosure wall came to be built around it that was up to 5m high and 3m thick (Karetsou 1981; Rutkowski 1986, 75–76, 82; Bradley 2000, 102–103, fig. 30). Stone buildings were added, and an altar built alongside the fissure, and the latter contained offerings. The fissure was being venerated long before its adaptation in Bronze Age. For Sardinia it is suggested that there are Neolithic/Bronze Age parallels with natural niches, alcoves or fissures as at Grutta I de Longu Fresu, Is Janas and Su Cannisoni, introduced next.

Grutta I de Longu Fresu: its shrine and art

The axis of this limestone cave is NNE–SSW. A few kilometres south-east of Seulo, it lies close to the stream 'Riu Su Longu Fresu'. In Sardo – the Sardinian language – the *Longu Fresu* is a type of tree locally said to be 'the plant of the dead'. The nearly horizontal, nearly straight, cave is about 15m long, 2–3m wide and 1m high. It has four natural niches or alcoves along its sides, two each side, formed long ago by the movement of falling water. The study of this cave has been rewarding.

At the far end of the main cave, and facing the cave entrance (15m away), half a human skull, and now anchored in calcite flowstone, had been positioned near a small, horizontal, longitudinal hollow which, once cleared of the small stones that closed it, also revealed more pieces of a human cranium and a long bone (Figs 13.1 and 13.2).

Analysis of a fragment of the skull by the Oxford University Radiocarbon Unit is radiocarbon dated to 5315±36 BP (4259–4042 cal BC OxA–X–2236–44). This is Middle-Late Neolithic.

Within a metre of this position, set in the northern wall of the cave, is a natural alcove or vertical niche extending from ground level to ceiling, which is suggestive of feminine symbolism. Its height is almost 1.2m and width between 0.3 and 0.5m. At its base is a 0.7m artificially-deepened pit possibly intended to allow falling water to drain away and prevent or reduce any flooding of the shrine in the cave.

To the left of this north-eastern niche is a rock-art painted panel 0.3m by 0.3m. The several marks were drawn by an adult finger using black natural paint (Fig. 13.3). Its remarkable preservation is due to being thinly covered with calcite flowstone that thickly but transparently covers it. Samples will be taken for dating.

To aid interpretation and explanation of the prehistoric markings, as shown in Figure 13.3, grey tones and white tonality have been added to the photograph (Fig. 13.4). Each stroke represents a continuous movement of the finger in the act of freehand drawing. One image appears to be an anthropomorphic horned-head figure (perhaps bearing a wooden carved mask which is even now typical

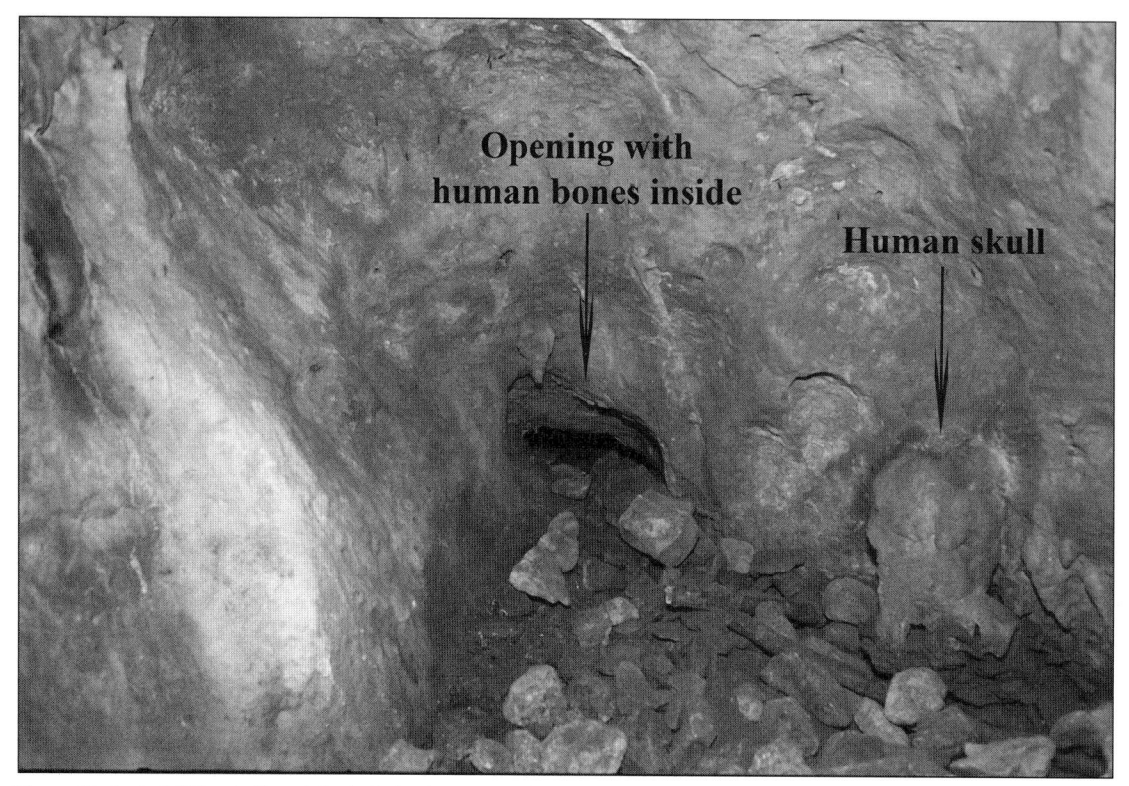

Figure 13.1: A child's skull and the hole in the cave wall that held additional pieces of skull, Grotta I de Longu Fresu.

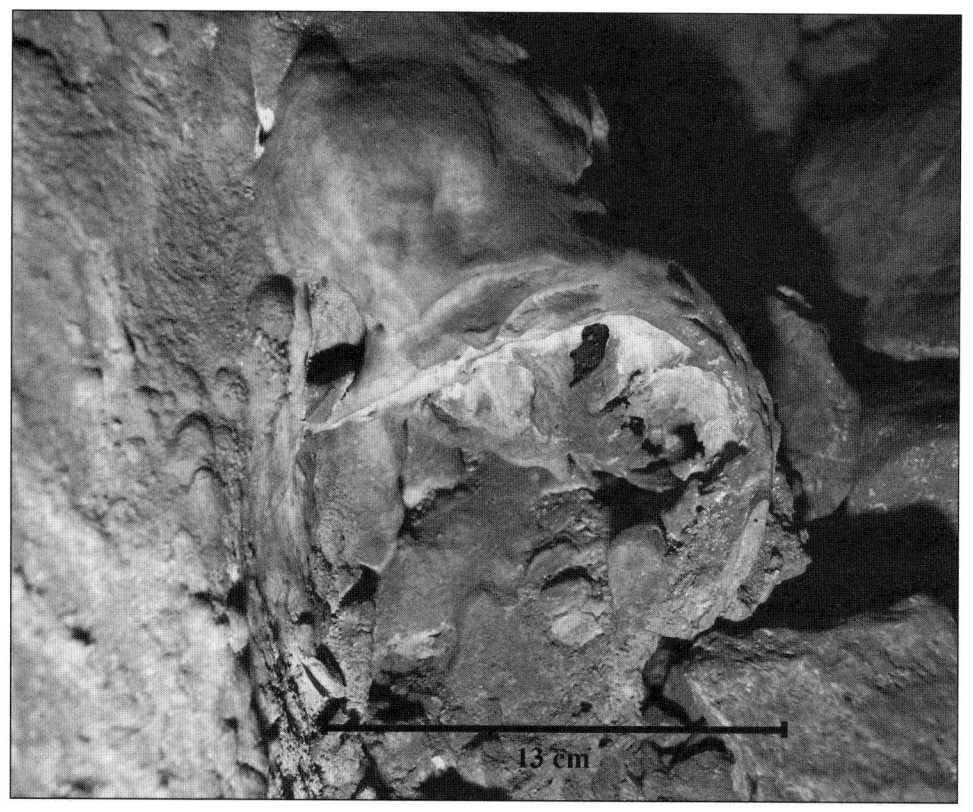

Figure 13.2: Close-up photograph of the stalagmite-coated skull dated to about 4250–4050 cal BC, Grotta I de Longu Fresu.

of surviving Sardinian village-pastoral and folklore traditions), holding a bow aimed towards a horned animal. The other artificial marks were also painted black but are less well-preserved and have not been interpreted.

In the north-western alcove of a similar size to the north-eastern one, pieces of human skull were found beneath the stones on the floor. Here, too, are features resulting from an ancient long-lasting waterfall and a pit dug at its base in the floor. The shape of the alcove again strongly evokes the idea of female symbolism.

The oval chamber at Is Janas

The major cave system known as Is Janas is located in the inner part of the Forest of Addolì, several kilometres south-east of Seulo. It has two principal branches. A high-ceilinged cave, 257m long sloping downwards from south to north, and of variable width between 2 and 6m, it is resplendent with many hundreds of fine stalactites and stalagmites. This is a show cave which has occasionally been opened to the public. A second long branch – about 150m long – runs from east to west. Part way along it, reached through a narrow 'squeeze' and passage several metres in length, is a roundish or oval-shaped chamber.

This space is about 20m in diameter and between 1 and 2m high. The room contains several hearths and is covered by a great quantity of stones, pottery sherds, lithic tools and animal bones scattered all about. The pottery is of the Ozieri culture (Late Neolithic) and Bonu Ighinu culture (Middle Neolithic), with dotted decoration and triangular zones (Fig. 13.5). This is present only inside a hearth near the entrance. An ongoing or later re-use in the Bronze Age also seems likely. Among the finds were sea shells, one with red ochre on the outside. Some of the base sherds from the hearth contained charred food residues.

Figure 13.5 shows a bone tool and some of the pottery. Remarkably it proved possible to match the impressions on these sherds to the tip of the bone tool that had patterned the pottery. What is more, the opposing, truncated, end of the bone tool was shown to have impressed another sherd.

From the big oval-shaped cavern, upon descending a 0.6m step, two small round rooms were encountered where burned human bones were found. A thick layer of ashes partially calcified with bones and plant remains, was present too.

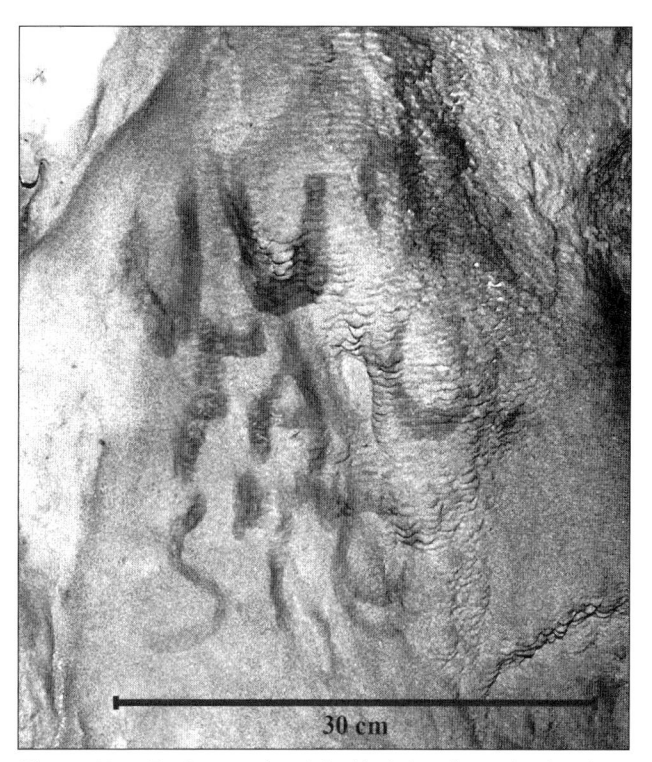

Figure 13.3: Rock art painted in black by 'fingertip drawing', Grotta I de Longu Fresu.

Figure 13.4: White and different shades of grey have been applied to the enlarged photograph to emphasise the various separate motifs, Grotta I de Longu Fresu.

'Sa Omu'e Is Ossus': the House of the Bones

In a limestone cliff west of Seulo is a big cavity, 'Sa Omu 'e is ossus' – which translates as 'The House of the Bones'. Because the tiny cave had served occasionally as a shepherd's shelter, even until recent times, a low dry-stone wall had been built at its

Figure 13.5: Bonu Ighinu Culture pottery sherd and the worked bone point used to impress the dots, found together at Is Janas Cave.

entrance. The cave's name is meaningful. The cave was found upon inspection to be full of bones. The deep, dusty, sandy cave deposits contained hundreds of bones of human and animal origin and numerous ceramic sherds. Within two years of its discovery in 2005, the majority of the bones were removed by persons unknown. The site is nearly above and not far from the major rock overhang considered next.

Su Cannisoni: a sacred rock overhang

This rockshelter (*riparo sotto roccia*) west of Seulo is marked 'Su Cannisoni' on modern maps although there is reason to suspect that the name is a variation of an earlier form, 'Su Cunnisoni'. The rock overhang is 20m long from north to south, 1 to 3m wide and almost 8m high (Fig. 13.6). A long natural vertical cleft (Fig. 13.7) at the southern end of the overhang resembles a vulva, and is pinkish in colour. A similar cleft, half the height and slightly pink, is about 15m distant in the northern section of the overhang. The name 'Cannisoni' or 'Cunnisoni' may relate to the Latin 'kunnus' or Sardo 'kunnu', meaning vulva in modern English. Whether reaching Sardinia via Latin or bypassing it in remoter antiquity, the archaic name could be an indicator of a prehistoric sacred place that was regarded as feminine and was appropriate for fertility cults. In writing about the Western Apache following discreet interviewing with tribal members, Basso (1996, 53–90) made worthwhile observations that could be widely applicable to prehistoric peoples and landscapes worldwide: 'Many place names evoke vivid images of places, and since these names were given by the ancestors as they explored and settled the land, they provide a path by which local people may reconstruct, imagine, and draw meaning from the past.' It is the same with symbolism. Symbolic thought is the ability, indeed knack, to use one thing to represent another. It originated in the Palaeolithic and continued as a prime component of intellectual, realistic and opportunistic imagery in Neolithic and Bronze Age landscapes: hence the suggestive, meaningful imagery evoked by the long pink fissures at Su Cannisoni.

Figure 13.6: The east-facing rockshelter, Su Cannisoni and the location of the Is Bittuleris Cave.

Discussion

Examination of the 15m long Grutta I de Longu Fresu in the mountains of the Barbagia di Seulo revealed natural features and placed deposits which have features in common with aspects of Neolithic constructions, like chambered long barrows, elsewhere in the world.

To begin with, the *position* of the Neolithic skull – at the far end of the cave, on the cave floor – is the same as the position of the skull found at the distal end of the Early Neolithic long barrow at West Kennet, Avebury, in southern England (Thurnam 1861, 416). This megalithic monument is a specially-designed tomb that seems as if it was a purpose-built artificial cave. Similarities of morphology between this tomb and Longu Fresu and the locations and types of placed deposits reward consideration. The two skulls were of young people. Many examples of megalithic cave-like constructions are known from Eurasia and even from more distant continents. For instance from Colombia, in the sixth century BC, in the Magdalena River valley and in the Department of Huila, there are monumental burial tombs of galleried barrow type, each with a statue of the Central American Jaguar god at the entrance (Dolmatoff 1972). Such transworld similarities of construction exemplify the principle of *convergence*. At great distances, on different continents, in different ages, architecture may follow a similar

Figure 13.7: One of two pinkish natural clefts beneath the overhang, Su Cannisoni.

evolution without any direct contact between the regions and cultures. This may be because the root concept developed from a common principle involving the human psyche in which dark moist caves were identified as if they were nature's womb within the Earth's body – in turn recognized as the 'feminine divine' in symbological terms (Cyriax 1921; Fowler and Blackwell 1999; see also Lewis-Williams 2002).

Another feature of the cave of Longu Fresu is that on the adjoining eastern wall is the painted art, described above. Similarly placed in the terminal cell of West Kennet long barrow at the far end of the gallery is fine sculptural art: a left-facing head that encompasses the entire megalith (Meaden 1999, 102–104).

At the foot of the north-western recess at Longu Fresu a human cranium was also found. Likewise, all the sub-chambers or side-cells at West Kennet held human bones and skulls (Piggott 1962; see Bayliss *et al.* 2007 for dating).

Sardinian galleried long barrows known as *tombe di giganti* also have similarities with long barrows from Wessex and *allées couvertes* from France. Several in Sardinia have a standing stone surviving at the entrance in the form of a tall carved female icon. At West Kennet the middle stone of the great façade is carved with a vulva over two metres in length (Meaden 1999, 107). This feature plainly heightens the feminine symbology of the entire monument. It is the much the same at the Neolithic 'Table des Marchands' in Brittany (Meaden 1997, 31).

A cross-cultural comparison between the chambered long barrows of Wessex, France, Corsica and Sardinia is appropriate. They can be viewed as artificial caves that could be repeatedly entered, just as real caves can. Most of them face between north-east and south-east. For West Kennet the planners arranged that the sun could penetrate to the end gallery in the weeks of the March and September equinoxes. Only then would sunlight illuminate the end chamber, as it still does today, where the child's skull was placed between the two orthostats that form the back wall. Thurnam, the finder, explained that it was 'the chief part of the skull of an infant about a year old' and lay with flint flakes and a heap of sherds (Thurnam 1861, 416, quoted by Piggott 1962, 7). One may also compare the side chambers of the Neolithic West Kennet barrow with the 'feminine' recesses of the Longu Fresu cave, and suggest that the West Kennet side chambers may be ancestor shrines on account of the early date of the human bones there (around 3800–3700 BC for West Kennet) if there was parallel Early Neolithic thinking in the two regions. Furthermore, the huge quantity of human bones in the small cave of 'Sa Omu 'e Is Ossus' speaks for placement and preservation within a womblike environment, while within the cliff face beneath this site, at Su Cannisoni, are located the two pink fissures described above.

The idea that long barrows duplicated the Earth Mother's womb had been recognized early in British archaeology by Cyriax (1921). The concept appears to be universal, and accords with known beliefs of today's Asian Indians whose Hindu temples are womb-temples of the Earth Mother. It is the same with the native Indians of America. The link between ancestors and Neolithic funerary monuments is thought of in terms of rebirth in the womb-tomb.

Despite the intervening distance, the farming communities who built this and other long barrows in Britain (and continental Europe) and the communities who used caves and *tombe di giganti* in Sardinia share a common, culturally-defined, time zone applicable to the Neolithic. Whatever purposes particular Neolithic monuments may have served, they and their landscapes were viewed as having symbolic, mystical and ancestral roles. Such philosophies and procedures augment a society's sense of history and affinity for the ancestral past in providing a stage for social, spiritual and ceremonial functions. By ritualizing caves and rockshelters in the landscape and forging an intimate relationship between people and shrines, the ancient peoples created a special and memorable native tradition. Their aim, as everywhere worldwide, was to improve their chances for prosperity, which they did by seeking to allay their natural concerns over fertility by devising gods and spirits and calling on divine aid for support.

Concluding remarks

Some of the numerous caves of the carbonitic plateaux of the Seulo area seem to have been used as burial places for the dead in Neolithic times, besides acting as shrines possibly dedicated to the Earth Mother. Sometimes the dead were buried and covered with stones. At other times bones lay on the cave floor and in one remarkable case some bones were confined to a 'stalagmite or flowstone natural coffin'. Other burials in the Seulo region have been found in secondary chambers and small caves. Those studied date from Neolithic to Bronze Age times, although the practice may have continued into later periods.

The work of researching the newly-found Seulo-region caves – especially ones with undisturbed archaeology – may aid an interpretation of the influence of prehistoric culture and sites generally in the region. In addition, through a study of bones and funerary practices we hope to understand better cultural rites and beliefs. This will hopefully permit comparisons with results known from elsewhere, and to uncover parallels and differences as to a tribal social past for peoples living in a close symbiotic relationship with the natural environment; and to learn whether particular caves were temporary shelters, winter abodes, homes, or ritual places.

To conclude, an arguably sacred landscape for the Neolithic and Bronze Age in the Barbagia di Seulo has been rediscovered. Partly known to the local inhabitants, it was there 'all the time' but had not previously been archaeologically detected among the prehistoric palimpsest of the mountainous, wooded landscape (Gradoli *et al.* 2006). Marcel Proust (1871–1922) observed that 'The real voyage of discovery lies not in seeking new landscapes, but in having new eyes'.

References

Albarella, U., Tagliacozzo, A., Dobney, K. and Rowley-Conwy, P. (2006) Pig hunting and husbandry in prehistoric Italy: a contribution to the domestication debate. *Proceedings of the Prehistoric Society* 72, 193–227.

Alderferfen, M. (1990) Late Preceramic ceremonial architecture at Asana, southern Perù. *Antiquity* 64, 479–493.

Antona, A. (1998) Le statuette di 'Dea Madre' nei contesti prenuragica: alcune considerazioni. In M. S. Balmuth and R. H. Tykot (eds.) *Sardinian and Aegean Chronology: Towards the Resolution of Relative and Absolute Dating in the Mediterranean*, 111–119. Studies in Sardinian Archaeology 5. Oxford, Oxbow Books.

Atzeni, E. (1978) La Dea Madre nelle culture prenuragiche. *Studi Sardi* 22, 1–51.

Basoli, P. (1992) Dipinti preistorici nel riparodi Luzzans (Ozieri, Sassari). In *Atti della XXVIII Riunione Scientifica dell' Istituto Italiano di Preistoria e Protostoria*, 495–506. Firenze, Istituto Italiano di Preistoria e Protostoria.

Basso, K. (1996a) *Wisdom Sits in Places: Landscape and Language among the Western Apache*. Walnut Creek (CA), Left Coast Press.

Basso, K. H. (1996b) Wisdom sites in places: notes on a western Apache landscape. In S. Feld and K. H. Basso (eds.), *Senses of Places*, 53–90. Santa Fe, SAR Press.

Bayliss, A., Whittle, A. and Wysocki, M. (2007) Talking about my generation: the date of the West Kennet long barrow. *Cambridge Archaeological Journal* 17/1, 85–101.

Bradley, R. (1991) Rock art and perception of the landscape. *Cambridge Archaeological Journal* 1, 77–101.

Bradley, R. (1997) *Rock Art and the Prehistory of Atlantic Europe*. London, Routledge.

Bradley, R. (2000) *An Archaeology of Natural Places*. London, Routledge.

Campbell, J. (1959) *Primitive Mythology: The Masks of God*. New York, Viking Press.

Carmichael, D. L., Hubert J., Reeves, B. and Schanche, A. eds. (1994) *Sacred Sites, Sacred Places*. London, Routledge.

Cyriax, T. (1921) Ancient burial places: a suggestion. *Antiquaries Journal* 28, 205–215.

D'Arragon, B. (1995) Comparisons of engraved anthropomorphic figures in the Alps and in Sardinia. In *North East West South 1995 International Rock Art Congress Proceedings. Symposium 12D: Rock Art and the Mediterranean Sea*. CD. Pinerolo: Centro Studi e Museo d'Arte Preistorica di Pinerolo

D'Arragon, B. (1997) Nuovo figure schematiche antropomorfe dalla Sardegna prenuraghica: le pitture rupestri della Grotta del Papa, Isola di Tavolara (SS–I). *Tracce* 9. <http://www.rupestre.net/tracce/TAVOLAR.htm>, accessed 4 April 2010.

Dolmatoff, G. R. (1972) San Agustin. <www.ancient-wisdom.co.uk/colombiasanagustin.htm>, accessed 16 April 2010.

Eliade, M. (1957) *The Sacred and the Profane*. San Diego (CA), Harcourt Brace Jovanovich.

Fowler, P. and Blackwell, I. (1999) *The Land of Lettice Sweetapple*. Stroud, Tempus.

Frazer, G. (1911) *The Golden Bough: A Study in Myth and Religion*. London, Macmillan.

Gheorghiu, D. ed. (2002) *New Approaches to the Archaeology of Art, Religion and Folklore*. British Archaeological Reports International Series 1020. Oxford, Archaeopress.

Gimbutas, M. (1989) *The Language of the Goddess*. London, Thames and Hudson.

Gimbutas, M. (1991) *The Civilisation of the Goddess*. San Francisco, Harper and Row.

Gosden, C. and Lock, G. (1998) Prehistoric histories. *World Archaeology* 30/1, 2–12.

Gradoli M. G., Dimitriadis, G. and Delogu, G. (2006) Il binomiouomo-territorio nella Barbagia di Seulo: prime segnalazioni dipitture parietali in grotte carsiche. In A. Guerci, S. Consigliere and S. Castagno (eds.) *Atti del XVI Congresso degli Antropologi Italiani (Genova, 29–31 ottobre 2005)*, 521–530. Milano, Edicolors Publishing.

Gradoli, G. and Meaden, G. T. (in prep.) Neolithic sacred sites in the mountains of Central Sardinia. In T. Meaden (ed.) *Archaeology of Mother Earth Sites and Sanctuaries through the Ages: Rethinking Symbols and Images, Art and Artefacts from History and Prehistory*. British Archaeological Reports. Oxford, Archaeopress.

Karetsou, A. (1981) The peak sanctuary at Mount Juktas. In R. Hägg and N. Marinatos (eds.) *Sanctuaries and Cults in the Aegean Bronze Age, Proceedings of the First International Symposium at the Swedish Institute in Athens, 12–13 May 1980*, 137–153. Stockholm, P. Astrom Forlag.

Lewis-Williams, D. J. (2002) *The Mind in the Cave: Consciousness And The Origins Of Art*. Thames and Hudson, London.

Lilliu, G. (1999) *Arte e Religione della Sardegna Prenuragica: Idoletti, Ceramiche, Oggetti d'Ornamento*. Sassari, Carlo Defino editore.

Lilliu, G. (2003) *La Cività dei Sardi dal Paleolitico al l'Età dei Nuraghi*. 3rd edition. Torino, Nuova ERI..

Lo Schiavo, F. (1978a) Figure antropomorfe nella Grotta del Bue Marino: Cala Gonone (Dorgali). In *Sardegna Centro-Orientale dal Neolitico alla Fine del Mondo Antico*, 53–55. Museo Civico Speleo-Archeologico. Mostra in occasione della XXII Riunione Scientifica dell'Istituto Italiano di Preistoria e Protostoria. Sassari, Soprintendenza Archeologica per le Provincie di Sassari e Nuoro.

Lo Schiavo, F. (1978b) Figurazioni antropomorfe nella grotta del Bue Marino: Cala Gonone (Dorgàli). In *Sardegna Centro–*

Orientale dal Neoliticoalla fine del Mondo Antico. Nuoro – Museo Civico Speleo-Archeologico. Mostra in Occasionedella XXII Riunione Scientifica dell'Istituto Italiano di Preistoria e Protostoria, 53–55. Sassari, Soprintendenza ai Beni Archeologici per le Provinciedi Sassari e Nuoro.

Lo Schiavo, F. (1980) La grotta del Bue Marino a Cala Gonone. In *Dorgali: Documenti Archeologici*, 39–45. Sassari, Chiarella.

McEwan, C. and Van de Guchte, M. (1992) Ancestral time and sacred space in Inca state ritual. In R. F. Townsend (ed.) *The Ancient Americas: Art from Sacred Landscapes*, 359–371. Chicago, Art Institute of Chicago.

Meaden, G. T. (1997) *Stonehenge: The Secret of the Solstice*. Second edition. London, Souvenir Press.

Meaden, G. T. (1999) *Secrets of the Avebury Stones*. London, Souvenir Press.

Meozzi, D. (2010) Bonorva chequered tomb (Sa Pala Larga Tomb **no.7)** <http://www.stonepages.com/scacchiera/>, accessed 25th April 2010.

Moravetti, A. (1980) Riparo sotto roccia con petroglifi in località Frattale (Oliena-Nuoro). In *Atti della XXII Riunione Scientifica nella Sardegna Centro-Settentrionale dell'Istituto Italiano di Preistoria e Protostoria*, 199–226. Firenze, Istituto Italiano di Preistoria e Protostoria .

Piggott, S. (1962) *West Kennet Long Barrow: Excavations 1955–56*. London, H. M. Stationery Office.

Rutkowski, B. (1986) *The Cult Places of the Aegean*. New Haven, Yale University Press.

Skeates, R. (this volume) Caves in need of context: prehistoric Sardinia.

Stefano, R. F. (2010) Protometaurina con spirali. <http://www.sardegnadigitallibrary.it/index.php?xsl=626&id=29469>, accessed 25 April 2010.

Taçon, P. S. C. (1999) Identifying ancient sacred landscapes in Australia: from physical to social. In W. Ashmore and A. B. Knapp (eds.) *Archaeologies of Landscape. Contemporary Perspective*, 33–57. Oxford, Blackwell Publishers Ltd.

Tanda, G. (1998) Cronologia dell'arte delle domus de janas. In M. S. Balmuth and R. H. Tykot (eds.) *Sardinian and Aegean Chronology: Towards the Resolution of Relative and Absolute Dating in the Mediterranean*, 121–139. Studies in Sardinian Archaeology V. Oxford, Oxbow Books.

Tanda, G. (2004) Dalla preistoria alle storie. In M. Brigaglia (ed.) *Storia della Sardegna*, 25–74. Cagliari, Editori Laterza.

Thurnam, J. (1861) Examination of a chambered long barrow at West Kennet, Wiltshire. *Archaeologia* 38, 405–421.

Tykot, R. H. (1994) Radiocarbon dating and absolute chronology in Sardinia and Corsica. In R. Skeates and R. Whitehouse (eds.) *Radiocarbon Dating and Italian Prehistory*, 115–145. Accordia Specialist Studies on Italy 3. London, Accordia Research Centre and The British School at Rome.

Ucko P. J. (1994) Foreword. In D. L. Carmichael, J. Hubert, J. B. Reeves and A. Schanche (eds.) *Sacred Sites, Sacred Places,* One World Archaeology 23, xiii–xxiii. London, Routledge.

Ugas, G. (1998) Considerazioni sulle sequenze culturali e cronologi chetral'Eneolitico e l'epoca Nuragica. In M. S. Balmuth and R. H. Tykot (eds.) *Sardinian and Aegean Chronology: Towards the Resolution of Relative and Absolute Dating in the Mediterranean*, 251–272. Studies in Sardinian Archaeology V. Oxford, Oxbow Books.

Van Dyke, R. M. (2008) Memory, place and the memorialization of landscape. In D. Bruno and J. Thomas (eds.) *Handbook of Landscape Archaeology*, 277–284. World Archaeological Congress Research Handbooks in Archaeology. Walnut Creek (CA), Left Coast Press.

Chapter 14

Notes from the Underground: Caves and People in the Mesolithic and Neolithic Karst

Dimitrij Mlekuž

Caves are not only unique sedimentary environments with good preservation of archaeological material, but as the archaeological record from caves testifies – also special places where distinct activities were performed. What makes caves special? What makes them different from open air locales? How do caves act back on humans? How do humans and caves mutually constitute each other and create a sense of self and belonging in the world? This chapter touches upon these themes using examples from the archaeological record of the Karst in northeast Italy and western Slovenia. By exploring the 'affordances' that caves provide we can focus on the social and contextual role they played in the practical tasks of past people. Caves are not passive backdrops for the activities that people perform, they are not natural places, and they do not satisfy the generic needs of people such as 'shelter'. We can understand caves as material culture where dwelling occurs. And, by focusing on the process of dwelling that they enable through the affordances they provide, they help us to challenge the dichotomies of the natural and built environment, or of the mundane

Introduction

Caves and rockshelters are special places in a landscape, which have often survived and are still visible today. As such they tend to be vastly over-represented in an archaeological record. The fact that they were used and regularly visited by past people signifies that they saw them as special and important places. Thus caves are not only unique sedimentary environments with good preservation of archaeological material, but also places where distinct activities were performed. In this chapter, I will focus on the following questions: What makes caves special? What makes them different to open air locales? How do caves act back on humans? How do humans and caves mutually constitute each other and create a sense of self and belonging in the world? I will explore these themes using examples from the archaeological record of the Karst in northeast Italy and western Slovenia. The Karst plateau (Kras in Slovenian, Karst in German and Carso in Italian) is a limestone landscape that rises above the Trieste bay (in the Adriatic Sea). It comprises the northwesternmost

tip of the Dinaric mountain range that extends along the East Adriatic coast. The area is covered by large dissolutional dolines and other classical karst features set in a landscape of broken rocks, patchy grass cover and stands of woodland. Although the region experiences heavy rainfall, there is a general lack of surface water. The porous limestone quickly absorbs water into crack and fissures, draining the surface. Soils – except in depressions – are thin and leached (terra rossa) and as a result of millennia of overuse some parts of the landscape are virtually barren. The area is pockmarked by caves and rockshelters (Fig. 14.1).

The Mesolithic and Neolithic archaeological record of the Karst consists almost exclusively of cave and rockshelter sites (Fig. 14.1). They are usually 'deep', with thick Holocene sedimentational sequences and long occupational histories, often extending back into the Early Mesolithic. For the Mesolithic (broadly between 9500–6000 cal BC), they are conventionally interpreted as temporary hunting camps for mobile hunter-gatherers, although we know little about open

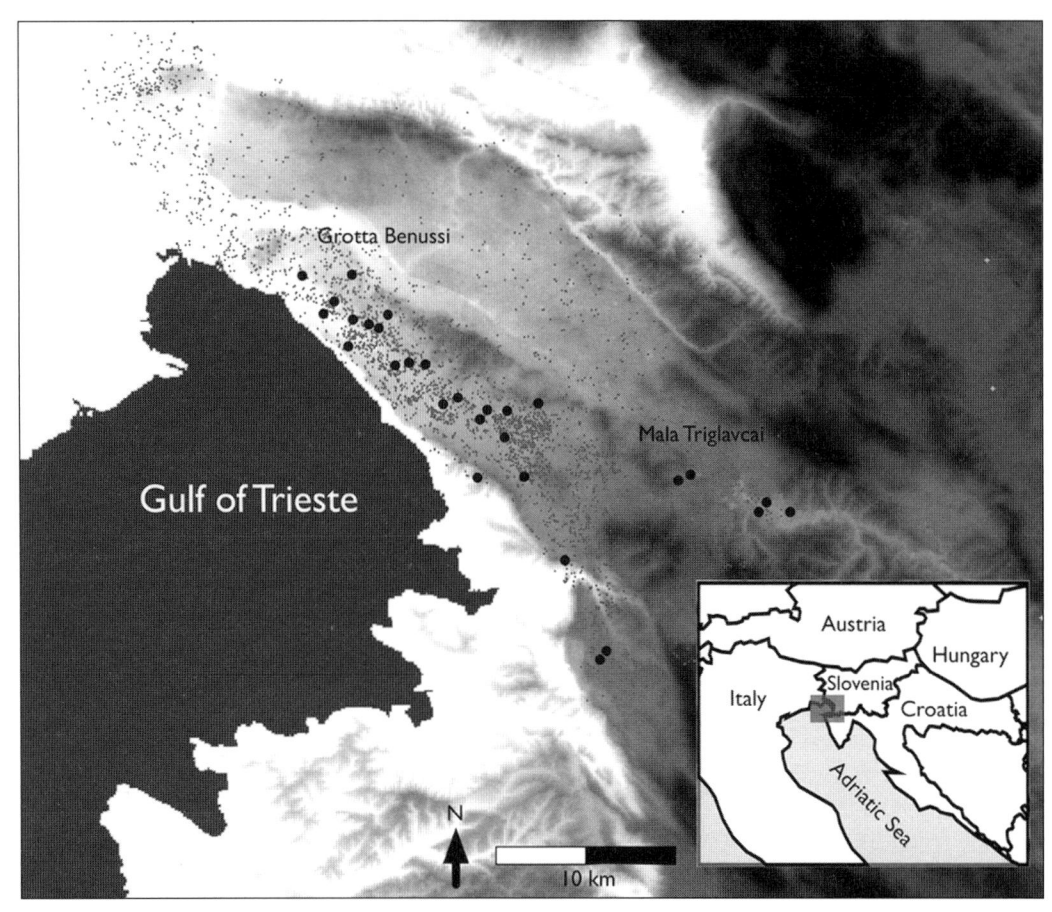

Figure 14.1: Karst. Limestone plateau overlooking the Trieste Gulf. The landscape is pockmarked with caves and rockshelters, many of them containing traces of prehistoric occupation.

air sites and other special places (Cremonesi 1981; Boschian and Montagnari Kokelj 1984; Biagi 1994). In the Neolithic (approximately 5500–3500 cal BC), pottery of the so-called Vlaška group and animal bones, the majority of which are ovicaprines, appear in the caves. This marks a new use of caves. Archaeological, geoarchaeological and archaeozoological data suggest that they were used as sheep pens for large flocks of ovicaprines (Boschian and Montagnari Kokelj 2000; Mlekuž 2005; Boschian 2006). In the Neolithic we have evidence of short, seasonal visits to the caves. The complementary seasonal patterns may suggest that cave sites in the Neolithic were not merely outstations of a larger pastoral system, with central sites elsewhere, but that they comprised a full yearly cycle of seasonal mobility. Thus we might see the Karst pastoralists as nomads moving from cave to cave (Mlekuž 2005). These can be interpreted as seasonal hunting or herding camps. These practices – with minor changes in intensity and scale – continued into the Bronze Age (Mlekuž 2007b).

Affordances

To approach the importance of caves in the lives of past people I shall use the concept of 'affordances', defined by the American environmental psychologist James Gibson (Gibson 1986; see also Ingold 1993). Gibson's 'direct perception' is direct in the sense that perception is not a computational activity of a mind within a body but an exploratory activity of an organism with its environment. From this perspective the environment is not a latent set of resources awaiting human exploitation, but part of the practice of dwelling in the world. Affordances can thus be defined as 'properties of the real environment as directly perceived by an agent in the context of practical action' (Ingold 1993, 64). But how does this relate to caves and rockshelters?

Mobile individuals move across the landscape during their daily schedule. As they move they perceive affordances along the tracks they walk. The landscape provides series of affordances, which can be exploited through choosing one locale instead of another. An encounter with affordance will lead to decisions about immediate and future actions.

For example, to a group of hunters an encounter with a cave can provide an affordance of shelter against rain or wind. To a herd of sheep it can offer shelter against the scorching midday sun and allow a herder to take a nap in its shade. During the night it can provide affordance to enclose the herd and protect sheep against predators. To a group that has lost its member a cave can provide an affordance of passage to the underworld, where they can deposit a dead body. The stones on the cave floor provide an affordance to build a stone wall in the entrance and to protect the dead body, with the stone wall denying predators the affordance to gnaw on the dead body. To a shaman the cave affords a literal entrance to the underworld, and experience of sensory deprivation and altered states of consciousness. The affordances are contextual, and relationally specific to individuals, rather than generic properties of the environment (Webster 1999). Every task people perform 'has its own thickness and temporal spread' (Gell 1992, 223). It makes sense only when related to those that have already been performed and those to be performed. Life is, thus, not just a succession of isolated tasks; it is a flow of tasks meaningfully related to one another. Tasks are implicitly or explicitly connected with another task, separated in time and space. Each task is made possible by a number of past tasks and future tasks that give it purpose. This network or 'referential system' (Gosden 1994, 15–18) of tasks unfolds over space and time. Things and substances move between tasks, carried as raw material, tools or as portable and attached affordance. Their transfer is often embedded in other activities such as hunting, visiting and herding. In this way a locale can become connected in a web of relations, flows and paths. The tasks that people perform and that involve affordances of caves are part of everyday life, which goes on elsewhere. Activities in a cave are always implicitly or explicitly connected with activities elsewhere, outside, in other locations, in other caves and in the landscape. Caves are part of landscape as their affordances are part of people's social life in the landscape. These tasks, movements and flows become habitual: part of the bulk of everyday social life. As people and things regularly stop at those special places in the landscape they become incorporated in caves and caves become embodied in persons inhabiting them, and this can provide additional affordances.

Caves

It is difficult to define what a cave is. White and Culver (2005, 81), for example, define a caves as 'a natural opening in the Earth, large enough to admit a human being, and which some human beings choose to call a cave.' This definition is close to my understanding of caves, as affordances of landscape. Caves provide affordances that other places in the landscape do not. Caves appear in many forms and shapes. They can be shallow rockshelters, deep narrow fissures or large underground cavities. Caves, as habitats, have certain environmental properties that distinguish them from the surrounding landscape. They tend to be cooler in the summer and warmer in the winter. In deep caves, the relatively limited penetration of sunlight creates zone of twilight and zones of darkness. However, the distinction between open air locale and rockshelter is sometimes blurred. An overhang can provide some of the affordances of the cave as well as the affordances of an open air locale.

The properties of a particular cave represent the affordances that can be used in the context of activities performed in a landscape and in the cave. Thus, properties such as size, access, openness, and light become important in understanding, which affordances caves provide in the context of practical tasks and actions. How many people or animals can fit in a cave? How well is it protected from the elements? How easily can it be accessed? How wide is the entrance? What kind of floor does it have? How dark is it in the cave?

Caves are not just passive backdrops to the activities that people perform; by providing the affordances to people who engage with them, they also act back. People and the material world are always conjoined in actions, and there is mutual constitution between people, things and places (Miller 1987; Knappett 2005; Latour, 2005). Through the performance of tasks, things, places and bodies are changed, and through this mutual constitution people are also changed. Tasks leave traces on matter, tools, places and bodies. Through repetition those traces accrete or layer one upon another. Through layering – a process of creating sediments, assemblages of traces that accrue over time, repair, adapt, modify or curate – life histories become sedimented and layered and biographies of objects, bodies and places are created (Gosden 1994; Knappett 2006). Things and places change, people become more skilled and older after each performed task, each day, and each seasonal course. Their bodies accumulate traces, skills, knowledge of how to perform body movements, gestures and postures that in turn constitute human beings. Rhythms of the daily or yearly engagement with the world are thus 'techniques of the self' (Foucault 1988; Warnier 2001): a way through which people constitute themselves, maintain or create their identities. There are many ways that caves can affect people and their bodies. One of them is through the affordance of bounding, amassing, cramping, and limiting movement.

How do caves act back on people?

The pioneering work of Edward Hall (1966) in the field of 'proxemics' emphasized the role of 'interpersonal distance' in the quality of social relations between people. Interpersonal distance can be conceptualized as a kind of non-verbal communication. It is not only a reflection of on-going relations between persons but can play an active role; by negotiating and adjusting the distance, people can maintain or change the quality of their interpersonal relations. Hall, as a cultural anthropologist, was interested in cultural frameworks that define and organize space; and, from a cross-cultural study of space-maintaining strategies, he outlined a typology of 'zones or spaces of interpersonal distance'.

Conceptualized as nested bubbles that surround persons, Hall defined several informal spaces based on types of sensory information available to the persons involved, like loudness of speech, olfactory cues, and body heat. The most intimate and the closest is intimate space, where involvement of the other person is unmistakable and characterized by strong and intense sensory inputs. The voice is usually held low or even to a whisper. Personal space is characterized by normal speech; minimally perceived olfactory inputs and extends approximately at arm's length around the person. Entry into this space is acceptable only for closest friends and acquaintances. Social and consultative spaces are the spaces in which people feel comfortable conducting routine social interactions with acquaintances as well as strangers. Public space is defined as the distance beyond, in which people perceive interactions as impersonal and relatively anonymous. However, this typology offers a rather static approach to interpersonal distance, and it might be more useful to conceptualize Hall's spaces as a continuum, as proposed by several authors (e.g. Aiello and Thompson 1980).

Studies in environmental psychology suggest some interesting physical determinants of interpersonal spacing. People keep more distance between themselves when indoors than when outdoors and personal space increases with reduction in room size (Cochran *et al.* 1984). Personal distance increases in darkness (Adams and Zuckerman 1991). Males have more need for personal space when ceiling height is low (Cochran and Urbanczyk 1982). Persons desire more space in a narrow room and persons exhibit more personal space in corners than in the centre of a room, and maintain closer distance when standing rather than while seated (Evans *et al.* 1996).

Thus the material world plays an important role in determining the quality of relations between people, which is expressed through interpersonal distance. Because interpersonal distance is tacit or habitual, persons usually become aware of the boundaries only when they are violated. Several studies have shown that, when an environmental setting forces people to interact in an inappropriate spatial zone, changes appear in body signs, such heart pulse-rate, and provoke feelings of discomfort, stress, threat, aggression, or fear. On the other hand, it can also provoke a desire for closer contact and intimacy. People tend to touch more – as with closeness in the dark (Andersen and Sull 1985).

People usually actively maintain their personal space. Physical setting of caves has therefore important consequences on interpersonal distance and on the quality and quantity of communication between persons. Enforced close interpersonal distance can lead to stronger responses than interaction at an appropriate distance. Caves which afford confined spaces, narrow passages, low ceilings and darkness can be places of more intense sociality. In this way they have an agency and act back on people.

Being in the cave

Objects and places evoke experiences. By moving along paths, and staying in familiar places, people evoke and create memories and weave stories, and create a sense of self and belonging. In this way, they embody places, creating a particular way of being-in-the world (Richardson 2004). Thus, for example, Miles Richardson (1982) in his ethnographic description of Cartago, Costa Rica, recounts how the experience of being in the plaza evokes the 'cultura', an appropriate and socially correct behaviour, while the experience of being in the market evokes behaviour connected with 'listo', smart, clever behaviour. Places become embodied in cultural forms. The material world, thus, has a crucial role to play in the production of people and in fixing relations between them.

The bodily process of being in a cave – moving through it, multi-sensuous engagement with its materiality and with other people in the cave – would have served as a focus for memory. The agency of a cave, in part, resides in its capacity to serve as a mnemonic device. Memories become embodied; places, meanings and memories are intertwined and create a 'sense of place' that rests on a social engagement with the landscape and is bound up with remembrance and time. Caves, thus, have their own agency. They act back on people who visit them. They bind, they separate, they direct movement, they cram people together, they lead into the underground, they hide from the sunshine, they seclude, they impose fear and a sense of discomfort, and they heighten the senses. They are material and real, their material presence – their affordances – can act as a foundation for interaction with us. They may profoundly affect the way we move, act, feel and relate to others. Thus the bodily experience of 'being-in-the-cave' is different than being-in-the-house or being-in-the-open. People who return from the cave is different persons than those who entered it. They have embodied the cave through the bodily experience of being in it.

Caves and the built environment

Outdoor camps of hunter-gatherers are open, with marked flexibility in spatial organization and social affiliation: people can move freely, and establish physical and social distance at will. In fact, social relations are mediated and regulated the spatial relations between huts. Personal space is easy to negotiate in outdoor camps, people can move closer or further away to a comfortable distance to maintain and negotiate personal space (Turnbull 1965).

With the appearance of surface level structures ('houses') in the Neolithic of Southeastern Europe, the built environment provided new ways in which social relations between people were created, reproduced and challenged (Bailey 1999, 2000). The materiality of houses provided a new way to produce people and to fix the relations between them. The house is, at the same time, a social institution and an artefact (Borić 2008). By building architecture, people build their social relations in material form. Architecture is a material setting in which relations between people are reproduced.

By contrast, in caves, the sociality of people is influenced by the materiality of the cave itself. A cave imposes its own sociality through a series of affordances provided to the people who dwell in it. That said, by strictly opposing caves to architecture we fall into the old trap of the nature/culture dichotomy (Bradley 2000; Ingold 2000). The boundary between caves as 'natural places' and architecture as 'built environment' is actually blurred (Bradley 2000).

Both houses and caves are containers for human activities (Ingold 2000, 185). The house is a kind of artificial cave, as it provides bounded and covered space, in the same way that a cave does; and the cave, as bounded and covered space, can be used in the same way as a built structure. Furhtermore, some structures have been built to deliberately resemble caves. Passage and chamber tombs might simultaneously resemble houses and caves (Barnatt and Edmonds 2002; Bradley 1998). Maltese monumental temples, with narrow passages, chambers and niches might have been made to evoke the experience of being in a cave (Trump and Cilia 2002). David Lewis-Williams (2004) sees Çatalhöyük houses as built caves, where the entrance into the house was a descent into darkness. Movement between the rooms forced people to bend low and crawl, as in caves. The resemblance is further supported by the presence of broken of stalactites, brought from the Taurus Mountain caves and deposited in houses. Even the architectural details, such as columns, platforms and ladders were built to enhance the resemblance of houses to limestone caves.

On the other hand, caves can be modified to resemble houses. By building walls – which organize space – a cave can be made into a house or home (Leitch and Smith 1997). Maybe instead of looking at structures and places, we should focus on acts of engaging with them: dwelling. We live in the world by dwelling. Heidegger (2000), in his essay entitled 'Building, Dwelling, Thinking', focuses on activities of 'building' and how these belong to our dwelling in the world – the way we are. What people build arises from their activity, from the context of their engagement with the environment. Schmitz (1997) sees dwelling as the cultivation of feelings in an enclosed space, as a bodily experience of being in an enclosed space. Therefore, building boundaries is instrumental in the process of dwelling as it creates envelopes and enclosures where we dwell (cf. Zaborowski 2005). As Tim Ingold puts it, building is the process that is continually going on, it is a work in progress. It is in a process of dwelling that we build, and we dwell through building. Making a place to live, or a home, is therefore the process of 'building, engagement with the environment' (Ingold 2000, 185–186). Thus every empty house is a cave and every cave can be made into home, when we dwell in it, making (or 'building') it intimate, familiar and adaptable.

Dwelling is a process that affects our bodies; houses and caves 'act back' on people and play an active role in the creation of people and relations between them. What distinguishes caves from houses or built environment is the role that the materiality of caves plays in the process. The house emerges, or, better, is built from, the relations of the people who build it together and with their environment, and the shape of the house reflects those relations. On the other hand, the shape of cave plays an active role in the way that people interact with each other and the environment. So even if cave can be modified and built, it still plays an active in the process of modification and building.

People inhabit the world shaped by their predecessors. We are 'thrown' into the world. As Chris Gosden (1994, 77) puts it, a 'world created by people will be a world into which their children will be socialized'. In this sense there is no difference between a house and a cave. We may live in a house built by ancestors, or in a cave, where ancestors have dwelt. The fact that a cave is natural makes it no different from a house for the people who dwell in it. As houses, caves can be seen as the works of ancestors, and their successors may have felt a responsibility to look after those places, to harness their power or even to transform them into something else (Bradley 1998, 21).

Secluded places or entrances to the otherworld: Mesolithic use of Karst caves

The evidence of Mesolithic presence on the Karst in northeast Italy and western Slovenia is based almost exclusively on records from caves and rockshelters. Caves and rockshelters were continuously used during the Mesolithic, as evidenced by middens, with large quantities of animal bones, human remains, charred organic matter, land-snails, shells and artefacts. Large quantities of ash and charcoal suggest that wood was commonly burnt inside or near the caves, and the presence of dispersed organic material in deposits suggests a substantial flow of substances (food, wood, litter, etc.) from the landscape into the caves (Boschian 1997; Boschian and Montagnari Kokelj 2000). This indicates that they were used as gatherings for – probably more than few – people and as important places in the landscape. Such an interpretation contrasts with the traditional one of caves as the hunting camps of mobile hunter-gatherers (Boschian and Montagnari Kokelj 1984; Biagi 1994).

Although the archaeological record imposes significant constraints on interpretative possibilities, much interpretation also reflects the limitations of our theoretical approaches. I will try to illustrate this briefly with reference to the example of Grotta Benussi (Pejca na Sedlu in Slovenian), an important Mesolithic site with evidence of frequent occupation or visits during the Mesolithic (Riedel 1975; Andreolotti and Gerdol 1972). Grotta Benussi is a narrow cavity with a low ceiling (Fig. 14.2). After passing through the low and narrow entrance between the rocks, one is enfolded by the darkness that leads into the limestone hill. One might ask what can a narrow cavity such as this offered, particularly compared to an open air camp. Was it about night shelter for only a hunting party? Why was the cave regularly visited over a period of more than a thousand years?

Robert Brightman (1993) describes a ritual eat-all feast among the Rock Cree North American Indians, known as 'wihkohtowin', which symbolically involves humans in the reproduction of game animals. Feasting

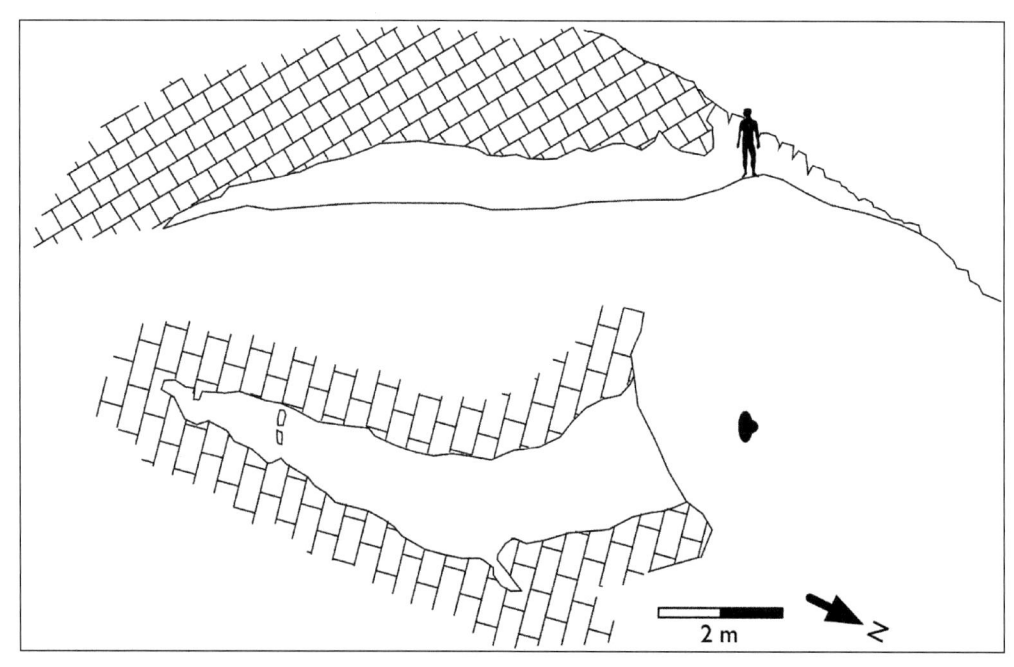

Figure 14.2: Grotta Benussi/Pejca na Sedlu.

is a manifestation of intense sociality between humans and between animals and humans, as reciprocity and communion. For the perspectives of both the hunters and the animals, the regeneration of life depends upon the maintenance of balance in the reciprocal give-and-take of vital forces. Ingestion of meat simultaneously feeds the spirits of animals and fills the human body with their essence. Animals give life to humans, and they return those gifts during ritual meals and by proper disposal of the bones. What is important is that the feast takes place indoors, in a sealed cabin. All flows of substances, people or even views, as well as the disposal and deposition of inedible remains, is strictly controlled in order not to offend the guardian spirits of the animals. There are important consequences of being inside and hidden from view. Crees conceive of themselves as subjects of surveillance reminiscent of the panopticon (Brightman 1993, 237).

Hunter-gatherers are engaged in social relations with non-human things and creatures. Animals and their guardian spirits are viewed as powerful and dangerous trading partners, and hunters sometimes think of themselves as locked in an advisory relationship with animals, who are conceived as powerful beings (Ingold 2000). Thus for hunter-gatherers, the environment is not just a passive backdrop, or a resource to be exploited, but a social arena shared with many sentient non-human beings. The enclosed space of a sealed cabin as in the case of Cree or a cave might provide affordance to commemorate and to commune, concealed from dangerous non-human creatures, during a potentially dangerous ritual.

There are also other possible routes of interpretation. David Lewis-Williams (2002) explores the role of caves in shamanistic contact with the other world. Caves can be seen as entrances to the underworld in a tiered animistic cosmos. The bodily and sensory experience of entering into the cave, and deeper into the darkness and underground, might be equated with the experience of flying, or passing, through the tunnel, vortex or cave that often appears in altered states of consciousness. As a result of this identification of the vortex leading to the other world with caves and subterranean passages, these natural features might have been regarded as entrances into the world beneath the earth. On the one hand, access to a cavern could therefore be seen as an entrance to the world of spirits and magical potency. And, on the other hand, dark caves, with their affordance of sensory deprivation, can be instrumental in attaining various altered states of consciousness, ranging from intense contemplation to visions, hallucinations, and out-of-body experiences. The experience of flow through a tunnel is often accompanied with visions of encounters with animals, which come from the vortex edge. Thus, by analogy, the cave wall can be seen as a permeable interface between people in the cave, in a lower level of the cosmos, and a spirit world that lay behind the walls.

However, even if the people who enter a cave do not seek experiences of altered states of conscious, as these are often reserved for shamans, caves still provide affordance to enter the liminal world beyond everyday experience. A cave, with its permeable walls, also provides affordance for the proper deposition of animal remains, where they can be returned to their guardian spirits.

Participation in communal gatherings in caves, through enforced interpersonal distance, darkness, unnatural postures, and a heightened sense of fear, thus affects the participants in powerful ways. This experience of entering a cave and returning back can be seen the experience of rites of passage (van Gennep 1960). Caves provide a setting for the three-fold process which comprises crossing the threshold, from outside to inside, life-changing trauma in the unfamiliar world, and ultimately return to society in a profoundly different status (cf. Whitehouse 2001).

I do not want to argue that Grotta Benussi or all caves were used in this way, but I want to argue that hunter gatherers' engagement with the environment involved more than just the exploitation of resources. Caves provided more than just shelter: they were places where the powerful experience of being in touch with the other world was felt.

Containers for animals, people and substances: Neolithic use of Karst caves

'Places gather' (Casey 1996, 24), but caves also hold, amass, contain and store. Caves and rockshelters provide the affordance of containment. They provide a physical envelope for a setting which separate outside from inside, and is excluded from the rest of the landscape. They can crowd people, animals, things and substances together, mix them, and hide them from view (cf. Warnier 2006).

There is plenty of evidence that Neolithic caves and rockshelters became containers for people, animals, things and substances (Boschian and Montagnari Kokelj 2000; Miracle and Forenbaher 2005; Mlekuž 2005). Caves became seasonal camps and pens for mobile herders and their flocks. However, caves were not only used as sheep pens but also as habitation places. The relative frequency of different body parts shows that ovicaprines were culled, processed and eaten on site. Deposition rates of bones are generally low and can be compared, for example, to the deposition rate of a single Navaho cohabitation group, suggesting that group size was small (Mlekuž 2005). Caves were regularly used both for penning animals and by camping pastoralists.

People and animals, each with their specific smells, sounds, food, and personal space, were kept in a same envelope or container of a cave. Thus the sociality between animals and humans in a cave was much denser than outside, in the open landscape.

Containment is a technology of power; power rests on agency to act directly upon the subjects or to make the subjects to act upon themselves (Warnier 2006). All these actions rest upon technology and include material culture such as fences, barriers and blockades. Containers in the form of corrals, fences and pens are the main material culture used by herders to control animals (Ingold 1980; Cribb 1991). Caves can be seen as a form of material culture associated with containment, often improved with fences or drystone walls that control and guide the acts of entering and leaving the enclosure of the cave. In this way the material culture of containment – the caves themselves – become embodied in persons through sensory motor conducts associated with containment, such as entering, leaving, maintaining the limits, forming a queue, and preventing the transit of substances (Warnier 2006). There is evidence from the Karst that caves were modified to be more effective containers. Numerous stone walls can be encountered in front of the caves and at least in one case (Grotta dell'Orso/Pečina pod Muzarji) it can be confirmed that the wall was built in prehistory (Guacci 1959). Karl Moser encountered the remains of a wattle fence in Grotta Moser/Pejca na Doleh (Moser 1903; Barfield 1972, 201). In the Mala Triglavca cave (Fig. 14.3), located at the edge of a minor doline, a drystone wall was built in front of the cave during in the Neolithic (as its stratigraphic position suggests). Containers have volumes and the volume of caves can be measured in terms of the number of sheep and people that can be enclosed. For example, in Mala Triglavca a flock of 120 sheep can fit with enough space for several people to sleep or perform daily activities. However, this would constitute rather cramped setting.

Containers of animals

Thus containment of live animals in caves marks different relations between humans and animals, relations which Tim Ingold (2000, 61–76) describe as 'domination'. Animals in the pastoral mode of production become a means of reproducing the social relations of pastoral production. The slaughter of domestic animals frees people from the obligations of sharing that apply to the hunted animals only. Reproduction and the multiplication of domestic animals make possible the accumulation of wealth. Thus the effect of drawing on domestic herds leads to the social fragmentation of human groups into autonomous, self-sufficient domestic units, where animals are members of the household and food resources at the same time (cf. Ingold 1980, 79–89). However, the incorporation of animals into the household was not always or only a tyrannical act of domination over hapless animals. The changes emerging from the incorporation of animals into the household are considerably more complex and contradictory, and include mastery, domination, objectification as well as care and nurture. Domestication practices brought humans and animals closer together in relations of not only control but also affinity, proximity and companionship. However, there would have been some unintended consequences of being contained in a cave together with animals.

Thus in the Karst caves (as well as in the eastern Adriatic hinterland), humans and sheep lived very close together, sharing living space, smells and sounds. The smell of smoke and cooking mixed with the smell of dung and sheep, people and animals, attended people and animals in their mundane tasks. There is evidence of both human and sheep milk

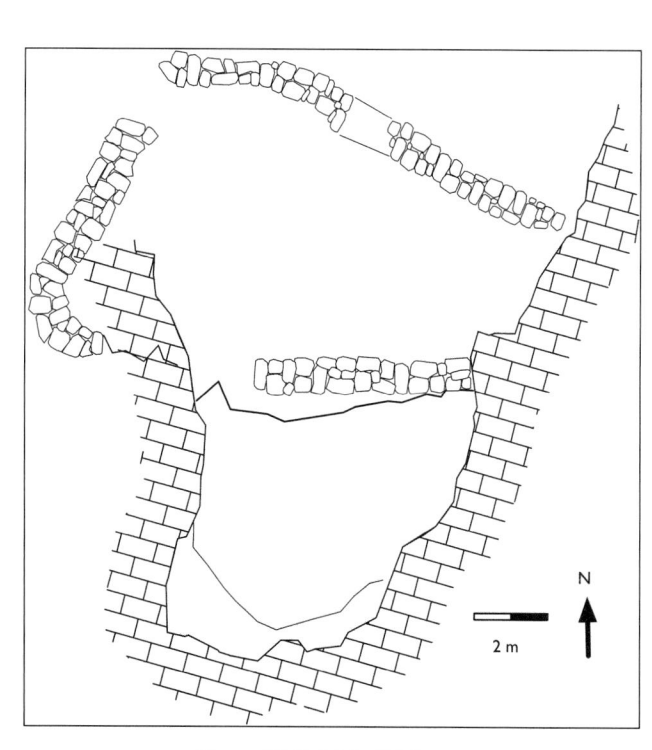

Figure 14.3: Mala Triglavca.

teeth shed on the sites (Štamfelj *et al.* 2004), and we can imagine children and lambs playing together or human children sucking milk directly from a sheep's udder.

I want to argue that this close contact and intimacy created more than just an association of animals and humans. It created a new kind of society, composed of sheep, goats and humans, interlocked together through the business of everyday life. All species involved were profoundly changed through their everyday contacts and interactions. Close everyday contact, mediated by the materiality of the cave, provided an opportunity for intimate and close contact between humans and animals. Sheep are gregarious, social animals: during socialization they establish a social order, they can recognize individual ovine faces and even human faces and remember them for years. When living close together with other species for prolonged periods they tend to bond, or create social links (Fisher and Matthews 2001; Estevez *et al.* 2007) – a feature that modern herders make use of when they socialize sheep dogs into herds. Through bonding with people (and other species) humans became incorporated within animal social organization and animals became part of the power and social relations of human households. A new hybrid society emerged, consisting of humans and non-humans alike. This new set of relations between people and animals brought about a different use of caves, which in turn influenced relations between people and animals. Caves as a material culture and as special places in a landscape thus played an active role in changing relationships between people and animals during the Neolithic.

Containers of substances

Through the process of everyday life people and animals would have incorporated themselves in the fabric of the cave. Caves are containers where different elements or substances of landscape, people and animals are brought together. In this way, by the combined agency of people, animals and cave itself in controlling the flow of substances and persons inside and outside the cave a history of the locale is created. Making use of containers constitutes of putting things, animals and humans together, and separating things that belong together from those which do not (Warnier 2006). Sediments, layers and features are created that pile up over time.

The sediments in the Karst caves are almost exclusively anthropogenic: they are literally residues of the dwelling of people and animals in the caves.

The stratified deposits can be extremely thick, as in the case of Podmol pri Kastelcu, where more than 8m of deposits have accumulated since the Early Neolithic (Turk *et al.* 1992). One of the most fascinating aspects of the Karst caves is the dung deposits that were periodically burnt. The long sequences of thin, layered lenses of white ash suggest that the process was repeated cyclically, probably over a long period (Brochier 1983, 1996, 2002; Brochier *et al.* 1992; Boschian and Montagnari Kokelj 2000; Boschian 2006). They appear in the form of 'heaps' (Fig. 14.4) which are mainly found near the cave walls, and which may be the result of cleaning the dung from the centre of the cave, burning it and of depositing the ash near the cave walls.

There are many practical reasons for burning dung. Probably one of the main reasons was to reduce the volume of manure deposits, or the disinfection of caves to protect the animals from parasites in the dung. However, we cannot assume that refuse disposal and site maintenance practices obey some universally applicable notion of functionality and hygiene (Brück 1999, 313). Dung is not necessarily a dirty, polluted substance, which has to be disposed of. The burning of dung can be understood as the symbolic manipulation of matter. The burning of dung is a process that transforms dung into a new substance: white ash. The process involves burning the dung, subjecting it to fire, which produces large quantities of smoke and heat, and can take quite a long time. During the long process of transformation, the material changes colour, texture, smell and volume. The regular and formalized nature of these depositional episodes suggests that they were an important part of occupational episodes in maintaining the floors of the living space in caves. Dung is a product of daily routines and is therefore a cultural construct. It is invested with particular meanings and powers, according to context and its state. Dung is also an animal product; it is literally a digested, condensed landscape brought into a cave by the agency of animals. Deposits of burnt herbivore dung can be understood as the culturally transformed substance of landscape stored in the container of the cave. Therefore, the proper manipulation, burning, and deposition of burnt dung can be seen as an important part of maintaining relations between people and animals, places and landscape (Mlekuž 2009). These practices have a clear temporal dimension. The dung takes a significant period to dry, and then to burn or smoulder. The burning marks a period when a cave is abandoned and empty. The act of burning is literally an act of temporally un-making: dismantling the camp in a cave. The regular deposition of dung

Figure 14.4: Heap of burnt animal dung from Neolithic layers of Mala Triglavca.

in the same place, near the cave wall, suggests that deposition practices were concerned mainly with the maintenance of the relation with previous occupations and the continuity of cave use. This is further supported by the fact that heaps of ashes appear to be undisturbed, and sometimes carefully preserved from trampling, by being covered by plate-like rocks. From this perspective, burnt animal dung can be seen as the 'stuff of memory', a material record of previous occupations and the activity of ancestors. The cave, then, is a container of material memories, preserved through the medium of burnt animal dung and other traces of previous occupations (Mlekuž 2009).

Another class of deposits consists of large stones, pottery and animal bones (Fig. 14.5). Even disarticulated human bones can often be found in these deposits. The metonymic qualities of many deposited objects (such as parts of cult vessels (Mlekuž 2007a), human and animal bones) suggest that people who deposited them were concerned mainly with the maintenance of relations with previous occupations and ancestors. Through the deposition of token deposits of human bones, people distributed themselves across the landscape and related themselves to specific locales. Those deposits were often placed near the cave walls, where, due to natural processes, they disappeared from the surface as they were transported deeper into the cave (Mlekuž et al. 2008). In this way, the material residues of dwelling in the cave were deposited.

Conclusions

Caves are not a passive backdrop for the activities that people perform, they are not natural places, and they do not satisfy the generic needs of people such as 'shelter'. By exploring the 'affordances' that caves provide we can focus on the social and contextual role they played in the practical tasks of past people. Activities that took place in caves were implicitly or explicitly connected with tasks elsewhere; these connections were materialized in substances and things that became embedded in the matrix of the caves. In this way caves grow: having their own biographies that consist of an unfolding of their relations with

Figure 14.5: Deposit of human and animal bones, pottery and large stones near cave wall. Neolithic layers of Mala Triglavca.

both human and non-human components of their environments. By focusing on the affordances that caves provide we can understand caves as material culture where dwelling occurs. And, by focusing on the process of dwelling that they enable through the affordances they provide, they help us to challenge the dichotomies of the natural and built environment, or of the mundane and the sacred. Caves provide affordances to people who engage with them, and they also 'act back' on people. Caves and rockshelters provide different sets of affordances to open air locales and built places. Caves afford a containment of things, substances and persons. They cram and amass together people, animals, things and even non-human beings, such as spirits. They envelop and separate people, animals and things from the outside landscape. And through sensory motor activities associated with containment they become embodied in subjects, and in this way provide new affordances for those subjects. Thus, at the end, we have reached a more social definition of a cave, as a place of intense sociality, shaped by the materiality of the cave itself.

References

Adams, L. and Zuckerman, D. (1991) The effect of lighting conditions on personal space requirements. *Journal of General Psychology* 118, 335–340.

Aiello, J. R. and Thompson, D. E. (1980) When compensation fails: mediating effects of sex and locus of control at extended interaction distances. *Basic and Applied Social Psychology*, 1(1), 65–81.

Andersen, P. A. and Sull, K. K. (1985) Out of touch, out of reach. *Western Journal of Speech Communication* 49, 57–72.

Andreolotti, S. and Gerdol, R. (1972) L'Epipaleolitico della grotta Benussi (Carso Triestino). *Atti e Memorie della Commissione Grotte 'E. Boegan'* 9, 59–103.

Bailey, D. W. (1999) The built environment: pit-huts and houses in the Neolithic. *Documenta Praehistorica* 26, 153–162.

Bailey, D. W. (2000) *Balkan Prehistory: Exclusion, Incorporation and Identity*. London and New York, Routledge.

Barfield, L. H. (1972) The first Neolithic cultures of North-Eastern Italy. In H. Schwabedissen (ed.) *Die Anfange des Neolithikums von Orient bis Nordeuropa. Teil VII. Westliches Mittlelmeerebiet und Britische Inseln. Fundamenta A/3*, 182–216. Köln and Wien, Böhlau Verlag.

Barnatt, J. and Edmonds, M. (2002) Places apart? Caves and monuments in Neolithic and earlier Bronze Age Britain. *Cambridge Archaeological Journal* 1, 113–129.

Biagi, P. (1994) Alcuni aspetti del Mesolitico nel Friuli e nel Carso Triestino. *Atti della XXIX Riunione Scientifica dell'Istituto Italiano di Preistoria e Protostoria*, 57–62. Firenze, Istituto Italiano di Preistoria e Protostoria.

Borić, D. (2008) First household and 'house societies' in European prehistory. In A. Jones (ed.) *Prehistoric Europe: Theory and Practice*, 109–142. Oxford, Wiley-Blackwell..

Boschian, G. (1997) Sedimentology and soil micromorphology of the late Pleistocene and early Holocene levels of Grotta dell'Edera (north-east Italy). *Geoarchaeology: An International Journal* 12, 227–233.

Boschian, G. (2006) Geoarchaeology of Pupićina Cave. In P. T. Miracle and S. Forenbaher (eds.) *Prehistoric Herders of Northern Istria. The Archaeology of Pupićina Cave. Volume 1*, 123–162. Pula, Arheološki Muzej Istre.

Boschian, G. and Montagnari Kokelj, E. (1984). Siti mesolitici del Carso Triestino: dati preliminari di analisi del territorio. In L. Ruaro Loseri (ed.) *Preistoria del Caput Adriae. Atti del Convegno Internazionale*, 40–50. Trieste, Udine, Istituto per l'Enciclopedia del Friuli-Venezia Giulia.

Boschian, G. and Montagnari Kokelj, E. (2000) Prehistoric shepherds and caves in the Trieste Karst (Northeastern Italy). *Geoarchaeology: An International Journal* 150(4), 331–371.

Bradley, R. (1998) Ruined buildings, ruined stones: enclosures, tombs and natural places in the Neolithic of south-west England. *World Archaeology* 30(1), 13–22.

Bradley, R. (2000) *The Archaeology of Natural Places*. London, Routledge.

Brightman, R. (1993) *Grateful Prey: Rock Cree Human-Animal Relationships*. Berkeley, University of California Press.

Brochier, J. É. (1983) Bergeries et feux de bois néolithiques dans le Midi de la France. *Quartär* 33–34, 181–194.

Brochier, J. É. (1996) Feuilles u fumiers? Observations sur rôle des poussièers sphérolitiques dans l'interprétation des dépôts archéologiques holocèenes. *Anthropozoologica* 24, 19–30.

Brochier, J. É. (2002) Les sédiments anthropiques: méthodes d'étude et perspectives. In C. Miskovsky (ed.) *Géologie de la Préhistoire: Méthodes, Techniques, Applications*, 459–477. Paris, Geopré.

Brochier, J. É., Villa, P. V., Giacomarra, M. G. and Tagliacozzo, A. T. (1992) Shepherds and sediments: geo-archaeology of pastoral sites. *Journal of Anthropological Archaeology* 11, 47–102.

Brück, J. (1999) Ritual and rationality: some problems of interpretation in European archaeology. *European Journal of Archaeology* 2(3), 313–344.

Casey, E. S. (1996) How to get from space to place in a fairly short stretch of time. In S. Field and K. H. Basso (eds.) *Sense of Place*, 13–52. Santa Fe (NM), School of American Research Press.

Cochran, C. D. and Urbanczyk, S. (1982) The effect of availability of vertical space on personal space. *Journal of Psychology* 111, 137–140.

Cochran, C. D., Hale, W. D. and Hissam, C. P. (1984) Personal space requirements in indoor versus outdoor locations. *Journal of Psychology* 117, 121–123.

Cremonesi, G. (1981) Caratteristiche economico-industriali del Mesolitico nel Carso. *Atti Della Società per la Praistoria e Protoistoria della Regione Friuli-Venezia Giulia* 4, 171–186.

Cribb, R. (1991) *Nomads in Archaeology*. Cambridge, Cambridge University Press.

Estevez, I., Andersen, I. L. and Nævdal, E. (2007) Group size, density and social dynamics in farm animals. *Applied Animal Behaviour Science* 103(3–4), 185–204.

Evans, G. W., Lepore, S. J. and Schroeder, A. (1996) The role of interior design elements in human responses to crowding. *Journal of Personality and Social Psychology* 70, 41–46.

Fisher, A. and Matthews, L. (2001). The social behaviour of sheep. In H. W. Gonyou and L. J. Keeling (eds.) *Social Behaviour in Farm Animals*, 211–245. Oxon, New York, CABI Publishing.

Foucault, M. (1988) Technologies of the self. In H. Gutman and L. H. Martin (eds.) *Technologies of the Self: A Seminar with Michel Foucault*, 16–49. Princeton (NJ), University of Princeton Press.

Gell, A. (1992) *The Anthropology of Time: Cultural Constructions of Temporal Maps and Images*. Oxford, Berg Publishers.

Gibson, J. J. (1986) *The Ecological Approach to the Visual Perception*. Hillshade (NJ), Lawrence Erlbaum Associates.

Gosden, C. (1994) *Social Being and Time*. Oxford, Blackwell.

Guacci, A. (1959) I muri della Grotta dell' Orso. *Tecnica Italiana* 24, 3–12.

Hall, E. T. (1966) *The Hidden Dimension*. Garden City (NY), Doubleday.

Heidegger, M. (2000) Bauen Wohnen Denken. In F.-W. von Herrmann (ed.) *Vorträge und Aufsätze (Gebundene Ausgabe)*, 139–156. Stuttgart, Klett-Cotta.

Ingold, T. (1980) *Hunters, Pastoralists and Ranchers*. Cambridge, Cambridge University Press.

Ingold, T. (1993) The temporality of landscape. *World Archaeology* 25(2), 152–174.

Ingold, T. (2000) *The Perception of the Environment: Essays in Livelihood, Dwelling and Skill*. New York, Routledge.

Knappett, C. (2005) *Thinking through Material Culture: An Interdisciplinary Perspective*. Philadelphia (Pa), University of Pennsylvania Press.

Knappett, C. (2006) Beyond skin: layering and networking in art and archaeology. *Cambridge Archaeological Journal* 16(2), 239–251.

Latour, B. (2005) *Reassembling the Social: An Introduction to Actor Network Theory*. Oxford, Oxford University Press.

Leitch, R. and Smith, S. (1997) Archaeology and the ethnohistory of cave dwelling in Scotland. In C. Bonsall and C. Tolan-Smith (eds.) *The Human Use of Caves*. British Archaeological Reports International Series 667, 122–126. Oxford, Archaeopress.

Lewis-Wiliams, J. D. (2002) *The Mind in the Cave: Consciousness and the Origins of Art*. London, Thames and Hudson.

Lewis-Williams, D. (2004) Constructing a cosmos: architecture, power and domestication at Çatalhöyük. *Journal of Social Archaeology* 4(1), 28–59.

Miller, D. (1987) *Material Culture and Mass Consumption*. Cambridge, Blackwell.

Miracle, P. T. and Forenbaher, S. (2005) Neolithic and Bronze-Age herders of Pupićina Cave, Croatia. *Journal of Field Archaeology* 30, 255–281.

Mlekuž, D. (2005) The ethnography of the Cyclops: Neolithic pastoralists in the eastern Adriatic. *Documenta Praehistorica* 32, 15–51.

Mlekuž, D. (2007a) 'Sheep are your mother': rhyta and the interspecies politics in the Neolithic of the eastern Adriatic. *Documenta Praehistorica* 34, 267–280.

Mlekuž, D. (2007b). Who were the Cyclopes? In M. Blečić, M. Črešnar, B. Hansel, A. Hellmuth, E. Kaiser and C. Metzer-

Nebelsick (eds.) *Scripta Praehistorica in Honorem Biba Teržan*, 69–82. Situla 44. Ljubljana, Narodni Muzej Slovenije.

Mlekuž, D. (2009) The materiality of dung: the manipulation of dung in Neolithic Mediterranean caves. *Documenta Praehistorica* 36, 219–226.

Mlekuž, D., Budja, M., Payton, R. and Bonsall, C. (2008) 'Mind the gap': caves, radiocarbon sequences, and the Mesolithic–Neolithic Transition in Europe – lessons from the Mala Triglavca rockshelter site. *Geoarchaeology: An International Journal* 23(3), 398–416.

Moser, K. (1903) Die Ausgrabungen in der Höhle 'Jama (Pejca) na Dolech' nächst der Eisenbahnstation Nabresina. Bericht über die Jahr 1902 in Österreich durchgeführten Arbeiten. *Mitteilungen der Anthropologischen Gesellschaft in Wien* 33, 69–99.

Richardson, M. (1982) Being-in-the-market versus being-in-the-plaza: material culture and the construction of social reality in Spanish America. *American Ethnologist* 9(2), 421–436.

Richardson, M. (2004) The artefact as abbreviated act: a social interpretation of material culture. In I. Hodder (ed.) *The Meaning of Things: Material Culture and Symbolic Expression*, 172–177. London and New York, Routledge.

Riedel, A. (1975) La fauna epipaleolitica della Grotta Benussi. *Atti e Memorie della Commissione Grotte 'E. Boegan'* 15, 123–144.

Schmitz, H. (1997) Der Mensch und die Grenze im Raum. In H. Schmitz (ed.) *Höhlengänge. über die Gegenwärtige Aufgabe der Philosophie*, 131–141. Lynkeus 7. Berlin, Akademie Verlag.

Štamfelj, I., Cvetko, E., Bitenc-Ovsenik, M. and Gašperšič, D. (2004) Identification of two human milk incisors from the archaeological sites Mala Trigiavca and Viktorjev spodmol. In I. Turk (ed.) *Viktorjev Spodmol in/and Mala Triglavca*, 221–240. Ljubljana, Založba ZRC.

Trump, D. H. and Cilia, D. (2002) *Malta, Prehistory and Temples*. Malta, Midsea Books.

Turk, I., Modrijan, Z., Prus, T., Culiberg, M., Šercelj, A., Perko, A., Dirjec, J. and Pavlin, P. (1992) Podmol pri Kastelcu – novo večfazno najdišče na Krasu, Slovenija. *Arheološki Vestnik*, 44, 45–96.

Turnbull, C. M. (1965) *Wayward Servants: The Two Worlds of African Pygmies*. London, Eyre and Spottinswoode.

van Gennep, A. (1960) *The Rites of Passage*. Chicago, University of Chicago Press.

Warnier, J.-P. (2001) A praxeological approach to subjectivation in a material world. *Journal of Material Culture* 6(5), 5–24.

Warnier, J.-P. (2006). Inside and outside: surfaces and containers. In C. Tilley, W. Keane, S. Kuechler, M. Rowlands and P. Spyer (eds.) *Handbook of Material Culture*, 186–195. London, Sage Publications.

Webster, D. S. (1999) The concept of affordance and GIS: a note on Llobera (1996). *Antiquity* 73(282), 915.

White, W. B. and Culver C. D. (2005) Cave, definition of. In D. C: Culver and W. B. White (eds.) *Encyclopedia of Caves*, 81–85. Burlington (MA), Oxford, Elsevier Academic Press.

Whitehouse, R. (2001) Tale of two caves: the archaeology of religious experience in Mediterranean Europe. In P. H. Biehl and F. Bertemes (eds.) *The Archaeology of Cult and Religion*, 161–167. Budapest, Archaeolingua.

Zaborowski, H. (2005) Towards a phenomenology of dwelling. *Communio: International Catholic Review* 32(3), 492–516.

Chapter 15

Cave Burials in Prehistoric Central Europe

Jörg Orschiedt

Burials within caves are common phenomena in European prehistory. Nevertheless, there are periods and regions where this kind of burial is absent or uncommon. Human remains within cave sediments have been found frequently during excavations. According to research traditions in some Central European regions these skeletons or isolated human remains were not interpreted as burials but as ritual human offerings and/or even remains of cannibalistic feasts. This kind of interpretation is usually favoured for post-Mesolithic periods when other forms of burials (e.g. in cemeteries) are more common. The fact that postglacial human remains from caves are sometimes overlooked by researchers leads to the fact that several of these remains are now being lost or have never been studied in detail. This paper analyses the appearance of burials within caves through time and tries to find evidence for cave burials in Central Europe during the Neolithic as well as the Bronze and Iron Ages.

Introduction

This paper deals with the occurrence of burials in cave sites within the Central European area in a chronological perspective. The area covered by this study includes the territory of Germany, Switzerland, Austria, Czech Republic and Poland. It highlights on the most important cases of primary burials form the Palaeolithic, Mesolithic, Neolithic, Bronze and Iron Age as well as isolated human remains with or without signs of manipulation. It aims towards an analysis of the burial data in caves for the Palaeolithic and Mesolithic period; the low number of burials reveals doubts about the interpretation of caves as a typical burial place for these periods. Another aspect is the discussion of the evidence of primary burials as the typical burial mode. Other forms of treatment of the deceased such as secondary depositions of bones or the 'non-interment burial' of exposing the dead above the ground should be considered as well as their archaeological potential. Therefore the evidence of burials in caves should be clarified, especially from the Neolithic to the Iron Age. The chapter deals with the paradigmatic interpretation of seeing cave sites with primary burials and isolated human remains from Post-Mesolithic periods as ritual sites with offerings of humans. Here, alternative perspectives are presented as a result of re-analyses of various sites.

The Palaeolithic evidence and the earliest burials in caves

Burials in caves and rockshelters are thought to be the typical way of treatment of the deceased in the Palaeolithic period. A closer look reveals that only a few individuals were buried within these sites in relation to the huge time span. It is rather tempting to see the beginning of a formal disposal of the deceased in the findings from Atapuerca, Sima de los Huesos, in Spain, where the bodies of a minimum number of 32 individuals were probably thrown down a shaft in several episodes between 300,000 and 200,000 BP. The discovery of a single handaxe in this context might reveal the first instance of a grave good (Arsuaga *et al.* 1997; Arsuaga and Martínez 2004, 110–113). The evidence of burials in the sense of primary burials within a clearly defined grave pit with or without grave goods shows that, for example, in the Middle Palaeolithic (MP), only about 40 Neanderthals and early modern humans in Europe and the Levant were buried in these locations. So the question arises, what happened to most of the deceased? There is no doubt that there must have been other forms of disposal and other places where this happened (Orschiedt 2005).

Unfortunately, the evidence for these kinds of practices is far more difficult to see and to reconstruct. Nevertheless, we do have evidence of primary burials in open air sites from the Upper Palaeolithic, especially 'rich' burials from the Gravettian and the Magdalenian (Wüller 1999). But here we face the same question, what happened to other individuals? Certainly the answer does not lie in rather spectacular finds like the Russian Sungir burials, exceptionally preserved in the permafrost soil for thousands of years with their grave goods and body ornaments, but in the larger number of isolated bones from open air sites, caves and rockshelters. Nevertheless, primary burials with rich equipment like Sungir are often seen as a blueprint for Palaeolithic burials, although they rather seem to be the exception.

In reality we have to think about burials above ground, as is the case in most recent hunter-gatherer populations, or more complex burial rituals such as secondary burials and other post mortem manipulations of the deceased. The fact that some of the human remains from the Middle and Upper Palaeolithic show signs of human manipulation, such as cut-marks, implies a complex behaviour towards the dead. However, these kinds of rather obvious manipulations have been seen in the past as evidence of cannibalism which has controversially been discussed (Orschiedt 1999a, 2005).

Most of the Palaeolithic human remains, either primary burials or supposed secondary burials or isolated bones, are from cave sites and rock shelters (Fig. 15.1). One of the reasons for this is certainly the excellent preservation in these places. Nevertheless, the lack of primary burials in Central Europe dating to the Middle and Upper Palaeolithic is striking compared to other areas (Orschiedt 2005). The fact that most

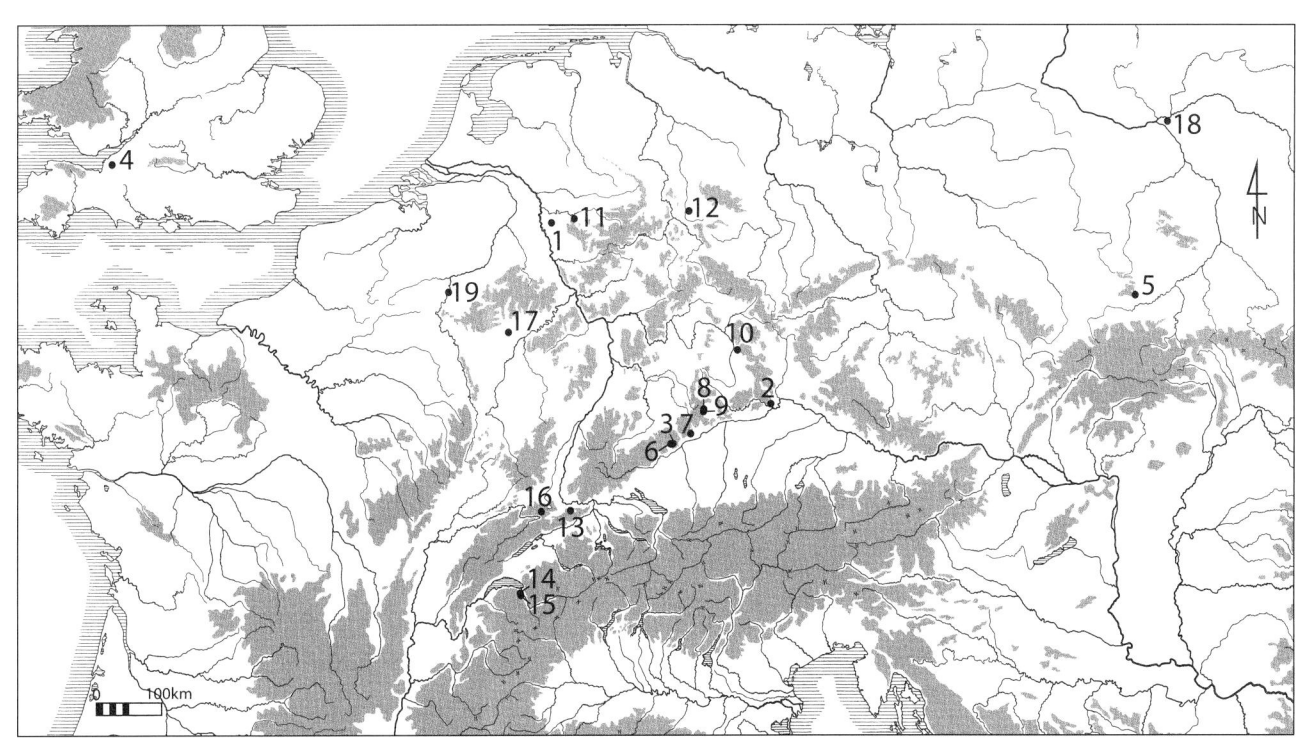

Figure 15.1: Palaeolithic and Mesolithic sites mentioned in the text. Palaeolithic: (1) Neandertal, near Düsseldorf; (2) Mittlere Klause Cave, near Essing close to Kehlheim; (3) Brillenhöhle, near Ulm; (4) Gough's Cave, Somerset, UK; (5) Maszycka Cave, Poland. Mesolithic (6) Höhlesbuckel site at Altental near Ulm; (7) Hohlenstein-Stadel, near Ulm; (8) Ofnet near Nördlingen; (9) Kaufertsberg near Nördlingen; (10) Büttnerloch near Forchheim close to Erlangen; (11) Blätterhöhle at Hagen; (12) Bettenroder Berg, near Göttingen; (13) Birsmatten Basisgrotte, Switzerland; (14) Abri de Vionnaz, Switzerland; (15) Châble-Croix, Switzerland; (16) Oberlarg Alsass, France; (17) Abri de Loschbour, Luxemburg; (18) Wieliszew Cave, Poland; (19) Grotte Margaux, Belgium.

of the larger cave sites have been used as sleeping places by cave bears and as hyena dens might have influenced the preservation of burials in this area. In contrast, the Levant region gives a different picture, as it shows evidence of several Middle Palaeolithic cave sites with burials of both modern humans and Neanderthals, like Skhul, Quafzeh, Amud, Tabun and Kebara. At most of these sites more than one and up to 15 individuals are present and most of them are primary burials (Orschiedt 2005). Their disturbance is not influenced by carnivores, which are missing in the faunal assemblages of this area, but by rockfall and sedimentation processes (Gamble 1999, 310–312; Orschiedt 2005).

However, the only supposed primary burial from the Middle Palaeolithic comes from the eponymous site at the Neander Valley, the Feldhofer cave. The specimen shows signs of post-mortem manipulation on the skullcap: a few cut marks on the frontal and the occipital (Czarnetzki 1977; Schmitz and Pieper 1992), which were probably caused by skinning activities and the removal of muscles on the neck, probably prior to a removal of the head. Almost nothing is known about the original position of the skeletal remains of this individual due to the circumstances of the find and the time of discovery prior to the discovery of archaeological excavation techniques. The remains themselves show defects that were caused during the excavation. These corresponding defects, (e.g. on the acetabulum of the pelvis and the caput of the femur,) are an indication of a preserved anatomical connection of at least the hip joint. One can hence suggest an anatomical connection of the postcranial remains and a possible primary burial in a grave pit within the shallow cave (Orschiedt 1999 *et al.* 39; Orschiedt 2005, 165).

The only primary burial within a cave site from the Upper Palaeolithic (UP) in Central Europe is a male individual of 30–40 years from Mittlere Klausenhöhle near Neuessing, Bavaria. There is no cultural attribution for this skeleton, which was found covered in red ochre but without datable grave goods (Wüller 1999, 45–55). An AMS date of 18,590±260 BP (OxA-9856) is in good agreement with an older conventional date (Street *et al.* 2006). The AMS dated finds from Mladec do not represent an early Upper Palaeolithic (Aurignacien) burial mode, nor do the finds from Pestera çu Oase in Romania. Both sites yielded human remains heavily affected by taphonomic factors such as carnivore activity. All of the remains were found in a secondary context (Svoboda 2000; Trinkaus *et al.* 2003; Wild *et al.* 2005; Teschler-Nicola 2006,).

Burial practices like secondary depositions of human remains are difficult to identify, as mentioned earlier. In most cases, the interpretation of such finds is complex and sometimes alternates between burials versus cannibalism, as in the case of Gough's Cave, Somerset in Great Britain. The fragmentary human remains dating to the late Palaeolithic (11,500–13,000 BP) found in the cave apparently show cut marks and perimortal breakage patterns. The find has been identified as possible evidence for a complex secondary burial ritual or cannibalism and more recently as evidence of nutritional cannibalism (Currant *et al.* 1989; Cook 1991a, 1991b; Andrews and Fernández-Jalvo 2003). A rather similar case from Germany, however, reveals evidence of a secondary burial within the Brillenhöhle in the Swabian Alb in southwest Germany. Although this find was interpreted in an earlier publication as evidence of cannibalism (Gieseler and Czarnetzki 1973), mainly due to its fragmentary nature, the occurrence of cut-marks and the position close to a fireplace, it has to be seen as a secondary burial from the Magdalenian (12,470±65 BP, OxA-11054) (Orschiedt 2002). The main argument that distinguishes the remains of a minimum of two adults and one infant from butchered animal bones is the high frequency of cut and scraping marks that indicates not only dismemberment and filleting activities but also an intensive cleaning of the bones (Fig. 15.2).

Another example from the same period is the remains from Maszycka cave in Poland (Kapica and Wiercinski 1995). Here, the remains from a minimum of 16 individuals reveal cut and scraping marks as well as perimortal breakage patterns. In contrast to Gough's Cave and the Brillenhöhle, where postcranial remains are present in larger numbers, up to 80 per cent of the remains from Maszycka cave consist of fragments of the cranial vault. This might imply the influence of taphonomic factors, as these areas belong to the most resistant parts of the body (Brain 1981, 312, 322). Nevertheless, the over-representation of skull fragments has been interpreted as an anthropogenic phenomenon

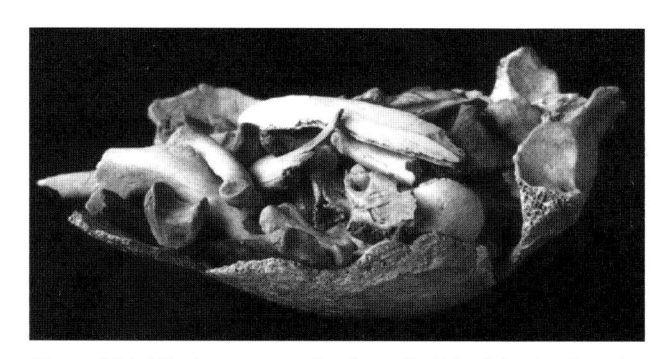

Figure 15.2: The human remains from the Magdalenian from the Brillenhöhle site. An example of a secondary burial inside the cave. (Photo: J. Orschiedt.)

and as a result of the ritual smashing of the skulls as well as the consumption of brain tissue (Kozlowski and Sachse-Kozlowska 1995, 170). As mentioned above, this scenario is a rather common interpretation widely ignoring other possibilities. The human remains and the archaeological material including the faunal remains are currently reinvestigated by the author and others in order to clarify this.

The Magdalenian and the late Palaeolithic, however, is a period where we have to be aware of a rather complex treatment of the deceased, which is sometimes difficult to understand, especially when confronted with older excavations and the lack of stratigraphic and taphonomic observations (Orschiedt 1999a).

From single graves to severed heads: Mesolithic cave burials

Burial modes in the Mesolithic are certainly complex and apart from single graves or cemetery-like sites in the open landscape and shell middens, caves and rock shelters still play a certain role as a final resting place for the deceased. But the evidence in Central Europe is still rather thin concerning the cave site burials, so the picture does not change much compared to the Palaeolithic (Fig. 15.1). The child burials of Abri Bettenroder Berg were seen as a typical example of such a primary burial. Based on stratigraphic observations and an AMS date on a bone not belonging to the skeleton, they were dated to the Mesolithic. However, a re-examination of the two burials revealed a much younger date. Both of them belong to the early Iron Age (*c.* 460 cal BC and 800 cal BC), and can be interpreted as deviant burials (Grote and Terberger 2011).

From other sites, such as the Höhlesbuckel site at Altental near Blaubeuren on the Swabian Alb, only little information is available. A single burial was discovered during construction works in 1949 within a small cave site, lying close to the cave wall below a cultural layer that contained stone artefacts. No description of the position of the body or other details is available. The skeletal remains have not been preserved except a calvarium which has been dated by AMS to the early Mesolithic (7570 cal BC, ETH-6668) (Haas 1991). This is one of the few burials from the Preboreal in Europe.

Other sites, like Büttnerloch near Forchheim close to Erlangen in Bavaria (Grünberg 2000, 37–38), Birsmatten Basisgrotte in Switzerland (Grünberg 2000, 291–293) or the Abri de Loschbour in Luxembourg (Grünberg 2000, 159–160), reveal the burial custom of single graves in caves or rock shelters.

A completely new way of treating the deceased is represented by the cases of cremation which appear in the Mesolithic. In Central Europe, various examples are known (Grünberg 2000, 51–54; Schunke 2004), but only three sites seem to be linked with rockshelters: the Abri de Vionnaz (Crotti 1993, 240), Châble-Croix in Switzerland (Pignat 2002, 112) with Wieliszew in Poland (Grünberg 2000, 171–172) and the rediscovered cremation from Loschbour (Toussaint *et al.* 2009). There is only preliminary information available, but it seems clear that these cremations represent single individuals.

Probably one of the most peculiar types of burial from cave sites in Central Europe is restricted to the area of southern Germany and Alsace. Sites like Ofnet, Hohlenstein-Stadel, Kaufertsberg and Oberlarg represent so-called head burials (Orschiedt 1998). Burials of this kind are characterised by one or a group of heads which were deposited in the entrance areas of caves or rock shelters. Kaufertsberg in Bavaria and Oberlarg from the Alsace yielded single individuals, while the site of Hohlenstein-Stadel, Bade-Wurttemberg contained three heads of two adults (one male, one female) and a child of about 2 years, potentially with some kind of family relationship (Fig. 15.3). It is obvious that at least the two adults and probably also the child were killed by blunt force trauma (Orschiedt 1999a, 131–135). The Ofnet cave site, however, represents the largest group of 34 individuals: nine females, 18 children and only five males (Orschiedt 1999a, 136–151). The heads were laid down in one large and one small nest-like structure inside two pits close to each other, covered with red ochre, and some of the individuals were equipped with large numbers of red deer canines and perforated snail shells. Here, too, eight individuals show signs of blunt force trauma. Most of these individuals, who certainly encountered a violent conflict, were laid down on the left side of the larger skull nest. It is important to state that all the individuals are represented by a more or less complete cranium, the lower jaw and several vertebrae of the neck. Some of these vertebrae show clear signs of cutting activities, which makes it obvious that the heads were separated from the trunk before natural decomposition. Both the Ofnet cave and the Hohlenstein-Stadel, dated by a series of radiocarbon dates between 6500 and 6000 cal BC (Hedges *et al.* 1989, 210–211) and 6800 to 6500 (Haas 1991) as well as Oberlarg dated around 6200 to 5900 cal BC (Newell *et al.* 1979, 126), and Kaufertsberg which is not yet dated due to contamination, represent a late Mesolithic horizon of the last hunter-gatherer population just before the first Neolithic settlements appeared.

Figure 15.3: The 'head burial' from Hohlenstein-Stadel in situ just after the discovery on July 26th 1937. (Institut für Urgeschichte und Quartärökologie, Universität Tübingen.)

A site that shows another, apparently even more complex kind of burial activity is the Blätterhöhle at Hagen, a cave site in Westphalia which was discovered in 2004 (Fig. 15.4). Due to the fact that archaeological investigations only started in 2006, the analyses are still ongoing and the results are preliminary. It seems clear that, aside from settlement activities below a rock shelter close to the cave entrance, the cave itself was used for burial purposes. The human remains were deposited in the cave together with three boar skulls, all of them are dated to the early Mesolithic around 8700 cal BC (Orschiedt *et al.* 2008). The situation resembles sites from Belgium from the same time period, such as Grotte Margaux and others (Cauwe 1998a; 1998b). These sites seem to represent a rather extensive treatment of the dead. The individuals buried here are only represented by single bones, as there are no anatomical connections present, so this kind of burial practice has to be regarded as a secondary collective deposition like an ossuary, which might have been deposited in a singular event or accumulated over a period of time.

Figure 15.4: Excavations at the Blätterhöhle at Hagen. The narrow space requires special excavation techniques. (Wippermann/ Historisches Centrum Hagen. Reproduced with permission.)

The Neolithic

During the Neolithic the use of caves was widely reduced to short occupations, although the exact kind of activities is sometimes hard to identify (Fig. 15.5). Usually, the Neolithic material is found mixed with other Holocene finds in the mostly disturbed upper layers. In most cases, clear stratigraphic situations are not preserved. Rare examples of Neolithic primary burials from caves and rockshelters without a Neolithic occupation horizon are, for instance, the child from Felsställe near Ulm (Kind 1987, 239–243) in southwest Germany and the double burial from Schellnäcker Wänd in Bavaria (Grünberg 2000, 51–52). Both burials have been dated to the late Neolithic by AMS dating (Haas 1991, 37; Grünberg 2000, 51–52).

The best known site is probably the Jungfernhöhle at Tiefenellern near Bamberg in Bavaria (Fig. 15.6). Excavated in the early 1950s, the cave contained the remains of a minimum of 41 individuals, mostly children and adult females. The human remains were mixed with animal bones and pottery from the early Neolithic (LBK culture), middle and late Neolithic, as well as Bronze and Iron Age finds up to Roman, Medieval and recent finds. Unfortunately, no archaeological layers could be identified. The excavation was done quite hastily, so we have to deal with a lack of documentation. The excavator published a monograph in which he stated that the human bodies were thrown into the cave after having been ritually killed, mutilated and cannibalised for ritual purposes (Kunkel 1955). Subsequently, for almost 40 years the site was interpreted as a rare example of an early Neolithic human sacrifice.

A reinvestigation of the human remains, however, was not able to prove these theories (Orschiedt 1999a, 164–178). In contrast to the claims of the first investigation, the human remains displayed neither cut marks nor blunt force trauma related to dismemberment, filleting or ritual slaughtering. It became apparent that the human remains were heavily selected; there is a clear dominance of skulls or skull fragments as well as long bones. The smaller bones of the postcranial skeleton, like hand and foot bones and vertebrae, are largely missing, as are the more fragile parts of the skull, like the skull base and the facial skeleton. The site was therefore reinterpreted as an early Neolithic burial site in which secondary burials were placed. Due to the lack of documentation and the fact that the remains were mixed with prehistoric and later pottery, a sample of human remains were AMS

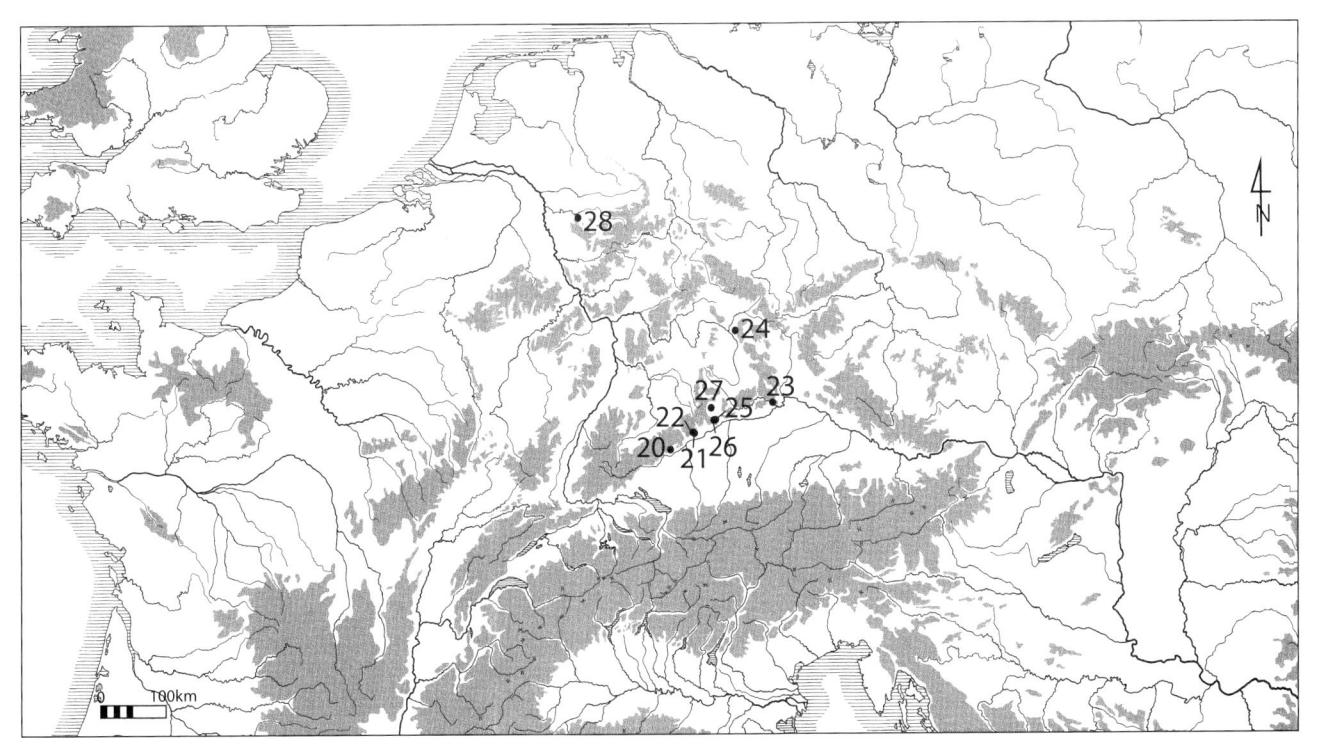

Figure 15.5: Neolithic sites mentioned in the text. (20) Felsställe near Ulm; (21) Hohlenstein-Stadel, near Ulm; (22) Vogelherd cave, near Ulm; (23) Schellnäcker Wänd, near Essing close to Kehlheim; (24) Jungfernhöhle at Tiefenellern near Bamberg; (25) Hanseles Hohl cave, near Nördlingen; (26) Hohler Stein near Nördlingen; (27) Kauftersberg 'Hexenküche', near Nördlingen; (28) Blätterhöhle at Hagen.

Figure 15.6: The 'Jungfernhöhle' at Tiefenellern near Bamberg. The entrance of the cave, which was thought to be a place of an early Neolithic human sacrifice of young females and children. (Photo: J. Orschiedt).

dated. Most of the dates are associated with the early Neolithic or LBK culture. Two dates around 350 cal BC indicate a late Neolthic use of the cave (Orschiedt 1999a, 164–167).

Another site in Bavaria, the Hanseles Hohl cave, might also represent a similar situation, although the information from it is rather sparse. Within a single cultural layer heavily damaged by rock fall, human remains associated with LBK pottery and spondylus shells were found. This site, too, might represent a secondary deposition in the early Neolithic (Orschiedt 1999a, 152–157). Other sites, like the Hohler Stein and Kauftersberg 'Hexenküche', both in Bavaria, are even less well investigated. Due to early excavations with their lack of documentation and most of all the subsequent loss of the human remains it seems almost impossible to judge the observations made by the excavators that the human remains were found mixed with animal bones and pottery from the Neolithic. Both sites have been identified as ritual sites where cannibalistic feasts were supposedly held (Orschiedt 1999a, 73).

A similar situation is found at the Hohlenstein-Stadel in Bade-Wurttemberg, southwest Germany. In the entrance area of the cave a round structure supported by a small wall was found below several large fireplaces. The human remains were not directly associated with Neolithic pottery, which was present within the disturbed upper layers. Three AMS radiocarbon dates between 4470–4230 cal BC (ETH-13321), 4460–4230 cal BC (ETH-13322) and 4360–4040 (ETH-13320) revealed a cultural association with the early phase of the late Neolithic. As in other cases, the find has been interpreted as the remains of a cannibalistic feast although no manipulations were mentioned. Re-examination showed that the human remains were dominated by skull fragments of the cranial vault and fragments of long bones with the smaller parts of the body underrepresented without the occurrence of cut marks or other signs of manipulation except some burned parts of the preserved bones. Here, too, the deposition of the human remains has to be interpreted as a secondary burial (Orschiedt 1999a, 179–188).

Late Neolithic cave burials are also present at the Vogelherd cave site, famous for the finds of figurines from the Aurignacian in which human remains were also present. In the entrance area an accumulation of human remains was found in the uppermost, non-Palaeolithic layer. AMS radiocarbon dating carried out during the re-investigation produced late and final Neolithic dates of 3890–3690 cal BC (ETH-

16684) and 2820–2540 cal BC (ETH-16683) (Orschiedt 1999b). The remains of a minimum of six individuals showed some carnivore damage. A photograph of the remains *in situ* showed the anatomical connection of a human torso, with other bones completely dislocated. Some of the cranial fragments revealed perimortal trauma which indicates that at least three of these individuals might have been killed and their bodies left on the surface. The remains were buried after they had already largely decomposed (Orschiedt 1999b). Additionally, the new dates on the human remains originally attributed to the Aurignacian layers were dated around 3800 and 3500, and 2500 cal BC rather surprisingly to the late and final Neolithic as well (Conard *et al.* 2004). According to the radiocarbon dates the human remains from Vogelherd cave might represent the same two time horizons with an unsolved stratigraphical discrepancy according to the observations made during the excavation. This might also be indicated by the fact that the some of the human remains formerly attributed to the Aurignacien show defect patterns and tooth marks caused by carnivore activity as well as those attributed to the Neolithic (Orschiedt 1999a, 96–99)

At the recently discovered Blätterhöhle at Hagen in Westphalia, aside from the Mesolithic burial (see above), a Neolithic collective burial was identified (Fig. 15.4). The human remains were found during the excavation of the cave, which was completely filled with sediments (Orschiedt *et al.* 2008). The human remains have been AMS radiocarbon dated to 3800–3200 cal BC and belong to the younger and late Neolithic. Within this time span, around 3500 cal BC, the first collective graves associated with the Funnel Beaker Culture and the Wartberg Culture were built in Westphalia and Hessen. The border of the distribution of megalithic tombs lies about 100–150km from the Blätterhöhle. This collective burial represents the first non-megalithic burial known so far outside this area. In the late Neolithic, the entrance of the cave site must already have been very narrow; we have to assume that this situation was intentionally used as a burial place for this reason, as it resembles the entrance situation of a megalithic tomb. Unfortunately, the human remains were affected by bioturbation mostly caused by badgers. The human remains were not directly associated with pottery or grave goods. Some fragments of undetermined and badly preserved pottery, however, were found within the cave itself and close to the rear wall of the collapsed rockshelter in front of the cave. The analysis of these fragments is still not finished but the pottery might be attributed to the late Neolithic.

Bronze Age cave burials

Several Bronze Age cave sites containing human remains are known in Central Europe (Fig. 15.7). There is a similar situation concerning the interpretation of these finds as in the Neolithic.

A very typical example of this is the Kyffhäuser caves close to Bad Frankenberg in the Harz region in Thuringia. The caves in this area are predominantly of a shaft-like form and are not easily accessible. The characteristic appearance of the very shallow caves excludes the possibility of their use as habitation areas. Finds from these sites dating to the latest Bronze Age (Urnfield Culture), including the human remains, are usually seen as ritual offerings. The excavation of the sites in the 1950s did not result in a scientific publication but in a popular book by Behm-Blanke (1958), *Höhlen, Heiligtümer, Kannibalen* (Caves, Sacred Places, Cannibals), which dominated the view of these and other cave sites for many years. The lack of a scientific publication, especially the detailed analysis of the human remains and their context, makes it difficult to decide what kind of activities were carried out, concerning the deposition of the human remains as well as pottery vessels, bronze and wooden objects, faunal remains and cereals. Several cases of manipulation have been mentioned unsystematically, such as cut marks and perimortal breakage patterns as well as traumatic lesions caused by blunt force trauma. Their interpretation, however, remains unclear, but the spectrum extends from ritual activities, including the deposition of the dead, with or without secondary burials or body parts, to the results of violent encounters in the Bronze Age (Behm-Blanke 1989; Walter 1998).

The most recent example is the still ongoing excavation at the Lichtensteinhöhle close to Dorste in Lower Saxony. Obtained with modern excavation techniques, although under difficult circumstances the result of this research sheds new light on the deposition of human remains in similar contexts. Additionally, the application of modern aDNA analyses proved that most of the individuals were related and paved the way for the interpretation of the site as a burial place, although the first interpretation had gone down the same traditional route of a site with ritual offerings. At this site, more than 30 individuals dating to the latest Bronze Age (Urnfield Culture) were deposited on the cave floor as primary and secondary burials. Together with the human remains, isolated finds of bronze objects, pottery and food remains were unearthed (Fig. 15.8). Most striking are huge layers of charcoal which indicate fires burning inside the narrow cave. No cases of intra vital trauma or any kind of post-

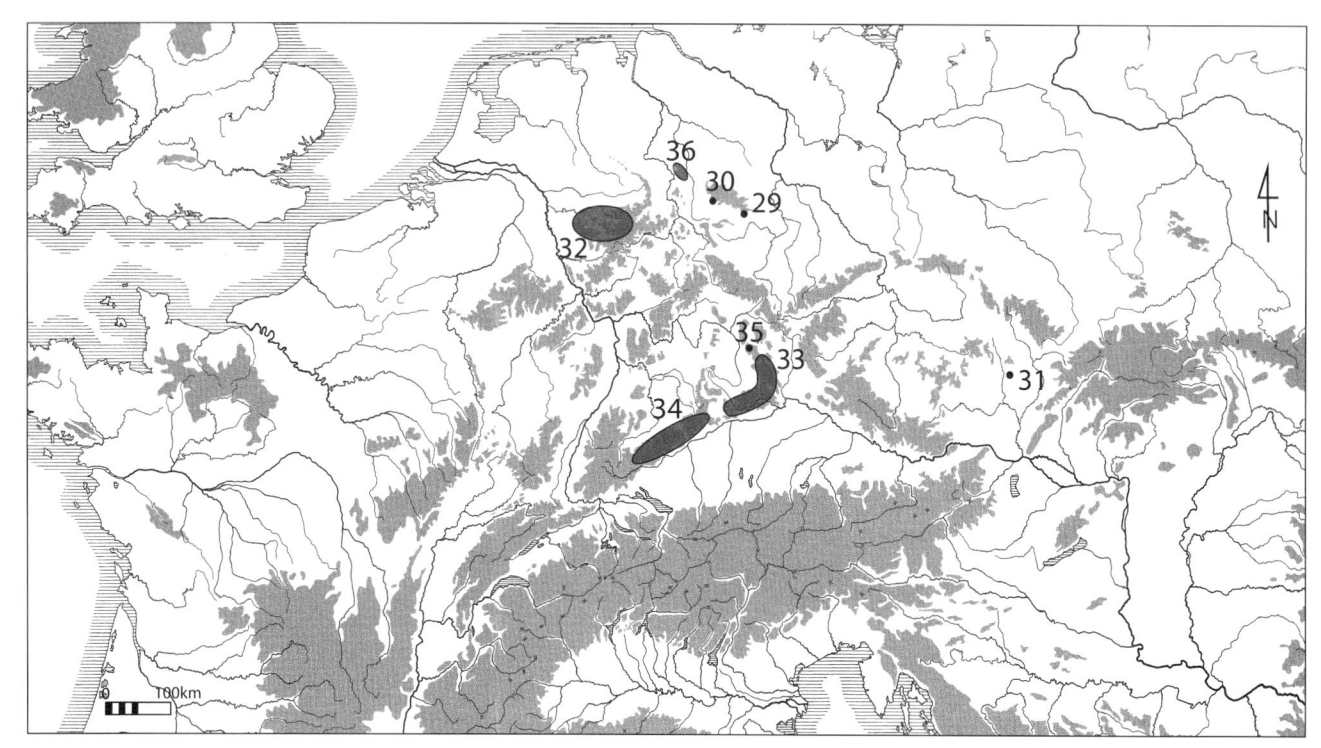

Figure 15.7: Bronze and Iron Age sites mentioned in the text. Bronze Age: (29) Kyffhäuser caves close to Bad Frankenhausen; (30) Lichtensteinhöhle close to Dorste. Iron Age: (31) Býčí skála cave, Czech Republic; (32) cave sites from the Sauerland region in southern Westphalia; (33) Franconian Alb in northern Bavaria; (34) Swabian Alb in Baden-Württemberg; (35) Dietersberg cave near Egloffstein; (36) Ith region in southern Lower Saxony.

Figure 15.8: Human remains in situ at the Lichtensteinhöhle. This Bronze Age site reveals new insights into the deposition of human bodies and skeletal remains in caves at that time period. (da Silva/Landkreis Osterode am Harz. Reproduced with permission.)

mortem manipulations have been reported (Flindt 1996; 1998).

In most cases, and especially where older excavations are concerned, Bronze Age finds of human remains within cave sites have been interpreted as human sacrifices with or without cannibalism. Stereotypical interpretations like these have also been applied to Iron Age contexts. It seems to be obvious that most researchers, even today, hesitate to see human remains in caves in a burial context. One of the most favoured arguments is the missing anatomical connection of the human remains. Usually the only accepted burial places for the Bronze and Iron Age are graveyards or mounds.

The Iron Age burial evidence

The most prominent example from the early Iron Age is the case of the Býčí skála cave in the Czech Republic (Fig. 15.7). Excavated in the 1880s, the site containing the remains of more than 40 individuals and numerous bronze objects from the eastern Halstatt culture was published as a chieftain's burial with the ritual killing and burial of young women (Wankel 1882). A reinvestigation accepted the ritual character of the site and interpreted it as a place of ritual sacrifices of humans and animals (Parzinger *et al.* 1995, 190). The idea of a burial site was refuted for lack of complete skeletons in anatomical connection accompanied with grave goods. This view ignores the fact that various sites exist with human remains from the Hallstatt or Latène Culture, which show some rather different aspects of the treatment of human remains in a burial context (Peter-Röcher 1997a, 318–319, 1997b, 1998; Orschiedt 1999a, 45–49, 81). Although any re-analysis of the human remains or demographic analysis is impossible due to the loss of most of the skeletal remains, the archaeological material of the site could well represent the grave goods of a cemetery of about 40 individuals (Angeli 1970, 149).

Cave sites from the Sauerland region in southern Westphalia are rich in Iron Age finds of pottery, metal objects and human remains. Human remains have been reported from a minimum of 13 out of 25 cave sites in this area. Most of them were associated with Iron Age material (Rothe 1983; Polenz 1991). Due to the fact that this information results from older excavations, mainly from the 1920s and 1930s, we have to be careful about the interpretation. Although some of the sites are easily accessible, any habitation activities are excluded. The deposition of human remains is seen as evidence of ritual offerings, again in most cases without any discussion. Unfortunately, most of the material was never analysed in detail and most of it seems to be lost today. A re-investigation of the remaining material is none the less desirable.

The same situation can be found on the Franconian and the Swabian Alb in Bavaria and Baden-Württemberg. The areas are rich in cave sites containing Iron Age material alongside human remains. Some of the human skeletal remains seem to be lost, but the lack of documentation causes the main problem in interpreting these sites. Here again the common interpretation is one of ritual offerings, excluding the occurrence of formal burials (Baum 1999, 80–81; Graf *et al.* 2008, 29–34). J. Biel however suggested the use of caves on the Swabian Alb as usual burial places for the late Hallstatt and early La Tène period (Biel 1987, 139–149). The Dietersberg cave near Egloffstein, which was seen as a typical 'Schachthöhle' or vertical cave, sheds some light on the treatment of human remains in an early Iron Age (late Hallstatt – early La Tène) context. Excavated in 1928, the site was interpreted as a ritual site where offerings of humans took place. The re-investigation by Norbert Baum revealed that, according to the demographic profile, this site has to be seen as a burial site (Baum 1999).

A further example is provided by the caves sites of the Ith region in southern Lower Saxony. The sites, which are usually easily accessible, represent a similar spectrum of archaeological material as the caves in the Kyffhäuser region (Geschwinde 1988). In most cases the fragmented and scattered human remains, some of which seem to be now lost, were not interpreted as burials or as the result of special burial customs, but as an evidence of ritual cannibalism (Geschwinde 1988, 105–106, 126–127; Schultz 1988, 129).

Conclusion

The start of a formal disposal of the dead in the Palaeolithic is usually linked with the first evidence of primary burials in the Middle Palaeolithic. Neanderthals and modern humans in the Near East (Levant region) dug the first graves before 100,000 BP. This evidence however might be misleading as it focuses only on the primary burials within a dug out grave pit. Other forms of interment or treatment of deceased people might be much older but are hard to proof archaeologically. The occurrence of primary burials in caves and rockshelters is dominant within the Middle Palaeolithic (MP). During the first phases of the Early Upper Palaeolithic (EAP) in Europe, the Chatelperronian and the Aurignacien most of the evidence for formal burials is missing. The picture changes in the Gravettain were burials from open air sites become more numerous and double and multiple burials as well as secondary treatment of human remains become more obvious. Caves and rockshelters are still important for primary and secondary burials; however, this might be related to the excellent conditions for preservation at these sites. The same is true for late Palaeolithic and Mesolithic burial sites. During the Mesolithic we have to note the occurrence of open air cemetery-like structures whose roots might lie in Late Palaeolithic times. The forms of disposal for the dead in the Mesolithic are complex, and it is probably misleading to suggest that primary burials is the dominating burial custom during this period

Scientific discussion prior to the 1990s seems to have excluded any other interpretation of human remains associated with Neolithic, Bronze and Iron Age material from cave sites than the ritual offerings of humans (Walter 1985, 77–81), although a non-ritual use of the caves is sometimes not excluded. The main argument against cave sites as burial sites for post-Mesolithic times usually refers to the elaborated burial customs which do not conform to cave burials or to the deposition of isolated human remains. It has to be noted, however, that, for example, human remains of the Urnfield Culture found in cave sites in West and Southern France are accepted as burials because of the absence of cemeteries (Schauer 1981, 412). In most cases, the absence of any modern osteoarchaeological analysis of the human remains, using differential diagnosis of perimortal breakage patterns, taphonomic or forensic analysis seems to produce stereotypical interpretations and self-fulfilling prophecies.

References

Andrews, P. and Fernández-Jalvo, Y. (2003) Cannibalism in Britain: taphonomy of the Creswellian (Pleistocene) faunal and human remains from Gough's Cave (Somerset, England). *Bulletin of the Natural History Museum London* 58 (suppl.), 59–81.

Angeli, W. (1970) *Zur Deutung der Funde aus der Byci-skala-Höhle. Krieger und Salzherren. Hallstattkultur im Ostalpenraum*. Exhibition catalogue. Mainz, Naturhistorisches Museum Wien.

Arsuaga, J. L., Martínez, I., Gracia, A., Carretero, J. M., Lorenzo, C., García, N. and Ortega, A. I. (1997) Sima de los Huesos (Sierra de Atapuerca, Spain) The site (hu970132). *Journal of Human Evolution* 33(2/3), 109–127.

Arsuaga, J. L., Martínez, I. and Trueba, J. (2004) *Atapuerca y la Evolucion Humana*. Barcelona, Fundació Caixa Catalunya.

Baum, N. (1999) Die Dietersberghöhle bei Egloffstein, Kr. Forchheim – von der Opferhöhle zum Bestattungsplatz. *Prähistorische Zeitschrift* 74(1), 79–121.

Behm-Blanke, G. (1958) *Höhlen, Heiligtümer, Kannibalen. Archäologische Forschungen im Kyffhäuser*. Leipzig, VEB Brockhaus.

Behm-Blanke, G. (1989) Heiligtümer, Kultplätze und Religion. In J. Herrmann (ed.) *Archäologie in der Deutschen Demokratischen Republik. Denkmale und Funde 1. Archäologische Kulturen, Geschichtliche Perioden und Volksstämme*, 166–176. Leipzig, Konrad Theiss.

Biel, J. (1987) *Vorgeschichtliche Höhensiedlungen in Südwürttemberg-Hohenzollern*. Forschungen und Berichte zur Vor- und Frühgeschichte in Baden-Württemberg 24. Stuttgart, Konrad Theiss.

Brain, C. K. (1981) *The Hunters or the Hunted? An Introduction to African Cave Taphonomy*. Chicago and London, University of Chicago Press.

Cauwe, N. (1998a) *La Grotte Margaux à Anseremme-Dinant. Étude d'une Sépulture Collective du Mésolithique Ancien*. Études et Recherches Archéologiques de l'Université de Liège 59. Liège, Université Liège.

Cauwe, N. (1998b) Sépultures collectives du Mésolithique au Néolithique. In J. Guilaine (ed.) *Sépultures d'Occident et Genèses des Mégalithismes*, 9–24. Paris, Errance.

Cook, J. (1991a) Preliminary report on marked human bones from the 1986–1987 excavation at Gough's Cave. In N. Barton, A. J. Roberts and D. A. Roe (eds.) *The Late Glacial in North-West Europe: Human Adaptation and Environmental Change at the End of the Pleistocene*. Council for British Archaeology Research Report 77, 160–168. Oxford, Alden Press.

Cook, J. (1991b) Preliminary report on marked human bones from the 1986–1987 excavations at Gough's Cave, Somerset, England. *Anthropologie* (Brno) 29(3), 181–187.

Conard, N. J., Grootes, P. M. and Smith, F. H. (2004) Unexpectedly recent dates for human remains from Vogelherd. *Nature* 430, 198–201.

Crotti, P. (1993) Das Mesolithikum. In *Die Schweiz vom Paläolithikum bis zum Frühern Mittelalter* 1, 203–241. Basel, Schweizerische Gesellschaft für Ur- und Frühgeschichte.

Currant, A. P., Jacobi, R. M. and Stringer, C.B. (1989) Excavations at Gough's Cave, Somerset 1986–87. *Antiquity* 63, 131–136.

Czarnetzki, A. (1977) Artifizielle Veränderungen an den Skelettresten aus dem Neandertal? In P. Schröter (ed.) *Festschrift 75 Jahre Anthropologische Staatsammlung München*, 215–219. München, Selbstverlag der Anthropologischen Staatsammlung.

Flindt, S. (1996) Die Lichtensteinhöhle bei Osterode, Landkreis Osterode am Harz. Eine Opferhöhle der jüngeren Bronzezeit im Gipskarst des südwestlichen Harzrandes – Forschungsstand und erste Grabungsergebnisse. *Die Kunde* N.F. 47, 435–466.

Flindt, S. (1998) Die Lichtensteinhöhle. In S. Flindt and C. Leiber (eds.) *Kulthöhlen und Menschenopfer im Harz, Ith und Kyffhäuser*. Archäologische Schriften des Landkreises Osterode am Harz 2, 50–80. Holzminden, Miskat.

Flindt, S. and Leiber, C. eds. (1998) *Kulthöhlen und Menschenopfer im Harz, Ith und Kyffhäuser*. Holzminden, Miskat.

Gamble, C. (1999) *The Palaeolithic Societies of Europe*. Cambridge, Cambridge University Press.

Geschwinde, M. (1988) *Höhlen im Ith. Urgeschichtliche Opferstätten im Südniedersächsischen Bergland*. Veröffentlichungen der Urgeschichtlichen Sammlungen des Landesmuseums Hannover 33. Hannover, Landesmuseum Hannover.

Gieseler, W. and Czarnetzki, A. (1973) Die menschlichen Skelettreste aus dem Magdalénien der Brillenhöhle. In G. Riek (ed.) *Das Paläolithikum der Brillenhöhle bei Blaubeuren (Schwäbische Alb)*. Forschungen und Berichte zur Vor- und

Frühgeschichte Baden – Württemberg 4(1), 165–168. Stuttgart, Müller & Gräff.

Graf, N., Graf, R. and Pasda, K. (2008) *Das Peterloch bei Woppental*. Beiträge zur Vorgeschichte Nordostbayerns 6. Nürnberg, Naturhistorische Gesellschaft Nürnberg e.V.

Grote, K. (1993) *Die Abris in Südlichen Leinebergland bei Göttingen. Archäologische Befunde zum Leben unter Felsschutzdächern in Urgeschichtlicher Zeit*. Veröffentlichungen der Urgeschichtlichen Sammlungen des Landesmuseums Hannover 43. Hannover, Landesmuseum Hannover.

Grote, K. and Terberger, T. (2011) Die prähistorischen Kinderbestattungen vom Abri Bettenroder Berg IX im Reinhäuser Wald bei Göttingen. *Archäologisches Korrespondenzblatt 41(2)*, 189–195.

Grünberg J. M. (2000) *Mesolithische Bestattungen in Europa*. Internationale Archäologie 40. Rahden/Westfalen, Verlag Marie Leidorf GmbH.

Haas, S. (1991) Neue Funde menschlicher Skelettreste und ihre Ergebnisse. In J. Hahn and C.-J. Kind (eds.) *Urgeschichte in Oberschwaben und der mittleren Schwäbischen Alb. Zum Stand neuerer Untersuchungen der Steinzeit-Archäologie* 37–38. Stuttgart, Konrad Theiss.

Hedges, R. E. M., Housely, R. A., Law, J. A. and Bronk, C. R. (1989) Radiocarbon dates from the Oxford AMS system: Archaeometry Datelist 9. *Archeometry* 31(2), 207–234.

Kapica, Z. and Wiercinski, A. (1995) Anthropological analysis of human skeletal remains from the Magdalenian period (younger Palaeolithic) recovered in Maszycka Cave, Olkusz Commune. In S. K. Kozlowski, E. Sachse-Kozlowska, A. Marshack, T. Madeyska, H. Kierdorf, A. Lasota-Moskalewska, G. Jabubowski, M. Winiarska-Kabacinska, Z. Kapica and A. Wiecinski (eds.) Maszycka Cave. A Magdalenian Site in Southern Poland. *Jahrbuch Römisch Germanisches Zentralmuseum* 1993 40(1), 245–251.

Kind, C.-J. (1987) *Das Felsställe. Eine jungpaläolithisch-frühmesolithische Abri-Station bei Ehingen-Mühlen, Alb-Donau-Kreis. Die Grabungen 1975–1980*. Forschungen und Berichte zur Vor- und Frühgeschichte in Baden-Württemberg 23, 365–372. Stuttgart, Konrad Theiss.

Kozlowski, S. K. and Sachse-Kozlowska, E. (1995) Magdalenian family from the Maszycka Cave. In S. K. Kozlowski, E. Sachse-Kozlowska, A. Marshack, T. Madeyska, H. Kierdorf, A. Lasota-Moskalewska, G. Jabubowski, M. Winiarska-Kabacinska, Z. Kapica and A. Wiecinski Maszycka Cave. A Magdalenian Site in Southern Poland. *Jahrbuch Römisch Germanisches Zentralmuseum* (1993) 40(1), 115–205.

Kunkel, O. (1955) *Die Jungfernhöhle bei Tiefenellern. Eine neolithische Kultstätte auf dem Fränkischen Jura bei Bamberg*. Münchner Beiträge zur Vor- und Frühgeschichte 5, 46–51. München, C.H. Beck.

Newell, R. R., Constandse- Westermann, T.-S. and Meiklejohn, C. (1979) The skeletal remains of Mesolithic man in western Europe: an evaluative catalogue. *Journal of Human Evolution* 8(1), 3–228.

Orschiedt, J. (1998) Ergebnisse einer neuen Untersuchung der Spätmesolithischen Kopfbestattungen aus Süddeutschland. In N. J. Conard and C.-J. Kind (eds.) *Aktuelle Forschungen zum Mesolithikum*. Urgeschichtliche Materialhefte 11, 147–160. Tübingen, Mo Vince.

Orschiedt, J. (1999a) *Manipulationen an Menschlichen Skelettresten: Taphonomische Prozesse, Sekundärbestattungen oder Anthropophagie*. Urgeschichtliche Materialhefte 13. Tübingen, Mo Vince.

Orschiedt, J. (1999b) Eine neolithische Sekundärbestattung aus dem Vogelherd bei Stetten, Baden-Württemberg. *Fundberichte aus Baden-Württemberg* 22, 161–172.

Orschiedt, J. (2002) Secondary burial in the Magdalenian: the Brillenhöhle in Southwest Germany. *Paléo* 14, 241–256.

Orschiedt, J. (2005) The head burials from Ofnet cave: an example of warlike conflict in the Mesolithic. In M. Parker-Pearson and I. J. Thorpe (eds.) *Warfare, Violence and Slavery in Prehistory*. British Archaeological Reports International Series 1374, 67–73. Oxford, Archaeopress.

Orschiedt, J., Auffermann, B. and Weniger, C.- G. (1999) *Familientreffen. Deutsche Neanderthaler 1856–1999*. Exhibition catalogue. Mettmann, Neanderthal Museum.

Orschiedt, J., Kegler, J., Gehlen, B., Schön, W. and Gröning, F. (2008) Die Blätterhöhle in Hagen (Westfalen): Vorbericht über die ersten archäologischen Untersuchungen. *Archäologisches Korrespondenzblatt* 38(1), 13–32.

Parzinger, H., Nekvasil, J. and Barth, F. E. (1995) *Die Býčí skála-Höhle: Ein Hallstattzeitlicher Höhlenopferplatz in Mähren*. Römisch-Germanische Forschungen 54. Mainz, Phillip von Zabern.

Peter-Röcher, H. (1997a) Menschliche Skelettreste in Siedlungen und Höhlen. Kritische Anmerkungen zu herkömmlichen Deutungen. *Ethnographisch Archäologische Zeitschrift* 38(3–4), 315–324.

Peter-Röcher, H. (1997b) Die Höhle von Býčí skála – Gaben an Götter und Ahnen. *Mitteilungen der Berliner Gesellschaft für Anthropologie, Ethnologie und Urgeschichte* 18, 47–56.

Peter-Röcher, H. (1998) Die Býčí skála-Höhle in Mähren: Opfer, Ahnenkult und Totenritual in der Hallstattzeit. *Das Altertum* 44, 3–30.

Pignat, G. (2002) Der Abri von Châble-Croix. In P. Curdy (ed.) *Die Ersten Menschen im Alpenraum: Von 50,000 bis 5000 v. Chr.*, 165–171. Exhibition catalogue. Zürich, Sitten, Walliser Kantonsmuseum.

Polenz, H. (1991) Opferhöhlen der vorrömischen Eisenzeit im südlichen Westfalen. In T. Hülsken, J. Niemeyer and H. Polenz (eds.) *Höhlen. Wohn- und Kultstätten des Frühen Menschen im Sauerland*, 33–71. Münster, Westfälisches Museumsamt.

Rothe, D. (1983) Ur- und frühgeschichtliche Funde in südwestfälischen Höhlen. *Karst und Höhle* 1982–83, 85–111.

Schauer, P. (1981) Urnenfelderzeitliche Opferplätze in Höhlen und Felsspalten. In H. Lorenz (ed.) *Studien zur Bronzezeit. Festschrift Wilhelm Albert von Brunn*, 403–418. Mainz, Philipp von Zabern.

Schmitz, R. W. and Pieper, P. (1992) Schnittspuren und Kratzer. Anthropogene Veränderungen am Skelett des Urmenschenfundes aus dem Neandertal- Vorläufige Befundaufnahme. *Rheinisches Landesmuseum Bonn* 2, 17–19.

Schultz, M.(1988) Osteologische Untersuchungen an den Skelettfunden aus den Ith-Höhlen. In M. Geschwinde (ed.) *Höhlen im Ith. Urgeschichtliche Opferstätten in Südniedersächsischen Bergland*, 127–149. Veröffentlichungen der Urgeschichtlichen Sammlungen des Landesmuseums Hannover 33. Hannover, Landesmuseum Hannover.

Schunke, T. (2004) Erstes mesolithisches Brandgrab. *Archäologie in Deutschland* 4, 57.

Street, M. Terberger, T. and Orschiedt, J. (2006) A critical review of the German Palaeolithic hominin record. *Journal of Human Evolution* 51, 551–579.

Svoboda, J. (2000) The depositional context of the Early Upper Palaeolithic human fossils from the Koneprusy (Zlatý kun) and Mladec caves, Czech Republic. *Journal of Human Evolution* 38, 523–536.

Teschler-Nicola, M. (2006) Taphonomic aspects of the human remains from the Mladeč Caves. In M. Teschler-Nicola (ed.) *Early Modern Humans at the Moravian Gate. The Mladeč Caves and their Remains*, 75–98. Wien, New York, Springer

Toussaint M., Brou L., Le Brun-Ricalens, F. and Spiers F. (2009) The Mesolithic site of Heffingen-Loschbour. A yet undescribed human cremation from the Rhine-Meuse- Schelde culture: anthropological, radiometric and archaeological implications. In P. Crombé, M. Van Strydonck and J., Sergant (eds) Chronology and Evolution within the Mesolithic of North-West Europe: Proceedings of an International Meeting, Brussels, May 30th–June 1st 2007, 239–260. Cambridge, Cambridge Scholars Publishing.

Trinkaus, E., Moldovan, O., Milota, S., Bılgar, A., Sarcina, L., Athreya, S., Bailey, S. E., Rodrigo, R., Mircea, G., Higham, T., Bronk Ramsey, C. and van der Plicht, J. (2003) An early modern human from the Pestera çu Oase, Romania. *Proceedings of the National Academy of Science* 100(20), 11231–11236.

Walter, D. (1985) *Thüringer Höhlen und Ihre Holozänen Bodenaltertümer*. Weimar, Museum für Ur- und Frühgeschichte Thüringens.

Walter, D. (1998) Die Höhlen im Kosackenberg bei Bad Frankenhausen. In S. Flindt and C. Leiber (eds.) *Kulthöhlen und Menschenopfer im Harz, Ith und Kyffhäuser*. Archäologische Schriften des Landkreises Osterode am Harz 2, 115–121. Holzminden, Miskat.

Wankel, H. (1882) *Bilder aus der Mährischen Schweiz und Ihrer Vergangenheit*. Wien, Holzhausen.

Wild, E. M., Teschler-Nicola, M., Kutschera, W., Steier, P., Trinkaus, E. and Wanek, W. (2005) Direct dating of Early Upper Palaeolithic human remains from Mladec. *Nature* 435, 332–335.

Wüller, B. (1999) *Die Ganzkörperbestattungen des Magdalénien*. Universitätsforschungen zur Prähistorischen Archäologie 57. Bonn, Habelt.

Chapter 16

Late Caucasian Neanderthals of Barakaevskaya Cave: Chronology, Palaeoecology and Palaeoeconomy

Galina Levkovskaya, Vasiliy Lyubin and Elena Belyaeva

This paper deals with materials from a Late Neanderthal site at a small cave named Barakaevskaya Cave (NW Caucasus), where anthropological remains of Neanderthal inhabitants (a child's mandible and several teeth) have been found. The lithic industry was attributed to a local variant of the Typical Mousterian. Multidisciplinary studies of the site allowed for the reconstruction of various aspects of palaenvironments and the paleoeconomy. In spite of the shallowness of the cultural layer (0.15–0.30m) it yielded evidence of a significant climatic shift. The Mousterian occupation of the site began under glacial conditions at the end of OIS 4 and continued during OIS 3 up to a very warm optimum correlated to the Moershooft interstadial dated to about 46,000–44,000 BP. Five rapid climatic oscillations are registered within this span. Certain changes in settlement patterns and subsistence strategies along this sequence have also been revealed. For the cold stage, the data suggest a very dense occupation within the limited and artificially protected space of the cave as well as low mobility in the surrounding landscape. Later, the Neanderthals occupied a greater area of the cave and became somewhat more mobile in relation to raw material procurement and hunting. In all periods the principal game was bison although the proportion to other hunted animals varied. To judge by intra-site activities that included very intensive lithic knapping and tool manufacturing (more than 26,000 chert artifacts, 109 bone retouchers and countless chips) as well as the treatment of carcasses and skins, the site was used as a base camp. The scanning electron microscope pollen data shows that Neanderthal occupation began when most plants were producing underdeveloped and dwarf pollen grains in conditions of their upper limits and law temperatures. Mousterian occupation of the cave was finished before the optimum of interstadial. Some phytoliths, diatom algae and some hundreds of pollen grains of land and water plants were found inside the cave, which is located at about 80m above the Gubs River. The pollen and palaeobotanical identified flora includes micro-remains of firewood, vitamins, medicine, oil, wickerwork, mat weaving, wood polishing, skin tanning, paint and melliferous plants. The research shows the important role of water and land plants in every day life of Neanderthals and the preferable use by them of the multifunctional plants. The diet of the Neanderthals included large quantities of meat, leaf, rut and stem vegetables, nuts, fruit and cereal starch plants. 79,600 pieces of small bones of domestic waste of the Neanderthals were found inside the cave (Baryshnikov 1994). 32 taxa of food plants were also identified. The comparing of palaeozoological and palaeobotanical data shows that the deficit of carbohydrates existied in the diet of Neanderthals. The cave was the place of the full cycle of flint tools making and their utilization: primary knapping, initial edging, final chapping, utilization, rechapping and reutilization (Lyubin 1994). Some zones of economical activity of Neanderthals are differentiated inside the cave on the bases of statistical data about distribution of retouchers, flint tools, chips and even phytolithhs, much quantity of which were found inside the fireplace.

Introduction

The Mousterian site of Barakaevskaya Cave is located in the NW Caucasus, in a canyon of the Gubs River, which belongs to the Kuban River system (Fig. 16.1). In the canyon a group of cave sites has been found that contains occupation levels with Middle Palaeolithic, Upper Palaeolithic and Neolithic industries. Neanderthal anthropological remains were found in the Gubs Canyon in two Mousterian cave sites (Barakaevskaya and Monasheskaya) and *Homo sapiens sapiens* in the Satanai rockshelter (Kharitonov and Romanova 1984; Amirkhanov 1986; Belyaeva *et al.* 1992; Autlev and Lyubin 1994; Zubov *et al.* 1994). Three cave sites (Barakaevskaya Cave, Monasheskaya Cave and Gubskaya Rockshelter) with rich Mousterian industries and a great number of animal bones can be defined as base camps. This paper is focused on the Mousterian site of Barakaevskaya Cave, which has yielded archaeological, anthropological, palaeozoological, pollen and scanning electron microscope (SEM) palaeobotanical/palynological materials in combination with palaeozoological materials. The data obtained through multidisciplinary studies of the site under consideration permit the examination of such complicated issues as palaeoenvironmental dynamics, chronostratigraphy and, in particular, settlement patterns and palaeoeconomy.

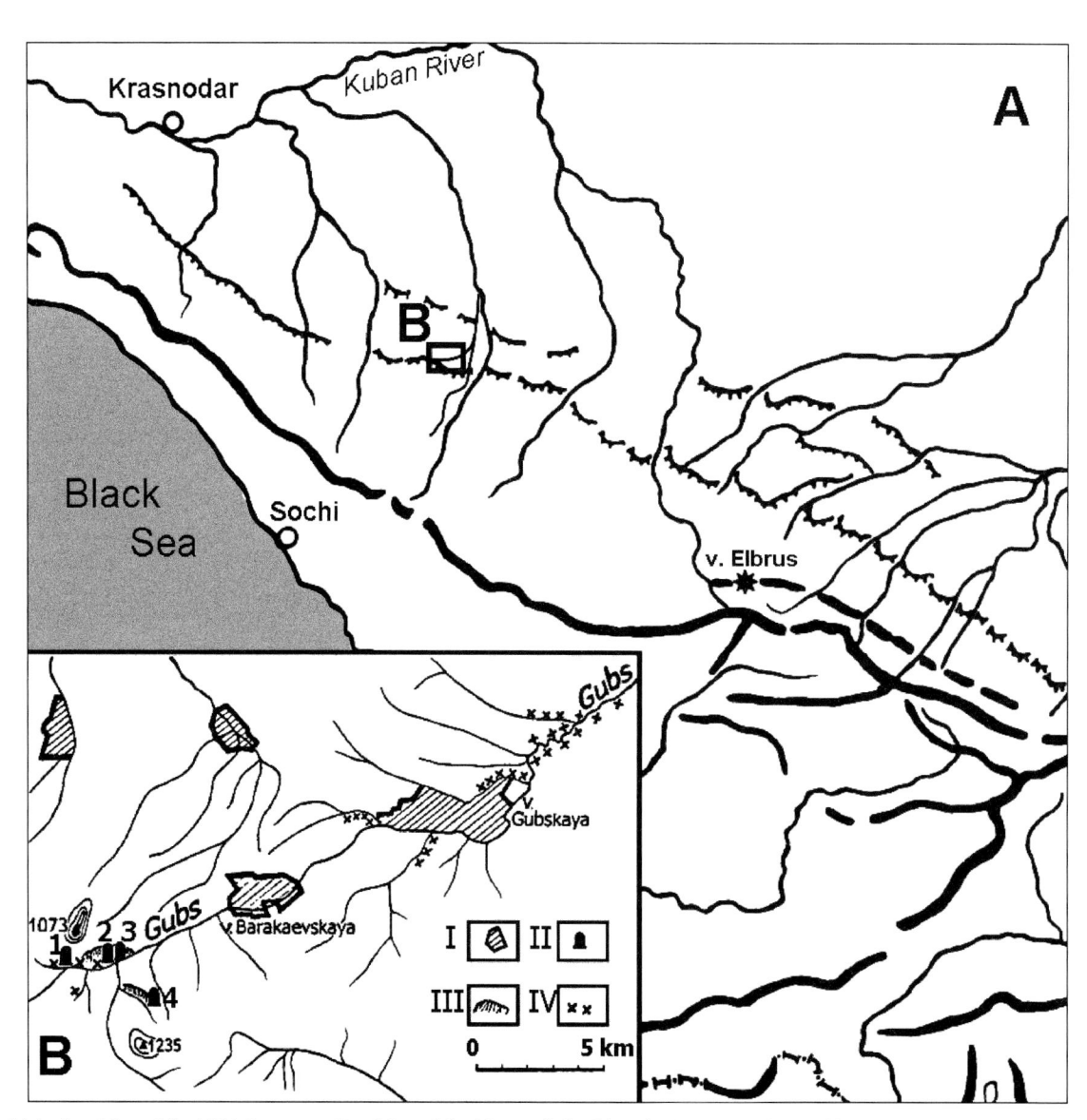

Figure 16.1: A – Map of the NW Caucasus; B – Map of the Upper Gubs River basin. I – modern villages; II – Mousterian cave sites; III – limestone cliffs; IV – surface occurrences of Mousterian lithics. 1 Barakaevskaya Cave, 2 Gubs Rockshelter N1, 3 Monasheskaya Cave, 4 Autlevskaya Cave.

A brief description of the Mousterian site at Barakaevskaya Cave

Location of the site

Barakaevskaya Cave is located in a limestone cliff extending along the northern side of the Gubs Canyon incised in spurs of the Scalisty ('rocky') ridge (Fig. 16.1). According to the geologist S. A. Nesmeyanov (1994), the Gubs Canyon began to form in the beginning of the Late Pleistocene. Parallel to the river, down-cutting in the walls of the canyon, there emerged several levels of erosional cavities and rock-shelters. A considerable number of them were more or less suitable for settlement and contain Palaeolithic and Mesolithic occupation levels. Barakaevskaya Cave is a small cave (Fig. 16.2) situated at a height of 834m above sea level and around 90m above the Gubs River bed, near the upper boundary of the modern broad-leaved forest zone. Just above the cliff with the cave lies a plateau covered with meadows and copses. In close proximity to Barakaevskaya Cave there is one of several karstic springs existing in the Gubs Canyon. The canyon is also rich in nodular small-sized dark chert. In spite of its mediocre knappable properties this local raw material predominated in all the Palaeolithic industries of the Gubs sites including that of Barakaevskaya Cave. Colored chert of high quality, transported from the lower reaches of the Gubs River, was also used to some extent.

History of the investigation

The Mousterian site at Barakaevskaya Cave was found in 1976 by P. U. Autlev, who discovered most sites of the Gubs River basin in the 1960–70s (Autlev and Lyubin 1994). In 1979–1981 the site at Barakaevskaya Cave was almost entirely excavated by V. P. Lyubin. It was established that under thin Holocene deposits

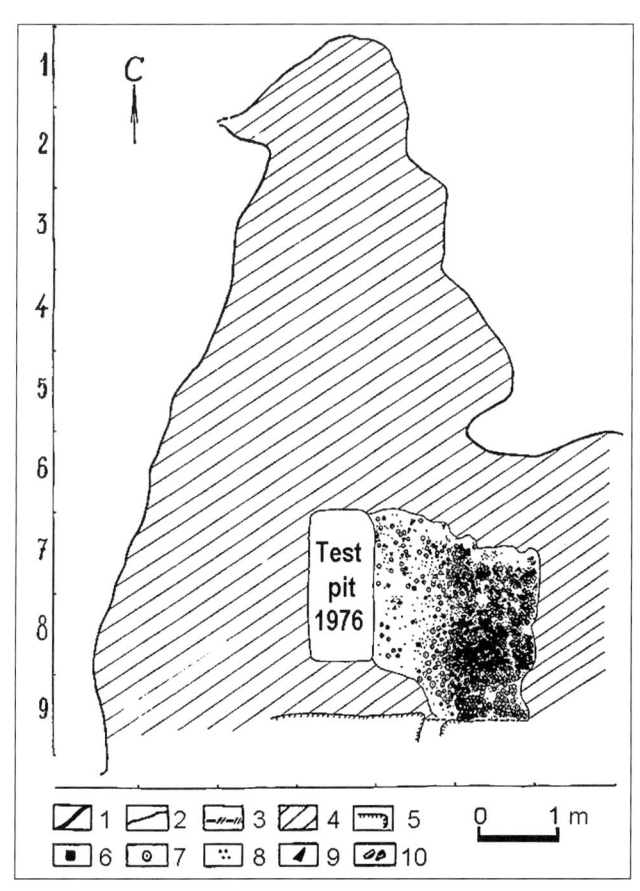

Figure 16.2: Mousterian site at Barakaevskaya Cave. Plan of horizon 4. 1 Outlines of the cave walls; 2 limits of the Mousterian deposits; 3 the southern border of the excavated area; 4 rock bed of the cave; 5 fallen blocks; 6 cores; 7 flakes and blades; 8 chips and chunks; 9 tools; 10 large bones (after Lyubin 1994).

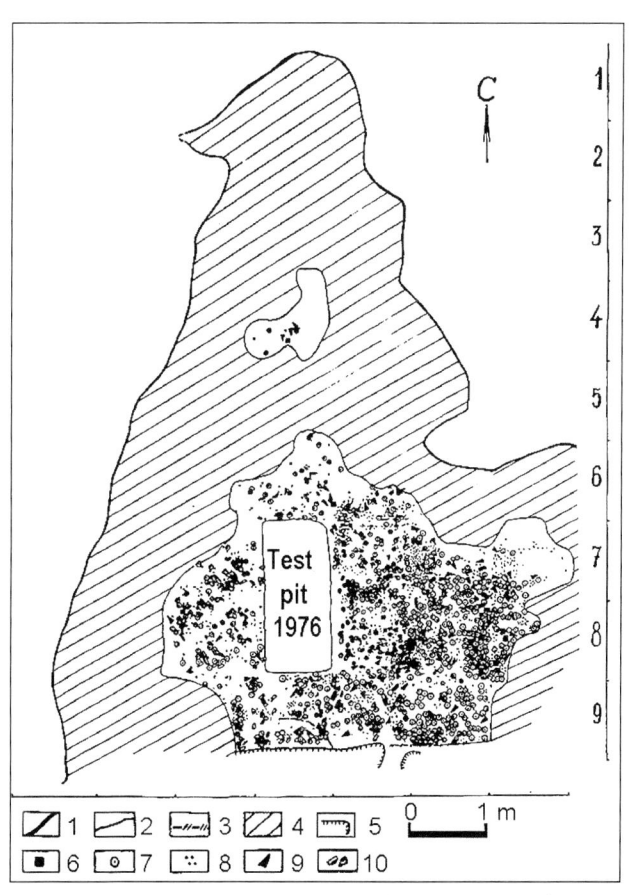

Figure 16.3: Mousterian site at Barakaevskaya Cave. Plan of horizon 3. 1 Outlines of the cave walls; 2 limits of the Mousterian deposits; 3 the southern border of the excavated area; 4 rock bed of the cave; 5 fallen blocks; 6 cores; 7 flakes and blades; 8 chips and chunks; 9 tools; 10 large bones (after Lyubin 1994).

with mixed archaeological material (Eneolithic-Middle Ages) there is only a single Mousterian (0.15–0.30cm thick) layer lying on the rock bed and covered with a travertine (stalagmitic) floor (0.01–0.07m) which seals off and presents a specific 'cap' of the Neanderthal level of occupation. The Mousterian cultural layer has been revealed primarily in an area of around 20m² in the southern part of the cave. The layer composed of loamy sediments and angular limestone clasts was covered with a travertine floor where phosphatic and calcitic cement predominated (Cherniakhovskiy 1994).

The Mousterian layer of Barakaevskaya Cave has yielded more than 20,000 chert artefacts and around 80,000 animal bones including several dozens bone retouchers (Filippov and Lyubin 1994; Lyubin and Autlev 1994). The lithic industry of Barakaevskaya Cave (Fig. 16.4, see the end of the article), like other Mousterian industries of the Gubs River canyon, can be defined as a local variant of Typical Mousterian (Lyubin 1977; Lyubin and Autlev 1994; Belyaeva 1999). Additionally, anthropological remains were found assigned to Neanderthal males: a child's mandible (Fig. 16.8, see the end of the article) and isolated teeth (Zubov *et al.* 1994). Different results of multidisciplinary research were published for Barakaevskaya Cave in 'The Neanderthals of Gups Canyon in the Northern Caucasus' (Lyubin 1994a).

Research methods

The Mousterian site at Barakaevskaya Cave has been studied comprehensively and using multidisciplinary

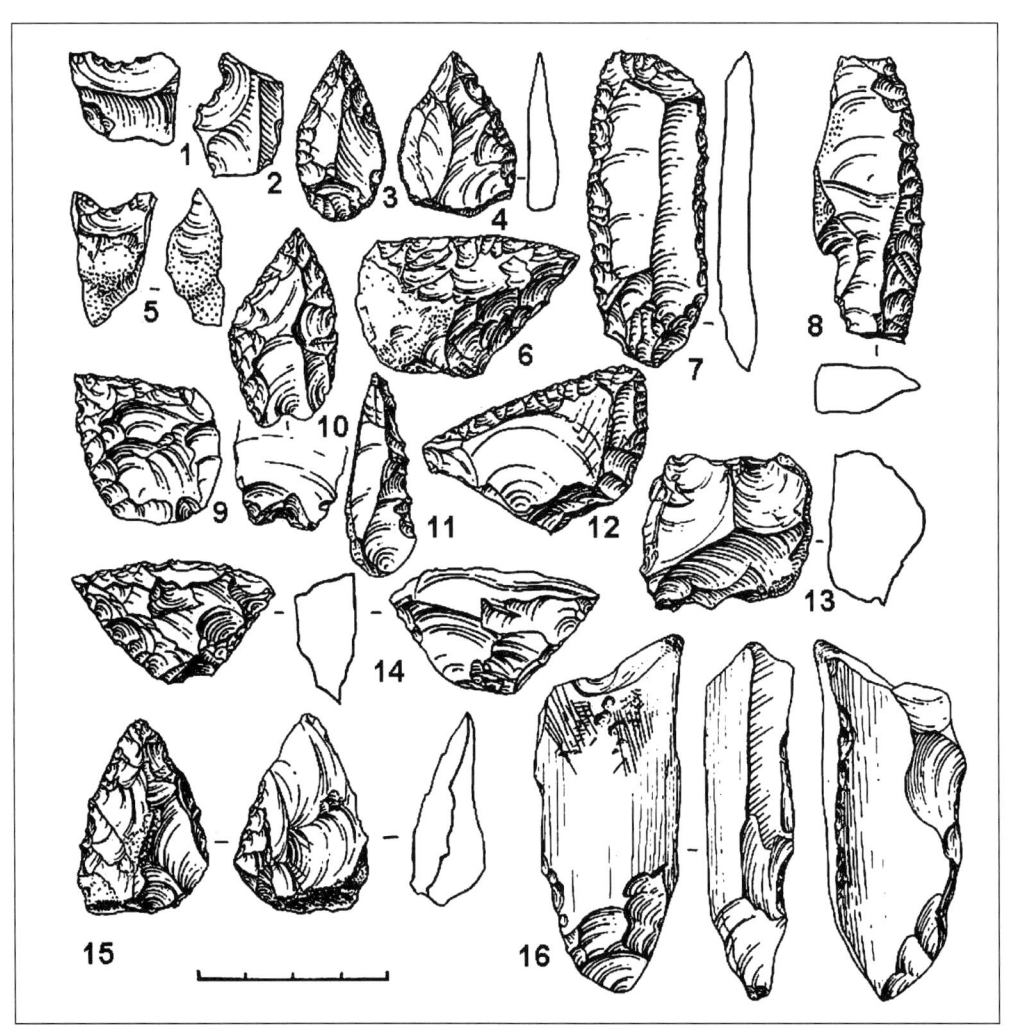

Figure 16.4: Mousterian site at Barakaevskaya Cave. Lithic industry: 1, 2, 5 notched tools; 6, 9, 12, 14 dejete scrapers; 7, 8 side-scrapers; 10, 15 convergent scrapers; 11 denticulated tool; 13 core; 16 bone retoucher. Note: the small size of most artefacts and a high intensity of retouch on them is a particular feature of the Barakaevskaya Mousterian layer.

approaches. Use-wear analysis, archaeological, anthropological, palaeozoological, geomorphological, microsedimentological and palynological research have been conducted. The methodological aspects of these studies have also been published (Lyubin 1994a). Some methodological innovations that allowed for the recovery of information about the plant gathering of the Neanderthals are described in the corresponding section of this article. In spite of the shallowness of the deposit, the thin (15–30cm) Mousterian layer was excavated according to four conventional artificial spits (horizons), to take into account possible environmental and industrial changes. All the sediments from each horizon were transported to the stream for wet-sieving. The 1–2mm mesh soil-sieves permitted the collection of numerous remains of small mammals and birds. Additionally, finer sieves (0.5mm) were used in the search for palaeocarpological plant remains and charcoals in control samples of all the excavated horizons.

Palaeoenvironments

Subdividing the Mousterian layer into four horizons permitted the observation of several rapid changes in pollen and faunal composition. On the basis of palynological and palaeozoological evidence, an attempt was made to reconstruct the palaeoenvironment dynamics during the period of Mousterian occupation of Barakaevskaya Cave. Studies of fauna have revealed a broad environmental trend, whereas the pollen analysis demonstrates both this same trend as well as rapid climatic oscillations within this trend. Nevertheless, both methods indicate a clear shift from very cold conditions with a predominance of subalpine communities (zone 1) or periglacial steppes (zone 2) to a warm interstadial with a low positioning of the forest belt. Both methods show that the area was not totally afforested throughout the period of the formation of the Mousterian layer (Table 16.1).

The type of climate reconstructed on the basis of sedimentological data

According to the sedimentological data (Lyubin 1994a, Figs 13–14) the lower part of the Mousterian layer of Barakaevskaya Cave (horizons 4 and 3), with evidence of desquamation (exfoliation) processes, was formed under conditions of a cold but rather continental type of climate and in correspondence with alternations of temperatures above and below zero. But the sedimentation of the Mousterian stalagmite 'floor' (capping the Neanderthal archaeological layer) is an indicator of wet but warm conditions.

Dynamics of palaeoenvironments

Pollen data allowed the differentiation of five rapid global climatic oscillations inside the span of the Barakaevskaya Mousterian sediments' formation. The five vegetational phases and pollen zones I–IV correspond with these rapid oscillations. They are indicators of some of the stages of climate improvements within one cold-warm cycle. The following rapid climatic and vegetational phases are represented (see Table 16.1, 16.6 and Fig. 16.9):

- *zone I* (excavated horizon 4) – extreme phase of the glacial or its cryoxerophilous phase with a rather continental type of the climate; conditions of the upper part of subalpine belt; a geo-botanical crisis for most flower plants at their upper limits (according to the SEM data, Fig. 16.5);
- *zone II* (excavated horizon 3) – first phase of transition from the glacial to interstadial; a combination of upper-mountain 'island' forests, periglacial *Poaceae* steppe and subalpine vegetation. The same type of the landscapes is reconstructed for the region on the bases of palaeozoological data (Baryshnikov 1994): combination of steppe and subalpine-alpine elements is registered in palaeozoological complex (see: Tables 16.1 and 16.6);
- *zone III* (excavated horizon 2) – beginning of the interstadial; combination of steppes, meadows and 'island' middle-mountain broad-leaved forests with large quantity of alder, hazel and with admixture of pine and birch.
- *zone IIIa* (excavated horizon 1) – thermoxerophilous phase of the interstadial; combination of steppes, meadows, dry bushes (*Ostrya carpinifolia, Carpinus orientalis*) and low-mountain broad-leaved polydominant (hornbeam-elm-hazel) forests with rare exotics which are absent in the modern flora of studied region. They were presented by *Castanea, Osmunda* (Colhis elements) and *Ostrya carpinifolia* (see Table 16.1) Appearance of some forest animals; domination of *Ovis orientalis* under *Capra caucasica* in palaeozoological complex.
- *zone IV* (the stalagmite cap of the Mousterian period)
- optimum of the interstadial; with a wet but more continental (than now) type of climate. Combination of steppes, meadows and 'island' low-mountain broad-leaved forests formed by hornbeam (*Carpinus*) and exotic walnut (*Juglans regia*). Now *Juglans* is present in cultural flora of Caucasus only.

Horizons of excavations	Pollen zones	Palynological data on vegetation and climate (after: Levkovskaya, 1994, with addition)		Palaeozoological data on vegetation and climate (after: Baryshnikov, 1994)
		Phytophases	Climate stages	
Pleistocene travertine overlaying the Neanderthal layer	IV	Rising of the vegetation belts. Forest-steppe with various "island" forests and meadows. Exotics of the Caucasus range.	IV. Low mountain broad-leaved forests with codomination of *Carpinus* and *Juglans regia* (exotic elements for natural flora of the region) and steppes.	Optimum of a warm interstadial (forest steppes with exotic *Juglans* societies).	Rising of the vegetation belts. *Ovis orientalis* became more common than *Capra caucasica*. The appearance of forest animals (*Apodemus sylvaticus*, *Sus scrofa*, *Cervus elaphus*, etc.) indicates a more afforested environment. *Bison priscus* is dominant in the fauna complex.
1	III a		IIIa. Low mountain forests with *Carpinus, Ulmus, Corylus*. Coniferous trees are rare. Exotics: *Castanea, Ostrya* (xerophyte), *Osmunda*.	Transition to optimum of interstadial	
2	III		III. Middle mountain polydominant broad-leafed forests. *Betula* and *Pinus* are present. Maximum of *Alnus, Corylus* and Poaceae.	Beginning of interstadial (appearance of broad-leaved trees).	Confluence of the subalpine and steppe belts (reconstructed conditions are mostly similar to those of the pollen zone II).
3	II	Combination of cold and warm forest-steppe landscapes. Upper mountain pioneer forests (*Betula, Alnus*) steppes, meadows and residual societies of phase 1 (*Bryales* mosses, *Pinus* and *Betula nana* – *Alnaster* yerniks)		Transition from cryoxerophilous stage of glacial to interstadial.	*Capra caucasica* became more frequent than *Ovis orientalis*. Appearance of *Spermophilus* cf. *musicus, Cricetus cricetus, Equus ferus* etc. indicates steppe landscapes.
4 (Neanderthal child's mandible)	I	Subalpine belt. Upper limits of *Pinus, Juniperus, Corylus, Alnus, Betula*. Combination of societies of *Juniperus, Pinus*, mosses (*Bryales*), meadows, dwarf shrub yernik formations (*Betula nana* – *Betula humilis* – *Alnaster*). Most plants produce dwarf pollen grains.		Extreme of glacial. Frost weathering of the cave roof. Geobotanical crisis due to lowering of the average summer temperatures by about 7.5°C.	Finds of alpine species: *Chionomys nivalis, Chionomys gud, Marmota paleocaucasica*. *Bison priscus* is dominant in the fauna complex.

Table 16.1: Barakaevskaya Cave site (the present broad-leaved forest belt). Reconstruction of palaeoenvironmental dynamics on the basis of pollen and palaeozoological data. (The palaeozoological reconstructions are based on 670 identified bones of animals (Baryshnikov: 1994, 69), and the pollen reconstructions on 2119 identified pollen grains).

The climatostratigraphy and chronology of the Barakaevskaya Mousterian layer

The climatostratigraphic basis of the chronology of the site

Palaeozoological and pollen data (Table 16.1) demonstrate that the Mousterian layer began to form during the subalpine phase. Even the upper part of the subalpine belt is reconstructed on the basis of SEM studies because not only trees but also most of the flower plants were under stress and mostly produced underdeveloped and dwarf pollen grains (Fig. 16.5).

The cave is currently located in the lower part of a broad-leaved forest belt. The elevation of Barakaevskaya Cave is 834m above sea-level, but the modern border of the subalpine and forest belts is located at 2400–2300m above the sea-level (Pavlov 1948, 362). There is a Caucasian temperature gradient: each 100m increase in altitude is related to a corresponding temperature decrease of 0.5°C (Povarnitsyn 1931). Thus the Mousterian lowering boundary of the subalpine belt was at about 1500m, and that of the summer average temperatures reached an average of approximately +7.5°C. Modern average summer temperatures in July in the Gubs canyon are approximately +20°C (Gvozdetskiy et al. 1966), but in the subalpine belt they are about +12.5°C (Pavlov 1948, 367). The same low average summer temperatures (between +12–14°C) are now typical for the most northern parts of Scandinavia according to the Agroclimatic Atlas of the World (see Holtsberg 1972, 35). The glacier sheet could have grown in this northern region in the conditions of wet climate and decreasing global temperatures of 7.5°C. This means that the Mousterian layer of Barakaevskaya Cave began to form during one of the glacial periods of the Upper Pleistocene. The Pleniglacial period (Wurm II glacial of the 18-0 stage 4, Kalininskiy glacial of the Russian Plain) dates (after Mellars 1996, 23; Gamble 1999, 189) to between (Mellars/Gamble) 75,000/71,000 – 60,000/58,000 BP, and the Interpleniglacial period to between 60,000/58,000 – 28,000/25,000 BP. According to stratigraphical scheme of Russian plain (Zarrina et al. 1980), the age of the Middle Valdai (Middle Wurm) glaciation was 57,000–50,000 BP and of the Late Valdai glaciation 23,000–17,000 BP. New data on the glaciation history (Mangerud et al. 2004, 111) show that the ages of the Upper Pleistocene glacier sheets of the northern region of Russia date to about 90,000–80,000, 60,000–50,000 and 20,000 BP. The glacial sheets of the span 90,000–80,000 and about 20,000 years had large sizes because they have been formed in uncontinal types of the cold climat. The glacial sheet advance about 60,000–50,000 was larger than during 90,000 in the Barents Sea region whereas in the area further to the east it was smaller due to the relatively continental type of its climate. The continental type of cold and moderately dry climate is reconstructed for the studied Gubs region on the basis of sedimentological and pollen data. It means that the Barakaevskaya cold phase corresponds to the glacial about 60,000–50,000 years. The age of Barakaevskaya horizons 4–3 (the end of the glacial phase about 60,000–50,000 years ago and the first phase of transition to the interstadial) is about 50,000 years BP.

Pollen data show that different horizons of the Neanderthal layer and the stalagmite 'cap' of the Mousterian layer were formed during the transition from the glacial phase to the optimum of the interstadial and the stalagmite cup – during this optimum. Hundreds of pollen diagrams and radiocarbon dates of the sediments of the last glacial-interglacial (Holocene) cycle show that the transition from the end of the Late Glacial period (about 10,000 BP) to the beginning of the Holocene optimum (about 6600 BP) could have continued for approximately 3400 years. This possible analogy shows that the five Barakaevskaya climatic phases of transition from the glacial period about 60,000–50,000 BP to the optimum of the interstadial could be connected with a span of approximately 3400 years. The age of the Barakaevskaya optimum could be about 46,600 BP.

The chronology of the Barakaevskaya Mousterian layer (in the context of some climatostratigraphical correlations)

The age of about 46,600 BP of the Barakaevskaya Mousterian optimum corresponds with the Moershooft interstadial (46,000–44,500 BP, Aitken and Stokes, 1997; 46,000–44,000 BP according to the standard pollen diagram of Moershooft interstadial, (Van der Hammen et al. 1967; Ran, 1990) and the rapid oscillation 12 of the stage 3 according to some isotope scales. The age of the oscillation 12 is about 46,000–44,000 BP on the isotope scales: the 18-0 GISP2 Greenland scale (Johnsen et al. 2001) and 13C stalagmite Villars cave scale (in France) (Genty et al. 2003).

The upper part of the Mousterian layer of Barakaevskaya Cave has the following radiocarbon dates (Belyaeva et al. 2009): 34,500±600 BP (Ki-10207), layer 2, horizon 1; and 35,200±600 BP (Ki-10208), layer 2, horizon 2.

But these dates are possibly more recent than the real age of the Barakayevskaya Mousterian layer, as is demonstrated by the following considerations.

First, the chronology of the Barakaevskaya cold phase (about 50,000 BP) and the optimum (approximately 46,500 BP) is correct in the chronological context of some palaeoenvironmental phases of the Kostenki 12 site from the Russian Plain. The Kostenki 12 phase of the glacial type of the climate (Levkovskaya *et al.* 2005a, 127 and Figure 2, zone 2) has the following IRSL/OSL dates (Anikovich *et al.*, 2005, 75; Hoffecker *et al.* 2008, 864) for one sample (UIC-917) obtained by S. L. Forman in his Chicago laboratory: 50,520±4,380, 51,330±4,950 and 52,440±3,850 years ago. The sediments of Glinde+Moershooft interstadials (in Kostenki 12: zones 3–8, the time of formation of palaesoils D, C and B and co-domination of elm forests on flood-terraces and herbs steppes, meadows on watersheds) have 15 IRSL/OSL dates of the span 50,520–43,470 years obtained for four samples (UIC-915, UIC-945, UIC-946 and UIC-947).

Second, two desquamation horizons were formed during Mousterian time in Gubs canyon according to the data for Monasheskaya cave layers IIIB and II (Belyaeva 1999, 46: fig. 15). The Barakaevskaya and lower Monasheskaya desquamations horizons are connected with the non-dry type of the cold climate because even the cryoxerophilous phases of glacial cycle in Barakaevskaya was moderately dry (see: Table 16.1). These both desquamation horizons are lying under the sediments which have been formed before some interstadial optima during which the *Juglans* appeared in the forests of Gubs canyon.

The upper Monasheskaya desquamation horizon (like the desquamation horizon of layers 4A, 4B and 4C of Matouzka cave site from the Kuban basin of North-West Caucasus) had formed during not only very cold but also extreme dry climatic phase of isotope stage 3 (Beliaeva 1999; Golovanova *et al.* 2006; Levkovskaya 2006). These both desquamation horizon are lying above the sediments which have been formed after one optimum with *Juglens* forests (in Barakaevskaya cave) and after two *Juglans* optima (in Monasheskaya).

The extremely dry cold climatic phase of isotope stage 3 had been correlated with H4 event at Caucasus and Russian Plain (Levkovskaya *et al.* 2007a, 2007b). Some problems of palaeogeography of this event and its correlations with archaeological layers were discussed at the Palaeolithic session of the European Association of Archaeologists in Croatia (Levkovskaya *et al.* 2007c).

The climate of H4 event was very cold (icebergs appeared near Spain), but it was not wet, so the continental glacial sheet was not formed in the northern parts of Eurasia due to its dry type of the climate (after Mangerud *et al.*, 2001). The continental glacial sheets developed in Eurasia between 90,000–80,000, 60,000–50,000 and 20,000 years ago (Mangerud *et al.* 2004) but the dry and cold extremum H4 was registred inside the span 50,000–30,000 years ago. The upper mountain dry pine forests are now dominants in the Caucasian regions with continental type of the climate, on the contrary to the wet spruce and fir forests that were typical for the upper mountain belt of Colchis geobotanical province with European maximum of precipitations.

Third, the Caucasian Mousterian cold and dry phase of the climate with domination of pine forests has the following uncalibrated radiocarbon dates: 37,250±500 BP (Ki-10209) from the upper desquamation horizon of Monasheskaya cave (Gubs canyon), 35,680±480 BP (Gr-631) from Malaya Vorontsovskaya cave site (Sochi Black Sea area of Caucasus) and 35,470±590 BP (LU-545) from Dziguta pit-bog (Sukhumi Black Sea area of Caucasus) TR/UR date 35,000±2,000 years was obtained for the upper Mousterian layer of Akhshtyr cave site from the Sochi area of Caucasis (Arslanov and Gej 1987; Lyubin, 1989; Levkovskaya, 1992, 2006; Chistyakov, 1996; Koulakov *et al.*, 2004; Beliaeva *et al.* 2009).

Fourth, the cold and the driest Middle Wurm phase of the Russian Plain (Bugrovskaya cryophase) was about 37,000–40,000 BP (see the stratigraphic scheme for the Russian Plain after Zarrina *et al.* 1980, Spiridinova 1991). It was the time of formation of the upper desquamation horizon of Mousterian sediments of Gubs canyon, because the upper desquamation layer of Monasheskaya cave (the Mousterian layer II) has radiocarbon date 37,250±500 BP (KI-10,209) (Beliaeva *et al.* 2009).

Thus, Barakaevskaya Mousterian layer with unreally young radiocarbon dates of the span 34,500–34,200 BP is older than *Juglans* optimum of Gubs canyon correlated with Moershooft interstadial (about 46,500–44,00 BP) and than the phase of extremely dry continental type of cold climate on the Russian plain (Bugrovskaya cryophase about 40,000–37,500 years BP) and in Caucasus (37,250–35,470 BP).

Archaeological materials from Barakaevskaya Cave do not contradict these inferences. The lithic industry from this site is clearly similar to that of Monasheskaya Cave but may reflect an earlier stage of the local Mousterian tradition.

Archeological evidence relating to settlement and land-use patterns

Settlement patterns

Judging by the distribution of the Mousterian cultural deposits at Barakaevskaya Cave, the Neanderthals settled primarily in the sunny southern half of the cave no more than 4–5m from the modern entrance arc. The main living space is distinguished by its considerable thickness of Mousterian deposits (0.25–0.30m) and a high density of finds over an area of 20m² (Fig. 16.2). A much thinner Mousterian deposit also exists in front of the cave mouth, immediately below some large fallen limestone blocks. In the Mousterian period, the cave was thus somewhat larger and the southern periphery of the site was also under the rock roof. A narrow, dark and damp northern part of the cavity appears to have been uncomfortable insofar as the thinnest Mousterian deposit with isolated finds was located there.

On the whole, however, the Mousterian layer of Barakaevskaya Cave was rich in a great number of cultural remains. In the excavation area 21,402 chert artefacts, 109 bone retouchers, and 79,600 animal bones dominated by splintered or trampled pieces have been found. Additionally, wet-sieving of the sediments yielded countless numbers of micro-chips and chunks of chert. In general, all the evidence from the Mousterian levels at Barakaevskaya Cave points to very intensive, long-term and almost continuous occupation, i.e. the site was used as a base camp (Lyubin 1994b). However, patterns of habitation at the site appear to have changed over time. The maximum density of finds (excluding chips 5800 bones and 1300 lithics/m²) as well as an abundance of ash indicating the use of fire was observed in the lowermost horizon 4, which corresponds to the most severe climatic conditions. The majority of the cultural remains of this level were concentrated in a small depression (4 m²) in the rock floor (Fig. 16.2). In the upper horizons, formed during warmer climatic conditions, the density of cultural remains decreased somewhat and they were more or less evenly distributed (Fig. 16.3). The changes in these settlement patterns may be explained primarily in terms of adaptations to climatic trends. In cold periods the Neanderthals probably preferred denser habitation within a confined part of the cave warmed by bonfires. Moreover, one may assume the presence of additional constructions to provide warmth in the limited space (Belyaeva 2004). In the later periods of Mousterian occupation during more favorable climatic conditions made it possible to use the entire inhabitable space of Barakaevskaya Cave and, most likely, some places outside the cave.

Judging by the chronostratigraphic correlations discussed above, the Neanderthal sites at Barakaevskaya Cave and Monasheskaya Cave appear to have been partly co-existent, although the uppermost levels of the latter, which show an increasing intensity of occupation, are younger than the Mousterian layer of Barakaevskaya Cave. Therefore, in the final stage of Neanderthal occupation in the Gubs micro-region, Monasheskaya Cave succeeded Barakaevskaya Cave as the main base camp. It may seem somewhat strange that Barakaevskaya Cave was left by the Neanderthals during favourable interstadial conditions. However, the travertine floor that covers the Mousterian layer, and which was formed through intensive dripping from the cave roof, suggests that in warm but moist climatic conditions the cave became too damp and, probably, less than comfortable for further habitation (Belyaeva 1999, 2004).

Land-use patterns and subsistence strategies

Important data concerning these questions have been obtained through detailed lithic analyses. As mentioned above, the Middle Paleolithic sites in the Gubs micro-region contain lithic assemblages attributed to a local variant of the Typical Mousterian, i.e. the most common type of Mousterian industries (Bordes 1961). These Mousterian industries belong to a single cultural tradition clearly defined by a unique complex of technological, morphological and typological characteristics (Lyubin 1977; Belyaeva 1999). At the same time, at all of the Gubs sites every lithic assemblage shows a number of special features that seem to reflect non-traditional factors of industrial variability, such as the intensity of occupation, raw material economy, set of activities, and so on (Belyaeva 1999; 2004).

It should be noted, first of all, that the lithic industry of Barakaevskaya Cave is distinguished from the other industries of the Gubs lithic tradition by the very small average size of all its components (Lyubin and Autlev 1994). Of 60 cores (0.3 per cent of the collection), almost all are very small and exhausted (Fig. 16.3). The majority of flakes (66.7 per cent) do not exceed 4cm, and tools are also dominated chiefly by small pieces (Fig. 16.4). It is noteworthy that the tool dimensions resulted not only from small-sized blanks but also from a high degree of reduction. For example, from a typological point of view, the convergent and dejete side scrapers from Monasheskaya Cave and those from Barakaevskaya Cave have very similar shapes. However, the latter are usually smaller but thicker and show a much more invasively stepped retouch that testifies to the continuous re-use and re-sharpening of these implements. All of this indicates a very economical and intensive utilization of raw materials. In addition, the Mousterian collection of Barakaevskaya Cave differs from that of Monasheskaya Cave in its smaller percentage of good multi-colored cherts, which were gathered in gravels along the lower reaches of the Gubs River, at a distance of 15–20km from the cave sites. Because the combined data point to a deficiency of both local and, in particular, imported cherts, the Mousterian inhabitants of Barakaevskaya Cave appear to have faced certain difficulties in the procurement and transportation of raw material. Given that all the sources of chert were not too distant from the site, one may suppose that the prevailing cold conditions compelled the Neanderthals to considerably limit their mobility and range of exploitable areas. This inference is in accordance with the evidence of an extraordinary degree of utilization of the fauna found at the site location, i.e. splitting of most bones to small pieces to extract marrow (Belyaeva 2004).

In spite of partly limited mobility in such cold conditions, the Neanderthal settlers of the Gubs area intensively exploited their surrounding landscape. Besides raw material procurement their subsistence strategy was aimed at providing themselves with game animals as well as plant foodstuffs (see below). Mousterian Neanderthals have been established as active hunters and no clear evidence of scavenging has been found. To judge by the very scarce remains of predators (wolf, fox and hyena) and only rare traces of their gnawing, the enormous accumulation of animal bones at Barakaevskaya Cave was primarily a result of Neanderthal activity. The predominance of adult and sub-adult animals in the mortality profiles for ungulates found in the Mousterian layer of the cave confirms that they were the game of Neanderthal hunters (Hoffecker and Baryshnikov 1998). During all the periods of their occupation in the Gubs area and in all climate conditions the principal game was steppe bison (*Bison priscus*). Bison was probably not hunted in the canyon but rather on the plateau above the cave. Among other game animals available in different periods were Caucasian mountain goats (*Capra caucasica* Guld. et Pal.), Asiatic mouflon (*Ovis orientalis* Gmel.), red deer (*Cervus elaphus* L.), giant deer (*Megaloceros giganteus* Blum.), chamois (*Rupicapra rupicapra* L.) wild boar (*Sus scrofa* L.) and horse (*Equus ferus* Boddaert). as well as some small mammals (Baryshnikov 1994).

Intra-site activities

As mentioned above, Barakaevskaya Cave is distinguished by an extraordinarily high density of cultural remains, including abundant by-products of primary stone flaking and tool production. Countless chert chips and a large number of intensively retouched tools associated with numerous bone retouchers confirm that, at least in cold periods, the cave was used not only as a simple living site but also as a real workshop with an entire cycle of lithic manufacturing (Filippov and Lyubin 1994; Lyubin 1994b; Belyaeva 1999). A great deal of evidence for the treatment of animal and plant resources has also been obtained.

Hunted animals were utilized to the maximum extent possible, including the extraction of marrow by splintering most bones, and many bones were also fragmented through trampling (Hoffecker and Baryshnikov 1998). Many bones were intentionally broken so that they could be reshaped and used as retouchers (Filippov and Lyubin 1994b). In addition, many hunted animals were procured for their skins, which were then transported to the site. The treatment of animal skins through scraping and cutting with some kind of tools was established by use-wear analysis (Schelinskiy 1992). It is also probable that the pattern of extremely heavy wear observed on the 20 Neanderthal teeth resulted partly from the processing of skins. Use-wear analysis of notched tools (Fig. 16.3), which are especially numerous (37.3 per cent) in the lithic assemblage of Barakaevskaya Cave, confirms that these implements were used for woodworking, and specifically for shaving small wooden poles (possibly darts). Many scrapers, denticulated pieces and some other implements could also have been used for wood and plant treatment. The use-wear analysis of the similar industry from Monasheskaya Cave has established that 65.2 per cent of the tools found there were intended for woodworking (Schelinskiy 1992). Of special interest is the discovery of a grindstone in the Mousterian layer of Barakaevskaya Cave. According to Schelinskiy (1994), this grindstone was intended for the production of mineral paints. It is also quite possible that such grindstones were also used for the processing of plant food. Certainly, the archeological data related to the exploitation of variable plant resources are rather scarce and this question should be considered more carefully on the basis of evidence obtained through pollen and palaeobotanical analyses.

Pollen and palaeobotanical evidence of Neanderthal plant gathering

There are no publications on the use of plants by the Caucasian Neanderthals. The Mousterian sediments of Barakaevskaya Cave are less than ideal deposits from which to obtain palynological and palaeobotanical information about the Neanderthals' use of plants. But the research carried out using some methodical innovations enabled the recovery of rich pollen complexes and some palaeobotanical micro-remains.

Methodological innovations

No charcoal or other plant macro-remains were found as a result of the excavation and flotation of the sediments. Only individual pollen grains were obtained after using the traditional methods of chemical treatment of the samples for pollen analysis, since most of them were covered with mineral colloid.

The information obtained about plant gathering from Barakaevskaya Cave is a result of four methodological research innovations and the accurate excavation of the cave, which allowed for the collection of pollen samples from four levels of the 15–30cm thick Mousterian layers.

First, a complicated method of chemical treatment of the sediments was used to destroy the colloids around the pollen grains.

The results of preliminary SEM researches of pollen complexes show that some horizons of Barakaevskaya cave sediments were characterized with domination of pollen grains surrounded with mineral colloids. The palynologists study light organic fractions. The heavy colloided forms like heavy mineral pieces disappeared from pollen complexes after centrifugation of the sediments with different heavy liquids. The palynologists of Eastern Europe used HCl acid: the cold one (according to the scheme of V. P. Gritchouk`s laboratory, personal communication) or the hot one (according to the method of E. A. Spiridonova, personal communication). The palynologists of most laboratories of Western Europe (laboratory of Long-term Palaeoenvironmental Prognoses of Oxford University and others) work with more effective for different types of sediments, but very toxic HF. We tried to use various acids to destroy fosphate-carbonate mineral colloids including pollen grains – HNO_3, HCl (cold), HCl (hot), but obtained rather small quantity of pollen grains. Only the alternate use of HF and HCl acids (according to the scheme of palynologist M. Carchiumaru, personal communication) in combination with our innovation (the repetition of this process three times for each sample) and use of KI hard liquid for centrifugation of the sediments enabled the collection of the pollen complexes (Levkovskaya 1994). Except of this, the unusually large quantity of preparations (about 30) was analised for each studied sample.

Second, the use of the SEM as additional resource for obtaining not only the pollen but also some palaeobotanical information relating to the Neanderthals' gathering of plants. Palaeobotanists and palynologists often use an SEM to study single biological taxa. The most important aspect of our innovation was the use of SEM analysis not only for single biological taxa, but also for pollen complexes with many taxa simultaneously. The liquids with pollen complexes were used for SEM studying. They were sprayed onto SEM tables. The research was carried out using a JSM-35 SEM microscope. An Aurum-Palladium admixture was used to dust the biological specimens (after drying) onto the SEM tables.

Various publications about the SEM morphology of phytoliths (e.g. Gol'eva 2001 and others) or the modern pollen grains of some plant families (*Asteraceae, Cyperaceae,* etc.) (e.g. Tarasevitch 1983; Meyer-Melikian *et al.* 2004) were used for identification of the Mousterian palaeoflora.

Third, statistical data about the abnormal morphology of pollen grains were used in order to reconstruct past geo-botanical stress. The original palynoteratical method was worked out and published with illustations in Krakov in the 'Proceedings of the 5th European Palaeobotanical and Palynological conference' (Levkovskaya 1999). It was developed (Levkovskaya 1984, 1999; 2005; Levkovskaya and Bogolubova 2005; Levkovskaya *et al.* 2005b) on the basis of statistics of different abnormalities of pollen grains, which had been collected during thirty years for fossil materials (Pleistocene and Holocene) and sub-fossil sediments from most modern geobotanical sub-zones of Western Siberia (Levkovskaya 1968, 1971) and the 30km zone around Chernobyl.

Fourth, the monograph *Plant Resources of Caucasus* (Grossgaim 1952) was used to interpret the plant gathering as indicated by the collected Mousterian palaeoflora. This monograph was based on ethnographic data and the results of chemical analyses of modern plants. It contains information on approximately 2500 Caucasian plants. Some more information about useful properties of the plants was used as well: some ethnographic data from Siberia and the Far East of the USSR (Mitlianskaya 1983) and materials from the *Atlas of Areas and Resources of the Medicinal Plants of USSR* (Tchikov 1980).

SEM results: indications of intensive Neanderthal plant use and geo-botanical stress

The SEM micrograph (Fig. 16.5) of the sample from the border of horizons 4 and 3 illustrates: (1) the palaeobotanical microremain – phytolith of *Poaceae*, of *Festucoid* type (Fig. 16.5.1; also Fig. 16.6.1); (2) a big quantity of immature and unidentified pollen grains of different taxa; all of which are badly developed (i.e. are devoid of sculpture, etc.); (3) a spore (Fig. 16.5.4), possibly of Jurassic age, from the limestone roof of the cave; (4) the identified pollen grains presented by individual palynomorphs of *Juniperus* (Fig. 16.5.2), *Asteraceae – Cirsium* type (Fig. 16.5.3), *Alnus* (Fig. 16.5.7) (it is represented by an immature grain); and *Cyperaceae* (Fig. 16.5.8), or by poliads with conglomerates of immature and unusually dwarf pollen grains of *Betula* and *Chenopodiaceae* (Figs 16.5.5

and 16.5.6) The conglomerates show that the process of development of pollen grains inside the anthers was not finished.

About a hundred contours of unidentified pollen grains are shown on the SEM micrograph. They are dominants in the complex. Parts of them are dwarf or ultra-dwarf forms. Most are devoid of sculpture that was not developed as a result of bad (stress) conditions for the generative spheres of the most plants of the area. According to E. N. Ananova (1966), some palynologists interpret the 'abortive' complexes with domination of underdeveloped palynomorphs as complexes of re-deposited forms. Though, the dwarf pollen grains or polyads of underdeveloped forms could not have appeared as a result of redepositional processes. Some criteria of differenciation of the complexes with domination of underdeveloped (immature) or re-deposited palynomorphs are published (Ananova 1966; Levkovskaya 1999; 2005).

Had we seen the normally developed pollen grains on the SEM micrograph, it would have been possible to obtain a rich list of plants useful for Neanderthals. But horizon 4 was formed in a sub-alpine belt in glacial conditions, and horizon 3 during the transition to the interstadial (Table 16.1). Thus, the plants existed in conditions of geobotanical stress due to low temperatures (Levkovskaya 1999, 2005) and produced mostly immature pollen grains. As a consequence, these grains cannot be identified.

The SEM micrograph provides the basis for two conclusions: The first concerns the palaeoenvironment. The Neanderthals existed during a period of geo-

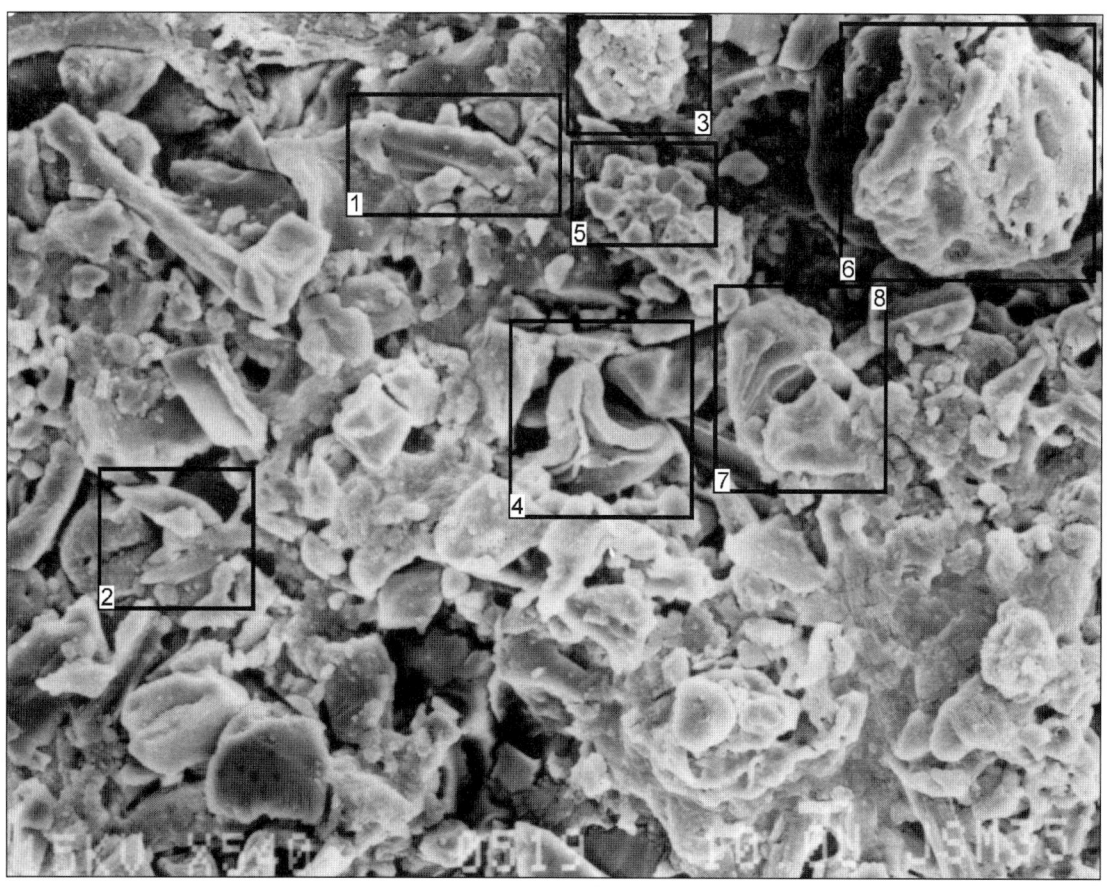

Figure 16.5: Barakaevskaya Cave site. SEM (x 540). Palynological-phytolith complex. Sediments with the Neanderthal child mandible. Zone I, subalpine belt (Table 16.1, Fig. 16.10). A. The unidentified forms: about 200 contours of immature pollen grains of different taxa that are devoid of sculpture and dwarf size (10–25 mkm). B. The identified forms: 1 phytolith of Poaceae *(Festoucoid type, Fig. 16.6); 2* Juniperus *sp.; 3* Asteraceae: Cirsium *type; 4 spore of Jurassic age (from the cave roof); 5, 6 conglomerates (polyades) of underdeveloped and dwarf forms of: cf.* Betulaceae *(5), cf.* Chenopodiaceae *(6); 7* Alnus *sp. (a badly-developed form); 8* Cyperaceae. *The stress state (Levkovskaya, 1999; Levkovskaya et al., 2005, 2011) of most plants (domination of immature and dwarf pollen) is registered at the upper limit of trees. This unusually rich complex is the pollen indicator of transportation of large quantity of different plants inside the cave as anthropochores, partly by winds and mostly as the result of plant gathering by Neanderthals.*

botanical stress of the man's generative spheres of most of the flowering plants at their upper limits, because the latter produced mostly immature, underdeveloped and dwarf pollen grains. The low temperatures of the upper part of the subalpine belt were the cause of this geo-botanical stress. The second conclusion concerns the palaeoeconomy. The cave was a place of intensive Neanderthal economic activity. The micrograph illustrates this complexity through the quantity of pollen grains, which is unusually large for cave sediments.

Palaeobotanical finds and interpretation of plant gathering

The conclusion about plant gathering of the Neanderthals is based in the article on the palaeo-botanical and pollen data. The information about palaeobotanical finds from Mousterian layer of Barakaevskaya cave site is presented in Table 16.5 and at some SEM-micrographs (Figs 16.5.1, 16.6.1, 16.6.2, 16.6.4 and 16.6.6). The palaeobotanical flora includes: (1) three taxa of phytoliths: *Poaceae* gen.$_1$ – *Festucoid* type; this identification was carried out by specialist on phytoliths P. Andersen (Figs 16.5.1, 16.6.1); *Poaceae* gen.$_2$ – *Eragrostis* type (Fig. 16.6.4) and unidentified phytolith (Fig. 16.6.2); (2) remains (valve) of *Diatom* algae of *Pennales* order (Fig. 16.6.6).

The mirco-sedimentological data (Chernyakhovskiy 1994) demonstrate that the phytoliths in Barakaevskaya Cave were found only in the area of the Neanderthals' fireplace. This indicates that they were the true objects of the Neanderthals' plant gathering, and that grasses of *Poaceae* could have been used for kindling the camp fire.

Three knifes with slides of cutting grasses of *Poaceae* or *Cyperaceae* were found in the Mousterian layer of Caucasian cave site Taglar though the specialist on the trasological method (Schelinskiy 1994b, 23) believed that they were used for cutting *Phragmites* (*Poaceae*) or *Scirpus* (*Cyperaceae*).

The *Diatom* algaes of *Pennales* order are the plants of the sweet ponds. The remains of *Diatom* algae of the *Pennales* order were also found in the Mousterian layers of the following Caucasian cave sites: Matouzka (North-West Caucasus), Kudaro I and Kudaro III (Southern Osetia), Vorontsovskaya (Sochy Black Sea area of Western Caucasus), though, these remains were not found in the Mousterian layer of Akhshtyr cave and in the most Early Palaeolithic layers of Treugolnaya cave (Levkovskaya 1980; 1992; 2006, 2007; Kulakov *et al.* 2007).

The finds of pollen grains of water plants in the Mousterian layer of Hortus cave in France (Lumley 1972) were the basis for the conclusion that Palaeolithic inhabitants of the cave transported water into it and made special vessels for doing this. Today, Hortus cave is located 200m above the river.

Barakaevskaya Cave is currently located about 80m above the Gubs River and at a long distance from the nearest water source and, as noted above, the cave itself was probably protected from the winds. Without vessels is not possible to transport the water (with *Diatom* algaes) to such considerable height and over such a long distance.

The palaeobotanical finds mentioned above (phytoliths and *Diatom* algaes) show that the Neandertals possibly transportated some plants inside the cave for special purposes.

Pollen flora of water plants and an interpretation of plant gathering

Pollen grains of the following water or shore-water plants are identified from Mousterian layer of Barakaevskaya cave site: *Menyanthes trifoliata* L., *Portulaca* L. (a shore-river and weed plant), *Typha* L., *Myriophyllum* cf. *spicatum* L., *Alisma* L. and *Cyperaceae*: *Bolboschoenus* type or *Cyperus* type (Table 16.5 and SEM-micrograph: Fig. 16.6.5).

The data on subfossil pollen spectra from hundreds of modern lakes, rivers and flood plains show that most pollen production of water plants gets into ponts directly near the plants and is not transported by wind for long distances (see as exemple: Khomutova 1965; Levkovskaya 1967, 1987; Kabailene 1969;).

Except of this pollen producing of water plants was small. Data on subfossil pollen spectra from 22 small lakes of Lithuania (Kabailene 1969) illustrate that not all lake samples contain pollen of water plants. That is typical even for Kubinskiy lake from Karelia (Khomutova, 1978) though about 50 per cent of its surface is covered by modern plants. The average quantity of pollen grains in Lithuania lakes – only 6.6 per cent and maximal – 16 per cent (in single sample). The materials on large Ladoga lake (Levkovskaya, 1967b) show that only 9 from 30 studied samples contain pollen grains of water plants and the maximal quantity of them – only 0.7 per cent.

The important methodical meaning has the graph published for Neolithic and Early Bronze age sites of Lubana lowland in Latvia (Levkovskaya 1987, 70: Fig. 24). It is included in this article (Fig. 16.10). It shows that the pollen producing of seven modern lakes of Lithuania and of Lubana lake in Latvia is less than quantity of water plants in archaeological layers buried within of these lake sediments when the foraging economy dominated. But the quantity of water plants

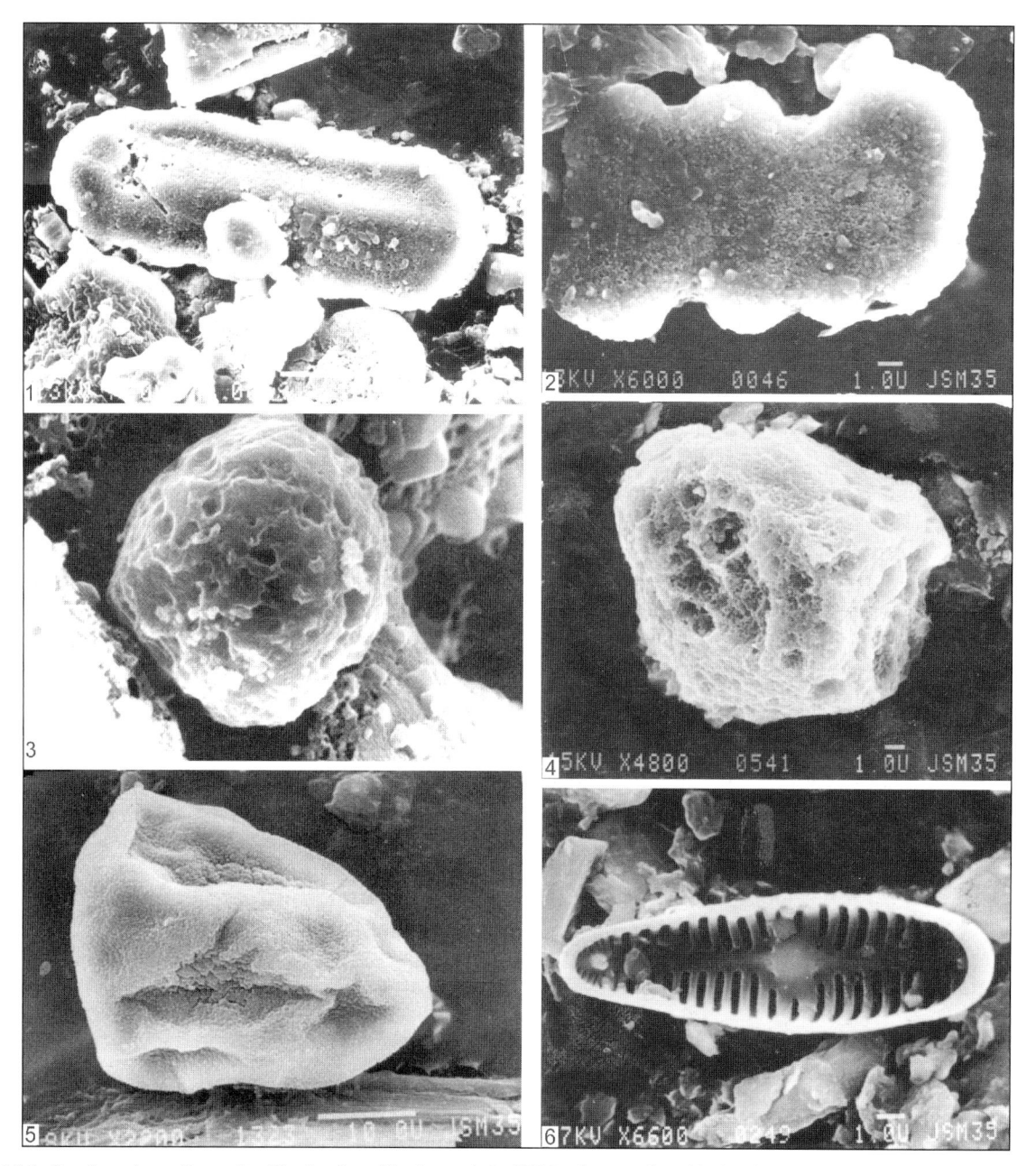

Figure 16.6: Barakaevskaya Cave site. The border of horizons 4–3. SEM-micrographs of individual pollen and palaeobotanical finds: 1 Fossil phytolith of Poaceae gen.1: Festucoid *type; X3000; 2 Unidentified phytolith; X6000; 3 Modern phytolith of* Eragrostis minor; *X7300; (after Golieva 2001, 105: table 2); 4 Fossil phytolith of* Poaceae gen.2: Eragrostis *type; X4800; 5 Pollen grain of* Cyperaceae: *of the* Shoenoplectus *type; X2800; 6 Diatom* algae of the Pennales *order; X6600.*

of archaeological layers is less since the beginning of the producting economy.

Pollen of water plants was not found in the subfossil samples collected from the alluvium of Gubs river, but they are dominanted in palynological complex of Mousterian layer 4 of Monasheskaya cave site of Gubs canyon (Beliaeva 1999, 67). 17 pollen grains of 5 taxa of water plants found in Mousterian layer of Barakaevskaya cave site. These caves are located at about 80 and 70m above the Gubs bed.

In the case of Barakaevskaya Cave, the role of winds and the processes of re-deposition of the Pleistocene sediments only minimally influenced the formation of the Mousterian pollen and palaeobotanical complexes of the inner part of the cave. The causes of this specificity of Barakaevskaya cave are as follows.

First, the entrance of the cave was rather small – approximately 6 by 3m (Figs 16.2–3); second, according to archaeological interpretations, the Neanderthals possibly closed the entrance of the cave in order to protect themselves from winds, frosts, heavy rains and animals; third, water re-deposition of sediments did not occur inside the dry cave; and, fourth, the Mousterian layer of Barakaevskaya Cave overlies the stone floor of it and there are no pollen grains that could have been re-deposited before Neanderthals have appeared there.

All mentioned above facts show that most pollen grains could appear inside the cave due to Neanderthal activity and not because of wind transportation. They are mostly the objects of plant gathering of Neanderthals or anthropochores because pollen of water plants could had been transported into the cave by people as a result of special collection them for different aims (eating and others) or accidentally: in clothes, hair, skins, on tools, bags, catches etc. There are no criteria for differentiation of these pieces of information from each other on the basis of pollen data only.

Pollen grains of *Myriophyllium* sp. are presented especially often among the pollen of water plants found inside of Palaeolithic Caucasian cave sites, that shown materials of cave sites Kudaro I (quantity of water plants reached 18 per cent), Kudaro III: layer 4 – up to 30 per cent, layer 3 – up to 14 per cent, Matouzka, Vorontsovskaya, Monasheskaya etc. in which pollen of water plants found at different levels etc. (Levkovskaya 1980: 137, 138, 142, 144, 146, 1992: 96, 98; 1994: 87; Beliaeva 1999: 105).

Myriophyllum cf. *spicatum* L belongs to a group of polishing plants (Grossgaim 1952, 335), because its leaves and stems contain silica. It could have been used by Neanderthals as a polishing plant for woodwork: haft handles, helve-hammers, shafts, hilts, etc. Such usage of this plant is plausible because 37.3 per cent of the Barakaevskaya lithic Mousterian assemblage must be used for woodworking (Schelinskiy 1975). According to V. P. Lyubin (Lyubin 1994, 163) the most important part of the economical activity of the Barakaevskaya Neanderthals was the woodworking.

Menyathes trifoliata L. belongs to the group of plants used as sources of vitamin C, pigments (green) and stomach-ailment medicine (Grossgaim 1952, 62, 178, 253, 364).

All the other water plants (*Alisma, Portulaca, Bolboschoenus* and *Typha*) are included in different categories of food plants.

Pollen grains of *Alisma* have been found in archaeological layers of different regions: in the Dniestr river basin – in the Ketrosi Mousterian layer (Levkovskaya 1983) and in the Lubana Lowland of Eastern Latvia in Early Neolithic layers of some peat bog sites (Levkovskaya 1987). It has been identified by Grossgaim as a root vegetable plant. Its roots could be used as food after drying, although the plant itself is poisonous.

Portulaca cf. *oleraceae* is a leaf and stem vegetable and vitamin plant. It contains about 179.7–300 mg per cent of vitamin C (Grossgaim 195, 43, 257).

The species of *Bolboschoenus* are starch vegetable plants. The nodules of these plants could have been used to flavour food (Grossgaim 1952, 66–67). It is not possible to identify the genera of the *Cyperaceae* due to the high level of morphological variations of the pollen grains in this family. The identification of the pollen *Bolboschoenus* is reliable in our case, due to the SEM research, which allowed for the viewing of small-scale detail in the sculpture of the pollen grains. The protuberances on the surface of the apertures of the illustrated SEM micrograph studied grain have different contours (Fig. 16.6.5). This is typical for *Bolboschoenus*, in contrast to *Cyperus* (Tarasevitch 1983, 94–99). Some species of *Bolboschoenus* include water, shore-water and swamp plants.

The species of *Typha* are vegetable plants. Their roots, leaves and stems are all edible. It is also the starch plant. Moreover *Typha* is a vitamin, basket and mat wickerwork, and spinning plant (Grossgaim 1952, 34, 46, 68, 263, 378, 381, 382).

It is clear that most pollen grains of water plants appeared inside the cave as the result of Neanderthals activity. The finds of water plants could be interpreted as anthropochores plus indicators of plant gathering simultaneously. But it's very difficalt to suppose that the water plants were transportated inside the dry Barakaevskaya cave at the hight about 80m accidentally. Thus the pollen grains of water plants found inside the Barakaevskaya cave site with specially closed by Neanderthals entrance possibly could be interpretated as the evidence of plant gathering.

Pollen flora of land plants from the Mousterian layer of Barakaevskaya Cave and an interpretation of plant resources

Tables 16.2–4 illustrate the pollen flora of land plants. It includes 44 taxa. They were identified from four excavated Mousterian horizons in the inner part of Barakaevskaya Cave. The flora of the Mousterian travertine (of the 'cap' of the Neanderthal layer) is given in these Tables 16.2–4 only for comparative purposes (it is not used in the calculations).

The right-hand sections of these Tables present an interpretation of the plant resources based on different identified trees (coniferous, evergreen, small-leaved and broad-leaved), bushes and herbs. They are listed with reference to the information provided about the useful properties of such plants in the monograph *Plant Resources of Caucasus* (Grossgaim 1952).

The identified Neandertal layer pollen flora of land plants includes four coniferous plants, three small-leaved trees, six bushes, two evergreen trees or bushes (Table 16.2), 16 broad-leaved trees (Table 16.3) and 15 herbs (Table 16.4).

Single pollen grains of terrestrial plants (as opposed to water plants) are sometimes transported over long distances by the wind. For example, single pollen grains of subtropical Colchis elements are found in the sub-fossil layers of modern Caucasian glaciers (Kartashova and Troshkina 1971). We do not risk using this information for making conclusions about plant gahtering of Neanderthals. These datas could be the bases of conclusions about the plant resources of the area of Gubs canyon and Barakaevskaya cave for the Neanderthal time.

However, as mentioned above, Barakaevskaya Cave had a small natural entrance and was intentionally closed by its inhabitants in order to protect themselves from different bad palaeoenvironmental influences. This means that the hundreds of pollen grains of the 44 taxa of land plants found inside the cave represent a combination of micro-remains of anthropochores, objects of plant gathering and a minimal quantity of grains transported by the wind.

Tables 16.2–5 show that the Neanderthals had rich plant resources available for different aspects of their life. The richness of the palaeoflora, however, changed over time. The total quantity of taxa of land plants in horizon 4 is 11 (subalpine conditions), in horizon 3 it is 20 (transition from stadial to interstadial conditions), in horizon 2 it is 29 (the beginning of the interstadial), in horizon 1 it is 21 (thermoxerophylous phase of the interstadial), and, at the time of stalagmite formation, there are 14 taxa (the optimum of the interstadial, when some pollen grains were destroyed as a result of chemically active water circulation). Thus, the richness of the plant resources available to the Neanderthals depended on the types of palaeoenvironments. There is no doubt that the real palaeoflora of all the phases included many more taxa, because most plants were indentified only till genera or even family and could belong to different species.

The Neanderthals had various plant resources available to them, even during the subalpine conditions when flora were extremely limited and plants existed in stress conditions, since the Neanderthals could use some plants for different purposes simultaneously (see: Tables 16.3–5 and 7). For example, *Juniperus* (Table 16.2) could have been used for firewood, vitamins, medicine, joinery, tanning and oil. Its conyberries are also eaten with meat as a spice (Grossgaim 1952, 51, 62, 71, 88, 175, 251, 296, 347, 397). It is especially important that *Juniperus* could have been used a source of vitamin C during winters (horizon 4) – the galberries of different species of *Juniperus* contain 73.0–129.7 mg per cent of vitamin C (Grossgaim 1952, 251).

The flora of the land plants does not include food starch plants and only two taxa of food cereal plants (Grossgaim 1952, 69) – *Amarantus* and *Chenopodium* cf. *album*. The latter was identified in the Mousterian layer of Monasheskaya cave (Lyubin *et al.* 1973) and for the Barakaevskaya section by a specialist on the morphology of *Chenopodiacieae*, M. M. Monoszon (1973). According to A. Grossgaim (1952), the seeds of this plant are used in baking.

Thus, the specific feature of the land plant flora is the small amount of plants with a large quantity of protein resources.

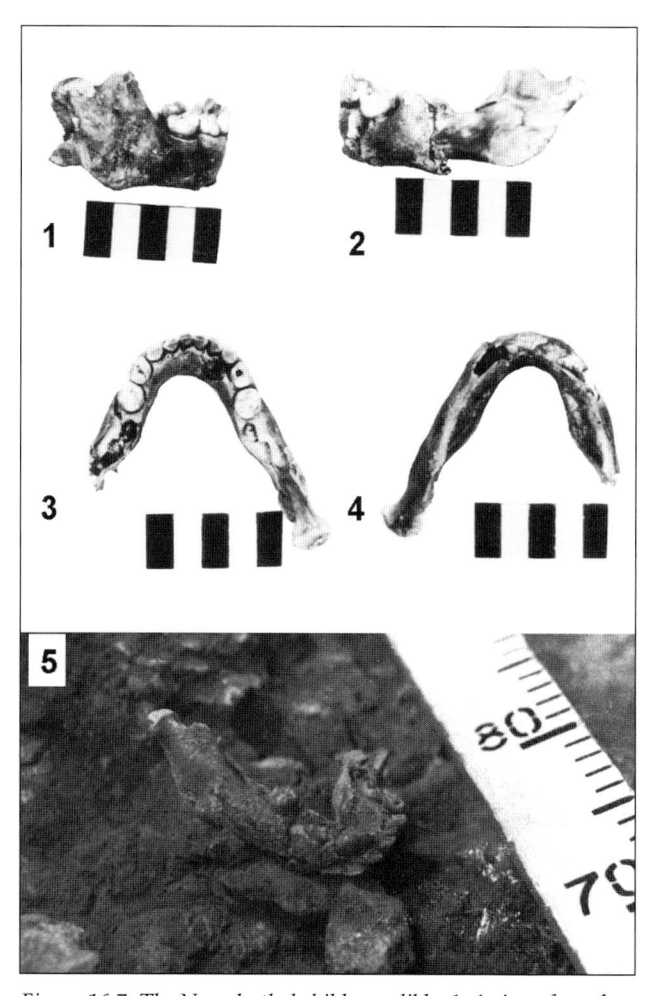

Figure 16.7: The Neanderthal child mandible. 1–4 views from four sides; 5 the position in the Mousterian deposits in situ.

The Neanderthals' diet: palaeozoological, palaeobotanical and pollen data

A. The palaeozoological data (after Baryshnikov 1994)

The palaeozoological data show that the Neanderthals consumed large quantities of meat. Inside Barakaevskaya Cave about 79,600 bones were found, most of which are splintered. The sizes of the bone fragments are mostly 1–2cm long, and only rare bones are larger than 12cm long. According to Baryshnikov's interpretation most of them are 'kitchen wastes'.

The focus of the Neanderthals' hunts was different mammals (Baryshnikov 1994). The Neanderthals' menu could include 25 taxa of mammals and 10 taxa of birds, although the remains of small animals inside the cave could also have been brought in by birds of prey.

The list of mammals that could be eaten by Neanderthals is presented in Table 16.6 which was published earlier (see Baryshnikov 1994, 75: Table 4). An especially important role here belongs to *Bison priscus* Boj., but *Capra caucasica* Guld. et Pall., *Ovis orientalis* Gmel., *Megaloceros giganteus* Blum. and *Equus ferus* Boddaert. were also often hunted.

The list of hunted birds (determination by A.V. Panteleev) includes: *Anas crecca* L. and *A. querquedula* L., *Columba livia* L., *Vtlanocorypha sp.*, *Lullula arborea* (L.), *Anthus trivialis* (L.), *Rhodopechys sanguinea* (Gould), *Corvus corax* L., *Passerifmes* (unidentified), and *Aves* (unidentified) (Baryshnikov 1994, 74: Table 3).

B. The pollen and palaebotanical data

The plant food resources of studied region were rich during the Mousterian span (before travertine formation). The group of food plants includes the following 32 taxa *Juniperus, Betula Litwinowii (B. pubescens), B. pendula, Berberis, Salix, Ephedra* cf. *distachia* (see Table 16.2), *Fagus* cf. *orientalis*, cf. *Pyrus, Quercus, Castanea, Acer, Corylus avellana, Ostrya carpinifolia, Malus orientalis, Fraxinus* (Table 16.3), *Amaranthus, Geranium, Polygonum viviparum, Chenopodium* cf. *album,* cf. *Cirsium, Sonchus, Silene,* and *Helianthemum* (Table 16.4). According to Grossgaim (1952), the seeds of some of these plants also belong to the group of food oil resources (see Tables 16.2–4). Exept of this to the group of oil food plants belong: *Picea orientalis, Buxus sempervierens* (Table 16.2), *Ulmus laevis* (Table 16.3), *Plantago major* (Table 16.4). Five taxa of food water plants (Table 16.5) belong to the food plants as well: *Menyanthes trifoliata, Typha, Alisma, Portulaca,* and cf. *Bolboshoenus.*

The flora of the land plants contains a small quantity of plants with a high concentration of carbohydrates, with the exception of *Amarantus sp.* The seeds of the cereal plant *Amarantus* sp., contain 41 per cent starch, 19 per cent protein and 8.0 per cent fat (Grossgaim 1952, 69).

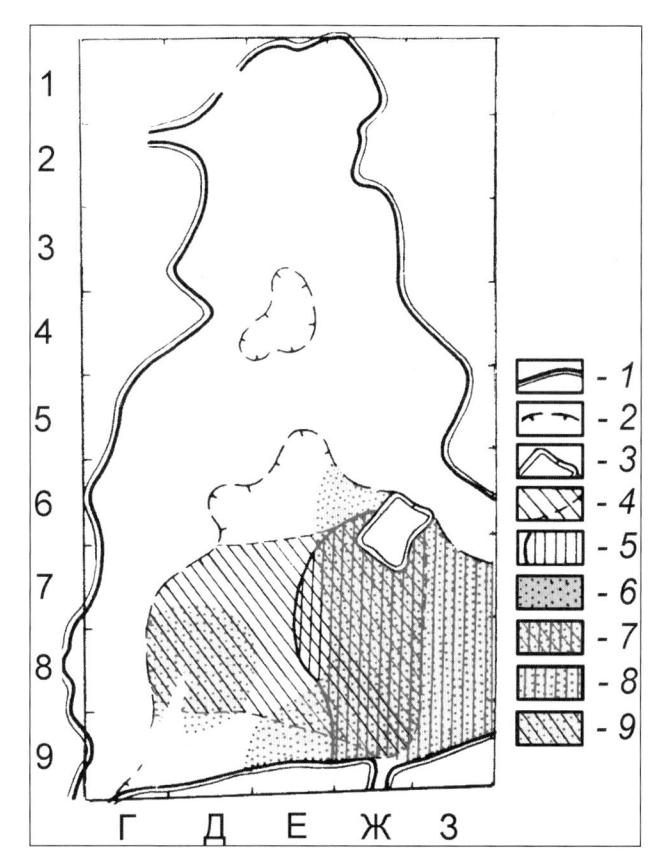

Figure 16.8: Barakaevskaya cave. The areas of different types of Neanderthal economic activity during formation of the lower part of Mousterian layer (Filippov and Lyubin 1994, 236: Fig. 55). Excavated Mousterian horizon 3. 1 – outlines of the cave walls; 2 – the limits of the Mousterian deposits; 3 – crumbling block; 4–9 – the parts of the cave with: 4 – major concentrations of retouchers, made mostly from tubular bison bones; 5 – major concentrations of stone tools; 6 – major concentration of stone chips; 7 – major concentration of retouchers, stone tools and chips; 8 – major concentration of stone tools and chips; 9 – major concentrations of retouchers, chips and isolated Neanderthal craftsman's workplace (square Г8, Д8) (Filippov and Lyubin 1994, 147). (Note: The graph is based on statistics on finds of 138 retouchers, 1.433 stone tools and 26.333 chips in two lower levels of excavation of Mousterian horizon 3 and 4 (Filippov and Lyubin, 1984, table 1, 145). The location of Neanderthal child's mandible is the northern part of the squares Ж-9 – Е-9. It is found about 3–4cm above the cave bottom (Lyubin, 1994, 153; Cherniakhovski, 1994, 64). The finds of Mousterian phytoliths are concentrated in place 3–7 – 3–8. According to Cherniakhovski (1994, 65) they are the evidence of Neanderthal fireplaces inside the cave).

The dry starch water plants, *Bolboschoenus* and *Typha*, and a land plant, *Amarantus* sp., could have been important sources of carbohydrates for the Neanderthals during cold seasons.

The rhizomes of *Typha* contain 44.2–47.5 per cent starch and sugar and could have been consumed after drying and grinding even in winter. A pestle with signs of *Typha* grinding has been found at the Italian Palaeolithic site, Bilancino, the age of which is about 30,000 years ago (Aranguren and Revedin 2008).

The groups of nut plants (*Corylus avellana, Fagus orientalis, Quercus, Castanea,*), fruit plants (*Berberis, Malus, Pyrus*), cereal plants (*Chenopodium* cf. *album, Amaranthus*) and especially of the starch root vegetable plants (*Typha,* cf. *Bolboshoenus*) were important in the life strategy of the Neanderthals. The seeds, fruits or roots of these groups of plants can be dried and preserved for cold seasons. The leaf and stem food plants (*Sonchus, Plantago major, Portulaca*, etc.) are now used in salads and were available mostly in the summer. The group of vitamin plants contains 32 taxa of land plants and three taxa of water plants (see Tables 16.2–5). Although the Neanderthals could have suffered from vitamin deficiencies during the winter. The coniferous plants – *Picea orientalis, Juniperus* and *Pinus silvestris* – could have been important sources of vitamins during the winter.

Thus, the plant diet of the Neanderthals could have included nut plants, fruit, cereal and starch plants, some spices (berry-cones of *Juniperus*, seeds of *Helianthemum*), and even delicacies (the stems of *Sonchus* – peeled and hand ground) (Grossgaim 1952, 45).

In addition, the Neanderthals (like some animals) could have used many more plants as food resources

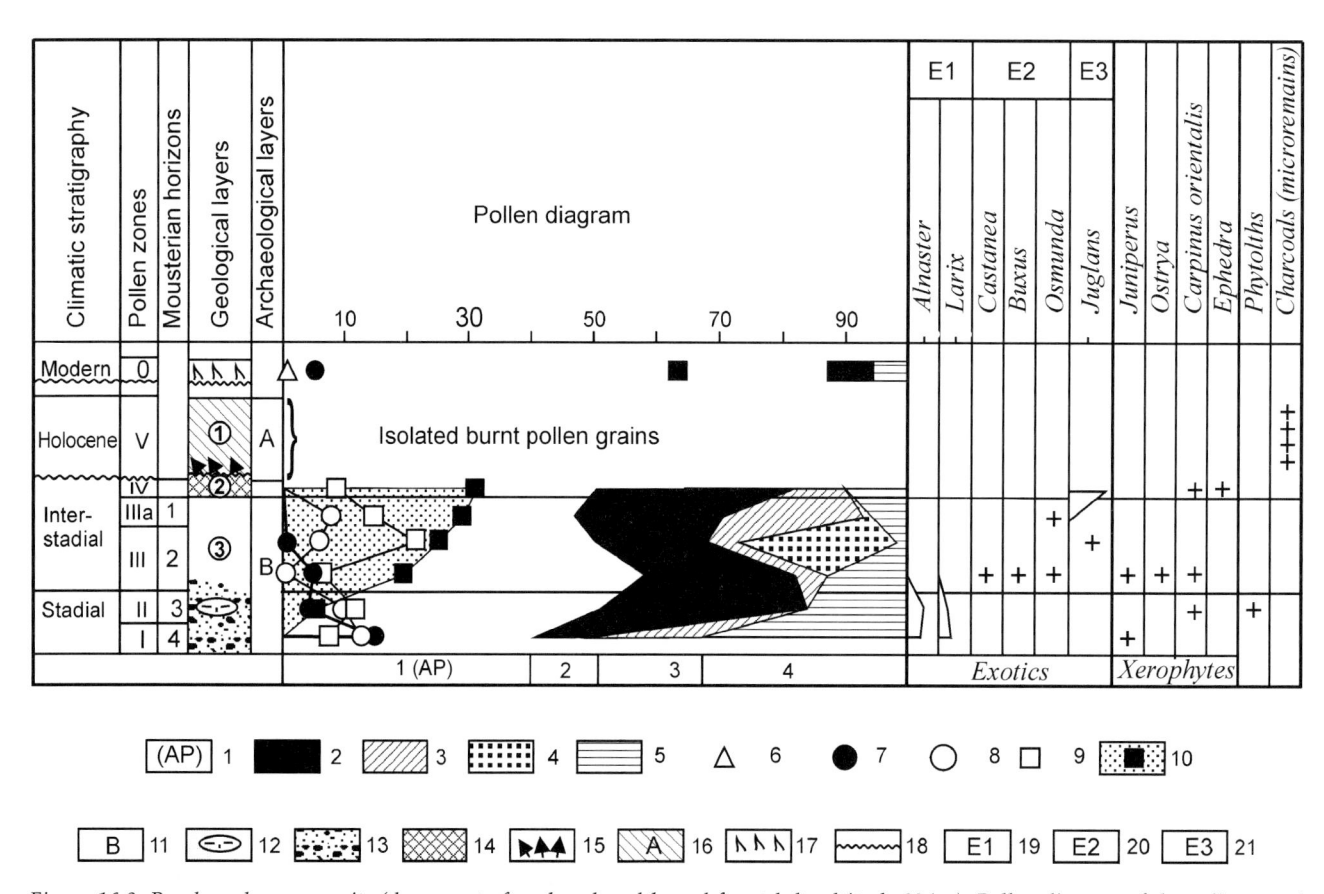

Figure 16.9: Barakaevskaya cave site (down part of modern broad-leaved forest belt, altitude 834m). Pollen diagram of the sediments. 1 the sum of arborael pollen (AP); 2 the sum of mesophilous grasses; 3 the sum of xerophilous grasses; 4 the sum of spores; 5 pollen of Picea; 6 pollen of Pinus; 7 pollen of Betula; 8 pollen of Alnus; 9 pollen of broad-leaved threes; 10 Mousterian layer with anthropological remain (with mandible of the Neanderthal child of about 8 years old); 11 Phosphate-carbonate travertine overlying the Mousterian layer; 12 Holocene ash sediments; 13 level of subfossil pollen spectra (sediments from the alluvium of Gubs river; the cave locates 80m above the river); 14 lenses with the artefacts of the Final Upper Palaeolithic (remains from erosion processes); 15 lenses with the artefacts of the Final Upper Palaeolithic; 16 Holocene ash sediments with pottery of several cultures; 17 pollen complex of the modern river alluvium (collected 90m below the cave entrance); 18 stratigraphical lacunas; 19 microtherm exotics with modern subalpine Carpathian area; 20-21: thermophilous exotics with modern areas in wet subtropical Caucasian centres: Colchis (20), Talysh (21).

than modern *Homo sapiens sapiens*. Two types of *Poaceae* phytoliths were found (Table 16.5) inside Barakaevskaya Cave: the *Festucoid*-type (Fig. 16.6.1) and the (?) *Eragrostis*-type (Fig. 16.6.4). According to A. A. Grossgaim (1952, 110), *Festuca* and *Eragrostis* are now important fodder plants for sheep and cattle. *Eragrostis* contains about 18 per cent protein. Possibly (?), the herbs of some *Poaceae* could be eaten not only by animals, but also by people. According to the anthropological conclusion (Zubov *et al.* 1994) and data of V. P. Lyubin (1994) all the finds of Neanderthal teeth (even the teeth of the child of about eight years old) are worn out, that could be the result of using a lot of plants as food resources.

The role of plants in the everyday life of the Neanderthals

No one can really know how the Neanderthals used plants in their everyday life. They used land and water plants. Only three identified taxa represent plants that are not multi-functional (see Tables 16.2–5). The plant resources of the Barakaevskaya Neanderthals were rich: 44 taxa of land plants and six taxa of water plants have been identified, but the Neanderthals could have used them for more than 300 aims (compare Tables 16.2–5 and 7) because most identified plants are multi-functional. The remains of firewood, food, vitamins, medicine, oil plants, pigments, and plants that could be used for joinery, basket weaving work, mats, wood polishing, spinning and hide tanning have been found inside Barakaevskaya Cave (see Tables 16.2–5 and 7). Some evidence of rotting plants (spores of *Trichia*) is also found in the cave. Most plants could have been transported into the cave by the Neanderthals during warm seasons on different purpose or as anthropochores. However, the stems of *Poaceae* (the dry ones) could have been cut even in the winter because there were of a small amount of snow around the cave in the reconstructed continental type of palaeoclimat. Many artefacts – about 33.7 per cent of the tools – from the Barakaevskaya Cave have signs of woodwork and plant stem cutting (Schelinskiy 1992).

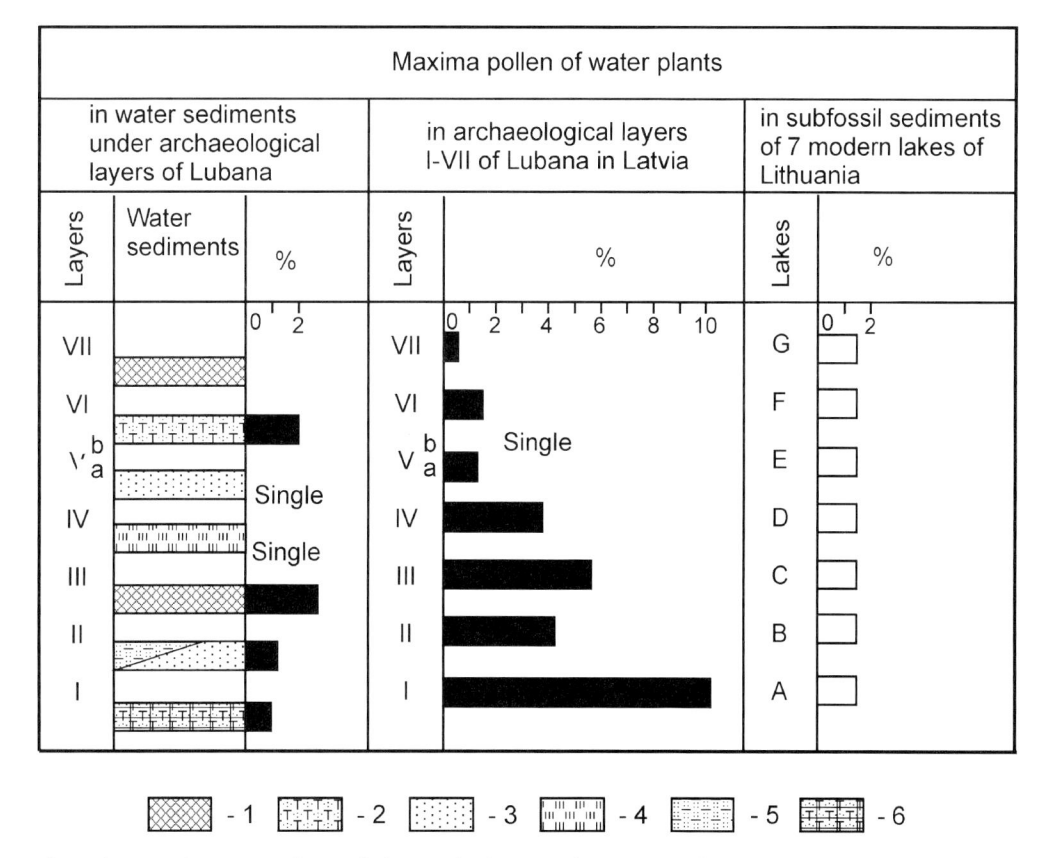

Figure 16.10: Pollen of water-plants as evidence of plant gathering at Lubana, Latvia (Levkovskaya 1987: 70); Pollen in: I–IV – Neolitic and V–VII – Early Bronze layers (Loze, 1969); 1–6 fossil lake sediments – the 'floors' of settlements during lake regressions; A–G modern Lithuanian lakes (Kabaliene 1969). The Neolithic (foraging economy) layers contain more pollen from water-plants than their 'floors', modern lakes sediments and layers of Bronze Age (producing economy).

The SEM micrograph (Fig. 16.5) presents a palynological-palaeobotanical complex with about 100 micro-remains of plants from a very small sample (not more than 15cm³). The total minimum volume of the excavated Mousterian Barakaevskaya layer is more than 3,000,000 cm³ (the minimum thickness of this layer is approximately 15cm and the excavated area was about 20m²–200,000cm²). This means that the Mousterian layer of Barakaevskaya Cave could contain more than 20 million pollen grains. Ultimately, the role of plant gathering in the everyday life of the Barakaevskaya Neanderthals was no less important than that of hunting.

Comparing the roles of palaeozoological and vegetation resources in the Neanderthals diet, we can say that Neanderthals diet could include a large amount of meat throughout the year. The role of vegetable food sources was significant only during the warm seasons. This means that their diet contained a large quantity of proteins and animal fats, but that they probably experienced a deficit of carbohydrates for several months, especially during the winters. The researchers shown the especially important role of water plants in the life strategy of Barakaevskaya Neanderthals because the identified flora of land plants includes only one plant (*Amaranthus*) which contains large quantity of carbohydrates opposite to some taxa of water plants which could be the recourse of carbohydrates for Neanderthal diet (*Typha, Alisma, Bolboshoenus*).

Conclusions

During a sufficiently lengthy period of the Middle Valday (Middle Weichselian, or Middle Wurm), Barakaevskaya Cave was a base camp of Late Neanderthals who developed a distinctive industrial tradition assigned to the Typical Mousterian. The Neanderthal occupation of Barakaevskaya Cave began in very cold conditions at the end of OIS 4 and continued during the first part of OIS 3 in a more favourable climate. The Neanderthals were able to adapt successfully to the varying changes in climate and landscape. The subsistence strategy of the Mousterian inhabitants of Barakaevskaya Cave included both hunting and gathering in the mosaic landscape surrounding the cave site in different climatic phases.

The plans of the cave (Figs 16.2–3) and the scheme of bone retouchers, flint tools and flake concentrations published by A. K. Filippov and V. P. Lyubin (1994, see Fig. 16.8) show that the entrance area of the cave (an area of 12–15m²) was the main living area for the Neanderthals because it was most dry and light. The eastern part of this area was the main place of their activities. It corresponds to the area of hollow into the stone floor of the cave. The traces of a Mousterian fireplace in horizon 4 and fragment of a large stone pestle near it have been found at area of hollow. According to V. E. Schelinskiy, this pestle was used to prepare mineral pigments (possibly of red ochre-haematite and limonite). The greatest number of archaeological finds was concentrated around the fireplace, while the phytolithhs were found directly in the sediments of the fireplace. The additional workplace of the Neanderthal 'master' existed possibly near the wall of the cave at the square Г8–Д8 were the individual place of rather high concentration of bone retouchers and flint-chips is registered.

According to V. P. Lyubin (1994, 163), the cave was the place of the full cycle of flint tools making and their utilization which includes: primary knapping, initial edging, final chapping, utilization, rechapping and reutilization processes. Thus, the cave was used not only as a living site but also as a workshop where a wide variety of activities took place.

The scanning electron microscope pollen data shows that Neanderthal occupation began when most plants were producing underdeveloped and dwarf pollen grains in conditions of their upper limits and law temperatures. Mousterian occupation of the cave was finished before the optimum of interstadial. Some phytolithhs, diatom algae and some hundreds of pollen grains of land and water plants were found inside the cave, which is located at about 90m above the Gubs River. The pollen and palaeobotanical identified flora includes micro-remains of firewood, vitamins, medicine, oil, wickerwork, mat weaving, wood polishing, skin tanning, paint and melliferous plants. The diet of the Neanderthals included large quantities of meat, leaf, rut and stem vegetables, nuts, fruit and cereal starch plants. 79,600 pieces of small bones of domestic waste of the Neanderthals were found inside the cave (Baryshnikov 1994). 33 taxa of food plants were also identified.

Research has shown the importance of the excavated Barakaevskaya Cave for the study of differents archaeological, palaeoeconomical (hunting and plant-gathering) and palaeoenvironmental problems.

The materials obtained for the Barakaevskaya cave could be especially important if combined with materials from the nearest Palaeolithic cave sites in the Gubs Canyon: Monasheskaya (with some Mousterian layers and rather long palaeoenvironmental dynamics), and the Upper Palaeolithic rockshelters – Gubskaya VII (Satanaj) in which the remains of *Homo sapiens sapiens* were found (Kharitonov and

Romanova 1984; Amirkhanov 1986) and Gubskaya I. The new researches will allow comparing the specificity of Late Neanderthals and *Homo sapiens sapiens* palaeoeconomy, to study the role of some palaeoenvironmental processes in their everyday life and correlate Mousterian archaeological complexes on the basis of accurate chronology.

Acknowledgements

In this article, V. P. Lyubin and E. V. Belyaeva are the co-authors of the essay sections entitled 'A brief description of the Mousterian site at Barakaevskaya Cave' and 'Archaeological evidence in relation to the settlement and land use patterns'. G. M. Levkovskaya is the author of the sections related to palaeoenvironments, climatostratigraphy, plant gathering and the Neanderthals' diet. The chronology of the Barakaevskaya Mousterian layer is written by all co-authors. One of the authors (G. M. Levkovskaya) is grateful to the Royal Society of the UK for their grant provided in 2006 that allowed her to work in the Oxford Laboratory of Reconstruction of Palaeoenvironments Prognosis. As a result of this visit, the Oxford pollen collections were used for the identification of the Mousterian Bakaraevskaya pollen flora. We would like to thank Kathy Willis, the head of the Oxford Laboratory, for her advices and the SEM specialist, L. A. Kartseva from Botanical Institut, Russian Academy of Science, for her help in using the SEM. In addition, we would like to thank P. Andersen, M. M. Monoszon and V. I. Tarasevitch for identification some taxa of the plants. The authors are grateful to specialist on different methods who had published different multidisciplinary data for Barakaevskaya cave or obtained some new materials that were used in the present article: to archaeologists P. U. Autlev, N. G. Lovpatche, palaeozoologist G. F. Baryshnikov, geomorphologist S. A. Nesmejanov, palaeopedologist A.G. Tcherniakhovskij, specialists on tracological method V. E. Schelinskiy and A. K. Phillipov, anthropologists A. A. Zubov, G. P. Romanova and V. P. Kharitonov and specialist on carst processes N. A. Gvozdetskij. We thank all colleagues and students who had worked at Barakaevskaya cave excavations.

POLLEN ZONES				LAND PLANTS. POLLEN FLORA FROM THE INNER PART OF THE CAVE	TYPE OF RESOURCE (BY GROSSGAIM 1952)									
I	II	III	IIIa	IV		firewood	food	vitamins	medicine	joinery	oil food plants	weaving plants	plants for mats	paints	tanning	melliferous herbs
EXCAVATED HORIZONS OF MOUSTERIAN LAYER			Travertine											
4	3	2	1												
					CONIFEROUS TREES:										
		+		+	Picea orientalis	*							*		*
+	+	+	+		Larix	*	*	*	*	*	*			*	*
+	+	+	+		Juniperus	*		*	*	*	*				*
+	+	+	+		Pinus sylvestris	*		*	*	*	*				
					SMALL LEAVED TREES:										
+	+	+			Betula Litwinowii (B. pubescens)		*	*	*	*	*			*	*	*
+	+	+	+		B. pendula	*	*	*	*	*	*	*		*	*	*
+	+	+	+	+	Alnus (incana+glutinosa)	*		*	*	*		*		*	*	*
					BUSHES:										
+		+			Berberis	*	*	*	*							
+	+	+			Betula fruticosa (humilis)	*	*	*	*	*				*	*	*
		+			Ligustrum	*	*							*	*	*
+		+			Salix	*	*	*		*		*				
+	+				Alnaster	*	*	*	*					*	*	*
		+			Cotinus	*		*	*	*				*	*	*
					EVERGREEN TREES AND BUSHES										
	+	+		+	Buxus sempervirens	*				*	*			*		*
				+	Ephedra procera	*			*							
		+			Eph. distachia		*	*	*						*	

Table 16.2: Some trees and bushes as plant resources of Neanderthals from the Barakaevskaya cave site. (The estimation of the useful properties of the fossil plants is given in Tables 2–5 on the basis of monograph of Grossgaim (1952), though many other publications could be used for same estimations. This monograph was chosen for the following reasons: 1. it contains information on 1350 useful plants of the Caucasian flora; 2. it is mostly based on the ethnographical information, though some results of laboratory studies are also used; and 3. it allows one to estimate useful properties of the taxa presented in the floristic Tables 2–5 although some of them are identified only by the genera. The monograph has examples of the common characteristics of different species within one genera (Grossgaim 1952, 68, 223).

Pollen zones				Excavated horizons of the Mousterian layer				Land plants. Pollen flora from the inner part of the cave	Type of resource (by Grossgaim 1952)										
I	II	III	III a	IV					firewood	food	vitamins	medicine	joinery	oil food plants	weaving plants	paints	tanning	melliferous herbs	spinning plants
				Travertine	1	2	3	4											
									Broad-leaved trees:										
						+			Fagus orientalis	*	*			*	*				*
				+	+	+	+		Ulmus	*		*		*	*		*	*	
				+	+	+			Carpinus orientalis	*				*			*		
				+	+	+			Carpinus betulus (C. caucasica)	*				*			*		
						+	+		cf. Pyrus	*	*	*		*		*	*		
					+	+	+		Quercus cf. robur	*	*	*	*	*		*	*		
					+	+			Castanea	*	*	*		*					
				+	+	+	+		Acer	*	*	*	*	*		*	*	*	
				+	+				Corylus avellana	*	*	*	*	*			*		
				+					Tilia caucasica	*	*	*	*	*	*		*	*	*
				+					Tilia cordata	*	*		*	*	*	*		*	*
				+					Juglans regia	*	*	*		*		*			
				+					Pistacia cf. mutica	*	*		*	*			*		
					+				Ostrya carpinifolia	*	*	*	*	*		*	*		
					+				Malus orientalis	*	*			*		*	*	*	
					+				Fraxinus	*	*			*		*		*	

Table 16.3: *Broad-leaved trees as Neanderthal plant resources from the Barakaevskaya cave site.*

Pollen zones				Land plants. Pollen flora from the inner part of the cave	Type of resource (by Grossgaim 1952)					
I	II	III	III a	IV		food	vitamins	medicine	oil food plants	paints	tanning	melliferous herbs
Excavated horizons of the Mousterian layer											
4	3	2	1	Travertine								
	+				Amaranthus	*	*					
	+		+		Colchicum			*				*
		+			Valeriana			*			*	
		+	+		Geranium	*	*			*		
	+	+	+		Polygonum viviparum	*	*					
					Polygonum aviculare		*			*		*
				+	Fagopyrum cf. sagitatum							
	+	+	+		Chenopodium cf. album	*	*		*			
	+	+	+		cf. Cirsium	*			*			*
+					Sonhus	*	*		*			
	+	+	+		Plantago major		*	*			*	
		+	+		Plantago lanceolata	*	*					
	+	+			Silene	*	*					
	+				Helianthemum			*				
	+				Arctostaphylos uva ursi			*			*	*

Table 16.4: Herbs as Neanderthal plant resources from the Barakaevskaya cave site.

ANTHROPOCHORES OR PLANT GATHERING INDICATORS (THE BORDER OF HORIZONS 4 AND 3)	USEFUL PROPERTIES OF THE PLANTS									INDICATORS OF		
	food	vitamin	medicine	polishing	weaving plants	plants for mats	spinning	paints	water transportation	plants rotting	burning of herbs
PHYTOLITHS FROM FIREPLACE (SEM):											
Non-identified phytoliths											*
Poaceae phytoliths (Festucoid-type?)											*
Poaceae phytoliths (Eragrostis-type?)											*
SPORES OF FUNGI SAPROPHYTE:											
Trichia										*	
Diatom algae (SEM)									*		
POLLEN OF WATER AND SHORE-WATER PLANTS FROM DRY CAVE:											
Menyanthes trifoliata	*	*	*								
cf. Bolboshoenus	*								*		
Portulaca	*	*							*		
Typha cf. latifolia	*	*			*	*	*		*		
Myriophyllum				*					*		
Alisma	*								*		

Table 16.5: Anthropochores or indicators of plant gathering from the Barakaevskaya cave site.

Taxa	\multicolumn{5}{c}{Mousterian layer}	Total
	1	2	3	4	Scree	
Lepus europaeus Pall.	1	1	-	-	-	2/2
Ochotona pusilla liubini Erb. et Baryshn.	2/2	-	1	-	-	3/3
Marmota paleocaucasica Baryshn.	-	-	-	1	-	1
Spermophilus cf. *musicus* Menet.	1	2/1	-	-	-	3/2
Apodemus sylvaticus L.	2/1	-	-	-	-	2/1
Ellobius talpinus Pall.	-	-	1	-	-	1
Cricetus cricetus L.	-	4/1	-	-	-	4/1
Arvicola chosaricus Alexandr.	1	8/2	2/1	-	-	11/4
Chionomys nivalis Martins	-	1	2/2	-	-	3/3
Chionomys gud Satun.	-	-	1	-	-	1
Chionomys sp.	-	2	1	-	-	3
Microtus arvalis L. s. l.	14/4	9/4	7/4	1	-	31/13
Arvicolinae indet.	2/1	-	-	-	-	2/1
Canis lupus L.	2/1	-	-	-	-	2/1
Vulpes vulpes L.	-	3/1	2/1	-	-	5/2
Martes foina Erxl.	1	1	-	-	-	2/2
Crocuta spelaea Goldf.	1	2/2	2/1	-	-	5/4
Equus ferus Boddaert s.l.	3/2	3/1	2/1	-	-	8/4
Sus scrofa L	2/1	-	-	-	-	2/1
Cervus elaphus L.	2/1	3/1	-	-	-	5/2
Megaloceros giganteus Blum.	5/1	7/1	1	1	11/1	25/5
Bison priscus Boj.	52/3	72/2	99/4	62/	2/1	290/11
Rupicapra rupicapra L.	1	-	-	-	-	1
Capra caucasica Guld. et Pall.	16/2	43/2	68/3	37/2	4/1	168/9
Ovis orientalis Gmel.	32/2	19/2	21/2	6/1	-	78/7
Indefinable bone fractions (mostly bison)	17262	27058	22893	11570	2	78812

Note: the numerators show the quantity of bone remains and the denominators – quantity of individuals.

Table 16.6: Palaeozoological complexes of mammals from various horizons of the Barakaevskaya Mousterian layer (after Baryshnikov 1994, 75: table 4)

Groups of plant resources	Anthropohors and plant-gathering evidences: acceptable (1), more realistic (2)		
	Land plants (1)	Water plants inside dry cave (2)	Sum
Food plants + oil food plants	27	5	32
Vitamin plants	32	3	35
Fire wood plants	25	-	25
Plants used for joinery	22	-	22
Plants used for tanning skins	22	-	22
Medicine plants	19	1	20
Paints	16	1	17
Melliferous plants	14	-	14
Plants used for weaving	4	1	5
Plants used for spinning	1	1	2
Plants used for making mats	1	1	2
Plants used for wood polishing	-	1	1
Variations of fossil flora (56 taxa) use	183	14	197

Table 16.7: The types of plants useful for people that were identified at Barakaevskaya Cave Mousterian layer under travertine (in the context of A. Grossgeim's (1952) data on modern plant resources of Caucasus).

References

Aitken, M. J. and Stokes, S. (1997) Climatostratigraphy. In R. Taylor and M. Aitken (eds.) *Chronometric Dating in Archaeology*, 1–30. New York and London, Plenum Press.

Amirkhanov, H. A. (1986) *Verhnij Paleolit Prikubanija*. Moscow, Nauka Press, Academy of Science of USSR.

Ananova, E. N. (1966) O nedorazvitoj pyltse v lednikovyh otlozhenijah. *Bjulleten' Komissii po Izucheniju Chetvertichnogo Perioda* 32, 18–22.

Aranguren, B. and Revedin. A. (2008) *Un Accampamento di 30.000 Anni fa a Bilancino (Mugello, Firenze)*. Firenze, Instituto Italiano di Preistoria e Protostoria.

Arslanov, H. A. and Gey, N. A. (1987) K paleogeografii i geohronologii stratigrafitsheskogo razreza srednego i pozdnego wurma Abhazii (pogrebennyj torf'annik, r. Dziguta). *Vestnik Leningradskogo Gosudarstvennogo Universiteta* 7/4 (n. 28), 107–108.

Autlev, P. U. and Lyubin, V. P. (1994) Istoria issledovania paleolita Gubskogo basseina. In V. P. Lyubin (ed.) *Neanderthaltsy Gupsskogo Ushchelya na Severnom Cavcase*, 12–20. Maykop, Meoty Press.

Baryshnikov, G. F. (1994) Ostatki pozvonochnyh iz Barakaevskoy moustierskoy stoyanki. In V. P. Lyubin (ed.) *Neanderthaltsy Gupsskogo Ushchelya na Severnom Cavcase*, 69–75. Maykop, Meoty Press.

Belyaeva, E. V. (1999) *Moustierskiy mir Gubskogo ushchelya (Severny Cavcas)*. St. Petersburg, Peterburgckoe Vostokovedenie.

Belyaeva, E. V. (2004) Middle Palaeolithic settlement in the Gubs river basin (Northwestern Caucasus). In N. Conard (ed.) *Settlement Dynamics of the Middle Palaeolithic and Middle Stone Age: Volume II*, 133–150. Tubingen, Kerns Verlag.

Belyaeva, E. V., Levkovskaya, G. M. and Kharitonov, V. M. (1992) Novye dannye o moustierskih obitatel'ah Gubskogo ushchelia. *Rossijskaya Arheologija* 3, 214–218.

Belyaeva, E. V., Leonova, E. V., Lyubin, V. P., Aleksandrovskij, A. L. and Aleksandrovskaja, E. I. (2009) Paleoekologicheskaja dinamika i obitanie tcheloveka v Gubskom mikroregione (Kubanskij Kavkaz) v srednem paleolite i mezolite. In V. A. Trifanov (ed.) *Adaptatsija Kul'tur Paleolita – Eneolita k Izmenenijam Prirodnoj Sredy na Severo-Zapadnom Kavkaze*, 27–46. St.-Petersburg, Insitut Istorii Materialnoj Kultury Rossijskoj Akademii Nauk (IIMK RAN), Teza Press.

Bordes, F. (1961) *Typologie du Paleolithique Ancien et Moyen*. Paris, Editions du Centre National de la Recherche Scientifique (CNRS).

Cherniakhovskiy, A. G. (1994) Sostav otlozhenij vypolnjavshih Barakaevskuju peshcheru. In V. P. Lyubin (ed.) *Neandertaltsy Gupsskogo Ushchelya na Severnom Kavkaze*, 64–69. Majkop, Meoty Press.

Fedele, F. G., Giaccio, B., Isaia, R. and Orsi, G. (2003) The Campainian ignimbrite eruption, Heinrich Event 4, and Palaeolithic change in Europe: a high-resolution investigation. In A. Robock and C. Oppenheimer (eds.) *Volcanism and the Earth's Atmosphere*. Geophysical Monograph 139, 329–356. Washington, D.C., American Geophysical Union.

Filippov, A. K. and Lyubin, V. P. (1994) Kostianue retushery iz moustierskogo sloya i prostranstvennoe razmeschenie kulturnyh ostatkov. In V. P. Lyubin (ed.) *Neanderthaltsy Gupsskogo Ushchelya na Severnom Cavcase*, 142–147. Maykop, Meoty Press.

Gamble, C. (1999) *The Palaeolithic Societies of Europe*. Cambridge, Cambridge University Press.

Genty, D., Blamart, D., Ouahdi, R., Glimour, M., Baker, A., Jouzel, J. and Van-Exter, S. (2003) Precise dating of Dansgaard-Oeschger climate oscillations in Western Europe from stalagmite data. *Nature* 421/20, 833–837.

Gol'eva, A. A. (2001) *Fitolity i ih Informatsionnaja Rol' v Izuchenii Prirodnyh i Arheologicheskih Obiektov*. Moscow, Syktyvkar and Elista, Institut geografii Rossijskoj Akedemii Nauk.

Golovanova, L. V., Doronichev, V. B., Levkovskaya, G. M., Lozovoj, S. P., Nesmejanov, S. A., Pospelova, G. A., Romanova, G. P. and Haritonov, V. P. (2006) *Peshchera Matuzka*. St. Petersburg, Ostrovitjanin.

Grossgaim, A. A. (1952) *Rastitelnye Bogatstva Kavkaza*. Moscow, Moskovskoe Obshchestvo Ispytatelej Prirody.

Gvozdetskiy, N. A., Dmitrienko, N. B. and Nefedjeva, E. A. (1966) *Kavkazskaja Gornaja Strana. Prirodnye Uslovija i Estestvennye Resursy SSSR. Kavkaz*. Moscow, Moskovskij Gosudarstvennyj Universitet.

Hoffecker, J. F. and Baryshnikov, G. F. (1998). Neanderthal ecology in the Northwestern Caucasus: faunal remains from the Borisovskoe Gorge sites. In J. J. Saunders, B. W. Styles, and G. F. Baryshnikov (eds.) *Quaternary Paleozoology in the Northern Hemisphere*, 187–211. Illinois State Museum Scientific Papers, 27. Springfield, Illinois State Museum.

Holtsberg, A. I. ed. (1972) *Agroklimaticheskij Atlas Mira*. Moscow-Leningrad, Gidrometeoizdat.

Johnsen, S. J., Dahl-Jensen, D., Gundestrup, N., Steffensen, J. P., Clausen, H. B., Miller, H., Masson-Delmotte, V., Sveinbjörnsdottir, A. E. and White, J. (2001) Oxygen isotope and palaeotemperature records from six Greenland ice-core stations: Camp Century, Dye-3, GRIP, GISP2, Renland and NorthGRIP. *Quarternary Science Reviews* 16, 299–307.

Kabailene, M. V. (1969) *Formirovanie Sporovo-Pyltsevyh Spektrov v Vodoemah Litvy i Metody Vosstanovlenija Paleorastitelnosti*. Vilnius, Akademija Nauk Litovskoe otdelenie.

Kartashova, G. G. and Troshkina, E. S. (1971) Osobennosti sporovo-pyltsevyh spektrov snezhno-firnovyh tolshch lednikov Tsentralnogo Kavkaza (na primere lednika Dzhankut). In S. S. Voskresenskiy and M. P. Grichuck (eds.) *Sporovo-Pyltsevoj Analiz pri Geomorfologicheskih Issledovanijah*, 48–57. Moscow, Moskovskij Gosudarstevennyj Universitet.

Kharitonov, V. M. and Romanova, G. P. (1984) Morfologitcheskie osobennosti cheloveka iz paleoliticheskoj stojanki v navese Satanaj. *Voprosy Antropologii* 74, 49–54. Moskva, Nauka Press, Akademija Nauk SSSR.

Khomutova, V. I. (1978) *Pyltsa i Spory v Donnyh Otlozheniah Kubenskogo Ozera. Part II. Trudy Instituta Ozerovedenia*, 102–210. Leningrad, Nauka Press, Akademija nauk SSSR, Leningradskoe otdelenie.

Kulakov, S. A., Baryshnikov, G. F. and Levkovskaya, G. M. (2007) Nekotorye rezultaty novogo izuchenija Akhshturskoj peshchernoj stojanki (zapadnyj Kavkaz). In H. A. Amirkhanov and S. A. Vasilyev (eds.) *Kavkaz i Pervonatchalnoe Zaselenie Chelovekom Starogo Sveta*, 65–81. St. Petersburg, Peterburgskoe Vostokovedenie.

Levkovskaya, G. M. (1967) *Zakonomernosti Raspredelenija Pyltsy i Spor v Sovremennyh Golotsenovyh Otlozheniyah Severa Zapadnoy Sibiri*. Aftoref. dissertatsii kand. geogr. nauk. Leningrad, Leningradskij Gosudarstvennyj Universitet.

Levkovskaya, G. M. (1967) O raspredelenii pyltsy i spor v poverhnostnom sloe donnyh otlozhenij Ladozhskogo ozera. In Geograficheskoe Obshchestvo SSSR (ed.) *Istorija ozer Severo-Zapada,* 140–145. Leningrad, Geograficheskoe obshchestovo SSSR.

Levkovskaya, G. M. (1973) Zonalnye osobennosti sovremennoy rastitelnosti I retsentnykh sporo-pyltsevykh spektrov Zapadnoj Sibiri // Metodicheskie voprosy palinologii. In Medvedeva, A. M. (ed.) *Trudy III Mezhdunarodnoy Palinolinologicheskoy Konferetsii,* 116–120. Moskva, Nauka.

Levkovskaya, G. M. (1980) Palinologicheskaja kharakteristika otlozhenij v peshcherakh Kudaro I i Kudaro III. In I. K. Ivanova and A. G. Cherniakhovsky (eds.), *Kudarskie Peshchernye Paleoliticheskie Stojanki v Jugo-Osetii. Voprosy Stratigrafii, Ekologii, Hronologii,* 128–151. Leningrad, Nauka Press (Aemija Nauk SSSR, Komissija pom Izucheniju Chetvertichnogo Perioda, Geologicheskij Institut)

Levkovskaya, G. M. (1983) O sobiratelstve i antropogennyh izmenenijah rastitelnosti v mustie. *Kratkie Soobshchenija Instituta Arheologii* 173, 73–76.

Levkovskaya, G. M., Berdovskaya, G. N. and Khomutova, V. I. (1983) Morfologicheskaya izmenchivost' pyltsy eli – vozmozhnyj istochnik oshibok pri paleogeograficheskih rekonstruktsiyakh (dannye po stoyanke Kostenki 14). In G. N. Papulov (ed.) *Materialy IV Vsesoyuznoy Palinologicheskoj Konferentsii (Tumen' 1981): 'Palinologiya i Paleogeografiya',* 53 – 57. Sverdlovsk, Uralskij Nauchnyj Tsentr AN SSSR.

Levkovskaya, G. M. (1987) *Prirodnaja Sreda i Pervobytnyj Tchelovek v Srednem Golotsene Lubanskoj Niziny (Vostochnaja Latvia).* Riga, Zinatne.

Levkovskaya, G. M. (1992) Palinologicheskaja kharakteristika mustierskoj stojanki Vorontsovskaja peshchera (Sochinskij rajon). In D. H. Mekulov (ed.) *Voprosy Arheologii Adygei,* 93–101. Majkop, Nauchno-issledovatelskij Institut Economiki, Jazyka, Literatury i Istorii.

Levkovskaya, G. M. (1994). Palinologicheskaya characteristica otlozheniy Barakaevskoy peshchery. In V. P. Lyubin (ed.) *Neanderthaltsy Gupsskogo Ushchelya na Severnom Cavcase,* 77–82. Maykop, Meoty.

Levkovskaya, G. M. (1999) Palynoteratical complexes as indicators of the ecological stress: past and present. *Proceedings of 5th European Palaeobotanical and Palynological Conference (Kraków, June 26–30, 1998). Acta Palaeobotanica* 2 (Suppl. 2), 643–648.

Levkovskaya, G. M. and Bogoljubova, A. N. (2005) Tipy palinoteratnykh 'otvetov' generativnoj sfery rastenij na klimaticheskie izmenenija vnutri plejstotsenovyh tsiklov: gljatsial-intergljatsial i stadial-interstadial. In S. A. Afonin and P. I. Tokarev (eds.) *Materialy XI Vserossijskoj Palinologicheskoj Konferencii 'Palinologija: Teorija i Praktika',* 132–133. Moscow, Paleontologicheskij Institut Rossijskoj Akademii Nauk.

Levkovskaya, G. M., Anikovich, M. V., Hoffecker, J. F., Holliday, V. T., Forman, S., Popov, V. V., Pospelova, G. A., Kartseva, L. A., Stegantseva, V. Ja. and Saniko, A. F. (2005a) Klimatostratigrafija drevnejshih paleoliticheskih sloev stojanki Kostenki 12 – Volkovskaja (pervye obobshchenija palinologicheskih, palinoteratnyh, paleozoologicheskih, paleopedologicheskih, paleomagnitnyh i SEM-paleobotanicheskih issledovanij). In M. V. Anikovich (ed.) *Problemy Rannej Pory Verhnego Paleolita Kostenkovsko-Borshchevskogo Rajona i Sopredelnyh Territorij. Trudy Kostenkovsko-Borshchevskoj Arheologicheskoj Ekspeditsii* 4, 93–130. St. Petersburg, Institut Istorii Materialnoj Kultury Rossijskoj Akademii Nauk.

Levkovskaya, G. M., Kartseva, L. A., Kolomets, O. D., Golubok, A. O., Rozshanov, V. V., Macko, V. P., Skvernjuk, I. I., Orehova, M. G. and Zaraj, G. (2005b) Spetsifika palinoteratnogo 'otveta' generativnoj sfery rastenij na chernobylskuju katastrofu. In S. A. Afonin and P. I. Tokarev (eds.). *Materialy XI Vserossijskoj Palinologicheskoj Konferencii 'Palinologija: Teorija i Praktika',* 134–136. Moscow, Paleontologicheskij Institut Rossijskoj Akademii Nauk.

Levkovskaya, G. M. (2006) Spetsifika shesti pozdneplejstotsenovyh termomerov i semi kriomerov rajona peshchernoj stojanki Matuzka. In L. V. Golovanova and V. B. Doronichev (eds.) *Peshchera Matuzka,* 54–71. St. Petersburg, Ostrovitjanin.

Levkovskaya, G. M. (2007) Palinologicheskaya kharakteristika otlozheniy Treugolnoj peshchery. In V. P. Doronichev, L. V. Golovanova, G. F. Baryshnikov, B. A. B. Blekwell, N. B. Garrut, G. M. Levkovskaya, A. N. Molot'kov, S. A. Nesmejanov, G. A. Pospelova and G. F. Hoffecker (eds.) *Treugolnaya peshchera. Rannij Palaeolit Kavkaza i Vostochnoj Evrop,* 73–81. St. Petersburg, Ostrovitianin.

Levkovskaya, G. M., Gambassini, P., Ronchitelli, A. M., Kolobova, K. and Anisutkin, N. K. (2007c). Middle/Upper Palaeolithic transitional time in Eurasia: cultural-historical, anthropological, palaeoecological and adaptation processes of the span 50–30 kyr BP. In A. Uglešić (ed.) *Final Programme and Abstracts. The 13th EAA Meeting Zadar, Croatia, 18th–23th September 2007,* 274–290. Zadar, University of Zadar.

Levkovskaya G. M., Lubin V. P., Kulakov S. A., Beliaeva E. A., Anisiutkin N. K., Vishniatskiy L. B., Nekhoroshev P. E., Chistyakov D. A., Housley R., Golovanova L. V., Bogolubova A. N. and Katamba N. A. (2007a) Archaeology-Palaeobotany-Palynology Database of the Neanderthal Epoche of Caucasus and Russian Plain: HE4 Event and Variations of its Palaeoenvirionments Industries and Types of Adaptations of Neanderthals). In A. Uglešić (ed.) *Final Programme and Abstracts. The 13th EAA Meeting Zadar, Croatia, 18th–23th September 2007,* 277–278. Zadar, University of Zadar.

Levkovskaya, G., Hoffecker, J., Anikovich, M., Popov, V., Lisizin, S., Anisutkin, N., Forman, S., Holliday, V. and Pospelova, G. (2007b) Time of H4 event in Kostenki-Borshevo region (archaeological, palaeoenvironmental and adaptation processes) In A. Uglešić (ed.) *Final Programme and Abstracts. The 13th EAA Meeting Zadar, Croatia, 18th–23th September 2007,* 283–284. Zadar, University of Zadar.

Lumley, H. de (1972) *La Grotte de l'Hortus (Valflaunés, Hérault), Memoir 1.* Marseille, Universite de Provence, CNRS.

Lyubin, V. P. (1977) *Moustierskie Cul'tury Cavcasa.* Leningrad, Nauka Press.

Lyubin, V. P. (1989) Paleolit Kavkaza. In P.I. Boriskovskii (ed.) *Paleolit Mira,* 9–142. Leningrad, Nauka Press.

Lyubin, V. P. ed. (1994a) *Neanderthaltsy Gupsskogo Ushchelya na Severnom Cavcase.* Maykop, Meoty Press.

Lyubin, V. P. (1994b) Itogi kompleksnogo izuchenia Barakaevskoy moustierskoy stoyanki. In V. P. Lyubin (ed.) *Neanderthaltsy Gupsskogo Ushchelya na Severnom Cavcase,* 151–164. Maykop, Meoty Press.

Lyubin, V. P., Autlev, P. U., Gritchuk, V. P., Goubonina, Z. P. and Monoszon, M. M. (1973) Mustierskaja stojanka v Goubskom navese 1 (Prikubanie). *Kratkie Soobsshenija Instituta Arheologii* 37, 54–62.

Lyubin, V. P. and Autlev P. U. (1994) Kamennaya industria. In V. P. Lyubin (ed.) *Neanderthaltsy Gupsskogo Ushchelya na Severnom Cavcase,* 99–140. Maykop, Meoty Press.

Mangerud, J., Astakhov, V. and Svendsen, J. I. (2004) The glaciation history of northern Russia. In M. V. Anikovich (ed.) *Kostenki i Rannjaja pora Verhnego Paleolita: Obshee i Lokal'noe. Mezhdunarodnaja Konferentsija. Putevoditel' i Tezisy,* 111–112. Voronezh, Istoki.

Mellars, P. (1996) *The Neanderthal Legacy: An Archaeological Perspective from Western Europe.* Princeton, Princeton University Press.

Meyer-Melikian, N. R., Bovina, I. Ju., Kosenko, Ja. V., Polevova, S. V., Severova, S. V., Tekleva, M. V. and Tokarev, P. I. (2004) *Atlas Pyltsevyh Zeren Astrovyh (Asteraceae).* Moscow, Tovarisshestvo Nauchnyh Izdanij KMK.

Mitlianskaya, T. B. ed. (1983) *Sel'skomu Uchitel'u o Narodnyh Hudozhestvennyh Remeslah Sibiri i Dal'nego Vostoka.* Moscow, Prosveshchenie.

Monoszon, M. M. (1973) *Opredelitel' Pyltsy Vidov Semejstva Marevyh.* Moscow, Nauka.

Nesmeyanov, S. A. (1994). Geomorphologicheskiy ocherk rayona Paleoliticheskih mestonahozhdenii Borisovskogo ushchelya na r. Gubs na Severnom Caucase. In V. P. Lyubin (ed.) *Neanderthaltsy Gupsskogo Ushchelya na Severnom Cavcase,* 22–36. Maykop, Meoty Press.

Pavlov, N. V. (1948) *Botanicheskaja Geografija SSSR.* Alma-Ata, AN Kazahskoj SSR.

Povarnitsyn, V. B. (1931) Tipy bukovyh lesov Djalabetskogo lesnogo massiva. In Sovet Izucheniju Proizvoditelnyh Sil (ed.) *Trudy Soveta po Izucheniju Proizvoditelnyh Sil* 2, 35–134. Leningrad, Botanitcheskij Institut Akademii Nauk Sojuza Sovetskih Sotsialisticheskih Respublik.

Ran E. Th. H. (1990) Dynamics of vegetation and environment during the middle pleniglacial in the Dinkel Vallye (the Netherlands). *Mededelingen Rijks Geologische Dienst* 44/3, 141–206.

Schelinskiy, V. E. (1975) Trasologicheskoe izuchenie funktsij kamennyh orudij Gubskoj mustierskoj stojanki v Prikubanie. In Akademija Nauk SSSR (ed.) *Kratkie Soobsshenija Instituta Arheologii* 141, 51–75. Moscow, Nauka Press.

Schelinskiy, V. E. (1992) Functsional'ny analyz orudii truda nizhnego paleolita Prikubania. In K. Mekulov, P. A. Ditler, and A. A. Sazonov (eds.) *Voprosy Archaeologii Adygei,* 194–209. Maykop, Meoty Press.

Schelinskiy, V. E. (1994a) Terochny kamen' iz moustierskogo kulturnogo sloya Barakaevskoy peshchery. In V. P. Lyubin (ed.) *Neanderthaltsy Gupsskogo Ushchelya na Severnom Cavcase,* 148–150. Maykop, Meoty Press.

Schelinskiy, V. E. (1994b) *Trasologija, Functsija Orudij Truda i Proizvodstvenno-Hoziastvennye Kompleksy Nizhnego i Srednego Paleolita (po Materialam Kavkaza, Kryma i Russkoj Ravniny).* Aftoreferat (thesis) dissertatsii doktora istorecheskih nauk. St.Petersburg, Institut Istorii Materialnoj Kultury Rossijskoj Akademii Nauk.

Spiridonova, E. A. (1991) *Evoljutsija Rastitel'nogo Pokrova Bassejna Dona v Verhnem Plejstotsene i Golotsene.* Leningrad, Nauka Akademija Nauk SSSR.

Tarasevitch, V. F. (1983) Morfologija pyltsy Cyperaceae. In A. E. Bobrov, L. A. Kuprijanova, M. F. Litvintseva and V. F. Tarasevich (eds.) *Spory Paporotnikoobraznyh i Pyltsa Golosemennyh i Odnodolnyh Rastenij Flory Evropejskoj Tchasti SSSR,* 90–111. Leningrad, Nauka.

Tchikov, P. S. ed. (1980) *Atlas Arealov i Resursov Lekarstvennyh Rastenij SSSR.* Moscow, Kartografija.

Tchistiakov, D. A. (1996). *Mustierskie Pamiatniki Severo-Vostochnogo Prichernomorija.* St. Petersburg, Evropejskij Dom.

Van der Hammen T., Maarleveld G. C., Vogel J. C. and Zagwijn W. H. (1967) Stratigraphy, climatic succession and radiocarbon dating of the last glacial in the Netherlands. *Geologie Mijnbouw* 46, 79–95.

Zarrina, E. P., Krasnov, I. I. and Spiridonova, E. A. (1980) Klimatostratigraficheskaja korreliatsija i khronologija pozdnego plejstotsena severo-zapada i tsentra Russkoj ravniny. In I. K. Ivanova (ed.) *Chetvertichnaja Geologija i Geomorfologija (Doklady Sovetskih Geologov k XXVI Mezhdunarodnomu Geologicheskomu Kongressu v Parizhe),* 99–120. Moscow, Nauka Press, Akademija Nauk SSSR.

Zubov, A. A., Romanova, G. P. and Kharitonov, V. M. (1994) Anthropologicheskiy analis nijney cheliusti rebenka-neanderthaltsa iz Barakaevskoy peshchery. In V. P. Lyubin (ed.) *Neanderthaltsy Gupsskogo Ushchelya na Severnom Cavcase,* 83–98. Maykop, Meoty Press.

Chapter 17

Interstratification in Layers of Unit III at Skalisty Rockshelter and the Origin of the Crimean Final Palaeolithic

Valery A. Manko

This article describes the cases of interstratification of materials representing Shankobien and Taubodrakian lithic industries in the Final Palaeolithic layers at Skalisty Rockshelter in the Crimean Mountains (Ukraine). Unit III at Skalisty Rockshelter contains 30 cultural layers, which correspond with the Epipalaeolithic stage and are dated to about 13,000–10,000 BP. Analysis of the flint industry in separate technocomplexes provides an opportunity to suggest that the site was visited by different teams of hunters who used different industries. One of these industries is called the Shankobien industry. Its technocomplexes are characterized by the presence of numerous types of geometric microliths, which were produced without using the microburin technique. The second type of industry is named the Taubodrakian industry. It is characterised by extensive use of the microburin technique in the manufacture of lunates, trapezes and triangles. The two industries also differ in the content of flint products and in a splitting technique used. Analysis of the flint technocomplexes shows that representatives of the two industries exploited the site alternately. The possibility of parallel co-use of the site by the representatives of different cultural traditions may be connected to the peculiarities of the economic strategies followed by the members of those cultures who inhabited the site in various seasons. This article also reviews the issues of origin of the two industries, which seems to be related to the geometric industries of the Eastern Mediterranean.

Introduction

As a result of slow sedimentation, visits by producers of different types of industries cannot be distinguished easily by means of archaeological methods; at such sites the cultural layer often turns to a heterogeneous mix that cannot be properly divided. In Ukraine, such a situation is common for most of the sites of the Ancient and Early Holocene (13,000–8000 BP). As an example, we can take a number of sites located in the steppe zone of Ukraine and in the Crimean Mountains.

We begin with the cave sites of Rogalik-Peredel'sk node (Lugansk region, Ukraine). Here, at the sites of Rogalik 2 and Peredel'sk 1 (Gorelik 2001), the following situation was encountered: artefacts characteristic of the Epigravettian and Osokorovian industries were found in a mixed state. So far, the reasons for such mixing have not been identified. Two possible interpretations can be considered: first, the technocomplexes comprise a single entity, i.e. the situation represents the synthesis of the two industries; second, the industries existed one after the other and the artefacts from the layer are mixed due to natural factors.

We can observe a similar situation at the cave site of Belolissya (Odessa region). Here, according to I. Sapozhnikov (2004), the technocomplex contains artefacts related to the Shankobien and Epigravettian industries. This raises the same questions as in the case of sites of the Rogalikso-Peredel'skiy node.

The problem of the site of Mirnoye (Odessa region) is also widely known, due to a number of artefacts attributed to the Kukrek and Grebeniky industries, which were found during earlier research. Analysis of the flint technocomplexes at the site of Mirnoye provoked debates as to the interrelation of the artefacts representing these two industries. The initiator of the excavations, V. N. Stanko, believed that we are dealing with a synthetic Kukrek-Grebeniky industry (Stanko 1982), while A. N. Sorokin suggested that the site was frequently visited by the makers of different industries (Sorokin 2002). It is possible that such visits were alternate.

According to the above, the problem of co-existence of the producers of various industries during the Final Palaeolithic-Mesolithic has been repeatedly raised in Ukrainian archaeology, but has not been solved so far because of the incompleteness of the archaeological sources.

We can hardly expect to solve this problem in the future either, as the sites located in the steppe zone are, as a rule, characterised by slow sedimentation, which hinders the conservation of cultural layers and the formation of sterile layers between different occupational phases at the site. However, we have other examples, like the cave sites located in the Crimean Mountains, where sometimes we encounter cases of excellent preservation of cultural layers associated with distinct visits to the sites. Here we are dealing with 'interstratification' which refer to those cases in which an ancient site was visited during a certain time period by different teams represented by different stone industries or technocomplexes. Interstratification is a rare phenomenon, which can be traced in archaeological excavations only when the cultural layer is perfectly preserved. A good example of such a situation is the Skalisty Rockshelter.

Interstratification at the Skalisty Rockshelter

Skalisty Rockshelter (Skalistoye village, Bakhchisarai district, the Crimean Autonomy) is situated in the second ridge of the Crimean Mountains on the bank of the river Bodrak (Fig. 17.1). The first excavations at the site were conducted in 1993–95. (Bibikov *et al.* 1994; Cohen *et al.* 1996). The urgent necessity of carrying out the excavations in 2004–2008 was associated with the impact of forces of nature, which threatened to ruin the site. In particular, water flows, occurring after heavy rains, washed out the upper layers of the site and destroyed the walls of the excavation conducted in the 1990s. At present, excavations have been carried out over an area of about 30 square meters, and have reached a depth of 6m. During the

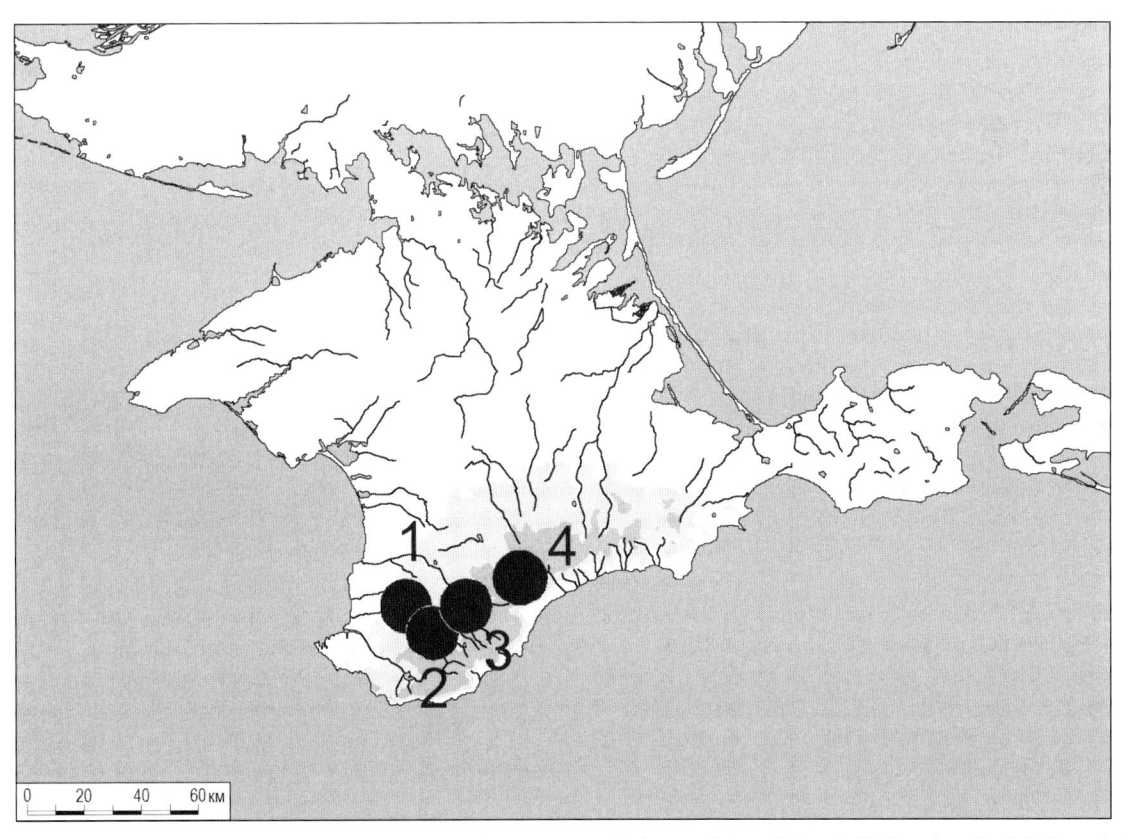

Figure 17.1: Epipalaeolithic sites in the Crimea Peninsula. 1. Fatma-Koba; 2. Szhan-Koba; 3. Vodopadny Rockshelter; 4. Skalisty Rockshelter.

course of the research, altogheter 38 cultural layers have been identified, 30 of these corresponding with the Epipalaeolithic. The layers provided finds such as artefacts of flint and of other types of stone and animal bones. Analysis of the flint technocomplexes (Manko and Yanevich 2006; Manko 2008) led to the conclusion that the flint technocomplexes of the site represent two types of industries, attributed to the Final Palaeolithic: the Shankobien and Taubodrakian industries (Figs 173–5).

All cases of interstratification are related to a unit of layers, Unit III (Fig. 17.2.3), assigned to the period 12,800–10,000 BP (uncalibrated), i.e. to the Ancient Holocene (Final Pleistocene) and the Final Palaeolithic. Both explicit and implicit instances of interstratification were noted. The oldest researched cultural horizon of the site consists of several cultural layers, assigned to cluster III-3-B. All the technocomplexes associated with this cluster have the characteristics of the Shankobien industry. This cultural horizon is overlapped by layer III-3-X-3, which is associated with the Taubodrakian industry. Unfortunately, layer III-3-X-3 was found only in the area in front of a shed. Under the shed, layer III-3-B is overlapped by layer III-3-A, which can be defined as mixed, containing artefacts of the two types of industries. Layer III-3-A overlaps layer III-3, which contains artefacts associated only with the Shankobien industry. Layer III-2-A-1, which is located above, contains only artefacts of the Taubodrakian industry. The stratigraphic layer III-2-A-1, located under a shed, is contemporaneous with layers III-3-X-2 and III-3-X-1, located on the ground in front of a shed. Their contemporaneity is proved by the fact that they are connected in a form of a layer between lenses of layers III-3-B and III-2-A, which are situated at the junction of the areas under the shed and the ones in front of it. Stratigraphy clearly shows that layer III-1, associated with the Shankobien industry, overlaps layer III-3-X-1. Thus, we admit the fact that the history of the Skalisty Rockshelter, in most cases associated with visits by producers of the Shankobien industry, provides us with at least two apparent cases of parallel site attendance by the producers of various industries.

It is impossible not to draw attention to the fact that the producers of different industries had completely different strategies for the use of the site.

Due to the dispersion of ash, the Shankobien industry layers have a specific colour (grey, dark brown), they contain multiple hearths (Fig. 17.2.2), pits, burnt bones and flints, which indicates that visits to the site were relatively long term. Analysis of the flint technocomplexes also confirms this thesis. The enormous number of chunks of raw materials in the Shankobien industry layers is especially illustrative. Sometimes it reaches 27,000 pieces per square meter, which provides evidence of the importance of primary flaking and confirms that they the flint workers used a strategy based on the continuous making of roughouts workpieces for work tools.

The layers containing the Taubodrakian industry are not stained with ash. None of the hearths are associated with the flint technocomplex of this industry (except for those cases where a sediment layer formed inside the hearth and contained no ash). The flint technocomplexes are characterised by a relatively low proportion specific volume of flakes, whose density does not exceed 1,000 per square meter. Bones and flints are not burnt. All these facts indicate that Taubodrakian industry producers visited the site for shorter periods than the ones of the Shankobien industry or used it less intensively.

Considerable differences in hunting strategy were identified. In total, six layers with representative faunal technocomplexes were investigated. Four of them are associated with a Shankobien industry, and the rest with a Taubodrakian industry (Table 17.1). Apparently, producers of the two cultural groups varied considerably: in the majority of Shankobien industry deposits red deer (*Cervus elaphus*) bones dominate, while in Taubodrakian industry deposits saiga (*Saiga tatarica*) bones prevail. The faunal complex of the Shankobien industry layer III-1 is an exception. Here, neither red deer bones nor saiga bones were found, while the roe was the main hunted species.

This correlation of animal bones may indicate that the producers of the Shankobien industry exploited primarily forest resources. The prevalence of saiga bones in the Taubodrakian industry deposits may indicate that their hunting strategy mainly used the steppe zones. The location of Skalisty Rockshelter perfectly ensured successful implementation of both strategies.

We can assume that the producers of the Shankobien industry used the rockshelter as a base camp and lived on-site for quite a long time. It is possible that the duration of their stay corresponded with the duration of the hunting season. The convenient site location provided a perfect view, enabling them to observe the movement of animals on the border of the forest, which covered the mountain slopes and Bodrak river valley. The ceiling of the rockshelter provided excellent protection from wind and rain. One can even argue that hunters of the Shankobien industry lived in the cave continuously. This is evidenced by an analysis of flint tools, including items relating to hunting, which comprise 50 to 70 per cent of the total number of tools. As a rule, such a proportion of geometric microliths is not characteristic for long-term settlement sites, for which a variety of economic activities is common. In

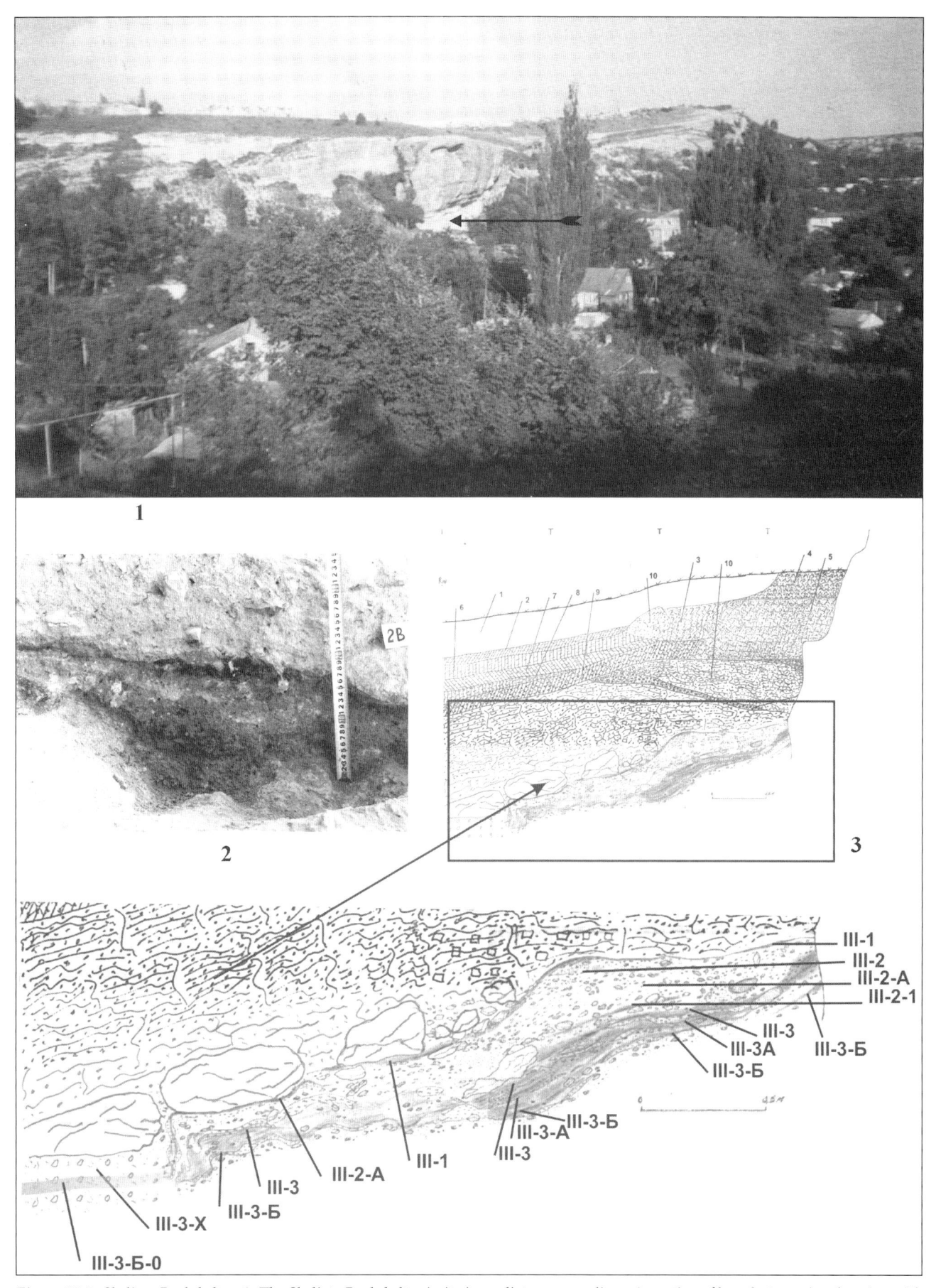

Figure 17.2: Skalisty Rockshelter. 1. The Skalisty Rockshelter in its immediate surroundings; 2. section of hearth; 3. section drawing with the relevant bottom layers enlarged.

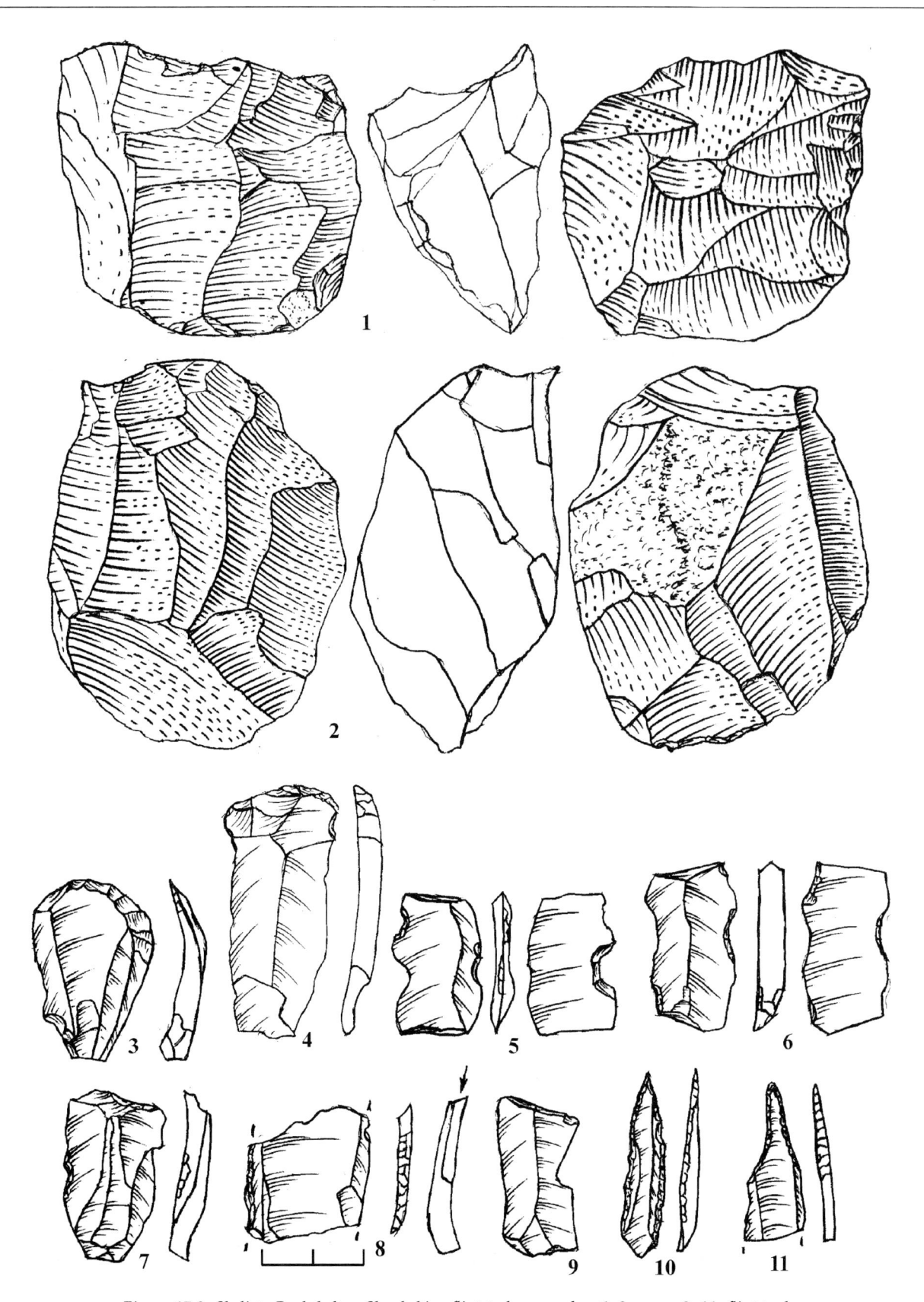

Figure 17.3: Skalisty Rockshelter. Shankobien flint technocomplex. 1–2. cores; 3–11. flint tools.

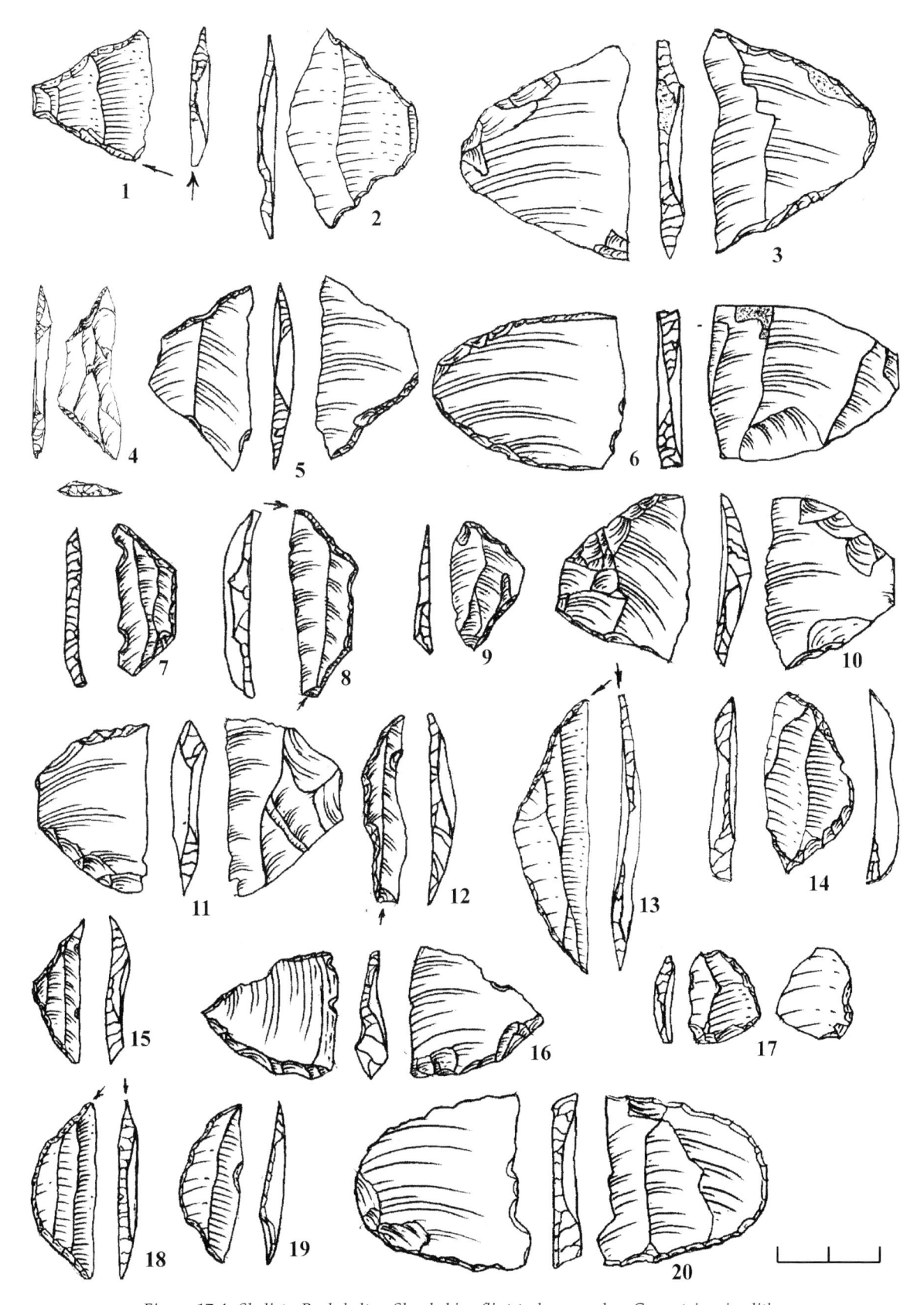

Figure 17.4: Skalisty Rockshelter. Shankobien flint technocomplex. Geometric microliths.

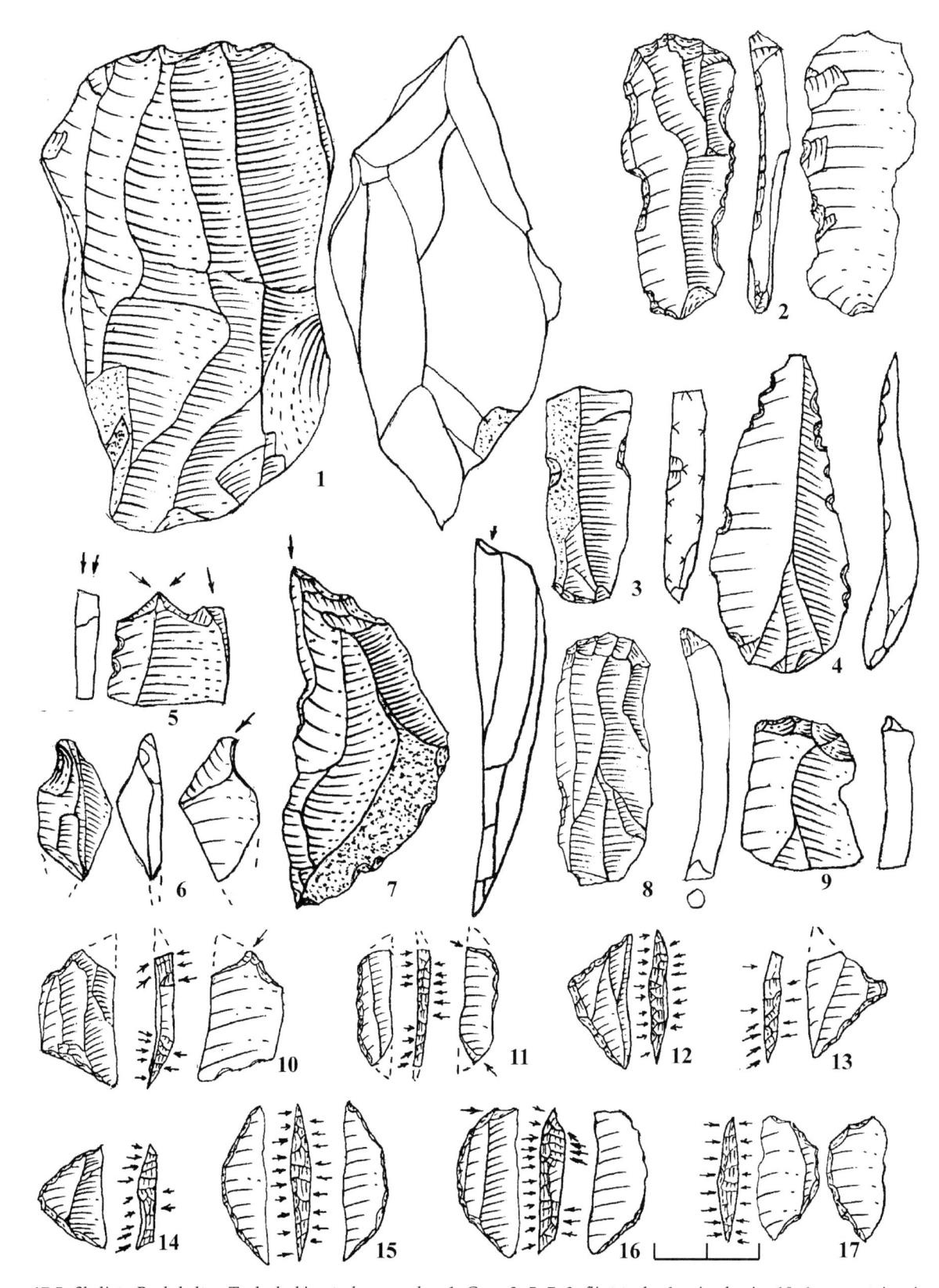

Figure 17.5: Skalisty Rockshelter. Taubodrakian technocomplex. 1. Core; 2–5, 7–9. flint tools; 6. microburin; 10–1. geometric microliths.

Species	Shankobien technocomplexes				Taubodrakien technocomplexes	
	III-1	III-3-Б	III-3-Б-0	III-3-Б-1	III-3-Х-1	III-3-Х-2
Lepus europaeus (Pallas)	14		2		3	
Capreolus capreolus	9					
Saiga tatarica		18	44	4	120	143
Lynx lynx			1			1
Canis lupus				1		1
Cervus elaphus		33	83	16	30	18
Equus hydruntinus			1			2
Sus scrofa		1		1		
Vulpes vulpes					1	3
Fish			4		8	
Birds	2		4		3	1
Not identified	508	815	978	637	433	244
All bones	533	867	1117	659	598	413

Table 17.1: Faunal remains from technocomplexes of Skalisty Rockshelter (studied by O. P. Zhuravlev).

such settlements, the proportion of items related to hunting is usually only 5–10 per cent of the Shankobien industry technocomplexes, so these sites cannot be interpreted as hunting places or as areas for butchering animal carcasses. Shalisty Rockshelter is located high above the valley. It is hard to imagine that the hunting took place right by the site, and it is unbelievable that animal carcasses were brought here from the valley. Such conclusions derive from the analysis of the faunal remains, among which there are no animal skulls, no anatomical groups and very few teeth. It appears that the site contained only those parts of carcasses that the hunters left after their own consumption. Geometric microliths with microwear traces were extracted from these bones at the site. The proportion of these microliths in the geometric technocomplexes reaches one third of the total number of tools.

Thus, in my opinion, the deposits associated with the Shankobien industry are derived from temporary hunting camps, focused on long-term hunting. The site was used as a place for hunting animals, for residence of the hunting team, for accumulation of flint raw materials, and for preparation of tools for hunting.

It appears that the producers of the Taubodrakian industry used the rockshelter to a lesser extent. Analysis of the cultural layers formed by hunters associated with the Taubodrakian industry demonstrates that their visits were short. Perhaps the site was used only to supply the raw flint for arrowheads used in hunting. Definitely, the hunters did not reside at the site, since staying at night in a mountain cave without a fire would have been impossible, even in summer.

Chronology of the Shankobien and Taubodrakian industries

Table 17.2 shows almost all the known absolute dates associated with the Shankobien and Taubodrakian industries.

The radiocarbon dates for Skalisty Rockshelter (Table 17.2.5–7) give us every reason to believe, that the earliest technocomplexes in Crimea were associated with the beginning of the Bølling environmental period, that is, with the very beginning of the Ancient Holocene.

Three dates from Skalisty Rockshelter (Table 17.2.3–4, 8) may be related to both the Taubodrakian and Shankobien industries. The dates of layers III-2 and III-3-A directly correspond with materials of the Taubodrakian industry. The dates of layer III-3 are close to the above-mentioned dates and contrast with the earlier dates, obtained while investigating the material from this layer. A similar date was obtained while screening the materials of the Shan-Koba site (layer VI.1), where a combination of microliths of the Taubodrakian and Shankobien industries was also found. The first case of interstratification of artefacts from Shan-Koba and Taubodrakian is connected with the Allerød environmental period.

Continuation of the co-existence of the two industries after the Allerød is also possible. There is no need to doubt that the Shankobien industry continued to exist in the Dryas 3 phase, and even in the Preboreal. This is evidenced by the radiocarbon dates of the Dryas 3 phase at Buran-Kaya-3 (Table 17.2.1–2), as well as by six dates of Preboreal date from Spahn-Koba (Table 17.2.11–16), and the Preboreal date from layer VI.1 of Shan-Koba

(Table 17.2.10). The lunates, similar to those of Shan-Koba, come from materials from the burial ground of Vasilievka III located in Dnieper Nadporozhie, dated to the Early Preboreal (Table 17.2.20–22).

There is also evidence for the co-existence of the two industries in a later period: in the Boreal. The dates of this period were obtained after examination of bone samples taken from 5–6 layers of the site of Fatma-Koba (Table 17.2.18–19). The deposits at Fatma-Koba primarily contained objects relating to the Taubodrakian industry, although elements of the Shankobien industry are present as impurities. Pure Shankobien technocomplexes are also found in the Boreal. For example, beyond Crimea, in the North-Western Black Sea Region, there is Belolissya, a Shankobien complex dated to the Boreal (Table 17.2.17).

The co-existence of the Taubodrakian and Shankobian traditions in the Boreal is confirmed not only by radiocarbon dates but also by relative chronology. Thus, layers V and IV of the site of Shan-Koba contain, in addition to Shankobien and Taubodrakian geometric microliths, such items as pencil-like cores, and blades with blunt edges. Their appearance is attributed to the Boreal, when the producers of the Kukrek flint

SITE, LAYER, MATERIAL	LAB. NO.	BP (UNCALIBRATED)
BURAN-KAYA-3		
1. Animal bone	Ki-6267	10,920±65
2. Animal bone	Ki-6267a	11,460±70
SKALISTY ROCKSHELTER		
3. III-2. Animal bone	OxA-5164	11,620±110
4. III-3. Animal bone	OxA-5165	11,750±120
5. III-3. Animal bone	OxA-4888	12,820±140
6. III-3-Б-1. Animal bone	Grn-	12,650±60
7. III-3. Ashes from a fire	Ki-13152	13,500±150
8. III-3-A. Ashes from a fire	Ki-13153	11,200±120
SHAN-KOBA		
9. VI.1. Animal bone	Ki-11086	11,260±190
10. VI.1. Animal bone	Ki-11085	9910±180
SHPAN-KOBA		
11. 3-2. Animal bone	KIA-3686	9760±60
12. 3-2. Animal bone	Ki-5824	9890±80
13. 3-4. Animal bone	KIA-3685	9930±60
14. 3-5. Animal bone	KIA-3684	9840±50
15. 3-5. Animal bone	Ki-5823	10,210±80
16. 3-6. Animal bone	KLA-3683	9940±50
BILOLISSYA		
17. Animal bone	Ki-10886	8900±190
FATMA-KOBA		
18. 5/6. Animal bone	Ki-10396	8520±150
19. 5/6. Animal bone	Ki-10395	8770±140
VASILIEVKA 3		
20. Bones from burial	OxA-3807	10,060±105
21. Bones from burial	OxA-3808	9980±100
22. Bones from burial	OxA-3809	10,080±100

Table 17.2: Radiocarbon dates from sites in the Black Sea region with Final Palaeolithic and Mesolithic layers.

industry appeared in Crimea. At present, it is difficult to establish whether a synthesis of these two industries took place in the later stages of their development, or if they continued to develop in isolation.

Origins of the Shankobien and Taubodrakian industries

The interstratification of the two types of industries represented at Skalisty Rockshelter made it possible not only to establish the fact that two populations with different flint industries resided at the same time in the Crimean Mountains, but also to justify the existence of the Taubodrakian industry.

The Shankobien industry became known in the 1920s and 30s, when G. A. Bonch-Osmolovsky initiated an investigation of the Shan-Koba site (Bonch-Osmolovsky 1934). Layer VI of Shan-Koba has long been a model for the characteristics of the ancient Shankobien industry. At the same time, all researchers of this technocomplex have underscored the fact that the stratification of this layer was the result of frequently repeated visits to the site. It was impossible to correctly separate episodes of formation of the cultural layer. For this reason, layer VI was divided into five horizons, in line with the depth of their location, in order to trace the dynamics of the industry. While assessing the developmental dynamics, the correlation of different types of geometric microliths was considered to be an important fact. Unfortunately, the researchers of layer VI of Shan-Koba did not realize that they had actually analysed the artefacts of two different industries. In contrast to the case of Skalisty Rockshelter, the formation of cultural layers at Shan-Koba was accompanied by slow sedimentation, which did not allow the researchers to establish the fact of interstratification. The emergence of pure technocomplexes at Skalisty makes it possible to describe – for the first time – the two different types of industries, to highlight their features, and raise the issue of their genesis.

The Shankobien industry and its genesis

The Shankobien industry is characterised by the use of one-platform cores of conical and prismatic shapes. The removal of short blades was carried out using a rigid hammerstone. This kind of technique was observed only in the oldest layer of Skalisty Rockshelter. Non-geometric technocomplexes of tools usually contain mainly endscrapers, burins on truncated faceted pieces, and perforators.

The Shankobien industry is the one that exhibits the richest variety of forms of geometric microliths. In particular, during the excavation of Skalisty Rockshelter, 18 types of microliths were identified (Table 17.3), although 11 out of these 18 were single samples. In fact, among the seven types, represented by large series, there are literally all of the forms of microliths that existed in the Final Palaeolithic in continental Ukraine. This typological diversity of the geometric microliths of Skalisty Rockshelter is accompanied by considerable variability in the correlation of length to width. Sets of geometric microliths are unstable; they form unique combinations in separate technocomplexes. For example, in cluster III-3-B-1 of Skalisty Rockshelter, the series of microlith types 5–6 are represented (Table 17.3), which is common for Rogalik-Osokorovka technocomplexes. In cluster III-3-B-2, a series of high microliths was found which is normally identified with the Zimovniki industry.

The aspect of application or absence of the microburin technique in Shankobien technocomplexes is also important. The answer to this question cannot be simple. On the one hand, the cluster of artefacts in layer VI in Shan-Koba contains a series of microburins. On the other hand, layer VI of Shan-Koba, as already mentioned, is a palimpsest, whose formation extended over time. Consequently, we do not know whether the microburins from layer VI are related to the Shankobien industry, or if they belong to the Taubodrakian component of the cultural layer.

Similar doubts appeared while analysing the Skalisty Rockshelter technocomplexes, where Shankobien technocomplexes were noticed. No microburins were found here, and there were no traces of the Taubodrakian technocomplexes, which are usually accompanied by large series of microburins. Thus, it is possible to state that the early Shankobien technocomplexes are not associated with the use of the microburin technique. However, at present we cannot make any conclusions on the basis of this, nor can we state the opposite as regards the later stages of development of the industry, attributed to the second half of the Ancient Holocene and to the Early Holocene. It is likely that the microburin technique could have finally emerged in the Shankobien industry, under the influence of contacts between its producers with representatives of the Taubodrakian industry.

As for the early stage of the Shankobien industry's development, well illustrated by the Skalistenskiy technocomplexes, we find only a sporadic appearance of pseudo-microburins, whose presence was not necessarily connected with deliberate use of microburin techniques.

No.	Type	Figure
	1. Geometric microliths of the Taubodrakien industry	
1.	Symmetrical trapeze with bipolar retouch on two sides	Fig. 17.1.1
2.	Symmetrical trapeze with bipolar retouch on three sides	Fig. 17.1.2
3.	Asymmetrical trapeze with bipolar retouch on three sides	Fig. 17.1.3
4.	Symmetrical triangle with bipolar retouch on two sides	Fig. 17.1.4
5.	Segment-like triangle with bipolar retouch on two sides	Fig. 17.1.5
6.	Symmetrical lunate with bipolar retouch on the arch	Fig. 17.1.6–7
7.	Asymmetrical lunate with bipolar retouch on the arch	Fig. 17.1.8
	2. Geometric microliths of the Shankobien industry	
1.	Symmetrical trapeze with backing retouch on two sides	Fig. 17.2.1–4
2.	Symmetrical trapeze with backing retouch on two sides and with flat retouch on the ventral surface	Fig. 17.2.3
3.	Symmetrical trapeze with alternate backing retouch on two sides	Fig. 17.2.5
4.	Symmetrical trapeze with inverse backing retouch on two sides	Fig. 17.2.6
5.	Symmetrical trapeze with backing retouch on three sides	Fig. 17.2.7
6.	Trapeze with backing retouch on three sides and with a notch on the long retouching side	Fig. 17.2.8
7.	Asymmetrical trapeze with backing retouch on two sides	Fig. 17.2.9
8.	Asymmetrical trapeze with backing retouch on two sides and with flat retouch on the ventral surface	Fig. 17.2.10
9.	Asymmetrical trapeze with inverse backing retouch on two sides	Fig. 17.2.11
10.	Asymmetrical trapeze with backing retouch on three sides	Fig. 17.2.12
11.	Symmetrical quasi-segment with backing retouch on two sides	Fig. 17.2.13
12.	Asymmetrical quasi-segment with backing retouch on two sides	Fig. 17.2.14
13.	Symmetrical triangle with backing retouch on two sides	Fig. 17.2.15
14.	Symmetrical triangle with backing retouch on two sides and with flat retouch on the ventral surface	Fig. 17.2.16
15.	Asymmetrical triangle with backing retouch on two sides	Fig. 17.2.17
16.	Symmetrical lunate with backing retouch on the arch	Fig. 17.2.18
17.	Asymmetrical lunate with backing retouch on the arch	Fig. 17.2.19

Table 17.3: Skalisty Rockshelter, Unit III. Type-list of geometric microliths.

The tools of non-geometric technocomplexes of the Shankobien industry have never been represented in significant series, due to the fact that, as a rule, geometric microliths comprise more than half of tool technocomplexes. The existence of series of points, which functioned as perforators, is the most distinctive feature of the non-geometric technocomplexes. These are bladelets with converging sharply retouched edges, forming 'darts' at the distal ends of the bladelets. The burins are on truncated faceted pieces and the angle burins are rarely dihedral. Usually there are more scrapers than burins. There are also technocomplexes in which burins are completely absent.

The flake production technique was also not standardized. In some technocomplexes there are two main types of cores: one-platform cores with a narrow flaking surface, and one-platform prismatic-like ones. In the lower layers of Skalisty Rockshelter there are also bi-direct two-platform cores. The main type of a debitage is a curved blade, which was removed with a help of a solid hammerstone. In the ancient technocomplexes of Skalisty Rockshelter the use of a soft hammerstone was established. The structure of the technocomplex of blade-like workpieces (blanks) is variable. The specific number of blades and fragments cut from them may range from 30 to 70 per cent of all laminated blanks. The specific number of microblades can vary from 5 to 35 per cent. As a rule, the twisted blanks are related to a group of microblades.

The origin of the Shankobien industry is often associated with the Epigravettian. Such an assumption seems to me unrealistic, since the relict Epigravettian technocomplexes simultaneously co-existed with the Shankobien technocomplexes (for instance, the Epigravettian technocomplexes of Leontyevka, Rogalik VII, Kamennaya Bakla II, and others). Moreover, there are not any intermediate technocomplexes between the Epigravettian and the Shankobien industries. There are no examples of points similar to those found in the Epigravettian technocomplexes among the oldest technocomplexes of Skalisty Rockshelter. A suggestion made by A. A. Yanevich about the possibility of the Shankobien industry's genesis being related to sites such as Vishennoye II (Yanevich 1992) is, in my opinion, erroneous. Archbacked blades are present in the Epigravettian technocomplex of Vishennoye II, however, the age of this technocomplex is not clear, and such microliths occur in the Epigravettian from the very beginning of its development. Consequently, their presence cannot be seen as evidence supporting the hypothesis that the geometric technocomplex of the Shankobien industry emerged out of it. Moreover, in the Epigravettian industry of Buran-Kayah, which immediately preceded the Shankobien industry, there were no microliths like that found (Yanevich 2000).

Since the Shankobien industry is the oldest industry with geometric technocomplexes in the Final Palaeolithic in Ukraine, one should recognize the impossibility of its genesis being linked to the other geometric Final Palaeolithic industries of the Ukraine. Accordingly, it is possible to make the following interpretative statements.

First, if we suggest that the genesis of the Shankobien industry can be linked to the Epigravettian, then it must relate to some of its variations, which are unknown at present. This seems to be entirely unlikely, as the example of the industry of Rogalik VII – which existed simultaneously with the Osokorovskaya industry – demonstrates that even in a brand-new environment the producers of the Epigravettian techniques did not adopt geometric microliths. Another example is provided by the sites of the Dyuvenzee cultural area. Here, the flint industry was characterized as Tardigravettian, where Epigravettian techniques remain throughout the whole Early Holocene. Given this, it is difficult to imagine that another industry was established based on the Epigravettian, while no features of the preceding industry were present. Accordingly, the appearance of the Shankobien industry in Ukraine can only be associated with migration; moreover, the producers of this new industry must have brought it to Ukraine in a complete form.

Second, it is impossible to suggest that these migrants could have been the producers of the Azillian industry from Western and Southern Europe. Here, the formation of geometric industries is undoubtedly related to the late Magdalenian, which is totally impossible as a source for the earliest geometric industries of Crimea (at least for the reason that the Magdalenian industries are odd, not only for Crimea but also for the steppes of Ukraine).

Third, the genesis of the Shankobien industry is most likely associated with migration of populations from the Eastern Mediterranean basin and Central Asia, where geometric microliths have been found in the deposits corresponding to the 20,000–12,000 BP period. This interpretation does not contradict the absolute chronology and corresponds to the distribution pattern of innovative technologies, which continued to exist at a later time, until the Neolithic.

Fourth, if the latter interpretation is correct, then the process of the spread of the Final Palaeolithic in Ukraine is also a process of pre-Neolithic replication, because it seems to be an absolute copy of the geometricisation process, which preceded the beginning of the food producing economy in the Fertile Crescent.

It is hardly possible to reliably determine to what industry the emergence of the Shankobien technocomplexes in Crimea is related. There are no analogs for this industry outside Ukraine. What is special about the Zarzian industry is its typological similarity with the Shankobien industry. Hhowever, it is characterized by the presence of specific types of points and lunates with helwan retouch along the arc. It is impossible to draw comparisons with the Early Natufian, where the lunates with helwan retouch are specific to the technocomplexes. Besides, a connection between the Zarzian and the Natufian industries is clear, as the microburin technique was present there. It is worth adding that most of the technocomplexes of these two industries are contemporary with the Shankobien industry. (For the chronology of the technocomplexes of the Near and Middle East, see Böhner and Schyle 2002–6.) Palegavra is the only exception among the sites of the Zarzian industry. Here, the beginning of the formation of the layer attributed to the Zarzian industry is dated to the Dryas 1 period. It should be noted that the typology of the geometric microliths of the Palegavra technocomplex is practically the same as that of the Shankobien technocomplexes: the geometric microliths are accompanied by perforators. Here it should be noted that, unfortunately, I have limited facilities to compare the technocomplexes of Palegavra and Skalisty Rockshelter, because, although

I have used the primary publication, it is not in a good condition (Braidwood and Howe 1960, 57–59, pl. 24). For this reason, I can not insist on the correctness of my observations.

The comparison of the Skalisty Rockshelter materials with the technocomplexes of geometric Kebara from Jordan, which are quite different from the technocomplexes of geometric Kebara from Israel, is highly illustrative. Unfortunately, I can only analyze the data of one technocomplex of this kind, associated with phase D of the site of Karaneh IV (Muheisen 1988, 265–269; Muheisen and Wada 1995, 75–95). The version of geometric Kebara, which was illustrated in the Karanehian complex, differs from the classic Kebarian technocomplexes in terms of the almost complete absence of trapezes with a retouched upper part, and in the use of geometric microliths. While geometric microliths in the classic geometric Kebara were used mainly as pricking points, in the Karaneh they were used as transverse points, just like the Shankobien technocomplexes, where pricking points were extremely rare. In the technocomplex of Karaneh IV, the most frequent are low symmetrical trapezes, series of lunates with sharply retouched arcs, and quasi-segments. Karaneh differs from the Skalisty technocomplex types only in terms of the absence of trapezes of high forms.

So, there are very strong analogs for the Shankobien industry. This, of course, cannot be a reason to exaggerate the significance of such analogies. Their presence is not informative enough to make conclusions about the genesis of the Shankobien industry, but at least it confirms the earlier expressed idea, that the Eastern Mediterranean Area and Middle East are those regions where the genesis of the Shankobien industry began.

The Taubodrakian industry and its genesis

The Taubodrakian industry was first described with reference to the Skalisty Rockshelter materials. The case of interstratification noted at the site became a reason for describing it: a specific Taubodrakian complex was found between the layers with the usual set of Shankobien microliths. Taubodrakian microliths do not differ from those of Shankobien typologically: all seven Taubodrakian types find a match in Shankobien technocomplexes (Table 17.3). The essence of the difference is: firstly, that Taubodrakian microliths were made using the microburin technique, and, secondly, that they have a particular design of lunate arcs and trapeze/triangle sides, and are covered with a sharp bi-direct retouch.

Geometric microliths of the Taubodrakian industry are small in size – their length rarely exceeds 2cm – but they were usually made of relatively massive pieces 3–7mm thick, whereas in Shankobien technocomplexes there are primarily microliths, made of pieces of 2–3mm thick.

This is not a complete list of differences. Taubodrakian microliths are accompanied by notched blades, which are usually missing in the Shankobien technocomplexes of Skalisty Rockshelter. On the other hand, in the Taubodrakian layers of the site, no perforators were found, while in the Shankobien technocomplexes there are series of them. The technocomplexes of scrapers and burins have similar features to those from the Shankobien technocomplexes.

The most prominent technocomplexes associated with the Taubodrakian industry are from Vodopadny Rockshelter (Fig. 17.6) and from layer 5/6 of Fatma-Koba (Fig. 17.7). Unfortunately, the stratigraphic position of the Final Palaeolithic layers of these sites is blurred. Moreover, the Taubodrakian findings are accompanied by classic Shan-Koba elements. Nevertheless, the presence of these technocomplexes, as well as the presence of the Taubodrakian component in layer VI of Shan-Koba, shows that the Taubodrakian industry was not a local phenomenon, which existed only in Skalisty Rockshelter.

The interstratification of the two types of industries clearly shows that they existed simultaneously, but this does not lead to a clear understanding of their genetic connection. It is possible that the available stratigraphic column of the Skalisty Rockshelter does not contain any layers related to the oldest manifestations of the two industries, where their characteristics could be traced simultaneously. It is possible that one of these industries was established on the basis of another, when there was a process of microburin technique degradation, or, conversely, its invention. If any of the alleged scenarios took place, then we are dealing with an example of two industries that had a common origin but then followed different directions. We have to admit that such a development was possible in the same, rather limited territory, and that this may well represent evidence of an informational isolation of the carriers of the two isolated traditions. At present, dealing with this problem, we can only make hypotheses and interpretations.

It is impossible not to mention the possibility that the genesis of the Taubodrakian industry took place outside Ukraine. Unlike the Shankobien technocomplexes, which have nothing but approximate analogues described above, the Taubodrakian technocomplexes do have analogues, which are almost absolute, or at least highly appropriate.

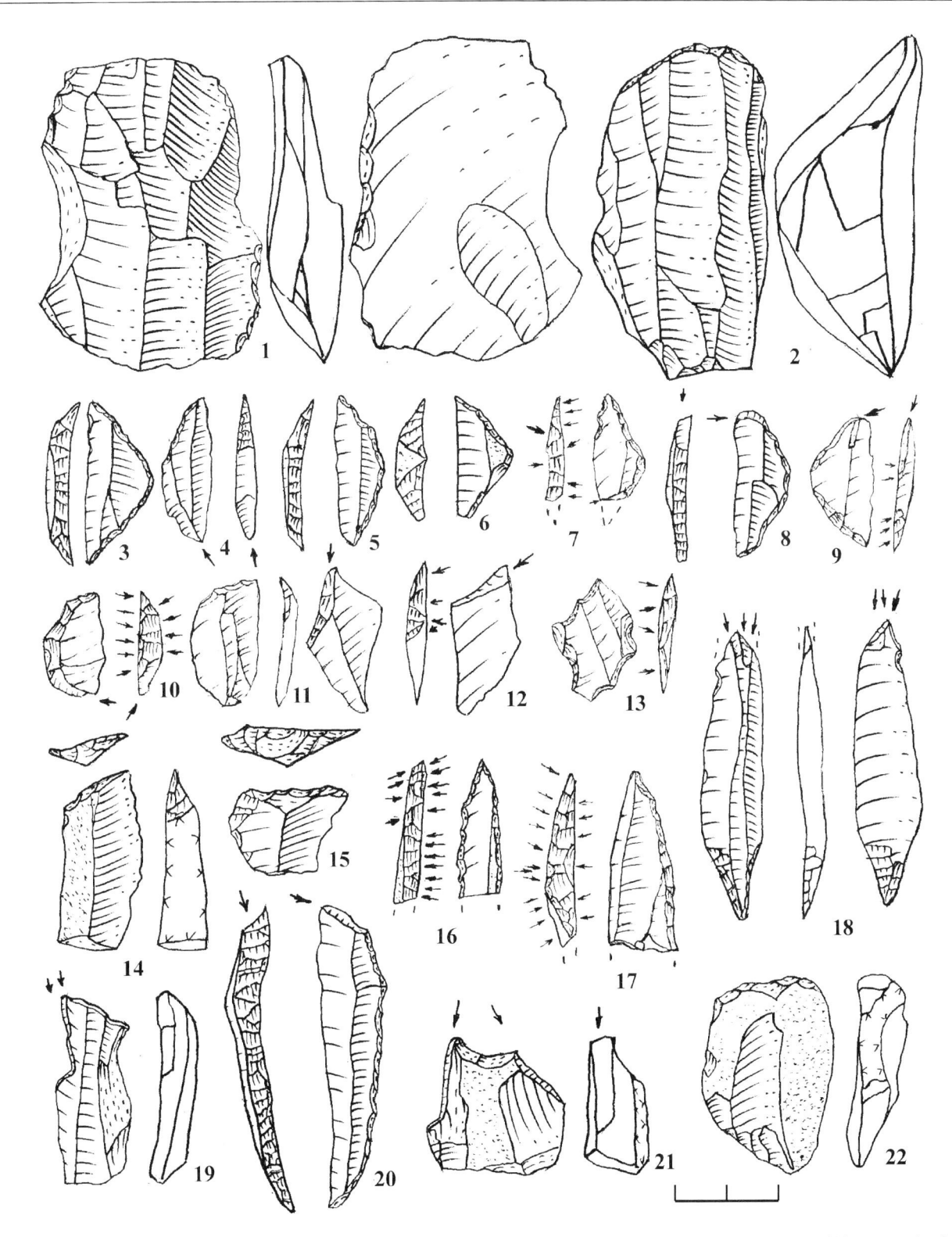

Figure 17.6: Vodopadny Rockshelter. Taubodrakian technocomplex. 1–2. cores; 3–11, 13, 17, 20. geometric microliths; 12. microburin; 14–16, 19, 21–22. flint tools; 18. tanged point.

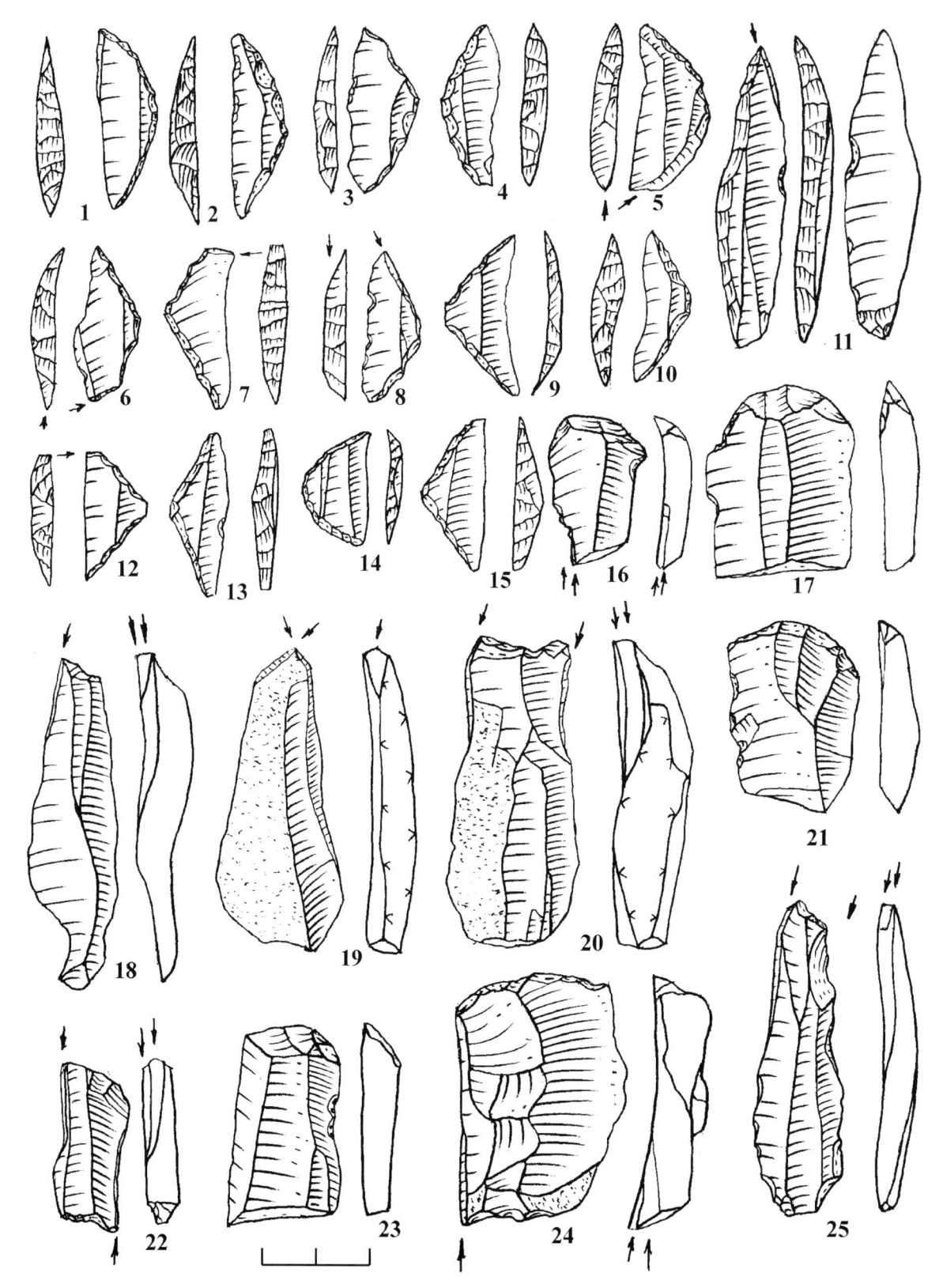

Figure 17.7: Fatma-Koba. Taubodrakian technocomplex. 1–14. geometric microliths; 15–25. flint tools.

The first striking fact is that the Taubodrakian and Late-Natufian technocomplexes are so similar. The examples of sites like Rosh Zin (Henry 1976), Rosh Horeysha (Marks and Larson 1977), and Nahal Seher VI (Goring-Morris and Bar-Yosef 1987), demonstrate the similarity of three characteristics: the flake production techniques, the flint technocomplexes' typologies, and the geometric microliths' manufacturing technologies. In the specified Late-Natufian technocomplexes a predominant role was given to lunates and triangles, made by means of the microburin technique and with bipolar retouch on the sides and along the arcs. Notched and denticulated pieces pay an important role in the technocomplexes. The flake production technique used a solid hammerstone, and the typology of cores is almost identical.

In the case of the Late-Natufian technocomplexes, we have every reason to admit that the two industries are almost completely identical. Nevertheless, this fact does not allow us to link the genesis of the Taubodrakian industry with the Late Natufian. This is because this industry existed in the Middle East only during the Dryas 3 period, which means that it was established much later than a similar industry in Crimea, the appearance of which is associated with the Allerød. In this case, we are dealing either with a clear example of convergence, which complied with general patterns of development of geometric systems, or with the fact that both industries developed out of the same source (or at least experienced the intensive impact of the same industry).

The Shankobien and Taubodrakian industries: a common origin?

The origin of the Shankobien and Taubodrakian industries from the same source is quite possible, because there is a highly probable suggestion for the industry – namely the industry of Karain B and Okuzini in Turkey – which might have played the role of an innovation provider, especially regarding the manufacture of geometric microliths.

In the Karain B (Albrecht 1988) the geometrization of the tool complex for missile weapons began in the Late Palaeolithic and was associated with layers 30–27, dated to 18,000–16,000 BP. This was precisely the time when distinct points with an arched back appeared, as well as the first lunates, including those with bipolar retouch. Layers 20–17 are attributed to the Epipalaeolitic, which is dated to around 16,000–12,000 BP. The Epipalaeolithic technocomplexes contain lunates and triangles with bipolar retouch along the arcs and on the sides; sometimes the elongated triangles and lunates have flat retouch on the ventral surface at the end with the bulb of percussion, as identified on one of the points from layer 5–6 of Fatma-Koba. Probably, the microburin technique was applied there. The endscrapers, including those of high forms, were present in the technocomplexes. Burins were rare. There are series of perforators and notched and denticulated pieces. The cores are usually of two types: one-platform diagonal-platform prismatic-like ones, and two-platform bi-direct ones.

To trace the development of the industry of Karain B, one may take the site of Okuzini as an example (Léotard *et al.* 1998). Its 12 layers were successfully investigated and grouped into four phases.

Phase 1 (layers XII–VII at Okuzini) is dated to around 17,000–14,000 BP. Even at that time, along with points with arched backs, there are lunates and triangles, finished with bipolar retouch. Starting from layer VIII, there are points and lunates with flat retouch on the ventral surfaces. In the same layer a low symmetrical trapeze with a retouched upper pat was present. The bi-direct cores were used from the beginning of phase 1.

Phase 2 at Okuzini corresponds to layers VI–V and is chronologically almost identical with the Oldest Dryas. In this phase there are the backed bladelets, but the geometric technocomplex is finally formed. Also lunates and trapezes with retouched upper parts, triangles, and low trapezes are present. In this phase, bipolar retouch is not so widely used, and among the microliths there are items with regular sharp retouch on thin workpieces. In layer VI the occurrence of notched blades was noted for the first time.

Phase 3 (layers IV–Ib) at Okuzini relates to the Ancient Holocene (Final Pleistocene) of Ukraine and, consequently, was contemporary with the Taubodrakian layers of the Skalisty Rockshelter. The main trends in the development of the Karain B industry stayed unchanged. The role of triangles becomes less important, while lunates dominated, including those denticulated on the base, and low trapezes with retouched and non-retouched upper parts. The production of microliths included both bipolar and regular sharp retouch. In the technocomplexes there are perforators, endscrapers and scrapers with retouch on the end and at the side.

Phase 4 (layer 1a) at Okuzini dates to the Early Holocene (Preboreal–Boreal). In this phase, the main array is represented by triangles with bipolar retouch, trapezes and lunates; bipolar retouch was applied to them very rarely. Among the cores there are one-platform items of conical and prismatic shapes. There is no evidence of application of the reduction technique, even at the later stage.

It should be noted that, at certain stages, the technocomplexes of sites such as Karain B and Okuzini have different degrees of similarity with the Taubodrakian industry; these are examples of both the domination of the bipolar retouch technique and its degradation to some extent. That is why it is correct to compare the Karain B industry with both types of geometric industries of the Skalisty Rockshelter. It is worth mentioning that the Shankobien and Taubodrakian industries went through the same stage of development, and that this stage was possibly based on the Karain B industry or on another similar industry.

It is important to define the influence of the Karain B industry in the Epipalaeolitic in the Middle East, Asia Minor and Central Asia. O. Bar-Yosef relates the Karain B and the Okuzini industries to the Zarzian industry of the Zagros (Bar-Yosef 1998). If this is true, then one can assume that both geometric cultures of the Crimean Final Palaeolithic emerged from the Zarzian industrial traditions, which explains their typological similarily.

We can offer another hypothesis concerning the common origin of two geometric industries in Crimea. It is possible that the two traditions, which were revealed in Crimea, had previously existed in Asia Minor. As mentioned above, in the Karain B and in the Okuzini industries the bipolar technique, used to shape the sides and arcs of geometric microliths, alternated: being either dominant in some periods, or disappearing completely in others. It is possible that the Karain B industry included a wide range of technological methods of microlith manufacture, but in the process of the industry's development any single feature could cause significant variability in the industry. This is likely to be true, given that the development of the above-mentioned Asia Minor industry lasted for nearly 10,000 years. If so, then the emergence of the Shankobien and Taubodrakian industries can be connected to the gradual migration of Karain B producers in Crimea, which endured, so that another wave of migration could coincide with the dominance of one or other technological method in producing geometric microliths.

Conclusion

The interstratification cases of the two industries, which show two contrasting manufacturing traditions for geometric microliths, indicate that in the Final Palaeolithic and Mesolithic in Crimea complex demographic processes took place. The simultaneous residence of producers of different flint industries in the same region indicates that they were relatively isolated; this is in fact a premise for the interpretation of two contemporary industries. The stability of two types of industry may suggest the existence of a complex system of land relations between different population groups, and regulation of the use of natural resources, usual hunting areas, and areas for hunting camps.

The presence in Crimea in the Final Palaeolithic of the two industries of common origin means that the process of exploration of Crimea by migrants was considerably extended in time, and was associated with several waves of migration. Given this, we may conclude that there was a significant demographic density in Asia Minor, in the Zagros Mountains. Such a situation led to a constant outflow of population towards the Crimean Mountains, and further to the steppe regions of Ukraine.

One may doubt such an interpretation, claiming that the area of the Crimean Mountains is too small to allow the two population groups to coexist. A. A. Yanevich, for example, has argued against this theory. However, the fact is that there is evidence confirming that two populations lived together in the Crimean Mountains – Shankobien and Swiderian populations – at least during the Dryas 3 phase and the Preboreal. In the late Preboreal, Shpan-Koba and Kukrek industries appeared in Crimea. This lets us presume that producers of several cultural traditions could inhabit Crimea at the same time, not hindering each other. Obviously, the natural resources of Crimea attracted these populations, ensuring the stable co-existence of different groups.

Seemingly, the first settlers, who had arrived from the area of the Fertile Crescent (its future name), paved the way for the later introduction of the Neolithic to South-East Europe. In fact, the process of primary settling in Crimea by the industry carriers, who used geometric microliths, was s process of pre-Neolithization. Subsequently, new waves of migrants from the Fertile Crescent brought to Eastern Europe food producing economic traditions.

Another informative fact is that the Shankobien and Taubodrakian industries became the basis for the formation of the first Neolithic industries in Eastern Europe at the turn of the eighth to seventh millennia BC. The Shankobien industry turned out to be a basis for the formation of the Neolithic culture of Tash-Air in the Crimean mountains and steppe. It is also necessary to assume that the Taubodrakian and Shankobien industries became the source for the formation of the Platovostavian Early Neolithic culture, distributed on the Kerch Peninsula, in the Southern Donbass and in the basin of the Lower Don river.

References

Albrecht, G. (1988) Preliminary results of the excavation in the Karain B cave near Antalya/Turkey: the Upper Palaeolithic assemblages and the Upper Pleistocene climatic development. *Paléorient* 14/2, 211–222.

Bar-Yosef, O. (1998) Öküzini – comparisons with the Levant. In M. Otte (ed.) *Anatolian Prehistory at the Crossroads of Two Worlds*. ERAUL (Études et Recherces Archaéologiques de l'Université de Liège) 85/2, 501–507. Liège, Université de Liège.

Bibikov, S. N., Stanko V. N. and Koen V. Ju. (1994) *Final'nyj Paleolit i Mezolit Gornogo Kryma*. Odessa, Vest.

Bonch-Osmolovskij, G. A. (1934) Itogi izuchenija krymskogo paleolita. In *Trudy II Mezhdunar. Konf. Associacii po Izucheniju Chetvertichnogo Perioda. Vyp. V. -M. -L.*, 114–183. Novosibirsk, Gos. Nauchno-Tehn. Gorno-Geologo-Neftjanoe Izd-Vo.

Böhner, U. and Schyle, D. (2002–6) ^{14}C radiocarbon CONTEXT database <http://context-database.uni-koeln.de>.

Braidwood, R. J. and Howe, B. (1960) *Prehistoric Investigations in Iraqi Kurdistan: Studies in Ancient Oriental Civilization*. Studies of Ancient Oriental Civilization, 31. Chicago, University of Chicago Press.

Cohen, V., Gerasimenko, N., Rekovetz, L. and Starkin, A. (1996) Chronostrathigraphie of Rockshelter Skalisty: implementations for the Late Glacial of Crimea. In M. Otte (ed.) *Prehistoire Européene* 9, 326–356. Liége, Université de Liège.

Gorelik, A. F. (2001) *Pamjatniki Rogalik-Peredel'skogo Rajona: Problemy Final'nogo Paleolita Jugo-Vostochnoj Ukrainy*. Kiev-Lugansk, RIO LIVD.

Goring-Morris, A. N. and Bar-Yosef, O. (1987) A Late Natufian campsite from the Western Negev, Israel. *Paléorient* 13/1, 107–112.

Henry. D. O. (1976) Rosh Zin: a Natufian settlement near Ein Avdat. In A. Marks (ed.) *Prehistory and Paleoenvironments in the Central Negev, Israel*. Volume I, 317–348. Dallas, Southern Methodist University Press.

Yanevich, A. A. (1992) Novaja final'nopaleoliticheskaja stojanka Vishennoe II v Krymu. V. In M. P. Olenkovskij (ed.) *Pizn'opaleolitichni Pam'jatki Centru Pivnichnogo Prichornomor'ja*. Herson.

Janevich, A. A. (2000) Buran-kajs'ka kul'tura gravetu Krimu. *Arheologija* 2, 20–30.

Léotard, J.-M., Bayón, L. and Kartal M. (1998) La Grotte d'Öküzini: évolution technologique et cynégétique. In M. Otte (ed.) *Anatolian Prehistory at the Crossroads of Two Worlds*. ERAUL (Études et Recherces Archaéologiques de l'Université de Liège) 85/2, 509–529. Liège, Université de Liège.

Manko, V. O. (2008) Kul'turnij shar III-3-H stojanki Grot Skeljastij. V. In N. O. Gavriljuk (ed.) *Arheologichni Vidkrittja v Ukraïni*, 234–238. Kiïv, Institut Arheologiï NAN Ukraïni.

Manko, V. O. and Yanevich O. O. (2006) The significance of Skalisty Rockshelter for Crimean prehistory. In V. Yanko-Hombach, I. Buinevich, P. Dolukhanov, A. Gilbert, R. Martin, M. McGann and P. Mudie (eds.) *Black Sea – Mediterranean Corridor during the Last 30 ky: Sea Level Change and Human Adaptation*, 117–119. Odessa, Astroprint.

Marks, A. E. and Larson, P. A. (1977) Test excavations at the Natufian Site of Rosh Horesha. In A. Marks (ed.) *Prehistory and Paleoenvironments in the Central Negev, Israel*. Volume II, 317–348. Dallas, Southern Methodist University Press.

Muheisen, M. (1988) Le gisement de Kharaneh IV: note sommaire sur la phase D. *Paléorient* 14/2, 265–269.

Muheisen, M. and Wada, H. (1995) An analysis of the microliths at Kharaneh IV, phase D, Square A20/37. *Paléorient* 21/1, 75–95.

Sapozhnikov, I. V. (2004) Mnogoslojnaja stojanka Mihajlovka (Beloles'e): problemy stratigrafii i datirovki. In G. M. Toshev (ed.) *Starozhitnosti Stepovogo Prichornomor'ja i Krimu* 11, 299–316. Zaporizhzhja, Zaporiz'kij Nacional'nij Universitet.

Sorokin, A. N. (2002). *Mezolit Zhizdrinskogo Poles'ja: Problema Istochnikovedenija Mezolita Vostochnoj Evropy*. Moskva, Nauka.

Stanko, V. N. (1983) *Mirnoe: Problema Mezolita Stepej Severnogo Prichernomor'ja*. Kiev: Naukova Dumka.